ARCH Models for
Financial Applications

ARCH Models for Financial Applications

Evdokia Xekalaki • Stavros Degiannakis

Department of Statistics
Athens University of Economics and Business, Greece

A John Wiley and Sons, Ltd, Publication

Registered office
John Wiley & Sons Ltd, The Atrium, Southern Gate, Chichester, West Sussex, PO19 8SQ, United Kingdom

For details of our global editorial offices, for customer services and for information about how to apply for permission to reuse the copyright material in this book please see our website at www.wiley.com.

Library of Congress Cataloguing-in-Publication Data

Xekalaki, Evdokia.
  ARCH models for financial applications / Evdokia Xekalaki, Stavros Degiannakis.
     p. cm.
  Includes bibliographical references and index.
  ISBN 978-0-470-06630-0 (cloth)
  1. Finance–Mathematical models. 2. Autoregression (Statistics) I. Degiannakis, Stavros. II. Title.
  HG106.X45 2010
  332.01'519536–dc22

                                                                                        2009052104

A catalogue record for this book is available from the British Library.

ISBN 978-0-470-06630-0 (H/B)

Set in 10/12pt Times by Thomson Digital, Noida, India
Printed and bound in Great Britain by TJ International, Padstow, Cornwall

To my husband and my son, a wonderful family

Evdokia Xekalaki

*To the memory of the most important person in my life, my father Antonis,
and to my mother and my brother*

Stavros Degiannakis

Contents

Preface

There has been wide interest throughout the financial literature on theoretical and applied problems in the context of ARCH modelling. While a plethora of articles exists in various international journals, the literature has been rather sparse when it comes to books with an exclusive focus on ARCH models. As a result, students, academics in the area of finance and economics, and professional economists with only a superficial grounding in the theoretical aspects of econometric modelling, while able to understand the basic theories about model construction, estimation and forecasting, often fail to get a grasp of how these can be used in practice.

The present book addresses precisely these issues by interweaving practical questions with approaches hinging on financial and statistical theory: we have adopted an interactional exposition of the ARCH theory and its implementation throughout. This is a book of practical orientation and applied nature intended for readers with a basic knowledge of time series analysis wishing to gain an aptitude in the applications of financial econometric modelling. Balancing statistical methodology and structural descriptive modelling, it aims to introduce readers to the area of discrete time applied stochastic volatility models and to help them acquire the ability to deal with applied economic problems. It provides background on the theory of ARCH models, but with a focus on practical implementation via applications to real data (the accompanying CD-ROM provides programs and data) and via examples worked with econometrics packages (EViews and the G@RCH module for the Ox package) with step-by-step explanations of their use. Readers are familiarized with theoretical issues of ARCH models from model construction, fitting and forecasting through to model evaluation and selection, and will gain facility in employing these models in the context of financial applications: volatility forecasting, value-at-risk forecasting, expected shortfall estimation, and volatility forecasts for pricing options.

Chapter 1 introduces the concept of an autoregressive conditionally heteroscedastic (ARCH) process and discusses the effects that various factors have on financial time series such as the leverage effect, the non-trading period effect, and the non-synchronous trading effect. Chapter 2 provides an anthology of representations of ARCH models that have been considered in the literature. Estimation and simulation of the models is discussed, and several misspecification tests are provided. Chapter 3 deals with fractionally integrated ARCH models and discusses a series of tests for testing the hypothesis of normality of the standardized residuals. Chapter 4 familiarizes readers with the use of EViews in obtaining volatility forecasts. Chapter 5 treats the case of ARCH models with non-normally distributed standardized innovations – in particular, models with innovations with Student t, beta, Paretian or Gram–Charlier type distributions, as well as generalized error distributions. Chapter 6 acquaints readers with the use of G@RCH in volatility forecasting. Chapter 7 introduces realized volatility as an alternative volatility measure. The use of high-frequency returns to compute volatility at a lower frequency and the prediction of volatility with ARFIMAX models are presented. Chapter 8 illustrates applications of volatility forecasting in risk management and options pricing. Step-by-step empirical applications provide an insight into obtaining value-at-risk estimates and expected shortfall forecasts. An options trading game driven by volatility forecasts produced by various methods of ARCH model selection is illustrated, and option pricing models for asset returns that conform to an ARCH process are discussed. Chapter 9 introduces the notion of implied volatility and discusses implied volatility indices and their use in ARCH modelling. It also discusses techniques for forecasting implied volatility. Chapter 10 deals with evaluation and selection of ARCH models for forecasting applications. The topics of consistent ranking and of proxy measures for the actual variance are extensively discussed and illustrated via simulated examples. Statistical tests for testing whether a model yields statistically significantly more accurate volatility forecasts than its competitors are presented, and several examples illustrating methods of model selection are given. Finally, Chapter 11 introduces multivariate extensions of ARCH models and illustrates their estimation using EViews and G@RCH.

The contents of the book have evolved from lectures given to postgraduate and final-year undergraduate students at the Athens University of Economics and Business and at the University of Central Greece. Readers are not expected to have prior knowledge of ARCH models and financial markets; they only need a basic knowledge of time series analysis or econometrics, along with some exposure to basic statistical topics such as inference and regression, at undergraduate level.

The book is primarily intended as a text for postgraduate and final-year under-graduate students of economic, financial, business and statistics courses. It is also intended as a reference book for academics and researchers in applied statistics and econometrics, and doctoral students dealing with volatility forecasting, risk evaluation, option pricing, model selection methods and predictability. It can also serve as a handbook for consultants as well as traders, financial market practitioners and professional economists wishing to gain up-to-date expertise in practical issues of financial econometric modelling. Finally, graduate students on master's courses holding degrees from different disciplines may also benefit from the practical orientation and applied nature of the book.

Writing this book has been both exciting and perplexing. We have tried to compile notions, theories and practical issues that by nature lie in areas that are intrinsically complex and ambiguous such as those of financial applications. From numerous possible topics, we chose to include those that we judged most essential. For each of these, we have provided an extensive bibliography for the reader wishing to go beyond the material covered in the book.

We would like to extend our thanks to the several classes of students whose queries and comments in the course of the preparation of the book helped in planning our approach to different issues, in deciding the depth to which chosen topics should be covered and in clearing up ambiguities. We are also grateful to the many people (university colleagues, researchers, traders and financial market practitioners), who by occasional informal exchange of views have had an influence on these aspects as well. Much of the material in this book was developed while the first author was on sabbatical leave at the Department of Statistics, University of California, Berkeley, which she gratefully acknowledges for providing her with a wonderful work environment. Moreover, it has been a pleasure to work on this project with Kathryn Sharples, Simon Lightfoot, Susan Barclay, Richard Davies, Heather Kay and Ilaria Meliconi at John Wiley & Sons, Ltd, copy editor Richard Leigh who so attentively read the manuscript and made useful editing suggestions, and Poirei Sanasam at Thomson Digital. Most importantly, we wish to express our gratitude to our families for their support.

May 2009   *Evdokia Xekalaki*
Stavros Degiannakis

Notation

$\circ$	Hadamard (elementwise) product
$A(L)$	Polynomial of ARCH
AD	Anderson–Darling statistic
AIC	Akaike information criterion
$\mathbf{B}$	Matrix of unknown parameters in a multivariate regression model.
$B(.,.)$	Cumulative distribution function of the binomial distribution
$B(L)$	Polynomial of GARCH
$B(t)$	Standard Brownian motion
$c(.)$	Smooth function on [0,1].
c_i	Autoregressive coefficients
$C(L)$	Polynomial of AR
$\mathbf{C}_t$	Matrix of conditional correlations
$C_t^{(\tau)}$	Call option at time t, with τ days to maturity
$C_{t+1|t}^{(\tau)}$	Call option at time $t+1$ given the information available at time t, with τ days to maturity
CGR	Correlated gamma ratio distribution
CM	Cramér–von Mises statistic
d	Exponent of the fractional differencing operator $(1-L)^d$ in FIGARCHmodels
$\tilde{d}$	Exponent of the fractional differencing operator$(1-L)^{\tilde{d}}$ in ARFIMAX models
$\tilde{\tilde{d}}$	Integer differencing operator
d_i	Moving average coefficients
dt_i	Duration or interval between two transactions, $dt_i \equiv t_i - t_{i-1}$
$D(L)$	Polynomial of MA
$DM_{(A,B)}$	Diebold–Mariano statistic
$ES_{t+\tau|t}^{(1-p)}$	Expected shortfall τ days ahead.
$ES_{t+1|t}^{(1-p)}$	Expected shortfall forecast at time $t+1$ based on information available at time t, at $(1-p)$ probability level.
$f(.)$	Probability density function
$f_a(.)$	a-quantile of the distribution with density function $f(.)$
$f_{(BG)}\left(x, y; \widehat{T}, \rho\right)$	Probability density function of the bivariate gamma distribution

$f_{(CGR)}\left(x;2^{-1}\,\widehat{T},\rho\right)$	Probability density function of the correlated gamma ratio distribution
$f_d(0)$	Spectral density at frequency zero
$f_{(GC)}(z_t;v,g)$	Probability density function of Gram–Charlier distribution
$f_{(GED)}(z_t;v)$	Probability density function of generalized error distribution
$f_{(GT)}(z_t;v,g)$	Probability density function of generalized t distribution
$f_{(SGED)}(z_t;v,\theta)$	Probability density function of skewed generalized error distribution
$f_{(skT)}(z_t;v,g)$	Probability density function of skewed Student t distribution
$f_{(t)}(z_t;v)$	Probability density function of standardized Student t distribution.
$F(.)$	Cumulative distribution function
$_2F_2(.,.;.,.;.)$	Generalized hypergeometric function
F_{il}	Filter
$FoEn_{(1,2)}$	Forecast encompassing statistic of Harvey *et al.* (1998)
$g(.)$	Functional form of conditional variance
g_{s_t}	Constant multiplier in SWARCH model
GED	Generalized error distribution
GT	Generalized t distribution
$\mathbf{H}_t$	Conditional covariance matrix of multivariate stochastic process, $\mathbf{H}_t \equiv V_{t-1}(\mathbf{y}_t)$
Hit_t	Index of VaR violation minus expected ratio of violations, $Hit_t = \tilde{I}_t - p$
HQ	Hannan and Quinn information criterion
$\mathbf{i}$	Vector of ones.
I_t	Information set
$\tilde{I}_t$	Index of VaR violations
JB	Jarque–Bera test
k	Dimension of vector of unknown parameters β
$\tilde{k}$	Number of slices
K	Exercise (or strike) price
$\tilde{K}$	Number of regimes in SWARCH model
$\overset{...}{K}$	Number of factors in the $\overset{..}{K}$ Factor ARCH(p,q) model
KS	Kolmogorov–Smirnov statistic
KS^*	Kuiper statistic
Ku	Kurtosis
l	Order of the moving average model
l_i	Eigenvalue
$l_t(.)$	Log-likelihood function for the tth observation
log	Natural logarithm
L	Lag operator
$L_T(.)$	Full-sample log-likelihood function based on a sample of size T
LR_{cc}	Christoffersen's likelihood ratio statistic for conditional coverage.

LR_{in}	Christoffersen's likelihood ratio statistic for independence of violations
LR_{un}	Kupiec's likelihood ratio statistic for unconditional coverage.
m	Number of intraday observations per day
$mDM_{(A,B)}$	Modified Diebold–Mariano statistic
$MGN_{(A,B)}$	Morgan–Granger–Newbold statistic
n	Dimension of the multivariate stochastic process, $\{\mathbf{y}_t\}$
n_{ij}	Number of points in time with value i followed by j
N	Total number of VaR violations, $N = \Sigma_{t=1}^{\tilde{T}} \tilde{I}_t$
$N(.)$	Cumulative distribution function of the standard normal distribution
NRT_t	Net rate of return at time t from trading an option.
p	Order of GARCH form
$p_{i,t}$	Filtered probability in regime switch models (the probability of the market being in regime i at time t)
p_t	Switching probability in regime switch models
$P(.)$	Probability
$P_t^{(\tau)}$	Put option at time t, with τ days to maturity
$P_{t+1\|t}^{(\tau)}$	Put option at time $t + 1$ given the information available at time t, with τ days to maturity
$\tilde{q}$	Order of BEKK$(p, q, \tilde{q})$ model.
q_t	Switching probability in regime switch models
rf_t	Rate of return on a riskless asset
R^2	Coefficient of multiple determination
RT_t	Rate of return at time t from trading an option
s_t	Regime in SWARCH model.
$\underset{\sim}{y}^*_{it}$	Risk-neutral log-returns at time t.
q	Order of ARCH form
$Q^{(LB)}$	Ljung–Box statistic
r_j	Autocorrelation of squared standardized residuals at j lags
skT	Skewed Student t distribution
S_t	Market closing price of asset at time t
$\tilde{S}(T)$	Terminal stock price adjusted for risk neutrality
SBC	Schwarz information criterion
$SGED$	Skewed generalized error distribution
SH	Shibata information criterion
Sk	Skewness
$SPA_{(i^*)}$	Superior predictive ability statistic for the benchmark model i^*
T	Number of total observations, $T = \tilde{T} + \breve{T}$
$\tilde{T}$	Number of observations for out-of-sample forecasting
$\breve{T}$	Number of observations for rolling sample
$\widehat{T}$	Number of observations for model selection methods in out-of-sample evaluation $v_t = \varepsilon_t^2 - \sigma_t^2$
$VaR_t^{(1-p)}$	VaR at $1-p$ probability level at time t

$VaR_{t+1\|t}^{(1-p)}$	VaR forecast at time $t+1$ based on information available at time t, at $(1-p)$ probability level
$VaR_{t+\tau\|t}^{(1-p)}$	VaR τ days ahead
$\widehat{VaR}_t^{(1-p)}$	VaR in-sample estimate at time t, at $1-p$ probability level
$vech(.)$	Operator stacks the columns of square matrix.
w	Vector of estimated parameters for the density function f
$\breve{w}$	Number of parameters of vector w
w_i	Weight
X	Transaction cost
y_t	Log-returns
$\mathbf{y}_t$	Multivariate stochastic process
$y_{t+1\|t(i)}$	One-step-ahead conditional mean at time $t+1$ based on information available at time t, from model i
$y_t^{(BC)}$	Box–Cox transformed variable of y_t
$z_{t+1\|t}$	One-step-ahead standardized prediction errors at time $t+1$ based on information available at time t.
α_c	Measure of cost of capital opportunity
β	Vector of unknown parameters in a regression model
$\gamma_d(i)$	Sample autocovariance of ith order
γ_t	Dividend yield
ε_t	Innovation process for the conditional mean of multivariate stochastic process, $\varepsilon_t \equiv \mathbf{y}_t - \boldsymbol{\mu}_t$
$\varepsilon_{t+1\|t}$	One-step-ahead standardized prediction errors at time $t+1$ based on information available at time t
$\tilde{\varepsilon}_t$	Innovation process in SWARCH model
θ	Vector of estimated parameters for the conditional mean and variance
$\theta_{(i)}^{(t)}$	Vector of estimated parameters for the conditional mean and variance at time t from model i
$\breve{\theta}$	Number of parameters of vector θ
κ	Order of the autoregressive model
μ	Instantaneous expected rate of return
$\mu(.)$	Functional form of conditional mean
μ_t	Predictable component of conditional mean
$\boldsymbol{\mu}_t$	Conditional mean of multivariate stochastic process, $\boldsymbol{\mu}_t \equiv E_{t-1}(\mathbf{y}_t)$
v	Tail-thickness parameter
π_{ij}	Percentage of points in time with value i followed by j, $\pi_{ij} = n_{ij}/\Sigma_j n_{ij}$
ρ	Correlation coefficient
σ	Instantaneous variance of the rate of return
$\sigma_{i,j,t}$	Conditional covariance between asset returns i and j at time t.
σ_{t+1}^2	True, but unobservable, value of the variance at time $t+1$

$\sigma^2_{t+1\mid t}$	One-step-ahead conditional variance at time $t+1$ based on information available at time t.
$\sigma^2_{t+1\mid t(i)}$	One-step-ahead conditional variance at time $t+1$ based on information available at time t, from model i.
$\sigma^2_{t+s\mid t}$	s-days-ahead conditional variance at time $t+s$ based on information available at time t
$\sigma^{2(\tau)}_{t+1}$	True, but unobservable, value of variance for a period of τ days, from $t+1$ until $t+\tau$
$\sigma^{2(\tau)}_{t+1\mid t}$	Variance forecast for a period of τ days, from $t+1$ until $t+\tau$, given the information available at time t
$\tilde{\sigma}^{2(\tau)}_{t+1}$	Proxy for true, but unobservable, value of variance for a period of τ days, from $t+1$ until $t+\tau$
$\bar{\sigma}^{(\tau)}_{t+1\mid t}$	Average standard deviation forecasts from $t+1$ up to $t+\tau$, given the information available at time t, $$\bar{\sigma}^{(\tau)}_{t+1\mid t} = \sqrt{\tau^{-1}\sum_{i=1}^{\tau}\sigma^2_{t+i\mid t}} = \sqrt{\tau^{-1}\sigma^{2(\tau)}_{t+1\mid t}}$$
$\hat{\sigma}^2_{t+1}$	In-sample conditional variance at time $t+1$ based on the entire available data set T
$\hat{\sigma}^{2(\tau)}_{t+1}$	In-sample conditional variance for a period of τ days, from $t+1$ until $t+\tau$, based on the entire available data set T
$\hat{\sigma}^{2(RV)}_{(un),t+1}$	In-sample realized volatility at time $t+1$ based on the entire available data set T
$\sigma^{2(RV)}_{t+1}$	Observable value of the realized variance at time $t+1$
$\sigma^{2(RV)(\tau)}_{t+1}$	Observable value of the realized variance for a period of τ days, from $t+1$ until $t+\tau$
$\sigma^{2(RV)}_{(un),t+1\mid t}$	One-day-ahead conditional realized variance at time $t+1$ based on information available at time t
τ	Point in time (i.e. days) for out-of-sample forecasting. Also days to maturity for options
υ_t	Vector of predetermined variables included in I_t
$\phi(.)$	Functional form of conditional variance in conditional mean in GARCH-M model
$\Phi(.;.;.)$	Confluent hypergeometric function
$\Phi(L)$	Polynomial of FIGARCH
$\varphi(t, a, \beta, \sigma, \mu)$	Characteristic function of stable Paretian distribution
$\chi^2_{(g)}$	Pearson's chi-square statistic
ψ	Vector of estimated parameters for the conditional mean, variance and density function, $\psi' = (\theta', w')$
$\psi(.)$	Euler psi function
$\breve{\psi}$	Number of parameters of vector ψ, $\breve{\psi} = \breve{\theta} + \breve{w}$
$\hat{\psi}^{(T)}$	Maximum likelihood estimator of ψ based on a sample of size T

$\bar{\Psi}_{(SE)(i)}^{(\tau)}$ — Mean squared error loss function for model i volatility forecasts τ days ahead, i.e. $\bar{\Psi}_{(SE)(i)}^{(\tau)} = \tilde{T}^{-1} \sum_{t=1\tilde{T}} \left(\sigma_{t+1|t(i)}^{2(\tau)} - \sigma_{t+1(i)}^{2(\tau)} \right)^2$

$\Psi_{(SE)t(i)}^{(\tau)}$ — Squared error loss function for models i volatility forecasts τ days ahead

$\bar{\Psi}^{(\tau)}$ — Average of a loss function for τ-days-ahead volatility forecasts, i.e. $\bar{\Psi}^{(\tau)} = \tilde{T}^{-1} \sum_{t=1}^{\tilde{T}} \Psi_t^{(\tau)}$

$\Psi_{t(i)}^{(\tau)}$ — Loss function that measures the distance between actual volatility over a τ-day period and model i volatility forecast over the same period

$\Psi_{t(i)}$ — Loss function that measures the distance between one-day actual volatility and its forecast by model i, i.e. $\Psi_{t(i)} \equiv \Psi_{t(i)}^{(1)}$

$\Psi_{t(i^*,i)}^{(\tau)}$ — Difference of loss functions (loss differential) of models i^* and i, $\Psi_{t(i^*,i)}^{(\tau)} = \Psi_{t(i^*)}^{(\tau)} - \Psi_{t(i)}^{(\tau)}$

$\bar{\Psi}_{(i^*,i)}^{(\tau)}$ — Sample mean loss differential of models i^* and i, $\bar{\Psi}_{(i^*,i)}^{(\tau)} = \tilde{T}^{-1} \sum_{t=1}^{\tilde{T}} \Psi_{t(i^*,i)}^{(\tau)}$

1

What is an ARCH process?

1.1 Introduction

Since the first decades of the twentieth century, asset returns have been assumed to form an independently and identically distributed (i.i.d.) random process with zero mean and constant variance. Bachellier (1900) was the first to contribute to the theory of random walk models for the analysis of speculative prices. If $\{P_t\}$ denotes the discrete time asset price process and $\{y_t\}$ the process of continuously compounded returns, defined by $y_t = 100 \log (P_t/P_{t-1})$, the early literature viewed the system that generates the asset price process as a fully unpredictable random walk process:

$$P_t = P_{t-1} + \varepsilon_t$$
$$\varepsilon_t \overset{i.i.d.}{\sim} N(0, \sigma^2),$$

(1.1)

where ε_t is a zero-mean i.i.d. normal process. Figures 1.1 and 1.2 show simulated $\{P_t\}_{t=1}^{T}$ and $\{y_t\}_{t=1}^{T}$ processes for $T = 5000$, $P_1 = 1000$ and $\sigma^2 = 1$.

However, the assumptions of normality, independence and homoscedasticity do not always hold with real data.

Figures 1.3 and 1.4 show the daily closing prices of the London Financial Times Stock Exchange 100 (FTSE100) index and the Chicago Standard and Poor's 500 Composite (S&P500) index. The data cover the period from 4 April 1988 until 5 April 2005. At first glance, one might say that equation (1.1) could be regarded as the data-generating process of a stock index. The simulated process $\{P_t\}_{t=1}^{T}$ shares common characteristics with the FTSE100 and the S&P500 indices.[1] As they are clearly

---

[1] The aim of the visual comparison here is not to ascertain a model that is closest to the realization of the stochastic process (in fact another simulated realization of the process may result in a path quite different from that depicted in Figure 1.1). It is merely intended as a first step towards enhancing the reader's thinking about or conceiving of these notions by translating them into visual images. Higher-order quantities, such as the correlation, absolute correlation and so forth, are much more important tools in the analysis of stochastic process than their paths.

ARCH Models for Financial Applications   Evdokia Xekalaki and Stavros Degiannakis
© 2010 John Wiley & Sons, Ltd

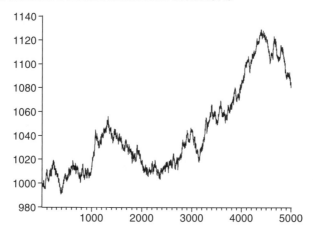

Figure 1.1   Simulated $\{P_t\}$ process, where $P_t = P_{t-1} + \varepsilon_t$, $P_1 = 1000$ and $\varepsilon_t \overset{i.i.d.}{\sim} N(0,1)$.

non-stationary, the autocorrelations presented in Figure 1.5 are marginally less than unity in any lag order. Figure 1.6 plots the distributions of the daily FTSE100 and S&P500 indices as well as the distribution of the simulated process $\{P_t\}_{t=1}^{T}$. The density estimates are based on the normal kernel with bandwidths method calculated according to equation (3.31) of Silverman (1986). S&P500 closing prices and the simulated process $\{P_t\}_{t=1}^{T}$ have similar density functions.

However, this is not the case for the daily returns. Figures 1.7 and 1.8 depict the FTSE100 and S&P500 continuously compounded daily returns, $\{y_{FTSE100,t}\}_{t=1}^{T}$ and

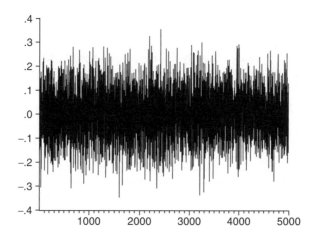

Figure 1.2   Simulated $\{y_t\}$ process, where $y_t = 100 \log(P_t/P_{t-1})$, $P_t = P_{t-1} + \varepsilon_t$, $P_1 = 1000$ and $\varepsilon_t \overset{i.i.d.}{\sim} N(0,1)$.

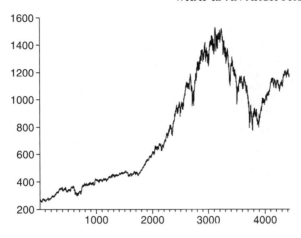

Figure 1.3   S&P500 equity index closing prices from 4 April 1988 to 5 April 2005.

$\left\{y_{SP500,t}\right\}_{t=1}^{T}$, while Figure 1.9 presents the autocorrelations of $\{y_t\}_{t=1}^{T}$, $\left\{y_{FTSE100,t}\right\}_{t=1}^{T}$ and $\left\{y_{SP500,t}\right\}_{t=1}^{T}$ for lags of order 1, ..., 35. The 95% confidence interval for the estimated sample autocorrelation is given by $\pm 1.96/\sqrt{T}$, in the case of a process with independently and identically normally distributed components. The autocorrelations of the FTSE100 and the S&P500 daily returns differ from those of the simulated process. In both cases, more than 5% of the estimated autocorrelations are outside the above 95% confidence interval. Visual inspection of Figures 1.7 and 1.8 shows clearly that the mean is constant, but the variance keeps changing over time, so the return series does not appear to be a sequence of i.i.d. random variables. A characteristic of asset returns, which is noticeable from the figures, is the volatility clustering first noted by

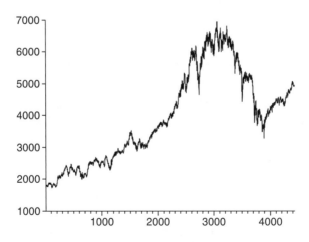

Figure 1.4   FTSE100 equity index closing prices from 4 April 1988 to 5 April 2005.

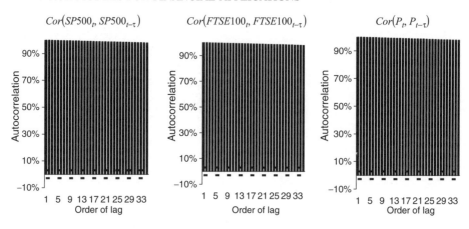

Figure 1.5   Autocorrelation of the S&P500 and the FTSE100 closing prices and of the simulated process $\{P_t\}_{t=1}^{T}$, for $\tau = 1(1)35$ lags. Dashed lines present the 95% confidence interval for the estimated sample autocorrelations given by $\pm 1.96/\sqrt{T}$.

Mandelbrot (1963): 'Large changes tend to be followed by large changes, of either sign, and small changes tend to be followed by small changes'. Fama (1970) also observed the alternation between periods of high and low volatility: 'Large price changes are followed by large price changes, but of unpredictable sign'.

Figure 1.10 presents the histograms of the stock market series. Asset returns are highly peaked (leptokurtic) and slightly asymmetric, a phenomenon observed by Mandelbrot (1963):

> The empirical distributions of price changes are usually too peaked to be relative to samples from Gaussian populations . . . the histograms of price changes are indeed unimodal and their central bells [recall] the Gaussian ogive. But, there are typically so many outliers that ogives fitted to the mean square of price changes are much lower and flatter than the distribution of the data themselves.

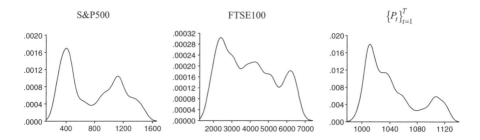

Figure 1.6   Density estimate of the S&P500 and FTSE100 closing prices, and of the simulated process $\{P_t\}_{t=1}^{T}$.

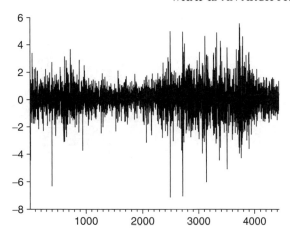

Figure 1.7   S&P500 equity index continuously compounded daily returns from 5 April 1988 to 5 April 2005.

According to Table 1.1, for estimated kurtosis[2] equal to 7.221 (or 6.241) and an estimated skewness[3] equal to −0.162 (or −0.117), the distribution of returns is flat (platykurtic) and has a long left tail relative to the normal distribution. The Jarque and Bera (1980, 1987) test is usually used to test the null hypothesis that the series is normally distributed. The test statistic measures the size of the difference between the skewness, Sk, and kurtosis, Ku, of the series and those of the normal distribution. It is computed as $JB = T\left(Sk^2 + \left((Ku-3)^2/4\right)\right)/6$, where T is the number of observations. Under the null hypothesis of a normal distribution, the JB statistic is χ^2 distributed

---

[2] Kurtosis is a measure of the degree of peakedness of a distribution of values, defined in terms of a normalized form of its fourth central moment by μ_4/μ_2^2 (it is in fact the expected value of quartic standardized scores) and estimated by

$$Ku = T\sum_{t=1}^{T}(y_t-\bar{y})^4 \Big/ \left(\sum_{t=1}^{T}(y_t-\bar{y})^2\right)^2,$$

where T is the number of observations and $\bar{y}$ is the sample mean, $\bar{y} = \sum_{t=1}^{T} y_t$. The normal distribution has a kurtosis equal to 3 and is called *mesokurtic*. A distribution with a kurtosis greater than 3 has a higher peak and is called *leptokurtic*, while a distribution with a kurtosis less than 3 has a flatter peak and is called *platykurtic*. Some writers talk about *excess kurtosis*, whereby 3 is deducted from the kurtosis so that the normal distribution has an excess kurtosis of 0 (see Alexander, 2008, p. 82).

[3] Skewness is a measure of the degree of asymmetry of a distribution, defined in terms of a normalized form of its third central moment of a distribution by $\mu_3/\mu_2^{3/2}$ (it is in fact the expected value of cubed standardized scores) and estimated by

$$Sk = \sqrt{T}\sum_{t=1}^{T}(y_t-\bar{y})^3 \Big/ \left(\sum_{t=1}^{T}(y_t-\bar{y})^2\right)^{3/2}.$$

The normal distribution has a skewness equal to 0. A distribution with a skewness greater than 0 has a longer right tail is described as *skewed to the right*, while a distribution with a skewness less than 0 has a longer left tail and is described as *skewed to the left*.

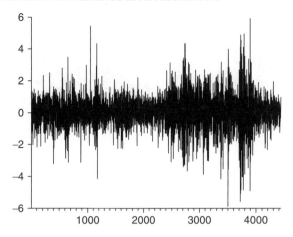

Figure 1.8   FTSE100 equity index continuously compounded daily returns from 5 April 1988 to 5 April 2005.

with 2 degrees of freedom. The two return series were tested for normality using *JB* resulting in a *p*-value that was practically zero, thus signaling non-validity of the hypothesis. Due to the fact that the *JB* statistic frequently rejects the hypothesis of normality, especially in the presence of serially correlated observations, a series of more powerful test statistics (e.g. the Anderson–Darling and the Cramér–von Mises statistics) were also computed with similar results. A detailed discussion of the computation of the aforementioned test statistics is given in Section 3.3.

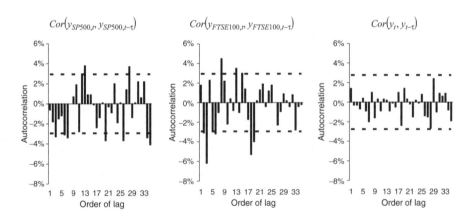

Figure 1.9   Autocorrelation of the S&P500 and FTSE100 continuously compounded daily returns and of the simulated process $\{y_t\}_{t=1}^T$, for $\tau = 1(1)35$ lags. Dashed lines present the 95% confidence interval for the estimated sample autocorrelations given by $\pm 1.96/\sqrt{T}$.

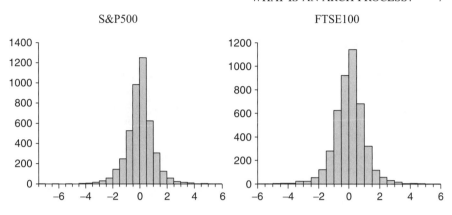

Figure 1.10   Histogram of the S&P500 and FTSE100 index log-returns.

In the 1960s and 1970s, the regularity of leptokurtosis led to a literature on modelling asset returns as i.i.d. random variables having some thick-tailed distribution (Blattberg and Gonedes, 1974; Clark, 1973; Hagerman, 1978; Mandelbrot, 1963, 1964; Officer, 1972; Praetz, 1972). These models, although able to capture the leptokurtosis, could not account for the existence of non-linear temporal dependence such as volatility clustering observed from the data. For example, applying an autoregressive model to remove the linear dependence from an asset returns series and testing the residuals for a higher-order dependence using the Brock–Dechert–Scheinkman (BDS) test (Brock et al., 1987, 1991, 1996), the null hypothesis of i.i.d. residuals was rejected.

Table 1.1   Descriptive statistics of the S&P500 and the FTSE100 equity index returns

	S&P500	FTSE100
Mean	0.034%	0.024%
Standard deviation	15.81%	15.94%
Skewness	−0.162	−0.117
Kurtosis	7.221	6.241
Jarque–Bera	3312.9	1945.6
[p-value]	[0.00]	[0.00]
Anderson–Darling	44.3	28.7
[p-value]	[0.00]	[0.00]
Cramér–von Mises	8.1	4.6
[p-value]	[0.00]	[0.00]

The annualized standard deviation is computed by multiplying the standard deviation of daily returns by $252^{1/2}$, the square root of the number of trading days per year. The Jarque–Bera, Anderson–Darling and Cramér–von Mises statistics test the null hypothesis that the daily returns are normally distributed.

1.2   The autoregressive conditionally heteroscedastic process

Autoregressive conditional heteroscedasticity (ARCH) models have been widely used in financial time series analysis and particularly in analysing the risk of holding an asset, evaluating the price of an option, forecasting time-varying confidence intervals and obtaining more efficient estimators under the existence of heteroscedasticity.

Before we proceed to the definition of the ARCH model, let us simulate a process able to capture the volatility clustering of asset returns. Assume that the true data-generating process of continuously compounded returns, y_t, has a fully unpredictable conditional mean and a time-varying conditional variance:

$$y_t = \varepsilon_t,$$
$$\varepsilon_t = z_t \sqrt{a_0 + a_1 \varepsilon_{t-1}^2}, \tag{1.2}$$

where $z_t \overset{i.i.d.}{\sim} N(0,1)$, z_t is independent of ε_t, $a_0 > 0$ and $0 < a_1 < 1$. The unconditional mean of y_t is $E(y_t) = E(z_t)E\left(\sqrt{a_0 + a_1 \varepsilon_{t-1}^2}\right) = 0$, as $E(z_t) = 0$ and z_t and ε_{t-1} are independent of each other. The conditional mean of y_t given the lag values of ε_t is $E(y_t | \varepsilon_{t-1}, \ldots, \varepsilon_1) = 0$. The unconditional variance is $V(y_t) = E(\varepsilon_t^2) - E(\varepsilon_t)^2 = E\left(z_t^2\left(a_0 + a_1 \varepsilon_{t-1}^2\right)\right) = a_0 + a_1 E(\varepsilon_{t-1}^2)$. As $E(\varepsilon_t^2) = E(\varepsilon_{t-1}^2)$, $V(y_t) = a_0(1-a_1)^{-1}$. The conditional variance is $V(y_t | \varepsilon_{t-1}, \ldots, \varepsilon_1) = E(\varepsilon_t^2 | \varepsilon_{t-1}, \ldots, \varepsilon_1) = a_0 + a_1 \varepsilon_{t-1}^2$. The kurtosis of the unconditional distribution equals $E(\varepsilon_t^4)/E(\varepsilon_t^2)^2 = 3(1-a_1^2)/(1-3a_1^2)$. Note that the kurtosis exceeds 3 for $a_1 > 0$ and diverges if a_1 approaches $\sqrt{1/3}$. Figure 1.11 plots the unconditional kurtosis for $0 \le a_1 < 1/\sqrt{3}$.

Both the unconditional and conditional means and the unconditional variance of asset returns remain constant, but the conditional variance has a time-varying character as it depends on the previous values of ε_t. Let us consider equation (1.2) as the true data-generating function and produce 5000 values of y_t. Figure 1.12

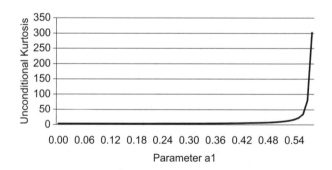

Figure 1.11   The unconditional kurtosis of $\varepsilon_t = z_t\sqrt{a_0 + a_1 \varepsilon_{t-1}^2}$ for $0 \le \alpha_1 < 1/\sqrt{3}$ and $a_0 > 0$.

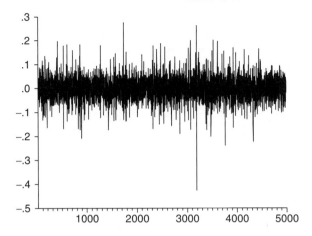

Figure 1.12   Simulated $\{y_t\}$ process, where $y_t = \varepsilon_t$, $\varepsilon_t = z_t\sqrt{a_0 + a_1\varepsilon_{t-1}^2}$, $z_t \stackrel{i.i.d.}{\sim}$ $N(0,1)$, $\varepsilon_0 = 0$, $a_0 = 0.001$ and $a_1 = 0.55$.

presents the simulated series y_t for $\varepsilon_0 = 0$, $a_0 = 0.001$ and $a_1 = 0.55$. Equation (1.2) produces a series with a time-varying conditional variance $V(y_t|\varepsilon_{t-1}, \ldots, \varepsilon_1) = 0.001 + 0.55\varepsilon_{t-1}^2$. Because of the non-constant variance, the series is highly leptokurtic. Figure 1.13 presents both the histogram (the series is obviously highly leptokurtic) and the conditional variance (revealing volatility clustering) of the simulated process. On the other hand, the y_t series standardized by its conditional standard deviation $y_t/\sqrt{V(y_t|\varepsilon_{t-1}, \ldots, \varepsilon_1)}$ is essentially the $z_t \stackrel{i.i.d.}{\sim} N(0,1)$ series. Therefore, the series itself is leptokurtic and has a non-constant variance, but its standardized version has a kurtosis equal to 3 and constant variance. Table 1.2

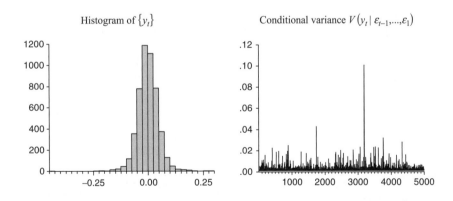

Figure 1.13   Histogram of the simulated $\{y_t\}$ process and conditional variance $V(y_t|\varepsilon_{t-1}, \ldots, \varepsilon_1) = a_0 + a_1\varepsilon_{t-1}^2$, where $y_t = \varepsilon_t$, $\varepsilon_t = z_t\sqrt{a_0 + a_1\varepsilon_{t-1}^2}$, $z_t \stackrel{i.i.d.}{\sim} N(0,1)$, $\varepsilon_0 = 0$, $a_0 = 0.001$ and $a_1 = 0.55$.

Table 1.2 Descriptive statistics of the simulated y_t and z_t series, where $y_t = \varepsilon_t$, $\varepsilon_t = z_t \sqrt{a_0 + a_1 \varepsilon_{t-1}^2}$, $z_t \sim N(0,1)$, $\varepsilon_0 = 0$, $a_0 = 0.001$ and $a_1 = 0.55$

	y_t	z_t
Mean	0.0006	0.0109
Standard deviation	0.046	0.999
Skewness	−0.091	0.014
Kurtosis	6.387	2.973
Jarque–Bera	2395.7	0.3
[p-value]	[0.00]	[0.86]
Anderson–Darling	12.3	0.2
[p-value]	[0.00]	[0.87]
Cramér–von Mises	1.69	0.04
[p-value]	[0.00]	[0.77]

presents the descriptive statistics of the y_t and z_t series. The kurtosis of the simulated y_t series is 6.387.

In the sequel, we provide a formal definition of the ARCH process.

Let $\{y_t(\theta)\}$ refer to the univariate discrete time real-valued stochastic process to be predicted (e.g. the rate of return of a particular stock or market portfolio from time $t-1$ to t), where θ is a vector of unknown parameters and $E(y_t(\theta)|I_{t-1}) \equiv E_{t-1}(y_t(\theta)) \equiv \mu_t(\theta)$ denotes the conditional mean given the information set I_{t-1} available at time $t-1$. The innovation process for the conditional mean, $\{\varepsilon_t(\theta)\}$, is then given by $\varepsilon_t(\theta) = y_t(\theta) - \mu_t(\theta)$ with corresponding unconditional variance $V(\varepsilon_t(\theta)) = E(\varepsilon_t^2(\theta)) \equiv \sigma^2(\theta)$, zero unconditional mean and $E(\varepsilon_t(\theta)\varepsilon_s(\theta)) = 0$, $\forall t \neq s$. The conditional variance of the process given I_{t-1} is defined by $V(y_t(\theta)|I_{t-1}) \equiv V_{t-1}(y_t(\theta)) \equiv E_{t-1}(\varepsilon_t^2(\theta)) \equiv \sigma_t^2(\theta)$. Since investors would know the information set I_{t-1} when they make their investment decisions at time $t-1$, the relevant expected return to the investors and volatility are $\mu_t(\theta)$ and $\sigma_t^2(\theta)$, respectively. An ARCH process, $\{\varepsilon_t(\theta)\}$, can be represented as:

$$\varepsilon_t(\theta) = z_t \sigma_t(\theta),$$
$$z_t \overset{i.i.d.}{\sim} f(w; 0, 1), \tag{1.3}$$
$$\sigma_t^2(\theta) = g(\sigma_{t-1}(\theta), \sigma_{t-2}(\theta), \ldots; \varepsilon_{t-1}(\theta), \varepsilon_{t-2}(\theta), \ldots; v_{t-1}, v_{t-2}, \ldots),$$

where $E(z_t) = 0$, $V(z_t) = 1$, $f(.)$ is the density function of z_t, w is the vector of the parameters of f to be estimated, $\sigma_t(\theta)$ is a time-varying, positive and measurable function of the information set at time $t-1$, v_t is a vector of predetermined variables included in I_t, and $g(.)$ is a linear or non-linear functional form of I_{t-1}. By definition, $\varepsilon_t(\theta)$ is serially uncorrelated with mean zero, but with a time-varying conditional variance equal to $\sigma_t^2(\theta)$. The conditional variance is a linear or non-linear function of lagged values of σ_t and ε_t, and predetermined variables $(v_{t-1}, v_{t-2}, \ldots)$ included in

I_{t-1}. For example, a simple form of the conditional variance could be

$$\sigma_t^2 = a_0 + a_1 \varepsilon_{t-1}^2, \qquad (1.4)$$

which will be referred to as ARCH(1) in the next chapter. In the sequel, for notational convenience, no explicit indication of the dependence on the vector of parameters, θ, is given when obvious from the context.

Since very few financial time series have a constant conditional mean of zero, an ARCH model can be presented in a regression form by letting ε_t be the innovation process in a linear regression:

$$
\begin{aligned}
y_t &= x_t' \beta + \varepsilon_t, \\
\varepsilon_t | I_{t-1} &\sim f(w; 0, \sigma_t^2), \\
\sigma_t^2 &= g(\sigma_{t-1}(\theta), \sigma_{t-2}(\theta), \dots; \varepsilon_{t-1}(\theta), \varepsilon_{t-2}(\theta), \dots; \upsilon_{t-1}, \upsilon_{t-2}, \dots),
\end{aligned}
\qquad (1.5)
$$

where x_t is a $k \times 1$ vector of endogenous and exogenous explanatory variables included in the information set I_{t-1} and β is a $k \times 1$ vector of unknown parameters.

Of course, an ARCH process is not necessarily limited to an expression for the residuals of a regression model. The dependent variable, $\{y_t\}$, can be decomposed into two parts, the predictable component, μ_t, and the unpredictable component, ε_t. Hence, another general representation of an ARCH process is:

$$
\begin{aligned}
y_t &= \mu_t + \varepsilon_t, \\
\mu_t &= \mu(\theta | I_{t-1}), \\
\varepsilon_t &= \sigma_t z_t, \\
\sigma_t &= g(\theta | I_{t-1}), \\
z_t &\overset{i.i.d.}{\sim} f(0, 1; w).
\end{aligned}
\qquad (1.6)
$$

Here $\mu(\theta | I_{t-1})$ and $g(\theta | I_{t-1})$ denote the functional forms of the conditional mean μ_t and the conditional standard deviation σ_t, respectively signifying that μ_t and σ_t are conditioned on the information available up to time I_{t-1} and depending on the parameter vector θ. The predictable component can be any linear or non-linear functional form of estimation. The conditional mean can also be expressed as a function of the conditional variance (see Section 1.6 for a detailed explanation). The form most frequently considered is an autoregressive moving average, or an ARMA(κ, l), process:

$$y_t = c_1 y_{t-1} + c_2 y_{t-2} + \cdots + c_\kappa y_{t-\kappa} + \varepsilon_t + d_1 \varepsilon_{t-1} + d_2 \varepsilon_{t-2} + \cdots + d_l \varepsilon_{t-l}, \quad (1.7)$$

or

$$\left(1 - \sum_{i=1}^{\kappa} c_i L^i\right) y_t = \left(1 + \sum_{i=1}^{l} d_i L^i\right) \varepsilon_t, \qquad (1.8)$$

where L is the lag operator.[4] Use of the AR(1) or MA(1) models is mainly imposed by the non-synchronous trading effect (see Section 1.5). The ARMA(κ, l) model can be

---

[4] That is, $L^2 y_t \equiv y_{t-2}$, $\sum_{i=1}^{3} (L^i) y_t = y_{t-1} + y_{t-2} + y_{t-3}$.

expanded to take into account the explanatory power of a set of exogenous variables that belong to I_{t-1}. In this case we may refer to an ARMA model with exogenous variables, or ARMAX(κ, l):

$$C(L)(y_t - x_t'\beta) = D(L)\varepsilon_t, \tag{1.9}$$

where $C(L) = (1 - \sum_{i=1}^{\kappa} c_i L^i)$ and $D(L) = (1 + \sum_{i=1}^{l} d_i L^i)$. If the series under study, $\{y_t\}$, exhibits long-term dependence, it is best modelled by a fractionally integrated ARMAX, or ARFIMAX($\kappa, \tilde{d}, l$) model,

$$C(L)(1-L)^{\tilde{d}}(y_t - x_t'\beta) = D(L)\varepsilon_t, \tag{1.10}$$

where $(1-L)^{\tilde{d}}$ is the fractional differencing operator and $\tilde{d} \in (-0.5, 0.5)$ is the fractional differencing parameter.[5] The ARFIMAX($\kappa, \tilde{d}, l$) specification was introduced by Granger (1980) and Granger and Joyeux (1980).

There is a plethora of formulations of the conditional mean in the literature. For example, Sarkar (2000) illustrated the ARCH model for a regression model in which the dependent variable is Box–Cox transformed. He referred to this as the Box–Cox transformed ARCH, or BCARCH, model:

$$\begin{aligned} y_t^{(BC)} &= \begin{cases} (y_t^{\lambda} - 1)\lambda^{-1}, & \lambda \neq 0, \\ \log(y_t), & \lambda = 0, \end{cases} \\ y_t^{(BC)}|I_{t-1} &\sim N(x_t'\beta, \sigma_t^2), \\ \varepsilon_t &= y_t^{(BC)} - x_t'\beta, \\ \sigma_t^2 &= g(\theta|I_{t-1}). \end{aligned} \tag{1.11}$$

Bickel and Doksum's (1981) extended form of the Box–Cox transformation allows negative values of y_t.

Feng et al. (2007) proposed an estimation method for semi-parametric fractional autoregressive (SEMIFAR) models with ARCH errors. A SEMIFAR ARCH model can be written as:

$$\begin{aligned} C(L)(1-L)^{\tilde{d}}\left((1-L)^{\tilde{\tilde{d}}}y_t - c(t/T)\right) &= D(L)\varepsilon_t, \\ \varepsilon_t &= z_t\sigma_t, \\ z_t &\sim N(0, 1), \\ \sigma_t^2 &= a_0 + a_1\varepsilon_{t-1}^2, \end{aligned} \tag{1.12}$$

where t/T is the rescaled time, $c(.)$ is a smooth function on $[0, 1]$, $\tilde{d} \in (-0.5, 0.5)$ is the fractional differencing parameter, and $\tilde{\tilde{d}} \in \{0, 1\}$ is the integer differencing operator. Beran (1995) first noted that the SEMIFAR class of models, is appropriate for simultaneous modelling of deterministic trends, difference stationarity and stationarity with short- and long-range dependence.

If one wishes to create a specification in which the conditional mean is described by an autoregressive model of order 1, the conditional variance is expressed by the simulated data-generating process in (1.4) and the standardized residuals are normally distributed, one may construct a model of the form:

---

[5] More information about fractionally integrated specification can be found in Chapter 3.

$$y_t = c_1 y_{t-1} + \varepsilon_t,$$
$$\varepsilon_t = z_t \sigma_t,$$
$$z_t \sim N(0, 1), \qquad (1.13)$$
$$\sigma_t^2 = a_0 + a_1 \varepsilon_{t-1}^2.$$

In this case, the vector of unknown parameters under estimation is $\theta' = (c_1, a_0, a_1)$. The elements of θ are now estimated simultaneously for the conditional mean and variance. However, the conditional mean can also be estimated separately. Artificial neural networks,[6] chaotic dynamical systems,[7] non-linear parametric and non-parametric models[8] are some examples from the literature dealing with conditional mean predictions. One can apply any model in estimating the conditional mean and may subsequently proceed to the estimation of the conditional variance (see, for example, Feng et al., 2007;[9] Pagan and Schwert, 1990[10]).

The rest of this chapter looks at the influence that various factors have on financial time series and in particular at the leverage effect, the non-trading period effect, the non-synchronous trading effect, the relationship between investors' expected return and risk, and the inverse relation between volatility and serial correlation.

1.3 The leverage effect

Black (1976) first noted that changes in stock returns often display a tendency to be negatively correlated with changes in returns volatility, i.e., volatility tends to rise in response to bad news and to fall in response to good news. This phenomenon is termed the *leverage effect* and can only be partially interpreted by fixed costs such as financial and operating leverage (see Black, 1976; Christie, 1982).

---

[6] For an overview of the neural networks literature, see Poggio and Girosi (1990), Hertz et al. (1991), White (1992) and Hutchinson et al. (1994). Thomaidis and Dounias (2008) proposed a neural network parameterization for the mean and a linear ARCH parameterization for the variance. Plasmans et al. (1998) and Franses and Homelen (1998) investigated the ability of neural networks to forecast exchange rates. They supported the view that the non-linearity found in exchange rates is due to ARCH effects, so that no gain in forecasting accuracy is obtained by using a neural network. Saltoglu (2003) investigated the ability of neural networks to forecast interest rates and noted the importance of modelling both the first and second moments jointly. Jasic and Wood (2004) and Perez-Rodriguez et al. (2005) provided evidence that neural network models have a superior ability compared to other model frameworks in predicting stock indices.

[7] Brock (1986), Holden (1986), Thompson and Stewart (1986) and Hsieh (1991) review applications of chaotic systems to financial markets. Adrangi and Chatrath (2003) found that the non-linearities in commodity prices are not consistent with chaos, but they are explained by an ARCH process. On the other hand, Barkoulas and Travlos (1998) mentioned that even after accounting for the ARCH effect, the chaotic structure of the Greek stock market is consistent with the evidence.

[8] Priestley (1988), Tong (1990) and Teräsvirta et al. (1994) cover a wide variety of non-linear models. Applications of SETAR and ARFIMA models can be found in Peel and Speight (1996) and Barkoulas et al. (2000), respectively.

[9] They proposed the estimation of the conditional mean and conditional variance parameters of the SEMIFAR ARCH model separately.

[10] For example, a two-step estimator of conditional variance with ordinary least squares is a consistent but inefficient estimator.

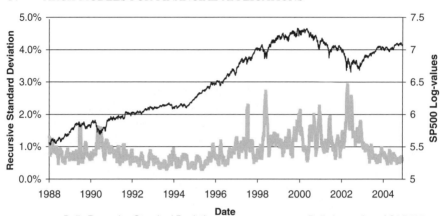

Figure 1.14   Daily log-values and recursive standard deviation of returns for the S&P500 equity index.

We can observe the phenomenon of leverage by plotting the market prices and their volatility. As a naive estimate of volatility at day t, the standard deviation of the 22 most recent trading days,

$$\sigma_t^{(22)} = \sqrt{\sum_{i=t-22}^{t} \left(y_i - \left(\sum_{i=t-22}^{t} y_i/22 \right) \right)^2 / 22}, \qquad (1.14)$$

is used. Figures 1.14 and 1.15 plot daily log-values of stock market indices and the relevant standard deviations of the continuously compounded returns. The periods

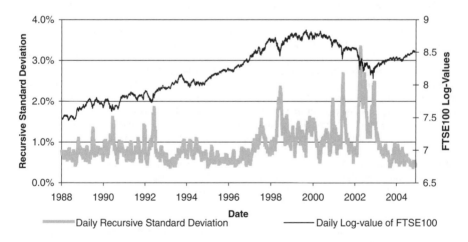

Figure 1.15   Daily log-values and recursive standard deviations of returns for the FTSE100 equity index.

Table 1.3   Mean and annualized standard deviation of the S&P500 and the FTSE100 equity index returns

	Monday	Tuesday	Wednesday	Thursday	Friday
S&P500					
Mean	0.061%	0.045%	0.045%	0.001%	0.021%
Std deviation	16.27%	16.13%	14.92%	15.59%	16.13%
Skewness	−0.895	0.315	0.450	0.153	−0.663
Kurtosis	10.494	6.039	6.332	5.767	7.017
FTSE100					
Mean	−0.014%	0.052%	−0.008%	0.033%	0.053%
Std deviation	16.60%	15.81%	14.94%	16.35%	15.93%
Skewness	−0.201	−0.200	−0.256	0.104	−0.069
Kurtosis	6.677	6.433	5.194	6.724	5.751

Annualized standard deviation is computed by multiplying the standard deviation of daily returns by $252^{1/2}$, the square root of the number of trading days per year.

of market drops are characterized by a large increase in volatility. The leverage effect is captured by a class of models that are explored in Section 2.4.

1.4   The non-trading period effect

Financial markets appear to be affected by the accumulation of information during non-trading periods, as reflected in the prices when the markets reopen following a close. As a result, the variance of returns displays a tendency to increase. This is known as the *non-trading period effect*. It is worth noting that the increase in the variance of returns is not nearly proportional to the market close duration, as would be anticipated if the information accumulation rate were constant over time. In fact, as Fama (1965) and French and Roll (1986) observed, information accumulates at a lower rate when markets are closed than when they are open. Also, as reflected by the findings of French and Roll (1986) and Baillie and Bollerslev (1989), the returns variance tends to be higher following weekends and holidays than on other days, but not by as much as it would be under a constant news arrival rate. Table 1.3 shows the annualized standard deviations of stock market returns for each day for the FTSE100 and S&P500 indices. In both cases, the standard deviation on Mondays is higher than on other days. The non-trading period effect is explored in the empirical example of Section 5.4.

1.5   The non-synchronous trading effect

The fact that the values of time series are often taken to have been recorded at time intervals of one length when in fact they were recorded at time intervals of another, not necessarily regular, length is an important factor affecting the return series, an effect known as the *non-synchronous trading effect*. For details, see Campbell et al. (1997, p. 84) and Tsay (2002, p. 176). For example, the daily prices of securities usually analysed are the closing prices. The closing price of a security is the price at which the last transaction occurred. The last transaction of each security is not implemented at the same time each day. So, it is falsely assumed that the daily prices are equally spaced at 24-hour intervals. The importance of non-synchronous trading was first recognized by Fisher (1966) and further developed by many researchers such as Atchison et al. (1987), Cohen et al. (1978), Cohen et al. (1979, 1983), Dimson (1979), Lo and MacKinlay (1988, 1990a, 1990b) and Scholes and Williams (1977).

Non-synchronous trading in the stocks making up an index induces autocorrelation in the return series, primarily when high-frequency data are used. To control this, Scholes and Williams (1977) suggested a first-order moving average (MA(1)) form for index returns, while Lo and MacKinlay (1988) suggested a first order autoregressive (AR(1)) form. According to Nelson (1991), 'as a practical matter, there is little difference between an AR(1) and an MA(1) when the AR and MA coefficients are small and the autocorrelations at lag one are equal, since the higher-order autocorrelations die out very quickly in the AR model'.

1.6 The relationship between conditional variance and conditional mean

1.6.1 The ARCH in mean model

Financial theory suggests that an asset with a higher expected risk would pay a higher return on average. Let y_t denote the rate of return of a particular stock or market portfolio from time t to time $t-1$ and rf_t be the return on a riskless asset (e.g. treasury bills). Then, the excess return (asset return minus the return on a riskless asset) can be decomposed into a component anticipated by investors at time $t-1$, μ_t, and a component that was unanticipated, ε_t:

$$y_t - rf_t = \mu_t + \varepsilon_t. \tag{1.15}$$

The relationship between investors' expected return and risk was presented in an ARCH framework by Engle et al. (1987). They introduced the *ARCH in mean*, or ARCH-M, model where the conditional mean is an explicit function of the conditional variance of the process in framework (1.5). The estimated coefficient on the expected risk is a measure of the risk–return trade-off. Thus, the ARCH regression model in framework (1.5) can be written as

$$y_t = x_t'\beta + \varphi\left(\sigma_t^2\right) + \varepsilon_t,$$
$$\varepsilon_t | I_{t-1} \sim f\left(0, \sigma_t^2\right), \tag{1.16}$$
$$\sigma_t^2 = g(\sigma_{t-1}, \sigma_{t-2}, \ldots; \varepsilon_{t-1}, \varepsilon_{t-2}, \ldots; \upsilon_{t-1}, \upsilon_{t-2}, \ldots),$$

where, as before, x_t is a $k \times 1$ vector of endogenous and exogenous explanatory variables included in the information set I_{t-1} and $\phi(\sigma_t^2)$ represents the risk premium, i.e., the increase in the expected rate of return due to an increase in the variance of the return.[11] Although earlier studies concentrated on detecting a constant risk premium, the ARCH-M model provided a new approach by which a time-varying risk premium could be estimated. The most commonly used specifications of the ARCH-M model are in the form

$$\phi(\sigma_t^2) = c_0 + c_1 \sigma_t^2 \qquad (1.17)$$

(Nelson, 1991; Bollerslev et al., 1994),

$$\phi(\sigma_t^2) = c_0 + c_1 \sigma_t \qquad (1.18)$$

(Domowitz and Hakkio, 1985; Bollerslev et al., 1988), or

$$\phi(\sigma_t^2) = c_0 + c_1 \log(\sigma_t^2) \qquad (1.19)$$

(Engle et al., 1987). A positive as well as a negative risk–return trade-off could be consistent with the financial theory. A positive relationship is expected if we assume a rational risk-averse investor who requires a larger risk premium during the times when the payoff of the security is riskier. On the other hand, a negative relationship is expected under the assumption that during relatively riskier periods the investors may want to save more. In applied research work, there is evidence for both positive and negative relationships. The relationship between the investor's expected return and risk is explored in the empirical example of Section 5.4. French et al. (1987) found a positive risk–return trade-off for the excess returns on the S&P500 composite portfolio, albeit not statistically significant in all the periods examined. Nelson (1991) observed a negative but insignificant relationship for the excess returns on the Center for Research in Security Prices (CRSP) value-weighted market index. Bollerslev et al. (1994) noted a positive, but not always statistically significant, relationship for the returns on the Dow Jones and S&P500 indices. Interesting studies employing the ARCH-M model were conducted by Devaney (2001) and Elyasiani and Mansur (1998). The former examined the trade-off between conditional variance and excess returns for commercial bank sector stocks, while the latter investigated the time-varying risk premium for real estate investment trusts. Theoretical results on the moment structure of the ARCH-M model were investigated by Arvanitis and Demos (2004) and Demos (2002).

---

[11] x_t is a vector of explanatory variables. The risk free rate of return, rf_t, could be incorporated as an explanatory variable, e.g. $x_t' = (rf_t)$ and $\beta = (1)$, or $rf_t + \mu_t = x_t'\beta + \varphi(\sigma_t^2) = (rf_t, \ldots)(1, \ldots)'$.

1.6.2 Volatility and serial correlation

LeBaron (1992) noted a strong inverse relationship between volatility and serial correlation for the returns of the S&P500 index, the CRSP value-weighted market index, the Dow Jones and the IBM returns. He introduced the *exponential autoregressive GARCH*, or EXP-GARCH(p,q), model in which the conditional mean is a non-linear function of the conditional variance. Based on LeBaron (1992), the ARCH regression model, in framework (1.5), can be written as:

$$y_t = x_t'\beta + \left(c_1 + c_2 \exp\left(-\sigma_t^2/c_3\right)\right)y_{t-1} + \varepsilon_t,$$
$$\varepsilon_t|I_{t-1} \sim f\left(0, \sigma_t^2\right), \tag{1.20}$$
$$\sigma_t^2 = g(\sigma_{t-1}, \sigma_{t-2}, \ldots; \varepsilon_{t-1}, \varepsilon_{t-2}, \ldots; \upsilon_{t-1}, \upsilon_{t-2}, \ldots).$$

The model is a mixture of the GARCH model and the exponential AR model of Ozaki (1980). For the data set LeBaron used, c_2 is significantly negative and remarkably robust to the choice of sample period, market index, measurement interval and volatility measure. Generally, the first-order autocorrelations are larger during periods of lower volatility and smaller during periods of higher volatility. The accumulation of news (see Section 1.4) and the non-synchronous trading (see Section 1.5) were mentioned as possible reasons. The stocks do not trade close to the end of the day and information arriving during this period is reflected in the next day's trading, inducing serial correlation. As new information reaches the market very slowly, traders' optimal action is to do nothing until enough information is accumulated. Because of the non-trading, the trading volume, which is strongly positive related with volatility, falls. Thus, we have a market with low trade volume and high correlation. The inverse relationship between volatility and serial correlation is explored in the empirical example of Section 5.4.

Kim (1989), Sentana and Wadhwani (1991) and Oedegaard (1991) have also investigated the relationship between autocorrelation and volatility and found an inverse relationship between volatility and autocorrelation. Moreover, Oedegaard (1991) found that the evidence of autocorrelation, for the S&P500 daily index, decreased over time, possibly because of the introduction of financial derivatives (options and futures) on the index.

2

ARCH volatility specifications

In this chapter, a number of univariate ARCH models are presented and their estimation is discussed. The main features of what seem to be most widely used ARCH models are described, with emphasis on their practical relevance. No attempt is made to cover the whole of the literature on the technical details of the models, which is very extensive. The aim is to give the broad framework of the most important models used today in financial applications. A careful selection of references is provided so that the interested reader can make a more detailed examination of specific topics. In particular, an anthology of representations of ARCH models that have been considered in the literature is provided.

2.1 Model specifications

In the literature, a large number of specifications of ARCH models have been considered for the description of the characteristics of financial markets. A wide range of proposed ARCH processes are covered in surveys such as Andersen and Bollerslev (1998c), Bera and Higgins (1993), Bollerslev et al. (1992, 1994), Degiannakis and Xekalaki (2004), Diebold and Lopez (1995), Gouriéroux (1997), Li et al. (2001) and Palm (1996). A good account of the state of the art up to 1995 can be found in Engle (1995). Theoretical or applied aspects of ARCH modelling have been covered in many textbooks: Asteriou (2006), Brooks (2002), Campbell et al. (1997), Chan (2002), Christoffersen (2003), Clements (2005), Cuthbertson and Nitzsche (2004), Enders (2003), Franses and van Dijk (2000), Gouriéroux and Jasiak (2001), Greene (1997), Hamilton (1994), Harris and Sollis (2003), Hill et al. (2000), Judge et al. (1985), Laurent and Peters (2004), Lütkepohl (2005), Lütkepohl and Krätzig (2004), Mills (1999), Taylor (2005), Tsay (2002) and Zivot and Jiahui (2006).

ARCH Models for Financial Applications   Evdokia Xekalaki and Stavros Degiannakis
© 2010 John Wiley & Sons, Ltd

Engle (1982) introduced the original form of $\sigma_t^2 = g(.)$, in equation (1.3), as a linear function of the past q squared innovations:

$$\sigma_t^2 = a_0 + \sum_{i=1}^{q} \left(a_i \varepsilon_{t-i}^2 \right). \tag{2.1}$$

For the linear ARCH(q) process to be well defined and the conditional variance to be positive, the parameters must satisfy $a_0 > 0$, $a_i \geq 0$, for $i = 1, \ldots, q$. An equivalent representation of the ARCH(q) process is given by

$$\sigma_t^2 = a_0 + A(L)\varepsilon_t^2, \tag{2.2}$$

where L denotes the lag operator and $A(L) = \left(a_1 L + a_2 L^2 + \cdots + a_q L^q \right)$. Defining $v_t = \varepsilon_t^2 - \sigma_t^2$, the model is rewritten as

$$\varepsilon_t^2 = a_0 + A(L)\varepsilon_t^2 + v_t. \tag{2.3}$$

By definition, v_t is serially uncorrelated with $E_{t-1}(v_t) = 0$, but neither independently nor identically distributed. The ARCH(q) model is interpreted as an autoregressive process in the squared innovations (see also Bera et al., 1992; Tsay, 1987) and is covariance stationary if and only if the roots of $\sum_{i=1}^{q} \left(a_i L^i \right) = 1$ lie outside the unit circle, or, equivalently, the sum of the positive autoregressive parameters is less than 1. If the process is covariance stationary, its unconditional variance is equal to $V(\varepsilon_t) \equiv \sigma^2 = a_0 \left(1 - \sum_{i=1}^{q} (a_i) \right)^{-1}$.

Also, by definition, the innovation process is serially uncorrelated, but not independently distributed. On the other hand, the standardized innovations are time invariant distributed. Thus, the unconditional distribution for the innovation process will have fatter tails than the distribution for the standardized innovations. For example, the kurtosis for the ARCH(1) process with conditional normally distributed innovations is $E(\varepsilon_t^4)/E(\varepsilon_t^2)^2 = 3(1-a_1^2)/(1-3a_1^2)$ if $3a_1^2 < 1$, and $E(\varepsilon_t^4)/E(\varepsilon_t^2)^2 = \infty$ otherwise, i.e., greater than 3, the kurtosis value of the normal distribution. Generally speaking, an ARCH process always has fatter tails than the normal distribution:

$$E(\varepsilon_t^4)/E(\varepsilon_t^2)^2 = E(\sigma_t^4 z_t^4)/E(\sigma_t^2 z_t^2)^2 = 3E(\sigma_t^4)/E(\sigma_t^2)^2 \geq 3E(\sigma_t^2)^2/E(\sigma_t^2)^2, \tag{2.4}$$

where the first equality comes from the independence of σ_t and z_t, and the inequality is implied by Jensen's inequality.

In empirical applications of the ARCH(q) model, a relatively long lag in the conditional variance equation is often called for, and to avoid problems of negative variance parameter estimates a fixed lag structure is typically imposed (see, for example, Engle, 1982, 1983; Engle and Kraft, 1983). To circumvent this problem, Bollerslev (1986) proposed a generalization of the ARCH(q) process to allow for past conditional variances in the current conditional variance equation, the generalized ARCH, or GARCH(p, q), model:

$$\sigma_t^2 = a_0 + \sum_{i=1}^{q} \left(a_i \varepsilon_{t-i}^2 \right) + \sum_{j=1}^{p} \left(b_j \sigma_{t-j}^2 \right) = a_0 + A(L)\varepsilon_t^2 + B(L)\sigma_t^2. \tag{2.5}$$

For $a_0 > 0$, $a_i \geq 0$, $i = 1, \ldots, q$ and $b_j \geq 0$, $j = 1, \ldots, p$, the conditional variance is well defined. Taylor (1986) independently proposed the GARCH model using a different acronym. Nelson and Cao (1992) showed that the non-negativity constraints on the parameters of the process could be substantially weakened, so they should not be imposed in estimation. Provided that the roots of $B(L) = 1$ lie outside the unit circle and the polynomials $1-B(L)$ and $A(L)$ have no common roots, the positivity constraint is satisfied if all the coefficients in the infinite power series expansion for $B(L)(1-B(L))^{-1}$ are non-negative. In the GARCH(1,2) model, for example, the conditions of non-negativity are that $a_0 \geq 0$, $0 \leq b_1 < 1$, $a_1 \geq 0$ and $b_1 a_1 + a_2 \geq 0$. In the GARCH(2,1) model, the necessary conditions require that $a_0 \geq 0$, $b_1 \geq 0$, $a_1 \geq 0$, $b_1 + b_2 < 1$ and $b_1^2 + 4b_2 \geq 0$. Thus, slightly negative values of parameters, for higher-order lags, do not result in negative conditional variance. Rearranging the GARCH(p,q) model, it can be presented as an autoregressive moving average process in the squared innovations of orders max (p,q) and p, ARMA (max $(p,q),p$), respectively:

$$\varepsilon_t^2 = a_0 + \sum_{i=1}^{q} \left(a_i \varepsilon_{t-i}^2 \right) + \sum_{j=1}^{p} \left(b_j \varepsilon_{t-j}^2 \right) - \sum_{j=1}^{p} \left(b_j v_{t-j} \right) + v_t. \tag{2.6}$$

The model is second-order stationary if the roots of $A(L) + B(L) = 1$ lie outside the unit circle, or equivalently if $\sum_{i=1}^{q} a_i + \sum_{j=1}^{p} b_j < 1$. Its unconditional variance is equal to $\sigma^2 = a_0 \left(1 - \sum_{i=1}^{q} a_i - \sum_{j=1}^{p} b_j \right)^{-1}$.

Very often, in connection with applications, the estimate for $A(L) + B(L)$ turns out to be very close to unity. This provided an empirical motivation for the development of the so-called integrated GARCH(p,q), or IGARCH(p,q), model by Engle and Bollerslev (1986):

$$\sigma_t^2 = a_0 + A(L)\varepsilon_t^2 + B(L)\sigma_t^2, \quad \text{for } A(L) + B(L) = 1, \tag{2.7}$$

where the polynomial $A(L) + B(L) = 1$ has $d > 0$ unit roots and max $(p,q)-d$ roots outside the unit circle.

Moreover, Nelson (1990a) showed that the GARCH(1,1) model is strictly stationary even if $a_1 + b_1 > 1$, as long as $E(\log(b_1 + a_1 z_t^2)) < 0$. Thus, the conditional variance in IGARCH(1,1) with $a_0 = 0$ collapses to zero almost surely, and in IGARCH(1,1) with $a_0 > 0$ is strictly stationary. Therefore, a process that is integrated in the mean is not stationary in any sense, while an IGARCH process is strictly stationary but covariance non-stationary.

Consider the IGARCH(1,1) model, $\sigma_t^2 = a_0 + a_1 \varepsilon_{t-1}^2 + (1-a_1)\sigma_{t-1}^2$, where $0 < a_1 < 1$. The conditional variance h steps in the future takes the form

$$E_t \left(\sigma_{t+h}^2 \right) = \sigma_{t+h|t}^2 = \sigma_t^2 + ha_0, \tag{2.8}$$

which looks very much like a linear random walk with drift a_0. A linear random walk is strictly non-stationary (no stationary distribution and covariance non-stationary) and it has no unconditional first or second moments. In the case of IGARCH(1,1), the conditional variance is strictly stationary even though its stationary distribution generally lacks unconditional moments. If $a_0 = 0$, equation (2.8) reduces to $\sigma_{t+h|t}^2 \equiv \sigma_t^2$, a bounded martingale which cannot take negative values. According to the martingale convergence theorem (Dudley, 1989), a bounded martingale must converge, and, in this case, the only value to which it can converge is zero. Thus, the stationary distributions for σ_t^2 and ε_t have moments, but they are all trivially zero. In the case of $a_0 > 0$, Nelson (1990a) showed that there is a non-degenerate stationary distribution for the conditional variance, but with no finite mean or higher moments. The innovation process ε_t then has a stationary distribution with zero mean, but with tails that are so thick that no second- or higher-order moments exist. Furthermore, if the variable z_t follows the standard normal distribution, Nelson (1990a) showed that

$$E\left(\log\left(b_1 + a_1 z_t^2\right)\right) = \log\left(2a_1\right) + \psi(1/2) + \left(2\pi b_1 a_1^{-1}\right)^{1/2}\Phi(0.5; 1.5; b_1/2a_1)$$
$$-(b_1/a_1)_2 F_2(1, 1; 2, 1.5; b_1/2a_1), \tag{2.9}$$

where $\psi(.)$ denotes the Euler psi function, with $\psi(1/2) \approx -1.96351$ (see Davis, 1965), $\Phi(.;.;.)$ is the confluent hypergeometric function (see Lebedev, 1972), and $_2F_2 (.,.;.,;.)$ is the generalized hypergeometric function (see Lebedev, 1972). Bougerol and Picard (1992) extended Nelson's work and showed that the general GARCH(p,q) model is strictly stationary and ergodic. Choudhry (1995), by means of the IGARCH(1,1) model, studied the persistence of stock return volatility in European markets during the 1920s and 1930s and argued that the 1929 stock market crash did not reduce stock market volatility. Using monthly stock returns from 1919 to 1936 in the markets of Czechoslovakia, France, Italy, Poland and Spain, Choudhry noted that in the GARCH(1,1) model the sum of a_1 and b_1 approaches unity, which implies persistence of a forecast of the conditional variance over all finite horizons.

J.P. Morgan (1996), in introducing their analytic methodology for the computation of value-at-risk, RiskMetrics™, considered the exponentially weighted moving average (EWMA) model,

$$\sigma_t^2 = 0.06\varepsilon_{t-1}^2 + 0.94\sigma_{t-1}^2, \tag{2.10}$$

which is a special case of the IGARCH(1,1) model for $a_0 = 0$ and $a_1 = 0.06$.[1]

Knight and Satchell (2002a) presented an alternative to the GARCH(1,1) model, the alternative GARCH, or ALT-GARCH, by replacing ε_{t-1}^2 by z_{t-1}^2:

$$\sigma_t^2 = a_0 + a_1 z_{t-1}^2 + b_1\sigma_{t-1}^2. \tag{2.11}$$

This minor modification helps to overcome the inability to obtain distributional results for either the conditional variance or the dependent variable. However, as Knight and Satchell (2002a) showed, the ALT-GARCH model did not produce a better fit to empirical data.

---

[1] Guermat and Harris (2002) derived the exponentially weighted maximum likelihood (EWML) procedure that potentially allows for time variation not only in the conditional variance of the return distribution, but also in its higher moments. The EWMA variance estimator can be obtained as a special case of the EWML.

2.2 Methods of estimation

2.2.1 Maximum likelihood estimation

In ARCH models, the most commonly used method in estimating the vector of
unknown parameters, θ, is the method of maximum likelihood (ML) estimation.
Under the assumption of independently and identically distributed standardized
innovations, $z_t(\theta) \equiv \varepsilon_t(\theta)/\sigma_t(\theta)$, in framework (1.6), let us denote their density
function by $f(z_t; w)$, where $w \in W \subseteq R^{\tilde{w}}$ is the vector of the parameters of f to be
estimated. So, for $\psi' = (\theta', w')$ denoting the whole set of the $\tilde{\psi} = \tilde{\theta} + \tilde{w}$ parameters[2]
that have to be estimated for the conditional mean, variance and density function, the
log-likelihood function for $\{y_t(\theta)\}$ is

$$l_t(y_t; \psi) = \log\left(f(z_t(\theta); w)\right) - \frac{1}{2}\log\left(\sigma_t^2(\theta)\right). \tag{2.12}$$

The full sample log-likelihood function for a sample of T observations is simply

$$L_T(\{y_t\}; \psi) = \sum_{t=1}^{T} l_t(y_t; \psi). \tag{2.13}$$

If the conditional density, the mean and the variance functions are differentiable for
each possible $\psi \in \Theta \times W \equiv \Psi \subseteq R^{\tilde{\psi}}$, the maximum likelihood estimator (MLE) $\hat{\psi}$
for the true parameter vector ψ_0 is found by maximizing equation (2.13), or
equivalently by solving the equation

$$\sum_{t=1}^{T} \frac{\partial l_t(y_t; \psi)}{\partial \psi} = 0. \tag{2.14}$$

If the density function does not require the estimation of any parameter, as in the case
of the normal distribution that is uniquely determined by its first two moments, then
$\tilde{w} = 0$. In such cases, equation (2.14) becomes[3]

$$\sum_{t=1}^{T} \left(\left(f(z_t(\theta))\right)^{-1} \frac{\partial f(z_t(\theta))}{\partial \theta} \left(-\frac{\partial \mu_t(\theta)}{\partial \theta} \sigma_t^2(\theta)^{-1/2} - 0.5\sigma_t^2(\theta)^{-3/2} \varepsilon_t(\theta) \frac{\partial \sigma_t^2(\theta)}{\partial \theta} \right) \right.$$
$$\left. - \frac{1}{2}\sigma_t^2(\theta)^{-1} \frac{\sigma_t^2(\theta)}{\partial \theta} \right) = 0. \tag{2.15}$$

---

[2] $\tilde{\theta}$ and $\tilde{w}$ denote the numbers of parameters of vectors θ and w, respectively.
[3] Equation (2.12) can be written as

$$\frac{\partial l_t(y_t; \theta)}{\partial \theta} = f(z_t(\theta))^{-1} \frac{\partial(f(z_t(\theta)))}{\partial \theta} \frac{\partial(z_t(\theta))}{\partial \theta} - \frac{1}{2}\sigma_t^2(\theta)^{-1} \frac{\sigma_t^2(\theta)}{\partial \theta}.$$

Note also that

$$\frac{\partial z_t(\theta)}{\partial \theta} = \frac{\partial\left((y_t - \mu_t(\theta))\sigma_t^2(\theta)^{-1/2}\right)}{\partial \theta} = -\frac{\partial \mu_t(\theta)}{\partial \theta}\sigma_t^2(\theta)^{-1/2} - \frac{1}{2}\sigma_t^2(\theta)^{-3/2}\frac{\partial \sigma_t^2(\theta)}{\partial \theta}\varepsilon_t(\theta).$$

Let us, for example, estimate the parameters of framework (1.6) for normal distributed innovations and the GARCH(p,q) functional form for the conditional variance as given in equation (2.5). The density function of the standard normal distribution is

$$f(z_t) = \frac{1}{\sqrt{2\pi}} \exp\left(-\frac{z_t^2}{2}\right). \tag{2.16}$$

For convenience, equation (2.5) is written as $\sigma_t^2 = \omega' s_t$, where $\omega' = (a_0, a_1, \ldots, a_q, b_1, \ldots, b_p)$ and $s_t = \left(1, \varepsilon_{t-1}^2, \ldots, \varepsilon_{t-q}^2, \sigma_{t-1}^2, \ldots, \sigma_{t-p}^2\right)$. The vector of parameters that have to be estimated is $\psi' = \theta' = (\beta', \omega')$, where β denotes a $k \times 1$ vector of unknown parameters as defined in equation (1.5). For normally distributed standardized innovations, z_t, the log-likelihood function in equation (2.12) is

$$l_t(y_t; \psi) = \log\left(\frac{1}{\sqrt{2\pi}}\right) - \frac{\left(y_t - x_t'\beta\right)^2}{2\sigma_t^2} - \frac{1}{2}\log\left(\sigma_t^2\right), \tag{2.17}$$

with x_t denoting a $k \times 1$ vector of endogenous and exogenous explanatory variables included in the information set I_{t-1}, and the full sample log-likelihood function in equation (2.13) becomes

$$L_T(\{y_t\}; \theta) = -\frac{1}{2}\left(T \log(2\pi) + \sum_{t=1}^{T} \frac{\left(y_t - x_t'\beta\right)^2}{\sigma_t^2} + \sum_{t=1}^{T} \log\left(\sigma_t^2\right)\right). \tag{2.18}$$

The first and the second derivatives of the log-likelihood for the tth observation with respect to the variance parameter vector are

$$\frac{\partial l_t(y_t; \beta, \omega)}{\partial \omega} = \frac{1}{2\sigma_t^2} \frac{\partial \sigma_t^2}{\partial \omega} \left(\frac{\varepsilon_t^2 - \sigma_t^2}{\sigma_t^2}\right), \tag{2.19}$$

$$\frac{\partial^2 l_t(y_t; \beta, \omega)}{\partial \omega \partial \omega'} = \left(\frac{\varepsilon_t^2 - \sigma_t^2}{\sigma_t^2}\right) \frac{\partial}{\partial \omega'}\left[\frac{1}{2\sigma_t^2} \frac{\partial \sigma_t^2}{\partial \omega}\right] - \frac{1}{2\sigma_t^4} \frac{\partial \sigma_t^2}{\partial \omega} \frac{\partial \sigma_t^2}{\partial \omega'} \frac{\varepsilon_t^2}{\sigma_t^2}, \tag{2.20}$$

where

$$\frac{\partial \sigma_t^2}{\partial \omega} = s_t + \sum_{i=1}^{p} b_i \frac{\partial \sigma_{t-1}^2}{\partial \omega}. \tag{2.21}$$

The first and second derivatives of the log-likelihood with respect to the mean parameter vector are

$$\frac{\partial l_t(y_t; \beta, \omega)}{\partial \beta} = \varepsilon_t x_t \sigma_t^{-2} + \frac{1}{2}\sigma_t^2 \frac{\partial \sigma_t^2}{\partial \beta}\left(\frac{\varepsilon_t^2 - \sigma_t^2}{\sigma_t^2}\right), \tag{2.22}$$

$$\frac{\partial^2 l_t(y_t; \beta, \omega)}{\partial\beta\partial\beta'}$$

$$= -\sigma_t^{-2} x_t x_t' - \frac{1}{2}\sigma_t^{-4}\frac{\partial\sigma_t^2}{\partial\beta}\frac{\partial\sigma_t^2}{\partial\beta'}\left(\frac{\varepsilon_t^2}{\sigma_t^2}\right) - 2\sigma_t^{-4}\varepsilon_t x_t \frac{\partial\sigma_t^2}{\partial\beta} + \left(\frac{\varepsilon_t^2 - \sigma_t^2}{\sigma_t^2}\right)\frac{\partial}{\partial\beta'}\left[\frac{1}{2}\sigma_t^{-2}\frac{\partial\sigma_t^2}{\partial\beta}\right],$$

$$(2.23)$$

where

$$\frac{\partial\sigma_t^2}{\partial\beta} = -2\sum_{i=1}^{q} a_i x_{t-i}\varepsilon_{t-i} + \sum_{j=1}^{p} b_j \frac{\partial\sigma_{t-j}^2}{\partial\beta}. \qquad (2.24)$$

The information matrix corresponding to ω is given by

$$I_{\omega\omega} = \frac{-1}{T}\sum_{t=1}^{T}\left(E\left(\frac{\partial^2 l_t(y_t; \beta, \omega)}{\partial\omega\partial\omega'}\right)\right) = \frac{1}{2T}\sum_{t=1}^{T}\left(\sigma_t^{-4}\frac{\partial\sigma_t^2}{\partial\omega}\frac{\sigma_t^2}{\partial\omega'}\right). \qquad (2.25)$$

The information matrix corresponding to b is given by

$$I_{\beta\beta} = \frac{-1}{T}\sum_{t=1}^{T}\left(E\left(\frac{\partial^2 l_t(y_t; \beta, \omega)}{\partial\beta\partial\beta'}\right)\right)$$

$$= \frac{1}{T}\sum_{t=1}^{T}\left(\sigma_t^{-2} x_t x_t' + 2\sigma_t^{-4}\sum_{i=1}^{q} a_i^2\varepsilon_{t-i}' x_{t-i} x_{t-i}'\varepsilon_{t-i} + \frac{1}{2}\sum_{j=1}^{p} b_j^2\left(\frac{\partial\sigma_{t-j}^2}{\partial\beta}\right)^2\right). \qquad (2.26)$$

The elements in the off-diagonal block of the information matrix are zero:

$$I_{\omega\beta} = \frac{-1}{T}\sum_{t=1}^{T}\left(E\left(\frac{\partial^2 l_t(y_t; \beta, \omega)}{\partial\omega\partial\beta'}\right)\right) = 0. \qquad (2.27)$$

So ω can be estimated without loss of asymptotic efficiency based on a consistent estimate of β, and vice versa. It should be noted at this point that although block-diagonality holds for models such as the GARCH, NARCH and Log-GARCH models, it does not hold for asymmetric models, such as the EGARCH, nor for the ARCH in mean models. In such cases, the parameters have to be estimated jointly.

Even in the case of the symmetric GARCH(p, q) model with normally distributed innovations, we have to solve a set of $\tilde\theta = k + p + q + 1$ non-linear equations in (2.15). Numerical techniques are employed to estimate the vector of parameters ψ.

2.2.2 Numerical estimation algorithms

The problem faced in non-linear estimation, as in the case of the ARCH models, is that there are no closed-form solutions. An iterative method therefore has to be applied to obtain a solution. Iterative optimization algorithms work by taking an initial set of values for the parameters, say $\psi^{(0)}$, and performing calculations based on these values to obtain a better set of parameter values $\psi^{(1)}$. This process is repeated until the likelihood function (2.13) no longer improves between iterations. If $\psi^{(0)}$ is a trial value of the estimate, then expanding $L_T(\{y_t\}; \psi)/\partial\psi$ and retaining only the first power of $\psi - \psi^{(0)}$, we obtain

$$\frac{\partial L_T}{\partial\psi} \approx \frac{\partial L_T}{\partial\psi^{(0)}} + \left(\psi - \psi^{(0)}\right)\frac{\partial^2 L_T}{\partial\psi^{(0)}\partial\psi'^{(0)}}. \tag{2.28}$$

At the maximum, $L_T/\partial\psi$ should equal zero. Rearranging terms, the correction for the initial value, $\psi^{(0)}$, obtained is

$$\left(\psi - \psi^{(0)}\right) = -\frac{\partial L_T}{\partial\psi^{(0)}}\left(\frac{\partial^2 L_T}{\partial\psi^{(0)}\partial\psi'^{(0)}}\right)^{-1}. \tag{2.29}$$

Let $\psi^{(i)}$ denote the parameter estimates after the ith iteration. Based on (2.29), the Newton–Raphson algorithm computes $\psi^{(i+1)}$ as

$$\psi^{(i+1)} = \psi^{(i)} - \left(\frac{\partial^2 L_T^{(i)}}{\partial\psi\partial\psi'}\right)^{-1}\frac{\partial L_T^{(i)}}{\partial\psi}, \tag{2.30}$$

where $\partial^2 L_T^{(i)}/\partial\psi\partial\psi'$ and $\partial L_T^{(i)}/\partial\psi$ are evaluated at $\psi^{(i)}$.

The scoring algorithm is a method closely related to the Newton–Raphson algorithm and was applied by Engle (1982) to estimate the parameters of the ARCH(p) model. The difference between the Newton–Raphson method and the method of scoring is that the former depends on observed second derivatives, while the latter depends on the expected values of the second derivatives. Thus, the scoring algorithm computes $\psi^{(i+1)}$ as

$$\psi^{(i+1)} = \psi^{(i)} + E\left(\frac{\partial^2 L_T^{(i)}}{\partial\psi\partial\psi'}\right)^{-1}\frac{\partial L_T^{(i)}}{\partial\psi}. \tag{2.31}$$

An alternative procedure (Berndt et al., 1974), which uses first derivatives only, is the Berndt, Hall, Hall and Hausman (BHHH) algorithm. The BHHH algorithm is similar to the Newton–Raphson algorithm, but, instead of the Hessian, $H^{(i)}$ (second derivative of the log-likelihood function with respect to the vector of unknown parameters), it is based on an approximation formed by the sum of the outer product of the gradient vectors for the contribution of each observation to the objective function. This approximation is asymptotically equivalent to the actual Hessian when evaluated at the parameter values which maximize the function. The BHHH algorithm computes $\psi^{(i+1)}$ as

$$\psi^{(i+1)} = \psi^{(i)} - \left(\sum_{t=1}^{T}\frac{\partial l_t^{(i)}}{\partial\psi}\frac{\partial l_t^{(i)}}{\partial\psi'}\right)^{-1}\frac{\partial L_T^{(i)}}{\partial\psi}. \tag{2.32}$$

When the outer product is near singular, a ridge correction may be used in order to handle numerical problems and improve the convergence rate. The BHHH algorithm is an optional method of model estimation in EViews. It can be modified by employing Marquardt's (1963) technique of adding a correction matrix to the sum of the outer product of the gradient vectors. The Marquardt updating algorithm is computed as

$$\psi^{(i+1)} = \psi^{(i)} - \left(\sum_{t=1}^{T} \frac{\partial l_t^{(i)}}{\partial \psi} \frac{\partial l_t^{(i)}}{\partial \psi'} - aI \right)^{-1} \frac{\partial L_T^{(i)}}{\partial \psi}, \tag{2.33}$$

where $\partial l_t^{(i)}/\partial \psi$ and $\partial L_T^{(i)}/\partial \psi$ are evaluated at $\psi^{(i)}$, I is the identity matrix and a is a positive number chosen by the algorithm. The effect of this modification is to push the parameter estimates in the direction of the gradient vector. The idea is that when we are far from the maximum, the local quadratic approximation to the function may be a poor guide to its overall shape, so we may be better off simply following the gradient. The correction may provide better performance at locations far from the optimum, and allows for computation of the direction vector in cases where the Hessian is near singular. The Marquardt optimization method is the default method of likelihood maximization in EViews.

Quasi-Newton methods use an initial estimate and computational history to generate an estimate $\tilde{H}^{(i)}$ of the Hessian matrix $H^{(i)}$ at each step i rather than performing the computational work of evaluating and inverting $\partial^2 L_T^{(i)}/\partial \psi \partial \psi'$. In general, $\psi^{(i+1)}$ is computed as

$$\psi^{(i+1)} = \psi^{(i)} - a\tilde{H}^{(i)^{-1}} \frac{\partial L_T^{(i)}}{\partial \psi}, \tag{2.34}$$

where a is a scalar appropriately chosen by the algorithm. In each iteration i, the inverse of the Hessian is approximated by adding a correction matrix $C^{(i-1)}$ to the approximate inverse of the Hessian used in the most recent step, $i-1$:

$$\tilde{H}^{(i)} = \tilde{H}^{(i-1)} + C^{(i-1)}. \tag{2.35}$$

Equation (2.35) can also be applied directly to the Hessian instead of its inverse. Different correction matrices have been suggested in the literature, but the Davidon–Fletcher–Powell (DFP) and Broyden–Fletcher–Goldfarb–Shanno (BFGS) algorithms are the most widely used. The DFP algorithm was proposed by Davidon (1959) and modified by Fletcher and Powell (1963). The BFGS method was developed by Broyden (1970), Fletcher (1970), Goldfarb (1970) and Shanno (1970). See, among many others, Broyden (1965, 1967), Fletcher (1987), Gill et al. (1981), Shanno (1985), Cramér (1986) and Thisted (1988). Details of the optimization algorithms are available in Quandt (1983), Judge et al. (1985) and Press et al. (1992a, 1992b).

If the likelihood maximization must be achieved subject to non-linear constraints in the vector of unknown parameters, i.e. lower and upper bounds on the parameters, algorithms such Lawrence and Tits's (2001) feasible sequential quadratic programming (SQPF) technique can be used.

Finally, simulated annealing is a global optimization method for non-smooth functions that distinguishes between different local maxima based on the probabilistic Metropolis criteria. Since the simulated annealing algorithm makes very few assumptions regarding the function to be optimized, it is quite robust with respect to non-quadratic surfaces. See Goffe et al. (1994) for details.

2.2.3 Quasi-maximum likelihood estimation

The assumption of normally distributed standardized innovations is often violated by the data. This has motivated the use of alternative distributional assumptions, presented in Chapter 5. Alternatively, the MLE based on the normal density may be given a quasi-maximum likelihood interpretation. Bollerslev and Wooldridge (1992), based on results by Weiss (1986) and Pagan and Sabau (1987), showed that the maximization of the normal log-likelihood function can provide consistent estimates of the parameter vector θ even when the distribution of z_t in non-normal, provided that

$$E(z_t|I_{t-1}) = 0,$$
$$E\left(z_t^2|I_{t-1}\right) = 1. \tag{2.36}$$

This estimator is inefficient, however, with the degree of inefficiency increasing with the degree of departure from normality. Therefore, the standard errors of the parameters have to be adjusted. Let $\hat{\theta}$ be the estimate that maximizes the normal log-likelihood function (2.13), based on the normal density function (2.16), and let θ_0 be the true value. Then, even when z_t is non-normal, under certain regularity conditions:

$$\sqrt{T}\left(\hat{\theta}-\theta_0\right) \xrightarrow{D} N\left(0, A^{-1}BA^{-1}\right), \tag{2.37}$$

where

$$A \equiv p\lim_{T \to \infty} T^{-1} \sum_{t=1}^{T} E\left(\frac{-\partial^2 l_t(\theta_0)}{\partial\theta\partial\theta'}\right), \tag{2.38}$$

$$B \equiv p\lim_{T \to \infty} T^{-1} \sum_{t=1}^{T} E\left(\frac{\partial l_t(\theta_0)}{\partial\theta}\right)\left(\frac{\partial l_t(\theta_0)}{\partial\theta}\right)', \tag{2.39}$$

for l_t denoting the correctly specified log-likelihood function. The matrices A and B can be consistently estimated by

$$\hat{A} = -T^{-1} \sum_{t=1}^{T} E\left(\frac{\partial^2 l_t(\hat{\theta})}{\partial\theta\partial\theta'}\Big|I_{t-1}\right), \tag{2.40}$$

$$\hat{B} = T^{-1} \sum_{t=1}^{T} E\left(\left(\frac{\partial l_t(\hat{\theta})}{\partial\theta}\right)\left(\frac{\partial l_t(\hat{\theta})}{\partial\theta'}\right)\Big|I_{t-1}\right), \tag{2.41}$$

where l_t is the incorrectly specified log-likelihood function under the assumption of normal density function (see Bollerslev and Wooldridge, 1992; Engle and González-Rivera, 1991). Thus, standard errors for $\hat{\theta}$ that are robust to misspecification of the family of densities can be obtained from the square root of the diagonal elements of

$$T^{-1}\hat{A}^{-1}\hat{B}\hat{A}^{-1}. \tag{2.42}$$

Therefore, even if the normality assumption of the standardized innovations does not hold, but the conditional mean and variance are correctly specified, then the standard error estimates will still be consistent.

Recall that if the model is correctly specified and the data are in fact generated by the normal density function, then $A = B$ and, hence, the variance–covariance matrix, $T^{-1}\hat{A}^{-1}\hat{B}\hat{A}^{-1}$, reduces to the usual asymptotic variance–covariance matrix in maximum likelihood estimation

$$T^{-1}\hat{A}^{-1}. \tag{2.43}$$

For symmetric departures from normality, the quasi-maximum likelihood estimator (QMLE) is generally close to the exact MLE. But, for non-symmetric distributions, Engle and González-Rivera (1991) showed that the loss in efficiency may be quite high.[4] In such a case, other methods of estimation should be considered. Lumsdaine (1991, 1996) and Lee and Hansen (1991, 1994) established the consistency and asymptotic normality of the QMLEs of the IGARCH(1,1) model. Lee (1991) extended the asymptotic properties to the IGARCH(1,1) in mean model, Berkes et al. (2003) and Berkes and Horváth (2003) studied the asymptotic properties of the QMLEs of the GARCH(p, q) model under a set of weaker conditions, and Baillie et al. (1996a) showed that the QMLEs of the FIGARCH(1,d,0) model are both consistent and asymptotically normally distributed.

2.2.4 Other estimation methods

Estimation methods other than maximum likelihood have appeared in the ARCH literature. Harvey et al. (1992) presented the unobserved components structural ARCH, or STARCH, model and proposed an estimation method based on the Kalman filter. These are state-space models or factor models in which the innovation is composed of several sources of error, where each error source has a heteroscedastic specification of the ARCH form. Since the error components cannot be separately observed given the past observations, the independent variables in the variance equations are not measurable with respect to the available information set, which complicates inference procedures.

Pagan and Hong (1991) considered a non-parametric kernel estimate of the expected value of squared innovations. Pagan and Schwert (1990) used a collection of non-parametric estimation methods, including kernels, Fourier series and two-stage least squares regressions. They found that the non-parametric methods provided good in-sample forecasts though the parametric models yielded superior out-of-sample forecasts. Gouriéroux and Monfort (1992) also proposed a non-parametric estimation

---

[4] Bai and Ng (2001) proposed a procedure for testing conditional symmetry.

method in order to estimate the GQTARCH model in equation (2.71). Bühlmann and McNeil (2002) proposed a non-parametric estimation iterative algorithm that requires neither the specification of the conditional variance functional form nor that of the conditional density function, and showed that their algorithm gives more precise estimates of the volatility in the presence of departures from the assumed ARCH specification.

Engle and González-Rivera (1991), Engle and Ng (1993), Gallant and Tauchen (1989) and Gallant et al. (1991, 1993), among others, combined parametric specifications for the conditional variance with a non-parametric estimate of the conditional density function. In a Monte Carlo study, Engle and González-Rivera (1991) found that their semi-parametric method could improve the efficiency of the parameter estimates up to 50 per cent over the QMLE, particularly when the density was highly non-normal and skewed, but it did not seem to capture the total potential gain in efficiency.

Another attractive way to estimate ARCH models without assuming normality is to apply the generalized method of moments (GMM) approach (for details, see Bates and White, 1988; Ferson, 1989; Mark, 1988; Rich et al., 1991; Simon, 1989). Let us, for example, represent the GARCH(p,q) model as $\sigma_t^2 = \omega' s_t$, where $\omega' = (a_0, a_1, \ldots, a_q, b_1, \ldots, b_p)$ and $s_t = (1, \varepsilon_{t-1}^2, \ldots, \varepsilon_{t-q}^2, \sigma_{t-1}^2, \ldots, \sigma_{t-p}^2)$. Under the assumption that

$$
\begin{aligned}
E\big((y_t - x_t'\beta)x_t\big) &= 0, \\
E\big((\varepsilon_t^2 - \sigma_t^2)s_t\big) &= 0,
\end{aligned}
\tag{2.44}
$$

the parameters could be estimated by GMM by choosing the vector $\theta' = (\beta', \omega')$ so as to minimize

$$
(g(\theta; I_{t-1}))' \, \hat{S}(g(\theta; I_{t-1})),
\tag{2.45}
$$

where

$$
g(\theta; I_{t-1}) =
\begin{bmatrix}
T^{-1} \sum_{t=1}^{T} (y_t - x_t'\beta)x_t \\
T^{-1} \sum_{t=1}^{T} \big((y_t - x_t'\beta)^2 - \omega' s_t\big)s_t
\end{bmatrix}
\tag{2.46}
$$

and the matrix $\hat{S}$ can be constructed by any of the methods considered in the GMM literature.

Geweke (1988a, 1988b, 1989) argued that a Bayesian approach, rather than the classical one, might be more suitable for estimating ARCH models due to the distinct features of these models. In order to ensure positivity of the conditional variance, some inequality restrictions should be imposed. Although it is difficult to impose such restrictions in the classical approach, under the Bayesian framework diffuse priors can incorporate these inequalities. Also, as the main interest is not in the individual parameters but rather in the conditional variance itself, in the Bayesian framework exact posterior distributions of the conditional variance can be obtained. For a summary of recent developments in Bayesian approaches, see Bauwens and Lubrano (1998, 1999).

Giraitis and Robinson (2000) estimated the parameters of the GARCH process using the Whittle estimation technique and demonstrated that the Whittle estimator is strongly consistent and asymptotically normal, provided the GARCH process has a marginal distribution with eight finite moments. Whittle (1953) proposed an estimation technique that works in the spectral domain of the process.[5] Moreover, Mikosch and Straumann (2002) showed that the Whittle estimator is consistent as long as the fourth moment is finite and inconsistent when the fourth moment is infinite. Thus, as noted by Mikosch and Straumann, the Whittle estimator for GARCH processes is unreliable as the ARCH models are applied in heavy-tailed data, sometimes without finite fifth, fourth, or even third moments.

Hall and Yao (2003) showed that for heavy-tailed innovations, the asymptotic distribution of quasi-maximum likelihood parameter estimators is non-normal and suggested percentile-t subsample bootstrap approximations to estimator distributions.

2.3   Estimating the GARCH model with EViews 6: an empirical example

So far, we have dealt with the ARCH(p) model and its generalization, the GARCH(p, q) model, and reviewed some properties of the FTSE100 and S&P500 indices. In this section, we estimate the first conditional variance models for the FTSE100 and S&P500 equity indices, the US dollar to British pound (\$/£) exchange rate, and gold bullion. The data were obtained from Datastream for the period from 4 April 1988 to 5 April 2005. The daily prices and log-returns of the dollar–pound exchange rate and gold bullion are presented in Figures 2.1–2.4. Volatility clustering, or alternation between periods of high and low volatility, is clearly visible in Figures 2.3 and 2.4, thus suggesting the presence of heteroscedasticity. We denote the FTSE100, S&P500, \$/£ and gold daily log-returns processes as $\{y_{FTSE100,t}\}_{t=1}^T$, $\{y_{SP500,t}\}_{t=1}^T$, $\{y_{\$/£,t}\}_{t=1}^T$ and $\{y_{Gold,t}\}_{t=1}^T$, respectively. Figures 1.10 and 2.5 present the histograms of $\{y_{FTSE100,t}\}_{t=1}^T$ and $\{y_{SP500,t}\}_{t=1}^T$, and of $\{y_{\$/£,t}\}_{t=1}^T$ and $\{y_{Gold,t}\}_{t=1}^T$, respectively.

In what follows, framework (1.6) is estimated in the case of a first order autoregressive conditional mean, a GARCH(p, q) conditional variance specification, and normally distributed innovations. In particular, we deal with the estimation of the process given by

$$
\begin{aligned}
y_t &= c_0(1-c_1) + c_1 y_{t-1} + \varepsilon_t, \\
\varepsilon_t &= \sigma_t z_t, \\
\sigma_t^2 &= a_0 + \sum_{i=1}^q \left(a_i \varepsilon_{t-i}^2\right) + \sum_{j=1}^p \left(b_j \sigma_{t-j}^2\right), \\
z_t &\overset{i.i.d.}{\sim} N(0,1).
\end{aligned}
\tag{2.47}
$$

---

[5] For further details of the Whittle estimation technique for ARMA processes, see Brockwell and Davis (1991).

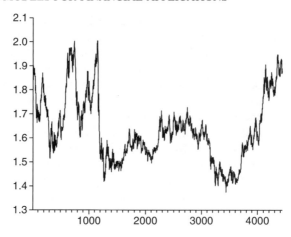

Figure 2.1 $/£ daily closing prices from 4 April 1988 to 5 April 2005.

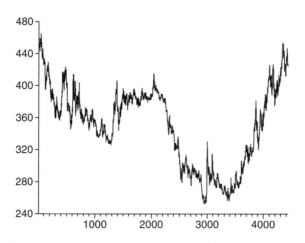

Figure 2.2   Gold bullion daily closing prices (US dollars per toy ounce) from 4 April 1988 to 5 April 2005.

The probability density function for the residuals is assumed to be the standard normal distribution with density function $f(z_t) = (2\pi)^{-1/2} \exp\left(-z_t^2/2\right)$. The AR(1) conditional mean is assumed to take into consideration the non-synchronous trading effect.[6] We consider estimating framework (2.47) for $p = 0, 1, 2$ and $q = 1, 2, 3$. Therefore, for each data set, we will estimate nine conditional variance specifications,

---

[6] An ARMA specification for the conditional mean, which includes a constant term, can be represented either as $\left(1 - \sum_{i=1}^{\kappa} c_i L^i\right) y_t = c_0 + \left(1 + \sum_{i=1}^{l} d_i L^i\right) \varepsilon_t$ or   as $y_t = c_0 + \varepsilon'_t$, $\left(1 - \sum_{i=1}^{\kappa} c_i L^i\right) \varepsilon'_t = \left(1 + \sum_{i=1}^{l} d_i L^i\right) \varepsilon_t$. The majority of packages, e.g. EViews and G@RCH from OxMetrics, estimate the second specification.

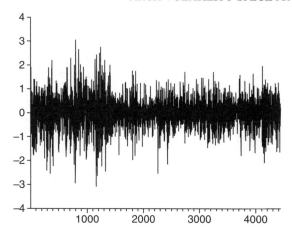

Figure 2.3 $/£ continuously compounded daily returns from 4 April 1988 to 5 April 2005.

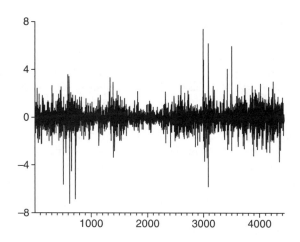

Figure 2.4   Gold bullion continuously compounded daily returns from 4 April 1988 to 5 April 2005.

leading to a total of 36 models. These will be estimated using EViews with the help of a program given in the appendix to this chapter. The program, named *chapter2_ GARCH.prg*, repeats the main code:

```
equation sp500_garch01.arch(1,0,m=10000,h) sp500ret c ar(1)
sp500_garch01.makegarch sp500_vargarch01
sp500_garch01.makeresid sp500_unresgarch01
sp500_garch01.makeresid(s) sp500_resgarch01
schwarz(1,1)=sp500_garch01.@schwarz
```

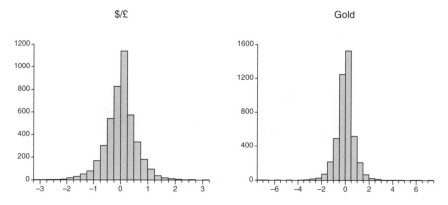

Figure 2.5   Histogram of $/£ and gold daily log-returns.

which computes the AR(1)-GARCH(p, q) model nine times for each index. The first line estimates an ARCH model with $p = 0$, $q = 1$. The conditional mean with *sp500ret* as the dependent variable has a constant (c) and first-order autoregressive form (AR(1)). The model (or equation in EViews) is named *sp500_garch01*. The option (h) takes into consideration the Bollerslev and Wooldridge robust quasi-maximum likelihood standard errors. The second line (*makegarch*) creates, for the model named *sp500_garch01*, the conditional variance. The Marquardt algorithm (equation (2.33)) is the default method for numerical maximization of the likelihood function. EViews provides the option of selecting the BHHH algorithm (equation (2.32)) by adding the option (b) in the model equation, i.e. *arch(1,0,b,m=10000,h)*. The conditional variance is saved in the *vargarch01* series. The third line (*makeresid*) creates, for the model named *sp500_garch01*, the residuals series named *unregarch01*, and the fourth line (*makeresid(s)*) constructs the standardized residuals series named *regarch01*. The fifth line saves the Schwarz (1978) information criterion of the *garch01* model in the matrix named *Schwarz*.

Based on the available data set of $T = 4435$ observations, the parameter vector $\psi^{(T)} = \left(c_0^{(T)}, c_1^{(T)}, a_0^{(T)}, a_1^{(T)}, \ldots, a_q^{(T)}, b_1^{(T)}, \ldots, b_p^{(T)}\right)$ is estimated. The conditional standard deviation estimations are computed as

$$\hat{\sigma}_{t+1} = \sqrt{a_0^{(T)} + \sum_{i=1}^{q} \left(a_i^{(T)} \hat{\varepsilon}_{t-i+1}^2\right) + \sum_{j=1}^{p} \left(b_j^{(T)} \hat{\sigma}_{t-j+1}^2\right)}, \qquad (2.48)$$

where

$$\hat{\varepsilon}_t = y_t - c_0^{(T)}\left(1 - c_1^{(T)}\right) - c_1^{(T)} y_{t-1}. \qquad (2.49)$$

Here, $\hat{\sigma}_{t+1}$ denotes the in-sample estimator of the conditional standard deviation at time $t+1$. The vector of the unknown parameters, $\psi^{(T)}$, has been estimated on the basis of all the T available observations.

We now look into how one would choose among the candidate models the model that best serves one's purposes. We start by applying Schwarz's (1978) Bayesian criterion (SBC) to select the model that fits the data best. The SBC is computed as

$$SBC = -2T^{-1}L_T\left(\{y_t\}; \hat{\psi}^{(T)}\right) + \tilde{\psi}T^{-1}\log(T),\qquad(2.50)$$

where $L_T(.)$ is the maximized value of the log-likelihood function, $\hat{\psi}^{(T)}$ is the MLE of ψ based on a sample of size T, and $\tilde{\psi}$ denotes the dimension of ψ.

A measure of the distance between the actual and estimated values could also serve as a useful method of model selection. However, in the area of volatility estimation, constructing such a measure is not a simple task. The actual variance is not observed and a proxy of the true but unobserved volatility must thus be employed. The daily squared log-returns or the squared demeaned log-returns are popular proxies of daily volatility. Although, the squared demeaned log-return, ε_t^2, constitutes an unbiased estimator of daily variance as $E_{t-1}\left(\varepsilon_t^2\right) = \sigma_t^2$, it is a noisy estimator of the unobserved volatility, as its conditional variance is

$$V_{t-1}\left(\varepsilon_t^2\right) = E_{t-1}\left(\left(z_t^2\sigma_t^2 - \sigma_t^2\right)^2\right) = \sigma_t^4 E_{t-1}\left(\left(z_t^2 - 1\right)^2\right) = \sigma_t^4(Ku-1).\qquad(2.51)$$

One should also keep in mind that the squared value of an ARCH process is the product of the actual variance times the square of a normally distributed process, $\varepsilon_t^2 = \sigma_t^2 z_t^2$. Hence, the squared residual, ε_t^2, as a proxy variable of σ_t^2 contains the noisy factor z_t^2 (see also Ebens, 1999).

The average squared distance between the in-sample estimated conditional variance and the squared residuals is computed by

$$\bar{\Psi}_{(SE)(i)} = T^{-1}\sum_{t=1}^{T}\Psi_{(SE)t(i)},\qquad(2.52)$$

where $\Psi_{(SE)t(i)} = \left(\hat{\varepsilon}_{t+1(i)}^2 - \hat{\sigma}_{t+1(i)}^2\right)^2$, $\hat{\varepsilon}_{t+1(i)}^2$ and $\hat{\sigma}_{t+1(i)}^2$ are the squared residuals and conditional variance, respectively, of model i, for $i = 1, \ldots, 9$. The program in the appendix, named *chapter2_MSE.prg*, repeats the main code:

```
matrix(4437,9) squared_dist_sp500
!k=0
for !p=0 to 2
    for !q=1 to 3
    !k=!k+1
    for !i=1 to 4437
squared_dist_sp500(!i,!k)=(sp500_vargarch{!p}{!q}(!i)-
sp500_unresgarch{!p}{!q}(!i)^2)^2
        next
```

```
mse(!k,1)=(@mean(@columnextract
(squared_dist_sp500,!k)))
next
next
```

which computes the $\bar{\Psi}_{(SE)(i)}$ function for each index. The first line creates a matrix with 4437 rows and nine columns named *squared_dist_sp500*. The seventh line computes the squared distance between the conditional variance series *sp500_ vargarch{!p}{!q}*, for *!p*=0,1,2 and *!q*=1,2,3, and squared residuals *sp500_unresgarch{!p}{!q}*. The ninth line saves in each *!k*th row of the first column of matrix *mse*, the average (*@mean*) of the values of each *!k*th column (*@columnextract*) of matrix *squared_dist_sp500*. Table 2.1 presents the values of the SBC criterion and mean squared error, $\bar{\Psi}_{(SE)(i)}$, function for each model. We do not proceed according to the $\bar{\Psi}_{(SE)(i)}$ criterion, as it measures the distance from a very noisy proxy of the unobserved variance. But we will return to the subject of proxy variables for conditional variance in Chapter 7 and refer to the construction of more accurate proxies of the true variance that are based on intraday information.

Table 2.1   SBC criterion and mean squared error, $\bar{\Psi}_{(SE)(i)}$, loss function for nine models

Model	S&P500	FTSE100	$/£	Gold
		SBC criterion		
ARCH(1)	2.790	2.788	1.714	2.258
ARCH(2)	2.746	2.726	1.704	2.254
ARCH(3)	2.730	2.687	1.698	2.233
GARCH(1,1)	**2.612**	**2.603**	**1.619**	2.151
GARCH(1,2)	2.614	2.604	1.620	**2.143**
GARCH(1,3)	2.612	2.605	1.622	2.145
GARCH(2,1)	2.614	2.604	1.620	2.144
GARCH(2,2)	2.614	2.606	1.622	2.144
GARCH(2,3)	2.614	2.608	1.624	2.145
		MSE function		
ARCH(1)	5.90848	5.06876	0.47302	4.20501
ARCH(2)	5.81506	4.75479	0.46873	**4.19529**
ARCH(3)	5.76106	4.59446	0.46629	4.24705
GARCH(1,1)	5.56599	4.38743	0.44669	4.33443
GARCH(1,2)	5.56883	4.37587	0.44670	4.23790
GARCH(1,3)	5.56158	**4.37381**	0.44669	4.23937
GARCH(2,1)	5.56682	4.38168	0.44670	4.28710
GARCH(2,2)	5.57681	4.37433	0.44669	4.25615
GARCH(2,3)	**5.56019**	4.37444	**0.44668**	4.23791

The best-performing models are shown in bold face.

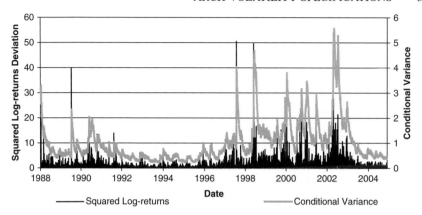

Figure 2.6   In-sample estimated conditional variance and squared log-returns of the AR(1)-GARCH(1,1) model for the S&P500 index.

In-sample information criteria such as the SBC have been widely used in the ARCH literature, despite the fact that their statistical properties in the ARCH context are unknown. Empirical studies in various areas of finance have provided evidence in favor of information criteria. In Chapter 10 we will refer analytically to various methods that evaluate the ability of models either in describing the data (in-sample evaluation) or in predicting future values of the variable under investigation (out-of-sample evaluation). For the moment, we proceed with model selection based on evaluation of the maximum value of the log-likelihood function. The model with the minimum value for the SBC criterion would be the best-performing one. On the basis of the results, in the majority of the cases, the best-performing model appears to be the AR(1)-GARCH(1,1).[7] Figures 2.6–2.9 plot the in-sample estimated conditional variance and squared log-returns of the best–performing models by the SBC criterion.

Table 2.2 presents the estimated parameters. In all cases, the autoregressive parameter is not statistically significant, providing no evidence of non-synchronous trading effect. The conditional variance parameters, however, are statistically signifi-cant. The sum of the coefficients on the lagged squared innovation and lagged conditional variance, $a_1 + \sum_{j=1}^{p} b_j$, is very close to unity, implying that shocks to the variance are highly persistent. Figure 2.10 presents the autocorrelations of squared log-returns, $\{y_t^2\}_{t=1}^{T}$. The highly significant autocorrelation function of squared log-returns, which decreases as the log order increases, indicates the systematic autocor-relation pattern of variance. Recall that the main characteristic of an ARCH process is that it allows the conditional variance, $\sigma_t^2 = E_{t-1}(\varepsilon_t^2)$, to depend on its lagged values, σ_{t-j}^2, or the previous values of the squared residuals, ε_{t-i}^2. A correctly specified conditional variance framework should not allow any autocorrelation pattern in the

[7] Engle (2004) described the GARCH(1,1) model as the 'workhorse of financial applications'. Hansen and Lunde (2005a) compared 330 ARCH models and concluded that, at least in the case of the Deutschmark–dollar exchange rate, no model beats the GARCH(1,1).

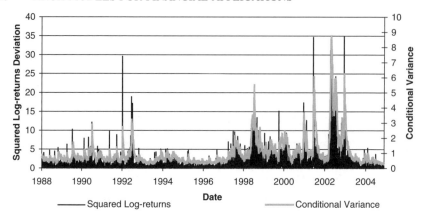

Figure 2.7   In-sample estimated conditional variance and squared log-returns of the AR(1)-GARCH(1,1) model for the FTSE100 index.

squared residuals. Thus, we test for any remaining ARCH effects in the variance equation via the autocorrelation of squared standardized residuals. According to Figure 2.11, which plots the autocorrelation of the squared standardized residuals, $\{\hat{\varepsilon}_t^2 \hat{\sigma}_t^{-2}\}_{t=1}^T$, for lags of order up to 35, almost all of the estimated autocorrelations are inside the 95% confidence interval.

The histograms of standardized residuals, conditional variances, conditional standard deviations and logarithmic conditional variances of the models with the minimum value in the SBC criterion are presented in Figures 2.12–2.15.

Table 2.3 provides the descriptive statistics of standardized residuals and logarithmic variance. The estimated skewness and kurtosis of the standardized residuals

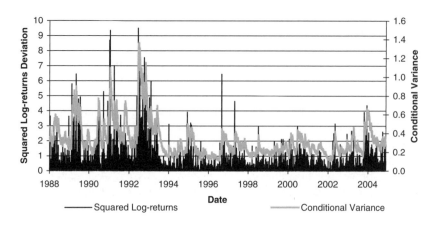

Figure 2.8   In-sample estimated conditional variance and squared log-returns of the AR(1)-GARCH(1,1) model for the $/£ exchange rate index.

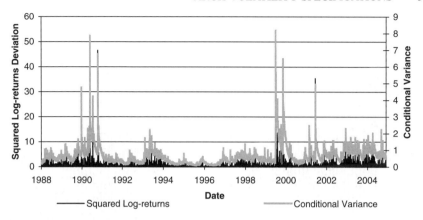

Figure 2.9   In-sample estimated conditional variance and squared log-returns of the AR(1)-GARCH(1,2) model for the gold price.

are clearly different from those of the normal distribution. The assumption of normality is rejected at any level of significance by a series of tests based on various statistics such as Anderson–Darling, Cramér-von Mises, and Jarque–Bera. The departure from normality justifies the use of the quasi-maximum likelihood estimation of standard errors. The computation of the variance–covariance matrix is an option available in EViews on the View button of the estimated model. By way of illustration, for the AR(1)-GARCH(1,1) model of the S&P500 index, Tables 2.4 and 2.5 present the estimated matrices $\hat{A}$ (equation (2.40)) and $\hat{B}$ (equation (2.41)), respectively. If the two matrices differ, the model might not be correctly specified, in which case the normal distribution would not be the true data-generating process of

Table 2.2   Estimated parameters of the models selected by the SBC criterion

Parameters	S&P500	FTSE100	$/£	Gold
c_0	0.04775	0.03688	0.00902	−0.01024
	(0.01213)	(0.01226)	(0.008)	(0.00929)
c_1	0.00876	0.02618	0.02416	−0.01353
	(0.01566)	(0.01582)	(0.01587)	(0.01776)
a_0	0.00502	0.01195	0.00332	0.00158
	(0.00157)	(0.00282)	(0.00116)	(0.00074)
a_1	0.04212	0.07058	0.03819	0.12964
	(0.00745)	(0.00984)	(0.00653)	(0.0429)
b_1	0.95291	0.91689	0.95147	−0.0948
	(0.00727)	(0.01026)	(0.00843)	(0.04363)
b_2	—	—	—	0.96449
				(0.00643)

Standard errors are given in parentheses.

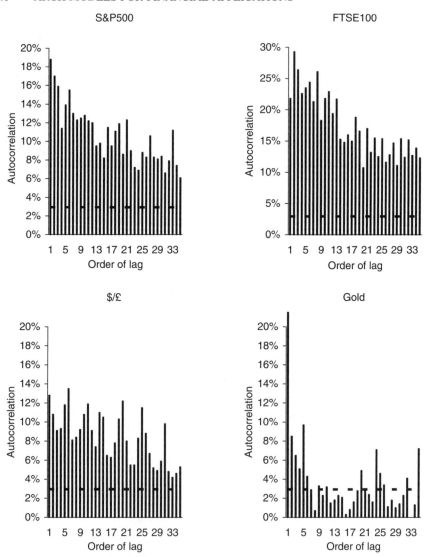

*Figure 2.10   Autocorrelation of squared log-returns. Dashed lines indicate the 95%
upper bound for the estimated sample autocorrelation, $1.96/\sqrt{T}$.*

$\{\varepsilon_t \sigma_t^{-1}\}_{t=1}^{T}$. The GARCH model captures some but not all of the leptokurtosis in the
unconditional distribution of residuals. The asymmetric and leptokurtic standardized
residuals indicate the importance of reconsidering the assumption that the standardized innovations are normally distributed.[8]

---

[8] In Chapter 5, the assumption of standard normally distributed innovations will be relaxed.

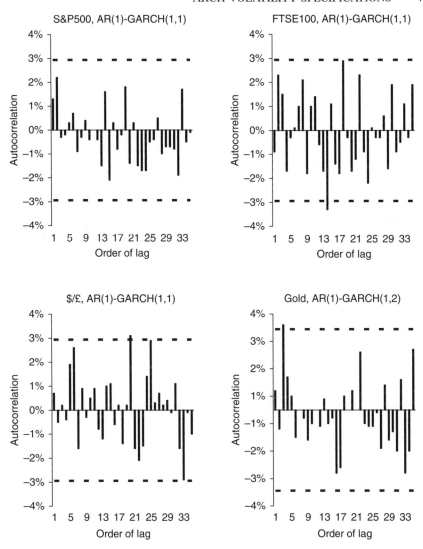

Figure 2.11   Autocorrelation of squared standardized residuals. Dashed lines indicate the 95% confidence interval for the estimated sample autocorrelations, $\pm 1.96/\sqrt{T}$.

The logarithmic transformation of conditional variance has a density that looks close to the normal one, in the sense that it is unimodal and reminds the Gaussian shape. In Chapter 7, we deal with the realised volatility, which is based on intraday log-returns, as a proxy of the unobserved variance. The statistical properties of the logarithmic realised volatility have been noted to be much closer to the normal one.

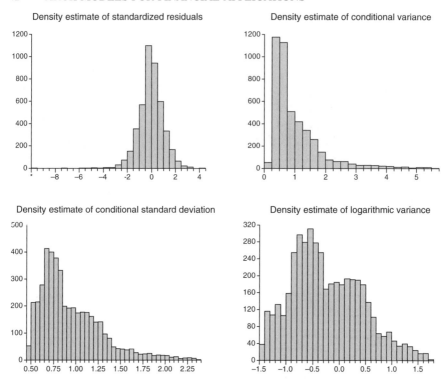

Figure 2.12   S&P500 index: distribution of standardized residuals, conditional variance, conditional standard deviation and logarithmic conditional variance of the AR(1)-GARCH(1,1) model.

To summarize the main findings, (i) even the simple GARCH model captures the autocorrelation pattern of conditional variance, (ii) an asymmetric and leptokurtic distribution of standardized residuals must be applied and (iii) the density of the logarithmic conditional variance is close to the normal distribution but statistically distinguishable from it.

2.4   Asymmetric conditional volatility specifications

The GARCH(p, q) model successfully captures several characteristics of financial time series, such as fat-tailed returns and volatility clustering. However, its structure imposes important limitations. The variance only depends on the magnitude and not the sign of ε_t, which is somewhat at odds with the empirical behaviour of stock market prices, where the leverage effect may be present. The models that have been considered so far are symmetric in that only the magnitude and not the positivity or negativity of innovations determines σ_t^2. In order to capture the asymmetry manifested by the

Density estimate of standardized residuals

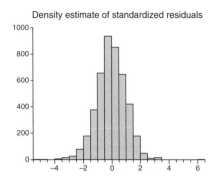

Density estimate of conditional variance

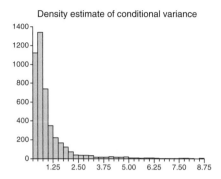

Density estimate of conditional standard deviation

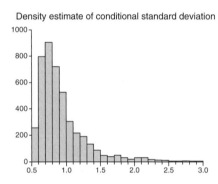

Density estimate of logarithmic variance

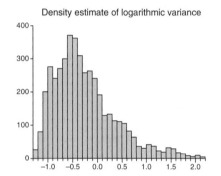

Figure 2.13   FTSE100 index: distribution of standardized residuals, conditional variance, conditional standard deviation and logarithmic conditional variance of the AR(1)-GARCH(1,1) model.

data, a new class of models, in which good news and bad news have different predictability for future volatility, was introduced.

The most popular model proposed to capture the asymmetric effects is Nelson's (1991) exponential GARCH, or EGARCH, model. He proposed the following form for the evolution of the conditional variance:

$$\log\left(\sigma_t^2\right) = a_0 + \sum_{i=1}^{\infty} \pi_i g\left(\frac{\varepsilon_{t-i}}{\sigma_{t-i}}\right), \quad \pi_1 \equiv 1, \tag{2.53}$$

and accommodated the asymmetric relationship between stock returns and volatility changes by making $g(\varepsilon_t/\sigma_t)$ a linear combination of $|\varepsilon_t/\sigma_t|$ and ε_t/σ_t:

$$g(\varepsilon_t/\sigma_t) \equiv \gamma_1(\varepsilon_t/\sigma_t) + \gamma_2(|\varepsilon_t/\sigma_t| - E|\varepsilon_t/\sigma_t|), \tag{2.54}$$

where γ_1 and γ_2 are constants. By construction, equation (2.54) is a zero-mean i.i.d. sequence (note that $z_t \equiv \varepsilon_t/\sigma_t$). Over the range $0 < z_t < \infty$, $g(z_t)$ is linear in z_t with slope $\gamma_1 + \gamma_2$ and over the range $-\infty < z_t \leq 0$, $g(z_t)$ is linear with slope $\gamma_1 - \gamma_2$. The first term of (2.54), $\gamma_1(z_t)$, represents the leverage effect, while the second term, $\gamma_2(|z_t| - E|z_t|)$, represents the magnitude effect as in the GARCH model. To make this

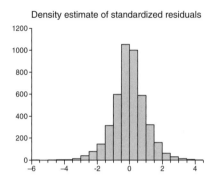

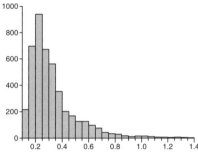

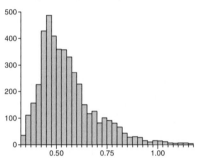

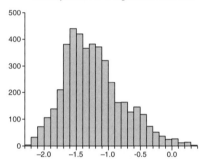

Figure 2.14 $/£ exchange rate index: distribution of standardized residuals, conditional variance, conditional standard deviation and logarithmic conditional variance of the AR(1)-GARCH(1,1) model.

tangible, assume that $\gamma_2 > 0$ and $\gamma_1 = 0$. The innovation in $\log(\sigma_t^2)$ is then positive (negative) when the magnitude of z_t is larger (smaller) than its expected value. Now assume that $\gamma_2 = 0$ and $\gamma_1 < 0$. In this case, the innovation in $\log(\sigma_t^2)$ is positive (negative) when innovations are negative (positive). Moreover, the conditional variance is positive regardless of whether the π_i coefficients are positive. Thus, in contrast to GARCH models, no inequality constraints need to be imposed for estimation. Nelson (1991) showed that $\log(\sigma_t^2)$ and ε_t are strictly stationary as long as $\sum_{i=1}^{\infty} \pi_i^2 < \infty$. A natural parameterization is to model the infinite moving average representation of equation (2.53) as an autoregressive moving average model:

$$\log(\sigma_t^2) = a_0 + \left(1 + \sum_{i=1}^{q} a_i L^i\right)\left(1 - \sum_{j=1}^{p} b_j L^j\right)^{-1} \qquad (2.55)$$
$$\times \left(\gamma_1(\varepsilon_{t-1}/\sigma_{t-1}) + \gamma_2(|\varepsilon_{t-1}/\sigma_{t-1}| - E|\varepsilon_{t-1}/\sigma_{t-1}|)\right),$$

or, equivalently,

$$\log(\sigma_t^2) = a_0 + (1 + A(L))(1 - B(L))^{-1} g(z_{t-1}). \qquad (2.56)$$

Density estimate of standardized residuals

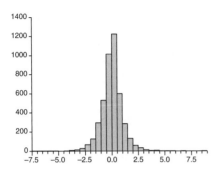

Density estimate of conditional variance

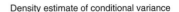
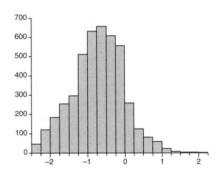

Density estimate of conditional standard deviation

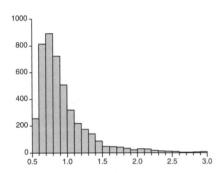

Density estimate of logarithmic variance

Figure 2.15   Gold bullion price: distribution of standardized residuals, conditional variance, conditional standard deviation and logarithmic conditional variance of the AR(1)-GARCH(1,2) model.

Theoretical results on the moment structure of the EGARCH model were investigated by He et al. (2002) and Karanasos and Kim (2003).

Another popular way to model the asymmetry of positive and negative innovations is the use of indicator functions. Glosten et al. (1993) presented the GJR(p, q) model,

$$\sigma_t^2 = a_0 + \sum_{i=1}^{q} \left(a_i \varepsilon_{t-i}^2 \right) + \sum_{i=1}^{q} \left(\gamma_i d(\varepsilon_{t-i} < 0) \varepsilon_{t-i}^2 \right) + \sum_{j=1}^{p} \left(b_j \sigma_{t-j}^2 \right), \qquad (2.57)$$

where γ_i, for $i = 1, \ldots, q$, are parameters that have to be estimated and $d(.)$ denotes the indicator function (i.e. $d(\varepsilon_{t-i} < 0) = 1$ if $\varepsilon_{t-i} < 0$, and $d(\varepsilon_{t-i} < 0) = 0$ otherwise). The GJR model allows good news ($\varepsilon_{t-i} > 0$) and bad news ($\varepsilon_{t-i} < 0$) to have differential effects on the conditional variance. Therefore, in the case of the GJR(0,1) model, good news has an impact of a_1, while bad news has an impact of $a_1 + \gamma_1$. For $\gamma_1 > 0$, the

Table 2.3   Descriptive statistics of standardized residuals and logarithmic variance

Index	S&P500	FTSE100	$/£	Gold
Model	AR(1)- GARCH(1,1)	AR(1)- GARCH(1,1)	AR(1)- GARCH(1,1)	AR(1)- GARCH(1,2)
Standardized residuals				
Mean	−0.0191	−0.0189	−0.0073	0.0163
Std. dev.	0.9992	0.9999	1.0006	1.0011
Skewness	−0.5172	−0.1603	−0.1434	0.0374
Kurtosis	6.9093	4.0043	4.6653	8.3205
In-sample estimated logarithmic variance				
Mean	−0.2341	−0.2444	−1.2296	−0.7087
Std. dev.	0.6666	0.6280	0.4580	0.6873
Skewness	0.4585	1.0428	0.6081	0.0337
Kurtosis	2.6640	4.0212	3.1263	3.1826

leverage effect exists. A semi-parametric extension of the GJR model was proposed by Yang (2006), the advantage of which is that it curtails the growth of volatility.

A similar way to model asymmetric effects on the conditional standard deviation was introduced by Zakoian (1990), and developed further in Rabemananjara and Zakoian (1993), by defining the threshold GARCH, or TARCH(p, q), model:

$$\sigma_t = a_0 + \sum_{i=1}^{q} \left(a_i \varepsilon_{t-i}^{+}\right) - \sum_{i=1}^{q} \left(\gamma_i \varepsilon_{t-i}^{-}\right) + \sum_{j=1}^{p} \left(b_j \sigma_{t-j}\right), \qquad (2.58)$$

where $\varepsilon_t^{+} \equiv \varepsilon_t$ if $\varepsilon_t > 0$, $\varepsilon_t^{+} \equiv 0$ otherwise and $\varepsilon_t^{-} \equiv \varepsilon_t - \varepsilon_t^{+}$.

Engle and Ng (1993) recommended the news impact curve as a measure of how news is incorporated into volatility estimates by alternative ARCH models. In their recent comparative study of the EGARCH and GJR models, Friedmann and

Table 2.4   The estimated matrix $\hat{A}$, for the AR(1)-GARCH(1,1) model of the S&P500 index

	c_0	c_1	a_0	a_1	b_1
c_0	1.643	−0.091	−21.688	−6.832	−10.426
c_1	−0.091	0.878	−0.221	0.399	0.384
a_0	−21.688	−0.221	2697.434	1038.786	1396.435
a_1	−6.832	0.399	1038.786	544.898	655.093
b_1	−10.426	0.384	1396.435	655.093	838.395

A consistent estimator of matrix A is $\hat{A}$ as given in equation (2.40). EViews performs maximum likelihood estimation of the variance covariance matrix, i.e. $\hat{\Sigma}_{\theta,MLE}$. Thus, $\hat{A}$ can also be computed as $\hat{A} = T^{-1} \hat{\Sigma}_{\theta,MLE}^{-1}$.

Table 2.5   The estimated matrix $\hat{B}$, for the AR(1)-GARCH(1,1) model of the S&P500 index

	c_0	c_1	a_0	a_1	b_1
c_0	6.720	−0.221	−498.154	−163.555	−248.104
c_1	−0.221	0.858	0.439	2.344	2.107
a_0	−498.154	0.439	46394.992	16062.837	23586.785
a_1	−163.555	2.344	16062.837	6180.670	8595.543
b_1	−248.104	2.107	23586.785	8595.543	12356.376

A consistent estimator of matrix B is $\hat{B} = T^{-1} \sum_{t=1}^{T} E\left(\left(\frac{\partial l_t(\hat{\theta})}{\partial \theta}\right) \left(\frac{\partial l_t(\hat{\theta})}{\partial \theta'}\right) |I_{t-1} \right)$. EViews performs quasi-maximum likelihood estimation of the variance–covariance matrix, i.e. $\hat{\Sigma}_{\theta,QMLE}$. Thus, $\hat{B}$ can also be computed as $\hat{B} = T\hat{A}\hat{\Sigma}_{\theta,QMLE}^{-1}\hat{A}$.

Sanddorf-Köhle (2002) proposed a modification of the news impact curve termed the 'conditional news impact curve'. Engle and Ng argued that the GJR model is better than the EGARCH model because the conditional variance implied by the latter is too high due to its exponential functional form. However, Friedmann and Sanddorf-Köhle (2002) argued that the EGARCH model does not overstate the predicted volatility.

The number of formulations presented in the financial and econometric literature is vast. In what follows, the best-known variations of ARCH modelling are presented.

Taylor (1986) and Schwert (1989a, 1989b) assumed that the conditional standard deviation is a distributed lag of absolute innovations, and introduced the absolute GARCH, or AGARCH(p, q), model,

$$\sigma_t = a_0 + \sum_{i=1}^{q} a_i |\varepsilon_{t-i}| + \sum_{j=1}^{p} b_j \sigma_{t-j}. \tag{2.59}$$

Geweke (1986), Pantula (1986) and Milhøj (1987) suggested a specification in which the log of the conditional variance depends linearly on past logs of squared innovations. Their model is the multiplicative ARCH, or Log-GARCH(p, q), model defined by

$$\log\left(\sigma_t^2\right) = a_0 + \sum_{i=1}^{q} a_i \log\left(\varepsilon_{t-i}^2\right) + \sum_{j=1}^{p} b_j \log\left(\sigma_{t-j}^2\right). \tag{2.60}$$

Schwert (1990) constructed the autoregressive standard deviation, or Stdev-ARCH (q), model,

$$\sigma_t^2 = \left(a_0 + \sum_{i=1}^{q} a_i |\varepsilon_{t-i}| \right)^2. \tag{2.61}$$

Higgins and Bera (1992) introduced the non-linear ARCH, or NARCH(p, q), model,

$$\sigma_t^\delta = a_0 + \sum_{i=1}^{q} a_i \left| \varepsilon_{t-i}^2 \right|^{\delta/2} + \sum_{j=1}^{p} b_j \sigma_{t-j}^\delta, \tag{2.62}$$

while Engle and Bollerslev (1986) proposed a simpler non-linear ARCH model,

$$\sigma_t^2 = a_0 + a_1 \left| \varepsilon_{t-1} \right|^\delta + b_1 \sigma_{t-1}^2. \tag{2.63}$$

In order to introduce asymmetric effects, Engle (1990) proposed the asymmetric GARCH, or AGARCH(p, q), model,

$$\sigma_t^2 = a_0 + \sum_{i=1}^{q} \left(a_i \varepsilon_{t-i}^2 + \gamma_i \varepsilon_{t-i} \right) + \sum_{j=1}^{p} b_j \sigma_{t-j}^2, \tag{2.64}$$

where a negative value of γ_i means that positive returns increase volatility less than negative returns. Moreover, Engle and Ng (1993) presented two more ARCH models that incorporate asymmetry for good and bad news, the non-linear asymmetric GARCH, or NAGARCH(p, q), model,

$$\sigma_t^2 = a_0 + \sum_{i=1}^{q} a_i (\varepsilon_{t-i} + \gamma_i \sigma_{t-i})^2 + \sum_{j=1}^{p} b_j \sigma_{t-j}^2, \tag{2.65}$$

and the VGARCH(p, q) model,

$$\sigma_t^2 = a_0 + \sum_{i=1}^{q} a_i (\varepsilon_{t-i}/\sigma_{t-i} + \gamma_i)^2 + \sum_{j=1}^{p} b_j \sigma_{t-j}^2. \tag{2.66}$$

Ding et al. (1993) introduced the asymmetric power ARCH, or APARCH(p, q), model, which includes seven ARCH models as special cases (ARCH, GARCH, AGARCH, GJR, TARCH, NARCH and Log-ARCH):

$$\sigma_t^\delta = a_0 + \sum_{i=1}^{q} a_i \left(\left| \varepsilon_{t-i} \right| - \gamma_i \varepsilon_{t-i} \right)^\delta + \sum_{j=1}^{p} b_j \sigma_{t-j}^\delta, \tag{2.67}$$

where $a_0 > 0, \delta > 0, b_j \geq 0, j = 1, \ldots, p, a_i \geq 0$ and $-1 < \gamma_i < 1, i = 1, \ldots, q$. The model imposes a Box and Cox (1964) power transformation of the conditional standard deviation process and the asymmetric absolute innovations. The functional form for the conditional standard deviation is familiar to economists as the constant elasticity of substitution (CES) production function. Ling and McAleer (2001) provided sufficient conditions for the stationarity and ergodicity of the APARCH (p, q), model. Brooks et al. (2000) applied the APARCH(1,1) model for 10 series of national stock market index returns. The optimal power transformation was found to be remarkably similar across countries.[9]

---

[9] Karanasos and Kim (2006) presented the APARCH(p, q), model as $\sigma_t^{\delta/2} = a_0 + \sum_{i=1}^{q} a_i \sigma_{t-i}^{\delta/2}$ $\times \left(\left| z_{t-i} \right| - \gamma_i z_{t-i} \right)^\delta + \sum_{j=1}^{p} b_j \sigma_{t-j}^{\delta/2}$ in order to study its autocorrelation structure.

Sentana (1995) introduced the quadratic GARCH, or GQARCH(p, q), model of the form

$$\sigma_t^2 = a_0 + \sum_{i=1}^{q} a_i \varepsilon_{t-i}^2 + \sum_{i=1}^{q} \gamma_i \varepsilon_{t-i} + 2 \sum_{i=1}^{q} \sum_{j=i+1}^{q} a_{ij} \varepsilon_{t-i} \varepsilon_{t-j} + \sum_{j=1}^{p} b_j \sigma_{t-j}^2. \qquad (2.68)$$

Setting $\gamma_i = 0$, for $i = 1, \ldots, q$, leads to the augmented ARCH model of Bera and Lee (1990). It does encompass all the ARCH models of quadratic variance functions, but it does not include models in which the variance is quadratic in the absolute value of innovations, as the APARCH model does.

Hentschel (1995) gave a complete parametric family of ARCH models. This family includes the most popular symmetric and asymmetric ARCH models, thereby highlighting the relation between the models and their treatment of asymmetry. Hentschel presents the variance equation as

$$\frac{\sigma_t^\lambda - 1}{\lambda} = \omega + a \sigma_{t-1}^\lambda f^\nu(\varepsilon_t) + \beta \frac{\sigma_{t-1}^\lambda - 1}{\lambda}, \qquad (2.69)$$

where $f(.)$ denotes the absolute value function of innovations,

$$f(\varepsilon_t) = |\varepsilon_t - \beta| - \zeta(\varepsilon_t - \beta). \qquad (2.70)$$

In general, this is a law of the Box–Cox transformation of the conditional standard deviation (as in the case of the APARCH model), and the parameter λ determines the shape of the transformation. For $\lambda > 1$ the transformation of σ_t is convex, while for $\lambda < 1$ it is concave. The parameter ν serves to transform the absolute value function. For different restrictions on the parameters in equations (2.69) and (2.70), almost all the popular symmetric and asymmetric ARCH models are obtained. For example, for $\lambda = 0$, $\nu = 1$, $\beta = 1$ and free ζ, we obtain Nelson's EGARCH model. However, some models, such as Sentana's quadratic model, are excluded.

Gouriéroux and Monfort (1992) proposed the qualitative threshold GARCH, or GQTARCH(p, q), model with the following specification:

$$\sigma_t^2 = a_0 + \sum_{i=1}^{q} \sum_{j=1}^{J} a_{ij} d_j(\varepsilon_{t-i}) + \sum_{j=1}^{p} b_j \sigma_{t-j}^2. \qquad (2.71)$$

Assuming constant conditional variance over various observation intervals, Gouriéroux and Monfort (1992) divided the space of ε_t into J intervals and let $d_j(\varepsilon_t)$ be 1 if ε_t is in the jth interval.

2.5 Simulating ARCH models using EViews

Let us assume that we wish to generate data from the GARCH(1,1) process, or $y_t = \varepsilon_t = z_t \sigma_t$ and $\sigma_t^2 = 0.0001 + 0.12 \varepsilon_{t-1}^2 + 0.8 \sigma_{t-1}^2$, with normally distributed standardized residuals. To this end, we proceed with the following steps:

1. Generate a series of, say, 3000 values from the standard normal distribution, i.e. $z_t \overset{i.i.d.}{\sim} N(0, 1)$.

2. Generate an equal number of values $\{\varepsilon_t\}_{t=1}^{3000}$ of the GARCH(1,1) process, by multiplying the collection $\{z_t\}_{t=1}^{3000}$ with the specific conditional variance form, or $\varepsilon_t = z_t\sqrt{\sigma_t^2}$, for $\sigma_t^2 = 0.0001 + 0.12\varepsilon_{t-1}^2 + 0.8\sigma_{t-1}^2$.

3. Generate the process of asset prices as $P_t = \exp(\log P_{t-1} + \varepsilon_t)$, based on the values $\{\varepsilon_t\}_{t=1}^{3000}$ of the innovation process and assuming $P_1 = 100$ as a starting point.

The simulation can be conducted in EViews using the following program (named *chapter2.simgarch11.prg*):

```
create chapter2.simgarch11 u 1 3000
series _temp = 1
!length = @obs(_temp)
delete _temp
    !a0 = 0.0001
    !a1 = 0.12
    !b1 = 0.8
series z = nrnd
series h
series e
series p
h(1) = !a0
e(1) = sqr(h(1))*z(1)
p(1) = 100
for !i = 2 to !length
    h(!i) = !a0 + (!a1*(e(!i-1)^2)) + (!b1*h(!i-1))
    e(!i) = sqr(h(!i))*z(!i)
    p(!i) = exp(log(p(!i-1)) + e(!i))
next
```

The first line of the program creates a workfile of 3000 observations for each time series. Lines 5–7 define the values of the parameters, lines 8–11 create the time series of z_t, σ_t^2, ε_t and P_t respectively, and lines 12–14 define the first values of the series. The *for* loop computes the values of the σ_t^2, ε_t and P_t simulated series, respectively.

Figure 2.16 plots the simulated processes, while Table 2.6 presents the relevant descriptive statistics. According to results obtained in the literature (e.g. Engle and Mustafa, 1992), the shocks to the variance, $v_t = E_t(\varepsilon_t^2) - E_{t-1}(\varepsilon_t^2) = \varepsilon_t^2 - \sigma_t^2$, generate a martingale difference sequence. These shocks are neither serially independent nor identically distributed. According to the BDS test, for the null hypothesis that the observations of the series are independent, only the process defined by z_t is independently distributed.[10] The test is presented for two correlated dimensions,

---

[10] To calculate the BDS test statistic in EViews, open the series you would like to test for and choose BDS Independence Test from the View button.

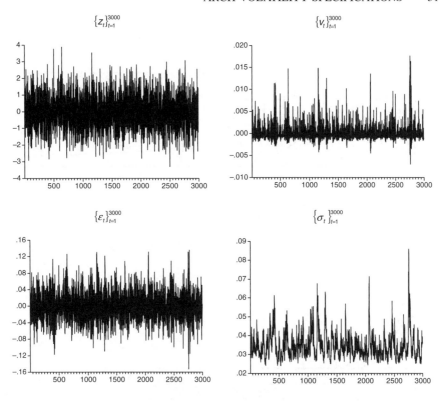

$\{z_t\}_{t=1}^{3000}$ $\{v_t\}_{t=1}^{3000}$

$\{\varepsilon_t\}_{t=1}^{3000}$ $\{\sigma_t\}_{t=1}^{3000}$

Figure 2.16   Simulated processes $z_t, \varepsilon_t, v_t, \sigma_t$.

Table 2.6   Descriptive statistics of the simulated processes $z_t, \varepsilon_t, v_t, \sigma_t$

	$\{z_t\}_{t=1}^{3000}$	$\{\varepsilon_t\}_{t=1}^{3000}$	$\{v_t\}_{t=1}^{3000}$	$\{\sigma_t\}_{t=1}^{3000}$
Mean	0.003320	0.000347	7.37E-07	0.034633
Median	0.006848	0.000218	−0.000572	0.033050
Maximum	3.883330	0.135524	0.017632	0.085937
Minimum	−3.313137	−0.152481	−0.006963	0.024033
Std. dev.	1.000943	0.035357	0.001890	0.007049
Skewness	0.080919	0.149335	3.090495	1.739608
Kurtosis	2.994521	3.502455	18.84448	8.198862
BDS test p-value	0.32	0.00	0.00	0.00

but it has been computed for higher values and the results are qualitatively unchanged. Figure 2.17 presents the autocorrelation of transformations of the processes defined by z_t, ε_t, v_t, σ_t. For the necessary computations one may use the following program, *chapter2.autocorrelation_sim.prg*:

$$Cor\left(\left|z_t\right|^d, \left|z_{t+\tau}\right|^d\right), \ d=0.5(0.5)3, \ \tau=1(1)100$$

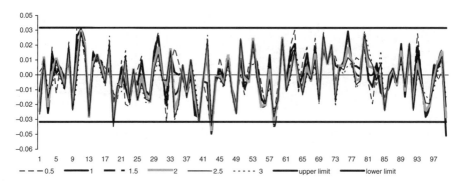

$$Cor\left(\left|\varepsilon_t\right|^d, \left|\varepsilon_{t+\tau}\right|^d\right), \ d=0.5(0.5)3, \ \tau=1(1)100$$

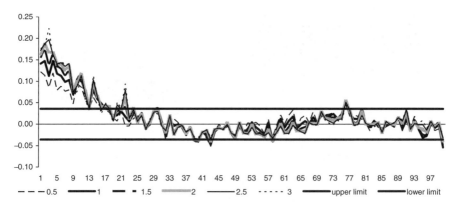

$$Cor\left(\left|v_t\right|^d, \left|v_{t+\tau}\right|^d\right), \ d=0.5(0.5)3, \ \tau=1(1)100$$

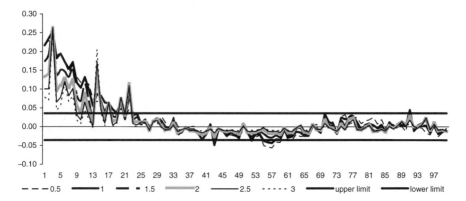

Figure 2.17   Autocorrelation of transformations of the processes $z_t, \varepsilon_t, v_t, \sigma_t$.

$Cor(\sigma_t^d, \sigma_{t+\tau}^d)$, $d = 0.5(0.5)3$, $\tau = 1(1)100$

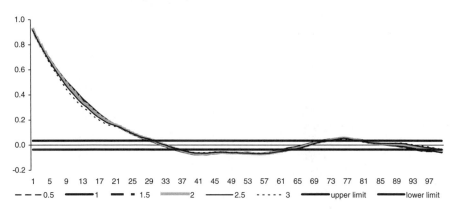

Figure 2.17 (Continued)

```
for !j=0.5 to 3 step .5
   !j100 = !j*10
   series d_abs_z_{!j100}
   d_abs_z_{!j100} = abs(z)^!j
next
matrix(100,6) corr_abs_z
for !i=1 to 100
   corr_abs_z(!i,1) = @cor(d_abs_z_5,d_abs_z_5(-!i))
   corr_abs_z(!i,2) = @cor(d_abs_z_10,d_abs_z_10(-!i))
   corr_abs_z(!i,3) = @cor(d_abs_z_15,d_abs_z_15(-!i))
   corr_abs_z(!i,4) = @cor(d_abs_z_20,d_abs_z_20(-!i))
   corr_abs_z(!i,5) = @cor(d_abs_z_25,d_abs_z_25(-!i))
   corr_abs_z(!i,6) = @cor(d_abs_z_30,d_abs_z_30(-!i))
next
```

The first loop creates six series with the values of $|z_t|^d$ for $d = 0.5(0.5)3$. The sixth line creates a matrix named *corr_abs_z* with 100 rows and six columns. The second loop computes the autocorrelations $Cor(|z_t|^d, |z_{t+\tau}|^d)$, for $d = 0.5(0.5)3$ and $\tau = 1(1)100$. In the first column of *corr_abs_z* the 100 autocorrelations $Cor(|z_t|^{0.5}, |z_{t+\tau}|^{0.5})$ are saved, in the second column the $Cor(|z_t|^1, |z_{t+\tau}|^1)$ are saved, etc. The half-length of the 95% confidence interval for the estimated sample autocorrelation equals $1.96/\sqrt{T} = 0.035785$, in the case of a process with independently and identically normally distributed components. The processes defined by v_t and ε_t are autocorrelated in around 20% of the cases. Ding and Granger (1996) and Karanasos (1996) gave the autocorrelation function of the squared errors for the GARCH(1,1) process, and Karanasos (1999) extended the results to the GARCH(p,q) model. He and Teräsvirta (1999a, 1999b) derived the autocorrelation function of the squared and absolute errors for a family of first-order ARCH processes.

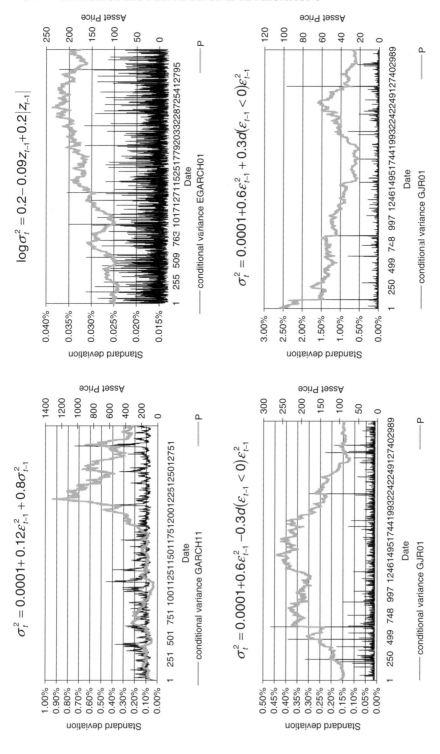

Figure 2.18 Simulated conditional variance $\{\sigma_t^2\}_{t=1}^{3000}$ and process $\{P_t\}_{t=1}^{3000}$ for $P_t = \exp(\log P_{t-1} + \varepsilon_t)$, $P_1 = 100$, $\varepsilon_t = z_t \sigma_t$ and $z_t \overset{i.i.d.}{\sim} N(0,1)$.

We now simulate the EGARCH(1,0) and GJR(0,1) conditional volatility models, considering the specifications $\log \sigma_t^2 = 0.2 - 0.09z_{t-1} + 0.2|z_{t-1}|$ and $\sigma_t^2 = 0.0001 + 0.6\varepsilon_{t-1}^2 \pm 0.3d(\varepsilon_{t-1} < 0)\varepsilon_{t-1}^2$, with $d(\varepsilon_{t-i} < 0) = 1$ if $\varepsilon_{t-i} < 0$, and $d(\varepsilon_{t-i} < 0) = 0$ otherwise. The simulated series were created by the EViews programs *chapter2. simgjr01.prg*[11] and *chapter2_simegarch01.prg*.

Figure 2.18 presents the simulated conditional variances along with the corresponding asset prices. The GARCH(1,1) process produces an asset price series whose volatility has the same magnitude in response to bad and good news. The other three processes create conditional volatility series with an asymmetric response in asset price changes. The $\log \sigma_t^2 = 0.2 - 0.09z_{t-1} + 0.2|z_{t-1}|$ and $\sigma_t^2 = 0.0001 + 0.6\varepsilon_{t-1}^2 + 0.3d(\varepsilon_{t-1} < 0)\varepsilon_{t-1}^2$ processes produce asset price series with higher volatility in market drops. On the other hand, $\sigma_t^2 = 0.0001 + 0.6\varepsilon_{t-1}^2 - 0.3d(\varepsilon_{t-1} < 0)\varepsilon_{t-1}^2$ relates market drops with a lower impact in conditional variance than market rises of the same magnitude. Simulated evidence for the one-step-ahead volatility, residuals, standardized residuals and shocks to the variance was provided by Degiannakis and Xekalaki (2007b).

2.6 Estimating asymmetric ARCH models with G@RCH 4.2 OxMetrics: an empirical example

In the paragraphs that follow, we will estimate five asymmetric ARCH specifications for the FTSE100 and S&P500 equity indices, the \$/£ exchange rate and the price of gold bullion. We use the same data set, for the period from 4 April 1988 to 5 April 2005, as in Section 2.3. The ARCH framework (1.6) with first-order autoregressive conditional mean, normally distributed innovations, and for various asymmetric volatility specifications is estimated:

$$
\begin{aligned}
y_t &= c_0(1-c_1) + c_1 y_{t-1} + \varepsilon_t, \\
\varepsilon_t &= \sigma_t z_t, \\
\sigma_t^2 &= g(\theta|I_{t-1}), \\
z_t &\overset{i.i.d.}{\sim} N(0,1).
\end{aligned}
\tag{2.72}
$$

We take into consideration Bollerslev's (1986) GARCH(p,q), Engle and Bollerslev's (1986) IGARCH(p,q), Nelson's (1991) EGARCH(p,q), Glosten et al.'s (1993) GJR(p,q) and Ding et al.'s (1993) APARCH(p,q) conditional volatility specifications. The asymmetric ARCH models are estimated for $p = q = 1$, an order that is used in the majority of empirical studies.

The models are estimated using the OxMetrics G@RCH package in each of the following forms:

GARCH(1,1),
$$
\sigma_t^2 = a_0 + a_1\varepsilon_{t-1}^2 + b_1\sigma_{t-1}^2;
\tag{2.73}
$$
IGARCH(1,1),
$$
\sigma_t^2 = \alpha_0 + \alpha_1\varepsilon_{t-1}^2 + (1-\alpha_1)\sigma_{t-1}^2;
\tag{2.74}
$$

---

[11] In order to estimate the GJR(0,1) model in the form $\sigma_t^2 = 0.0001 + 0.6\varepsilon_{t-1}^2 + 0.3d\,(\varepsilon_{t-1} < 0)\,\varepsilon_{t-1}^2$, replace *!gamma1=-0.3* with *!gamma1=0.3*.

EGARCH(1,1),

$$\log \sigma_t^2 = \alpha_0(1-b_1) + (1+\alpha_1 L)\left(\gamma_1 \frac{\varepsilon_{t-1}}{\sigma_{t-1}} + \gamma_2\left(\left|\frac{\varepsilon_{t-1}}{\sigma_{t-1}}\right| - E\left|\frac{\varepsilon_{t-1}}{\sigma_{t-1}}\right|\right)\right) + b_1 \log \sigma_{t-1}^2;$$

(2.75)

GJR(1,1),

$$\sigma_t^2 = a_0 + \alpha_1 \varepsilon_{t-1}^2 + \gamma_1 d(\varepsilon_{t-1} < 0)\varepsilon_{t-1}^2 + b_1 \sigma_{t-1}^2,$$

(2.76)

with $d(\varepsilon_t < 0) = 1$ if $\varepsilon_t < 0$ and $d(\varepsilon_t < 0) = 0$ otherwise;

and

APARCH(1,1),

$$\sigma_t^\delta = a_0 + \alpha_1 \left(|\varepsilon_{t-1}| - \gamma_1 \varepsilon_{t-1}\right)^\delta + b_1 \sigma_{t-1}^\delta.$$

(2.77)

Note that the EGARCH(1,1) specification can be expanded as

$$\log \sigma_t^2 = a_0(1-b_1) + \gamma_1 \frac{\varepsilon_{t-1}}{\sigma_{t-1}} + \gamma_2\left(\left|\frac{\varepsilon_{t-1}}{\sigma_{t-1}}\right| - E\left|\frac{\varepsilon_{t-1}}{\sigma_{t-1}}\right|\right)$$
$$+ a_1\gamma_1 \frac{\varepsilon_{t-2}}{\sigma_{t-2}} + a_1\gamma_2\left(\left|\frac{\varepsilon_{t-2}}{\sigma_{t-2}}\right| - E\left|\frac{\varepsilon_{t-2}}{\sigma_{t-2}}\right|\right) + b_1 \log \sigma_{t-1}^2.$$

Therefore, the EGARCH(1,1) form considers the logarithmic conditional variance as a function of one lag of itself and two lags of standardized residuals. In case we wish to estimate the conditional variance as a function of one lag of standardized residuals, we can estimate the EGARCH(1,0) specification, which has the form

$$\log \sigma_t^2 = \alpha_0(1-b_1) + \gamma_1 \frac{\varepsilon_{t-1}}{\sigma_{t-1}} + \gamma_2\left(\left|\frac{\varepsilon_{t-1}}{\sigma_{t-1}}\right| - E\left|\frac{\varepsilon_{t-1}}{\sigma_{t-1}}\right|\right) + b_1 \log \sigma_{t-1}^2.$$

The functional forms of the in-sample conditional standard deviation estimators, $\hat{\sigma}_{t+1}$, are specified as follows:

GARCH(1,1),

$$\hat{\sigma}_{t+1} = \sqrt{a_0^{(T)} + a_1^{(T)} \hat{\varepsilon}_t^2 + b_1^{(T)} \hat{\sigma}_t^2};$$

(2.78)

IGARCH(1,1),

$$\hat{\sigma}_{t+1} = \sqrt{a_0^{(T)} + a_1^{(T)} \hat{\varepsilon}_t^2 + \left(1 - a_1^{(T)}\right)\hat{\sigma}_t^2};$$

(2.79)

EGARCH(1,1),

$$\hat{\sigma}_{t+1} = \sqrt{\exp\left(a_0^{(T)}\left(1-b_1^{(T)}\right) + \left(1+\alpha_1^{(T)}L\right)\left(\gamma_1^{(T)}\frac{\hat{\varepsilon}_t}{\hat{\sigma}_t} + \gamma_2^{(T)}\left(\left|\frac{\hat{\varepsilon}_t}{\hat{\sigma}_t}\right| - E\left|\frac{\hat{\varepsilon}_t}{\hat{\sigma}_t}\right|\right)\right) + b_1^{(T)}\log\hat{\sigma}_t^2\right)},$$

(2.80)

with $E\left|\hat{\varepsilon}_t\hat{\sigma}_t^{-1}\right| = \sqrt{2/\pi}$ as $E|z_t| = \sqrt{2/\pi}$ under the assumption that $z_t \sim N(0,1)$;
GJR(1,1),

$$\hat{\sigma}_{t+1} = \sqrt{a_0^{(T)} + a_1^{(T)}\hat{\varepsilon}_t^2 + \gamma_1^{(T)}d(\hat{\varepsilon}_t < 0)\hat{\varepsilon}_t^2 + b_1^{(T)}\hat{\sigma}_t^2};$$

(2.81)

and APARCH(1,1),

$$\hat{\sigma}_{t+1} = \left(a_0^{(T)} + a_1^{(T)}\left(|\hat{\varepsilon}_t| - \gamma_1^{(T)}\hat{\varepsilon}_t\right)^{\delta^{(T)}} + b_1^{(T)}\hat{\sigma}_t^{\delta^{(T)}}\right)^{1/\delta^{(T)}}.$$

(2.82)

Laurent and Peters (2002a, 2002b, 2004, 2006) provide very instructive technical material about various versions of the G@RCH package, which gives step-by-step instructions on such basic matters as installation and loading of G@RCH, as well as technical details such as numerical methods of likelihood maximization.

The estimation of the five asymmetric model specifications of the four index series under consideration is effected through the use of the programs *chapter2.sp500.ox*, *chapter2.ftse100.ox*, *chapter2.usuk.ox* and *chapter2.gold.ox*. Program *chapter2. sp500.ox* is given in the appendix to the chapter. The rest of the programs consist of the same steps with obvious modifications. The core G@RCH program that carries out the necessary estimations is given below:[12]

```
#import <packages/Garch42/garch>
main()
{
decl garchobj;
garchobj = new Garch();
garchobj.Load("Chapter2Data.xls");
garchobj.Select(Y_VAR, {"SP500RET",0,0});
garchobj.SetSelSample(-1, 1, -1, 1);
garchobj.CSTS(1,1);
garchobj.DISTRI(0);
garchobj.ARMA_ORDERS(1,0);
garchobj.ARFIMA(0);
garchobj.GARCH_ORDERS(1,1);
garchobj.ARCH_IN_MEAN(0);
garchobj.MODEL(1);
garchobj.MLE(2);
garchobj.Initialization(<>);
garchobj.DoEstimation(<>);
garchobj.Output();
garchobj.STORE(1,1,1,0,0,"sp500_GARCH11",0);
delete garchobj;
}
```

G@RCH reads the data from the Excel file named *chapter2Data.xls* (line 6). It computes for the log-returns of the S&P500 index (line 7) a first-order autoregressive conditional mean (*garchobj.ARMA_ORDERS(1,0);*) and a GARCH conditional variance specification (*garchobj.MODEL(1);*) for $p = q = 1$ (*garchobj.GARCH_ ORDERS(1,1);*). The command *garchobj.MODEL(.),* for values 0, 1, ..., 11, estimates the EWMA, GARCH, EGARCH, GJR, APARCH, IGARCH, FIGARCH, FIGARCHC, FIEGARCH, FIAPARCH, FIAPARCHC and HYGARCH conditional

---

[12] The models can be estimated either by the program presented here or by its extended version that is available in the appendix to this chapter. The programs for the estimation of the other indices are available on the CD-ROM accompanying this book.

volatility specifications, respectively.[13] The command *garchobj.MLE(2)* takes into consideration the Bollerslev and Wooldridge robust quasi-maximum likelihood standard errors. For MLE(2), the QMLE estimate of the variance–covariance matrix, also termed in G@RCH the Sandwich formula, is computed according to equation (2.42). The command *garchobj.MLE(0)* estimates the variance–covariance matrix based on the inverse of the Hessian (i.e. the second derivatives of the log-likelihood function) according to equation (2.43). For MLE(1), the estimation of the variance–covariance matrix is based on the outer product of the gradients and is computed as $T^{-1}\hat{B}^{-1}$. By way of illustration, for the AR(1)-EGARCH(1,1) model of the S&P500 index, Table 2.7 presents the variance–covariance matrices computed by the three estimation methods of G@RCH.[14] The *garchobj.STORE (1,1,1,0,0,"sp500_GARCH11",0)* command saves the residuals, squared residuals and conditional variance in the *res_sp500_GARCH11.xls*, *sqRes_sp500_GARCH11.xls* and *condV_sp500_GARCH11.xls* Excel files.

We apply the SBC and Akaike's (1973) information criterion (AIC) to select the models that best fit the data sets. In G@RCH, the SBC criterion is computed as in (2.50). The AIC is computed as:

$$AIC = -2T^{-1}L_T\left(\{y_t\}; \hat{\psi}^{(T)}\right) + 2T^{-1}\breve{\psi}, \qquad (2.83)$$

where $L_T(.)$ is the maximized value of the log-likelihood function, $\hat{\psi}^{(T)}$ is the MLE of ψ based on a sample of size T, and $\breve{\psi}$ denotes the dimension of ψ. EViews and G@RCH use the same definitions of SBC and AIC, but other packages may compute information criteria in a slightly different way. Even the very early versions of EViews contained information criteria that omitted inessential constant terms of the likelihood function. Table 2.8 presents the values of the SBC and AIC criteria for each model. In both cases, the model corresponding to the minimum value of the criterion is judged to be the best-performing one. For the S&P500 index the AR(1)-EGARCH(1,1) model is selected, whereas in all the other cases the AR(1)-APARCH(1,1) model is preferred. G@RCH computes the SBC and AIC as well as the Hannan–Quinn and the Shibata information criteria, by setting the command *garchobj.TESTS(0,1)*. For values *(1,0)*, tests prior to model estimation are performed for the dependent variable, namely, Ljung and Box's (1978)Q test (see Section 2.7.1) and Engle's (1982) Lagrange multiplier test (Section 2.7.3), whereas for values *(0,1)*, tests following model estimation are conducted, namely tests on the residuals of the model.

Table 2.9 presents the estimated parameters of the AR(1)-EGARCH(1,1) and AR (1)-APARCH(1,1) models. We note that the non-synchronous trading effect is not present in the estimated models (coefficient c_1). The leverage effect is present in the S&P500 index, as coefficient γ_1 is significantly different from zero. In the case of the

---

[13] The FIGARCH, FIGARCHC, FIEGARCH, FIAPARCH, FIAPARCHC and HYGARCH specifications belong to the family of the fractionally integrated ARCH models, which are presented in the next chapter.

[14] The program *chapter2.sp500.VarCovar.ox* presents the variance–covariance matrices computed based on (i) the inverse of the Hessian, (ii) the outer product of the gradients, and (iii) its quasi-maximum likelihood estimation.

Table 2.7 AR(1)-EGARCH(1,1) model variance–covariance matrices for the S&P500 index

Variance–covariance matrix based on the inverse of the Hessian

Parameters	c_0	c_1	a_0	a_1	b_1	γ_1	γ_2
c_0	1.45E-04	-1.74E-05	-7.77E-04	-8.11E-05	7.27E-06	7.47E-06	8.87E-06
c_1	-1.74E-05	2.53E-04	-3.37E-05	1.41E-05	-5.34E-07	-2.35E-05	-8.54E-06
a_0	-7.77E-04	-3.37E-05	1.23E-02	4.56E-04	4.19E-04	1.43E-04	1.26E-04
a_1	-8.11E-05	1.41E-05	4.56E-04	1.27E-02	-8.48E-05	1.60E-03	-1.46E-03
b_1	7.27E-06	-5.34E-07	4.19E-05	-8.48E-05	5.20E-06	3.66E-06	-2.37E-06
γ_1	7.47E-06	-2.35E-05	1.43E-04	1.60E-03	3.66E-06	3.27E-04	-2.26E-04
γ_2	8.87E-06	-8.54E-06	1.26E-04	-1.46E-03	-2.37E-06	-2.26E-04	3.13E-04

Variance–covariance matrix based on the outer product of the gradients

Parameters	c_0	c_1	a_0	a_1	b_1	γ_1	γ_2
c_0	1.48E-04	-1.26E-05	-7.10E-04	-1.40E-04	6.34E-06	-3.03E-06	2.15E-05
c_1	-1.26E-05	2.93E-04	-1.48E-04	1.18E-04	-6.75E-07	-5.85E-06	-2.53E-05
a_0	-7.10E-04	-1.48E-04	8.51E-03	1.42E-03	2.06E-05	2.38E-04	-2.49E-05
a_1	-1.40E-04	1.18E-04	1.42E-03	7.17E-03	-3.09E-05	9.56E-04	-1.08E-03
b_1	6.34E-06	-6.75E-07	2.06E-05	-3.09E-05	2.34E-06	5.82E-07	-1.29E-06
γ_1	-3.03E-06	-5.85E-06	2.38E-04	9.56E-04	5.82E-07	1.88E-04	-1.38E-04
γ_2	2.15E-05	-2.53E-05	-2.49E-05	-1.08E-03	-1.29E-06	-1.38E-04	2.59E-04

Quasi-maximum likelihood estimate of the variance–covariance matrix

Parameters	c_0	c_1	a_0	a_1	b_1	γ_1	γ_2
c_0	1.52E-04	-2.66E-05	-1.04E-03	1.50E-04	1.12E-05	4.85E-05	-2.20E-05
c_1	-2.66E-05	2.22E-04	1.79E-04	-2.09E-04	-1.30E-06	-6.32E-05	2.20E-05
a_0	-1.04E-03	1.79E-04	2.08E-02	-4.78E-03	5.73E-05	-4.81E-04	6.27E-04
a_1	1.50E-04	-2.09E-04	-4.78E-03	2.79E-02	-2.20E-04	3.06E-03	-2.33E-03
b_1	1.12E-05	-1.30E-06	5.73E-05	-2.20E-04	1.29E-05	1.63E-05	-8.74E-06
γ_1	4.85E-05	-6.32E-05	-4.81E-04	3.06E-03	1.63E-05	6.29E-04	-4.14E-04
γ_2	-2.20E-05	2.20E-05	6.27E-04	-2.33E-03	-8.74E-06	-4.14E-04	4.31E-04

Table 2.8   Values of the SBC and AIC criteria for five models

Model	S&P500	FTSE100	$/£	Gold
	AIC criterion			
AR(1)-GARCH(1,1)	2.605210	2.595462	1.611359*	2.143217
AR(1)-IGARCH(1,1)	2.606124	2.597809	1.615456	2.142781
AR(1)-EGARCH(1,1)	**2.579729**	2.584482*	1.623787	2.121844*
AR(1)-GJR(1,1)	2.589309	2.585768	1.611592	2.127675
AR(1)-APARCH(1,1)	2.582430*	**2.583719**	**1.609807**	**2.124257**
	SBC criterion			
AR(1)-GARCH(1,1)	2.612421	2.602673	1.618570*	2.150428
AR(1)-IGARCH(1,1)	2.611893	2.603578	1.621225	2.148550
AR(1)-EGARCH(1,1)	**2.589825**	2.594577*	1.633882	2.131939*
AR(1)-GJR(1,1)	2.597962	2.594421	1.620245	2.136329
AR(1)-APARCH(1,1)	2.592525*	**2.593814**	**1.619903**	**2.134352**

*Model failed to converge.
The models that minimize the criterion are shown in bold face.

Table 2.9   Estimated parameters of the models selected by the *SBC* and *AIC* information criteria

Parameters	S&P500	FTSE100	$/£	Gold
c_0	0.02913	0.02182	0.00825	−0.00171
	(2.36)	(1.70)	(1.04)	(−0.17)
c_1	0.01430	0.02905	0.02515	0.00150
	(0.96)	(1.77)	(1.57)	(0.08)
a_0	0.05339	0.01166	0.00155	0.00415
	(0.37)	(3.89)	(1.64)	(1.83)
a_1	−0.31165	0.05635	0.02207	0.04941
	(−1.87)	(6.36)	(2.52)	(3.73)
b_1	0.98717	0.93958	0.95260	0.95467
	(274.90)	(93.54)	(90.54)	(76.76)
γ_1	−0.10187	0.45583	0.01442	−0.41640
	(−4.06)	(3.94)	(0.27)	(−3.59)
γ_2	0.11846	—	—	—
	(5.70)			
δ	—	1.29030	2.96884	1.43148
		(5.98)	(6.22)	(6.22)

The coefficients to standard error ratios are given in parentheses.
For the S&P500 index, the model is AR(1)-EGARCH(1,1), while for the FTSE100, the $/£, and the gold bullion price, the model is AR(1)-APARCH(1,1).

APARCH model, the asymmetric parameter γ_1 is also significant in all cases except that of the $/£ exchange rate. The leverage effect was also found statistically significant in Ebens's (1999) and Giot and Laurent's (2004) studies, but Andersen et al. (1999a) did not find such supporting evidence in the exchange rates markets. The Box–Cox power transformation was correctly imposed in the ARCH specification as δ differs from one or two in all cases. A negative value of $\varepsilon_t(\varepsilon_t < 0)$ indicates an actual return that is less that the expected one, as $\varepsilon_t = y_t - \mu_t \Rightarrow y_t < \mu_t$. In the EGARCH (1,1) model, a negative estimate of γ_1 provides evidence that bad news has a greater impact on the conditional variance than good news. In the case of the APARCH(1,1) specification, a positive value of γ_1 indicates a greater impact of negative residuals on the conditional volatility. The leverage effect exists in the case of the S&P500 and FTSE100 indices, whereas in the case of gold there is evidence for an asymmetric relationship between innovations and conditional volatility. However, when $\gamma_1 < 0$, a greater impact on volatility comes from positive residuals rather than from the negative ones. Figures 2.19–2.22 present the annualized in-sample estimated standard deviation, $\sqrt{252}\hat{\sigma}_{t+1}$, along with the logarithm of the daily closing prices in order to underscore the tendency of log-returns to be negatively correlated with changes in volatility. It is obvious, from Figures 2.19 and 2.20 referring to the S&P500 and FTSE100 indices, that periods of market drops are characterized by a high increase in volatility.

The asymmetric relationship between unexpected returns and conditional variance can also be plotted by the estimated news impact curve introduced by Pagan and Schwert (1990) and further explored by Engle and Ng (1993). The goal is to plot the volatility at time t against the impact of news, or unexpected return, at time $t-1$. We fix last period's volatility at σ, the unconditional standard deviation of log-returns, and examine the implied relationship between σ_t^2 and ε_{t-1}. The news impact curve for

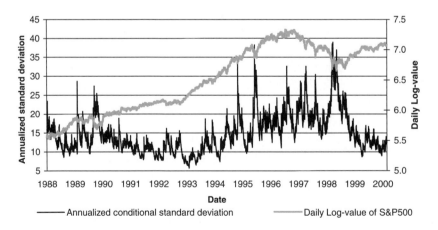

Figure 2.19   Annualized estimated standard deviation, $\sqrt{252}\hat{\sigma}_{t+1}$, and daily log-values for the S&P500 equity index.

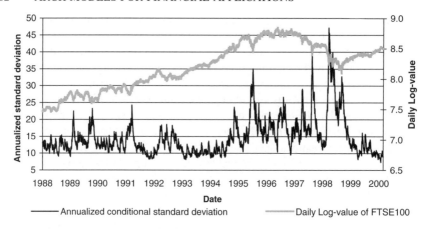

Figure 2.20 Annualized estimated standard deviation, $\sqrt{252}\hat{\sigma}_{t+1}$, and daily log-values for the FTSE100 equity index.

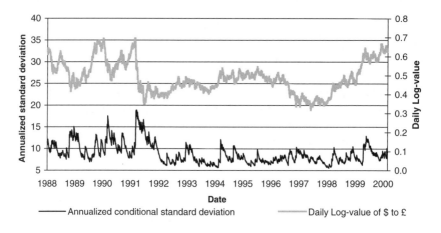

Figure 2.21 Annualized estimated standard deviation, $\sqrt{252}\hat{\sigma}_{t+1}$, and daily log-values for the $/£ exchange rate.

the EGARCH(1,1) model is computed by

$$
\sigma_t^2 = \exp\left(a_0^{(T)}\left(1-b_1^{(T)}\right) + b_1^{(T)} \log \sigma^2 + \left(1+\alpha_1^{(T)}\right)\left(-\gamma_2^{(T)}\sqrt{2/\pi}\right)\right)
$$
$$
\times \exp\left(\left(1+\alpha_1^{(T)}L\right)\left(\gamma_1^{(T)}\frac{\varepsilon_{t-1}}{\sigma} + \gamma_2^{(T)}\frac{|\varepsilon_{t-1}|}{\sigma}\right)\right),
\tag{2.84}
$$

where σ is the unconditional log-return standard deviation. For the APARCH(1,1) model the news impact curve is given by

$$
\sigma_t^2 = \left(a_0^{(T)} + b_1^{(T)}\sigma^{\delta^{(T)}} + \alpha_1^{(T)}\left(|\varepsilon_{t-1}|-\gamma_1^{(T)}\varepsilon_{t-1}\right)^{\delta^{(T)}}\right)^{2/\delta^{(T)}}.
\tag{2.85}
$$

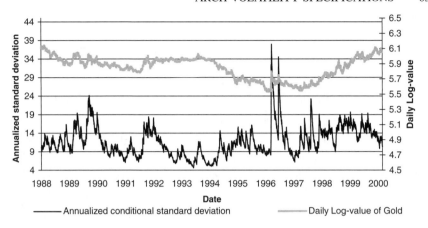

Figure 2.22   Annualized estimated standard deviation, $\sqrt{252}\,\hat{\sigma}_{t+1}$, and daily log-values for the gold price.

Figure 2.23 presents the news impact curves. In the case of the S&P500 and FTSE100 indices, the conditional volatility asymmetric modeling allows good news to generate less volatility than bad news. In the $/£ exchange rate case, the news impact curve is almost symmetric as the asymmetric parameter is statistically indistinguishable from zero. Finally, in the gold case, the good news has greater impact on volatility.

The standardized residuals, $\hat{\varepsilon}_t\hat{\sigma}_t^{-1}$, and their squared values, $\hat{\varepsilon}_t^2\hat{\sigma}_t^{-2}$, of the four models do not obey the standard assumptions of absence of autocorrelation and heteroscedasticity. In Table 2.10, we present the Box and Pierce (1970) Q statistic for

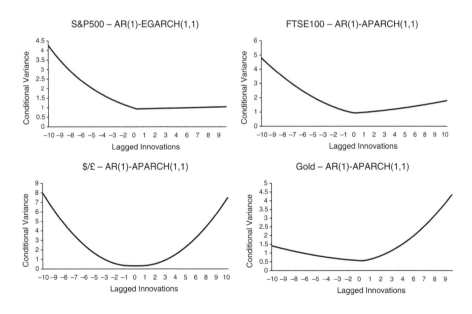

Figure 2.23   News impact curves.

Table 2.10   Descriptive statistics of standardized residuals and Q statistic of squared standardized residuals

Index	S&P500	FTSE100	$/£	Gold
Model	AR(1)-EGARCH(1,1)	AR(1)-APARCH(1,1)	AR(1)-APARCH(1,1)	AR(1)-APARCH(1,2)
		Standardized residuals		
Mean	0.0006	−0.0007	−0.0053	0.0017
Std. dev.	1.0025	1.0001	0.9999	0.9998
Skewness	−0.4596	−0.1347	−0.1325	0.0507
Kurtosis	6.3776	4.0246	4.6064	8.5209
	Q statistic and p-values (given in parentheses) of squared standardized residuals			
Lags	Q statistic	Q statistic	Q statistic	Q statistic
10	3.0558	11.320	6.5633	32.122*
	(0.93)	(0.18)	(0.58)	(0.00)
15	7.5972	20.515	8.4773	33.902*
	(0.87)	(0.083)	(0.81)	(0.00)
20	9.7798	29.392*	13.369	38.857*
	(0.94)	(0.04)	(0.77)	(0.00)

*The null hypothesis that the squared residuals are not autocorrelated is rejected.

the null hypothesis that there is no autocorrelation up to order 10, 15 and 20 computed on $\hat{\varepsilon}_t^2 \hat{\sigma}_t^{-2}$. The hypothesis of no autocorrelation up to lag i is rejected if the Q statistic is greater than the appropriate percentile of the chi-square distribution with $i-p-q$ degrees of freedom. The command *garchobj.BOXPIERCE(<i>)* computes the Q statistic for i lags. Figures 2.24–2.27 plot the histograms of standardized residuals and logarithmic conditional standard deviations, while Table 2.10 presents the descriptive statistics of the standardized residuals. Their skewness and kurtosis are clearly

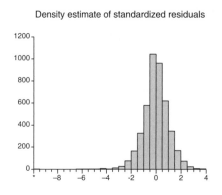
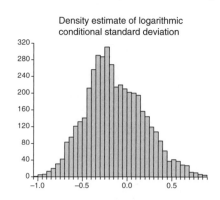

Figure 2.24   S&P500 index: distribution of standardized residuals and logarithmic conditional standard deviation of the AR(1)-EGARCH(1,1) model.

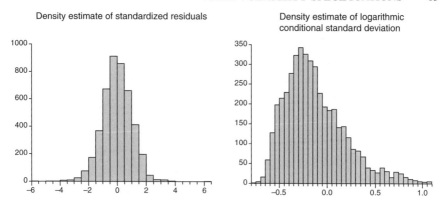

Figure 2.25 FTSE100 index: distribution of standardized residuals and logarithmic conditional standard deviation of the AR(1)-APARCH(1,1) model.

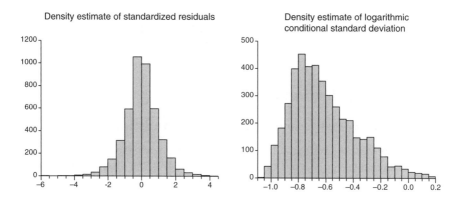

Figure 2.26 $/£ exchange rate index: distribution of standardized residuals and logarithmic conditional standard deviation of the AR(1)-APARCH(1,1) model.

different from that of the normal distribution, implying that a normality assumption would not be plausible. The asymmetric and leptokurtic standardized residuals indicate the importance of reconsidering the assumption that the standardized innovations are normally distributed. The logarithmic transformation of conditional volatility has a unimodal density with a shape far from being Gaussian. In summary, (i) asymmetric ARCH models capture the tendency of unexpected returns to be negatively correlated with changes in returns volatility, but (ii) the assumption of normally distributed standardized residuals is not satisfied.

The observant reader will notice that although we estimate the AR(1)-GARCH(1,1) model based on the same data set, G@RCH provides slightly different parameter estimates than EViews. The main reason is the different techniques that are used to maximize the log-likelihood function. EViews maximizes the likelihood function with the Marquardt and the BHHH algorithms. On the other hand, G@RCH uses the

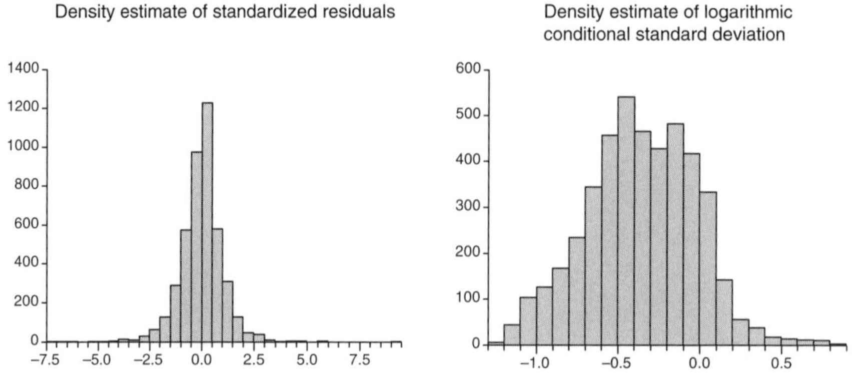

Density estimate of standardized residuals

Density estimate of logarithmic conditional standard deviation

Figure 2.27   Gold price: distribution of standardized residuals and logarithmic conditional standard deviation of the AR(1)-APARCH(1,2) model.

BFGS[15] (equations (2.34) and (2.35)) and the simulated annealing[16] algorithms. When using constrained parameters, the SQPF algorithm, which is similar to algorithm (18.7) of Nocedal and Wright (1999), is used. For details on imposing stationarity and positivity constraints, see Laurent and Peters (2006, p. 83).

2.7   Misspecification tests

2.7.1   The Box–Pierce and Ljung–Box Q statistics

EViews computes the Ljung and Box (1978) Q statistic to test the null hypothesis that the squared residuals are not autocorrelated up to order i. The Ljung–Box statistic is computed as

$$Q^{(LB)} = T(T+2) \sum_{j=1}^{i} \frac{r_j^2}{T-j}, \qquad (2.86)$$

where

$$r_j = \frac{T}{T-j} \frac{\sum_{t=j+1}^{T} \left(\hat{\varepsilon}_t^2 - \bar{\varepsilon} \right) \left(\hat{\varepsilon}_{t-j}^2 - \bar{\varepsilon} \right)}{\sum_{t=1}^{T} \left(\hat{\varepsilon}_t^2 - \bar{\varepsilon} \right)^2}, \text{ for } \bar{\varepsilon} = T^{-1} \sum_{t=1}^{T} \varepsilon_t^2,$$

denotes the EViews autocorrelation estimate of squared standardized residuals at j lags. Theoretically, the Ljung–Box statistic is asymptotically chi-square distributed with degrees of freedom equal to the number of autocorrelations. If the conditional mean has an ARMA specification, the degrees of freedom should be adjusted to represent the number of autocorrelations less the number of AR and MA terms previously estimated.

---

[15] See Doornik (2001, p. 248) and Laurent and Peters (2006, p. 26) for further details.
[16] See Charles Bos's homepage, http://www.tinbergen.nl/~cbos/index.html?content=/~cbos/software/maxsa.html, for further details.

The G@RCH package provides the Box and Pierce (1970) Q statistics for testing whether the squared residuals are autocorrelated up to order i. As Tse (2002) noted, many studies assume the Q statistics as chi-square distributed, although its distribution differs from the χ_i^2 distribution even asymptotically. Li and Mak (1994) and Ling and Li (1997) derived the asymptotic distribution of the Q statistics in the univariate and multivariate cases, respectively. Monte Carlo results on the finite-sample distributions of the Li–Mak and Ling–Li tests were reported by Tse and Zuo (1997) and Tse and Tsui (1999), respectively.

2.7.2 Tse's residual based diagnostic test for conditional heteroscedasticity

Tse (2002) proposed the residual based diagnostic (RBD) statistic from the auxiliary regression:

$$E\left(\frac{\hat{\varepsilon}_t^2}{\hat{\sigma}_t^2}\right) - 1 = a_1 \frac{\hat{\varepsilon}_{t-1}^2}{\hat{\sigma}_{t-1}^2} + \cdots + a_i \frac{\hat{\varepsilon}_{t-i}^2}{\hat{\sigma}_{t-i}^2} + \varepsilon_t^*. \tag{2.87}$$

As the regressors are not observed but estimated, the inference procedure based on the usual ordinary least squares (OLS) inference is invalid. Thus, the appropriate procedure is to correct for the asymptotic variance of the OLS. Tse (2002) derived the asymptotic distribution of $a_1, \ldots, a_i$ and showed that the RBD statistic that tests the null hypothesis, $H_0: a_1 = \cdots = a_i = 0$, is asymptotically chi-square distributed. The G@RCH package provides the RBD statistic. For example, the command *garchobj.RBD(<5>)* tests the null hypothesis $H_0: a_1 = \cdots = a_5 = 0$. Tsui (2004) studied the size and power of various diagnostic statistics for conditional heteroscedasticity models and provided evidence that the Tse (2002) and Li and Mak (1994) diagnostics are more powerful.

2.7.3 Engle's Lagrange multiplier test

Engle's (1982, 1984) Lagrange multiplier procedure tests the null hypothesis of no ARCH effect in a model's residuals. The test procedure is to regress the squared standardized residuals[17] on a constant and i lagged values of the squared standardized residuals

$$\frac{\hat{\varepsilon}_t^2}{\hat{\sigma}_t^2} = a_0 + a_1 \frac{\hat{\varepsilon}_{t-1}^2}{\hat{\sigma}_{t-1}^2} + \cdots + a_i \frac{\hat{\varepsilon}_{t-i}^2}{\hat{\sigma}_{t-i}^2} + \varepsilon_t^*, \tag{2.88}$$

for $\varepsilon_t^* \sim N(0, \sigma_{\varepsilon^*}^2)$, and use the squared multiple correlation times the number of observations, TR^2, an asymptotically chi-square distributed variable with i degrees of freedom, as a test statistic. The null hypothesis, $H_0: a_1 = \cdots = a_i = 0$ is tested against the alternative that at least one $a_i \neq 0$. EViews reports the values of two test statistics, namely, the value of the TR^2 statistic and that of an F statistic, which is

---

[17] Packages base their estimations on the squared standardized residuals but Engle proposed the use of squared residuals.

an omitted variable test for the joint significance of all lagged squared residuals. The exact finite-sample distribution of the F statistic under the null hypothesis is not known. In EViews, the command $eq1.archtest(4)$ tests the null hypothesis $H_0: a_1 = \cdots = a_4 = 0$ for the squared standardized residuals of the eq1 equation. G@RCH presents the F statistic. For example, the command $garchobj.ARCHLAGS$ $(<10>)$ tests the null hypothesis $H_0: a_1 = \cdots = a_{10} = 0$. As the volatility parameters must be positive, Demos and Sentana (1998) proposed a one-sided version of the test that is more powerful than the test based on TR^2. Lumsdaine and Ng (1999) showed that if the assumption of a correctly specified conditional mean does not hold, Engle's Lagrange multiplier test over-rejects conditional homoscedasticity, and they proposed the use of recursive residuals to improve the fit of a first-stage conditional mean regression. Hacker and Abdulnasser (2005) extended Engle's Lagrange multiplier test for ARCH effects to multivariate cases.

2.7.4 Engle and Ng's sign bias tests

Engle and Ng (1993) assessed whether there is asymmetry in the volatility of the residuals of a model by deriving the sign bias test (SBT), the negative sign bias test (NSBT) and the positive sign bias tests (PSBT) based on the following three auxiliary regressions:

$$\frac{\hat{\varepsilon}_t^2}{\hat{\sigma}_t^2} = a_0 + a_1 d(\hat{\varepsilon}_{t-1} < 0) + \varepsilon_t^*, \tag{2.89}$$

$$\frac{\hat{\varepsilon}_t^2}{\hat{\sigma}_t^2} = a_0 + a_1 d(\hat{\varepsilon}_{t-1} < 0)\hat{\varepsilon}_{t-1} + \varepsilon_t^*, \tag{2.90}$$

$$\frac{\hat{\varepsilon}_t^2}{\hat{\sigma}_t^2} = a_0 + a_1 (1 - d(\hat{\varepsilon}_{t-1} < 0))\hat{\varepsilon}_{t-1} + \varepsilon_t^*, \tag{2.91}$$

where $d(\varepsilon_t < 0) = 1$ if $\varepsilon_t < 0$, and $d(\varepsilon_t < 0) = 0$ otherwise. The SBT is used for testing whether squared standardized residuals can be predicted by the dummy variable $d(\varepsilon_{t-1} < 0)$. The NSBT is utilized to test whether large and small negative shocks have different impacts on volatility, while the PSBT is employed to test whether large and small positive shocks have different effects on volatility. The SBT, NSBT and PSBT tests are the t-ratios of parameter a_1 in equations (2.89), (2.90) and (2.91), respectively. The tests can be jointly formulated by defining the regression:

$$\frac{\hat{\varepsilon}_t^2}{\hat{\sigma}_t^2} = a_0 + a_1 d(\hat{\varepsilon}_{t-1} < 0) + a_2 d(\hat{\varepsilon}_{t-1} < 0)\hat{\varepsilon}_{t-1} + a_3 (1 - d(\hat{\varepsilon}_{t-1} < 0))\hat{\varepsilon}_{t-1} + \varepsilon_t^*, \tag{2.92}$$

and testing whether $H_0: a_1 = a_2 = a_3 = 0$. The joint test is conducted by computing the TR^2 statistic from (2.92), which is chi-square distributed with three degrees of freedom.

Engle and Ng (1993) derived the forms of the test statistics by assuming that the volatility model under the null hypothesis is correctly specified and is a special case of the model under the alternative hypothesis

$$\log\left(\sigma_t^2\right) = \log\left(\sigma_{0t}\left(a_0' z_{0t}\right)\right) + a_\alpha' z_{at}, \tag{2.93}$$

where $\sigma_{0t}(a_0'z_{0t})$ is the volatility model under the null, z_{0t} is the vector of explanatory variables of the model hypothesized under the null, a_0 is the parameter vector under the null, and a_α is the parameter vector corresponding to z_{at}, which is the vector of missing explanatory variables. The general form of the auxiliary regression is defined as

$$\frac{\hat\varepsilon_t^2}{\hat\sigma_t^2} = z_{0t}^* a_0 + z_{at}^* a_\alpha + \varepsilon_t^*, \qquad (2.94)$$

where $z_{0t}^* = \sigma_{0t}^{-1}\partial\sigma_t^2/\partial a_0$ and $z_{at}^* = \sigma_{0t}^{-1}\partial\sigma_t^2/\partial a_\alpha$, both evaluated under the null hypothesis, and $a_0 = a_\alpha = 0$. The Lagrange multiplier test statistic for testing $H_0: a_a = 0$ in (2.93) can be computed as the TR^2 statistic from (2.94). To perform the three sign bias tests, we allow z_{at} to include the variables $d(\hat\varepsilon_{t-1} < 0)$, $d(\hat\varepsilon_{t-1} < 0)\hat\varepsilon_{t-1}$ and $(1-d(\hat\varepsilon_{t-1} < 0))\hat\varepsilon_{t-1}$. In G@RCH, the command *garchobj. SBT(1)* provides the SBT, NSBT and PSBT diagnostic tests.

2.7.5 The Breusch–Pagan, Godfrey, Glejser, Harvey and White tests

Breusch and Pagan (1979) and Godfrey (1978) provided a test for the null hypothesis of homoscedasticity against heteroscedasticity of the form $\sigma_t^2 = \sigma^2(x_t'\beta)$, where x_t is a vector of independent variables. The test statistic is computed by an auxiliary regression, where the squared standardized residuals are regressed on x_t and a constant. Glejser (1969) and Harvey (1976) proposed testing for heteroscedasticity on the basis of a test statistic of the form of $\sigma_t^2 = (\sigma^2 + x_t'\beta)^i$, for $i = 1, 2$, and $\sigma_t^2 = e^{x_t'\beta}$, respectively. Glejser's statistic is computed by an auxiliary regression, where the absolute standardized residuals are again regressed on x_t and a constant. Harvey's test statistic is computed in terms of an auxiliary regression, where the logarithms of the squared standardized residuals are also regressed on x_t and a constant. White (1980) considered the case of testing a null hypothesis of no heteroscedasticity against heteroscedasticity of an unknown form. The test statistic is computed on the basis of an auxiliary regression, where the squared standardized residuals are regressed on a constant and all possible cross-products of the explanatory variables in the conditional mean. The heteroscedasticity tests discussed above are available in EViews 6 with the command *hettest*. For example, the command *model1.hettest(bpg) x1 x2* regresses the squared standardized residuals of the model1 equation on $x_{1,t}$, $x_{2,t}$ and a constant.

2.7.6 The Wald, likelihood ratio and Lagrange multiplier tests

The test procedures of this section are employed for drawing inferences on the parameters of an ARCH model. Bollerslev and Wooldridge (1992) constructed robust Wald and Lagrange multiplier statistics and Busch (2005) built a robust likelihood ratio test. For a general outline of such tests, see Bollerslev et al. (1994, p. 2982) and Brooks (2002, p. 490). Using EViews, one can compute the values of the Wald and likelihood ratio statistics on the estimated coefficients of an ARCH model by choosing from the package's menu: View $\rightarrow$ Coefficient Tests $\rightarrow$ Wald-Coefficient Restrictions or View $\rightarrow$ Coefficient Tests $\rightarrow$ Omitted Variables-Likelihood Ratio... or View $\rightarrow$ Coefficient Tests $\rightarrow$ Redundant Variables-Likelihood Ratio....

2.8 Other ARCH volatility specifications

2.8.1 Regime-switching ARCH models

Engle and Mustafa (1992) noted that the volatility consequences of the 1987 crash on option prices disappeared more rapidly than ARCH models suggested. The high volatility persistently implied by ARCH models did not agree with the volatility implied by stock options. Diebold (1986) was the first to suggest that the ARCH volatility persistence might be a consequence of the failure to model structural changes that occurred. In estimating ARCH models for stock returns and exchange rates, Lamoureux and Lastrapes (1990b) and Kim and Kon (1999) found that the inclusion of a deterministic shift in the conditional variance's scale parameter decreases volatility persistence substantially. Mikosch and Stărică (2004) obtained similar results, as a deterministic shift in the unconditional variance forced the GARCH(1,1) estimated parameters $a_1 + b_1$ to unity.

Driven by the above findings, Cai (1994) and Hamilton and Susmel (1994) proposed another class of ARCH models, the regime-switching ARCH models, a natural extension of regime-switching models for the conditional mean, introduced by Hamilton (1988, 1989). These models allow the parameters of the ARCH process to come from one of several different regimes, with transitions between regimes governed by an unobserved Markov chain.

Let $\tilde{\varepsilon}_t$ be the innovation process and let s_t denote an unobserved random variable that can take the values $1, 2, \ldots, \tilde{K}$. Suppose that s_t can be described by a Markov chain, $P(s_t = i_t | s_{t-1} = i_{t-1}, s_{t-2} = i_{t-2}, \ldots, \tilde{\varepsilon}_{t-1}, \tilde{\varepsilon}_{t-2}, \ldots)$. The idea is to model the innovation process, $\tilde{\varepsilon}_t$, as $\tilde{\varepsilon}_t \equiv \sqrt{g_{s_t}} \varepsilon_t$, where ε_t is assumed to follow an ARCH process. So the underlying ARCH variable, ε_t, is multiplied by the constant $\sqrt{g_1}$ when the process is in the regime presented by $s_t = 1$, is multiplied by $\sqrt{g_2}$ when $s_t = 2$, and so on. The factor for the first stage, g_1, is normalized at unity with $g_{s_t} \geq 1$ for $s_t = 2, 3, \ldots, \tilde{K}$. The idea is thus to model changes in regime as changes in the scale of the process. The $\tilde{K}$th-state, qth-order Markov switching ARCH, or SWARCH$(\tilde{K}, q)$, model is

$$\tilde{\varepsilon}_t \equiv \sqrt{g_{s_t}} \varepsilon_t,$$

$$\sigma_t^2 = a_0 + \sum_{i=1}^{q} \left(a_i \varepsilon_{t-i}^2 \right). \tag{2.95}$$

Hamilton and Susmel (1994) also proposed another regime-switching model that captures the leverage effect. In particular, they constructed the $\tilde{K}$th-state, qth-order Markov switching ARCH with Leverage effect, or SWARCH-L$(\tilde{K}, q)$, model,

$$\tilde{\varepsilon}_t \equiv \sqrt{g_{s_t}} \varepsilon_t,$$

$$\sigma_t^2 = a_0 + \gamma_1 d(\varepsilon_{t-1} < 0)\varepsilon_{t-1}^2 + \sum_{i=1}^q \left(a_i \varepsilon_{t-i}^2\right), \tag{2.96}$$

for $d(\varepsilon_{t-i} < 0) = 1$ if $\varepsilon_{t-i} < 0$, and $d(\varepsilon_{t-i} < 0) = 0$ otherwise, which is an extension of the GJR$(0, q)$ specification.

The SWARCH model cannot be extended to include the GARCH model because of the path dependence on the Markov switching GARCH model, as each conditional variance depends not just on the current regime but on the entire past history of the process. Gray (1996) considered a similar approach for the case of GARCH models leading to what he termed the generalized regime-switching (GRS) model that allowed not only the scaling parameter but also all the conditional variance parameters to be regime-dependent. Gray overcame the problem of path dependence by allowing each conditional variance to depend not on the entire past history of the process, but only on the current regime. For two regimes and under the assumption of conditional normality within each regime, the GRS model can be represented in the form

$$y_t = p_{1,t}\mu_{1,t} + \left(1 - p_{1,t}\right)\mu_{2,t} + \varepsilon_t,$$

$$\sigma_{i,t}^2 = a_{i,0} + a_{i,1}\varepsilon_{t-1}^2 + b_{i,1}\sigma_{t-1}^2,$$

$$\sigma_{t-1}^2 = p_{1,t-1}\left(\mu_{1,t-1}^2 + \sigma_{1,t-1}^2\right) + \left(1 - p_{1,t-1}\right)\left(\mu_{2,t-1}^2 + \sigma_{2,t-1}^2\right)$$

$$- \left(p_{1,t-1}\mu_{1,t-1} + \left(1 - p_{1,t-1}\right)\mu_{2,t-1}\right)^2, \tag{2.97}$$

for $i = 1, 2$, where $\mu_{i,t}$ and $\sigma_{i,t}^2$ are the conditional mean and conditional variance, respectively, at time t of regime i, and $p_{1,t} = P(s_t = 1|I_{t-1})$ describes the probability of the market being at regime 1 at time t given the information at $t-1$. The filtered probability is equal to

$$p_{1,t} = (1 - q_t)P(s_{t-1} = 2|I_{t-1}) + p_t P(s_{t-1} = 1|I_{t-1}), \tag{2.98}$$

with switching probabilities $q_t = P(s_t = 2|s_{t-1} = 2)$ and $p_t = P(s_t = 1|s_{t-1} = 1)$.[18] Moreover, $P(s_{t-1} = 2|I_{t-1}) = f_{2,t-1}(1 - p_{1,t-1})(f_{1,t-1}p_{1,t-1} + f_{2,t-1}(1 - p_{1,t-1}))^{-1}$, $P(s_{t-1} = 1|I_{t-1}) = f_{1,t-1}p_{1,t-1}(f_{1,t-1}p_{1,t-1} + f_{2,t-1}(1 - p_{1,t-1}))^{-1}$ and $f_{i,t} = f(y_t|s_t = i)$ is the density function of regime i at time t.

Haas et al. (2004) noted that although attractively combining Markov switching with GARCH effects, neither Gray's (1996) specification nor the modification proposed by Klaassen (2002), leads to an analytic expression for the covariance structure of the squared process and to covariance stationarity conditions. They thus proposed the $\tilde{K}$th-state Markov switching GARCH, or MSG($\tilde{K}$) model, which is analytically tractable and establishes stationarity conditions.

---

[18] Gray proposed either constant transition probabilities, $q_t = q_{t'}$ and $p_t = p_{t'}$, $\forall t \neq t'$, or time-varying transition probabilities. In the latter case, the transition probabilities can be estimated as a function of the cumulative normal distribution, $p_t = F(\theta_1|I_{t-1})$ and $q_t = F(\theta_2|I_{t-1})$, where θ_i, for $i = 1, 2$, are parameters to be estimated.

Representative applications of regime-switching ARCH specifications in stock returns and foreign exchange rates have been presented by Turner et al. (1989), Engel and Hamilton (1990), Pagan and Schwert (1990), Engel (1994), Dueker (1997), Vigfusson (1997), Assoe (1998), Fong (1998), Rydén et al. (1998), Billio and Pelizzon (2000), Bollen et al. (2000), Chaudhuri and Klaassen (2000), Maheu and McCurdy (2000), Susmel (2000), Dewachter (2001), Francq et al. (2001), Perez-Quiros and Timmermann (2001), Ang and Bekaert (2002), Brunetti et al. (2003), Beine et al. (2003), Guidolin and Timmermann (2003), Ahrens and Reitz (2004), Bhar and Hamori (2004), Gau and Tang (2004) and Li and Lin (2004).

2.8.2 Extended ARCH models

Fornari and Mele (1995) introduced the volatility-switching ARCH, or VSARCH(p, q), model,

$$\sigma_t^2 = a_0 + \sum_{i=1}^{q} a_i \varepsilon_{t-i}^2 + \gamma d(\varepsilon_{t-1} > 0) \frac{\varepsilon_{t-1}^2}{\sigma_{t-1}^2} + \sum_{j=1}^{p} b_j \sigma_{t-j}^2, \qquad (2.99)$$

where $d(\varepsilon_t > 0)$ is an indicator factor that equals 1 if $\varepsilon_t > 0$ and -1 if $\varepsilon_t < 0$, and $\varepsilon_t^2/\sigma_t^2$ measures the difference between the forecast of the volatility at time t on the basis of the information set at time $t-1$, σ_t^2, and the realized value ε_t^2. As Fornari and Mele (1995) noted, the volatility-switching model is able to capture a phenomenon that has not been modelled before. It implies that asymmetries can become inverted, with positive innovations inducing more volatility than negative innovations of the same size when the observed value of the conditional variance is lower than expected. Fornari and Mele (1996) constructed a mixture of the GJR and VSARCH models, the asymmetric volatility-switching ARCH, or AVSARCH(p, q), model, and estimated it for $p = q = 1$:

$$\sigma_t^2 = a_0 + a_1 \varepsilon_{t-1}^2 + b_1 \sigma_{t-1}^2 + \gamma d(\varepsilon_{t-1} > 0) \varepsilon_{t-1}^2 + \delta \big((\varepsilon_{t-1}^2/\sigma_{t-1}^2) - k \big) d(\varepsilon_{t-1} > 0). \qquad (2.100)$$

The first four terms are the GJR(1,1) model, except that $d(\varepsilon_t > 0)$ is a dummy variable that equals 1 (if $\varepsilon_t > 0$) or -1 (if $\varepsilon_t < 0$) instead of 0 or 1, respectively. The last term captures the reversal of asymmetry observed when $\varepsilon_{t-1}^2/\sigma_{t-1}^2$ reaches k, the threshold value. Note that the AVSARCH model is able to generate kurtosis higher than the GARCH or GJR models.

Hagerud (1996), inspired by the smooth transition autoregressive (STAR) model of Luukkonen et al. (1988), proposed the smooth transition ARCH model. In the STAR model, the conditional mean is a non-linear function of lagged realizations of the series introduced via a transition function. The smooth transition GARCH(p, q) model has the form

$$\sigma_t^2 = a_0 + \sum_{i=1}^{q} (a_i + \gamma_i F(\varepsilon_{t-i})) \varepsilon_{t-i}^2 + \sum_{j=1}^{p} b_j \sigma_{t-j}^2, \qquad (2.101)$$

where $F(.)$ is either the logistic or the exponential transition function, the two most commonly used transition functions for STAR models (for details see Teräsvirta, 1994). The logistic function considered is

$$F(\varepsilon_{t-i}) = (1 + \exp(-\theta\varepsilon_{t-i}))^{-1} - 0.5, \quad \text{for } \theta > 0, \tag{2.102}$$

and the exponential function is

$$F(\varepsilon_{t-i}) = 1 - \exp\left(-\theta\varepsilon_{t-i}^2\right), \quad \text{for } \theta > 0. \tag{2.103}$$

The two resulting models are termed logistic and exponential smooth transition GARCH, or LST-GARCH(p, q) and EST-GARCH(p, q), models, respectively. The smooth transition models allow for the possibility of intermediate positions between different regimes. For $-\infty < \varepsilon_t < \infty$, the logistic transition function takes values in $-0.5 \leq F(.) \leq 0.5$ and generates data where the dynamics of the conditional variance differs depending on the sign of innovations. On the other hand, the exponential function generates a return process for which the dynamics of the conditional variance depends on the magnitude of the innovations, as for $|\varepsilon_t| \to \infty$ the transition function equals 1, and when $\varepsilon_t = 0$ the transition function equals 0. Thus, contrary to what happens in the case of the regime-switching models, the transition between states is smooth as the conditional variance is a continuous function of innovations. A model similar to the LST-GARCH model was independently proposed by González-Rivera (1996). Recently, Nam et al. (2002) provided an application of a smooth transition ARCH model with a logistic function in the form

$$\sigma_t^2 = a_0 + a_1\varepsilon_{t-1}^2 + a_2\sigma_{t-1}^2 + \left(b_0 + b_1\varepsilon_{t-1}^2 + b_2\sigma_{t-1}^2\right)F(\varepsilon_{t-1}),$$
$$F(\varepsilon_{t-1}) = (1 + \exp\left(-\theta\varepsilon_{t-1}\right))^{-1}, \tag{2.104}$$

which they termed the asymmetric non-linear smooth transition GARCH, or ANST-GARCH, model. Nam et al. explored the asymmetric reverting property of short-horizon expected returns and found that the asymmetric return reversals can be exploited for contrarian profitability.[19] Note that when $b_0 = b_2 = 0$, the ANST-GARCH model reduces to González-Rivera's specification. Lubrano (1998) suggested an improvement over these transition functions, introducing an extra parameter, the threshold c, which determines at which magnitude of past innovations the change of regime occurs. His generalized logistic and exponential transition functions are given by

$$F(\varepsilon_{t-i}) = \frac{1 - \exp\left(-\theta\varepsilon_{t-i}^2\right)}{1 + \exp\left(-\theta(\varepsilon_{t-i}^2 - c^2)\right)} \tag{2.105}$$

---

[19] Contrarian investment strategies are contrary to the general market direction. Contrarian profitability is viewd differently under the two competing hypotheses: the time-varying rational expectation hypothesis and the stock market overreaction hypothesis. For details see Chan (1988), Chopra et al. (1992), Conrad and Kaul (1993), DeBondt and Thaler (1985, 1987, 1989), Lo and MacKinlay (1990b), Veronesi (1999) and Zarowin (1990).

and

$$F(\varepsilon_{t-i}) = 1 - \exp\left(-\theta(\varepsilon_{t-i} - c)^2\right), \tag{2.106}$$

respectively.

Engle and Lee (1993) proposed the component GARCH model in order to investigate the long-run and short-run movement of volatility. The GARCH(1,1) model can be written as

$$\sigma_t^2 = \sigma^2 + a_1\left(\varepsilon_{t-1}^2 - \sigma^2\right) + b_1\left(\sigma_{t-1}^2 - \sigma^2\right), \tag{2.107}$$

for $\sigma^2 = a_0(1 - a_1 - b_1)^{-1}$ denoting the unconditional variance. The conditional variance in the GARCH(1,1) model shows mean reversion to the unconditional variance, which is constant at all times. By contrast, the component GARCH, or CGARCH(1,1), model allows mean reversion to a time-varying level q_t. The CGARCH(1,1) model is defined as

$$\begin{aligned} \sigma_t^2 &= q_t + a_1\left(\varepsilon_{t-1}^2 - q_{t-1}\right) + b_1\left(\sigma_{t-1}^2 - q_{t-1}\right), \\ q_t &= a_0 + pq_{t-1} + \varphi\left(\varepsilon_{t-1}^2 - \sigma_{t-1}^2\right). \end{aligned} \tag{2.108}$$

The difference between the conditional variance and its trend, $\sigma_t^2 - q_t$, is the transitory or short-run component of the conditional variance, while q_t is the time-varying long-run volatility. Combining the transitory and permanent equations, the model reduces to

$$\begin{aligned} \sigma_t^2 &= (1 - a_1 - b_1)(1 - p)a_0 + (a_1 + \phi)\varepsilon_{t-1}^2 - (a_1 p + (a_1 + b_1)\phi)\varepsilon_{t-2}^2 \\ &\quad + (b_1 - \phi)\sigma_{t-1}^2 - (b_1 p - (a_1 + b_1)\phi)\sigma_{t-2}^2, \end{aligned} \tag{2.109}$$

which shows that the CGARCH(1,1) is a restricted GARCH(2,2) model. Moreover, because of the existence of the leverage effect, Engle and Lee (1993) considered combining the component model with the GJR model to allow shocks to affect the volatility components asymmetrically. The asymmetric component GARCH, or ACGARCH(1,1), model becomes

$$\begin{aligned} \sigma_t^2 &= q_t + a_1\left(\varepsilon_{t-1}^2 - q_{t-1}\right) + \gamma_1\left(d(\varepsilon_{t-1} < 0)\varepsilon_{t-1}^2 - 0.5q_{t-1}\right) + b_1\left(\sigma_{t-1}^2 - q_{t-1}\right), \\ q_t &= a_0 + q_{t-1} + \varphi\left(\varepsilon_{t-1}^2 - \sigma_{t-1}^2\right) + \gamma_2\left(d(\varepsilon_{t-1} < 0)\varepsilon_{t-1}^2 - 0.5\sigma_{t-1}^2\right), \end{aligned} \tag{2.110}$$

where $d(.)$ denotes the indicator function (i.e. $d(\varepsilon_{t-i} < 0) = 1$ if $\varepsilon_{t-i} < 0$, and $d(\varepsilon_{t-i} < 0) = 0$ otherwise).

Ding and Granger (1996) proposed the N-component GARCH(1,1) model:

$$\begin{aligned} \sigma_t^2 &= \sum_{i=1}^{N} w_i \sigma_{i,t}^2, \\ \sigma_{i,t}^2 &= \sigma^2(1 - a_i - b_i) + a_i \varepsilon_{t-1}^2 + b_i \sigma_{i,t-1}^2, \end{aligned} \tag{2.111}$$

for $i = 1, \ldots, N$, where σ^2 is the unconditional variance of ε_t, and $\sum_{i=1}^{N} w_i = 1$. This model assumes a conditional variance which is a weighted sum of N components. As Ding and Granger (1996) noted, a two-component GARCH(1,1) model can be designed such that $\sigma_{1,t}^2$ can capture the short-term fluctuation and $\sigma_{2,t}^2$ can model the long-term, gradual movements in volatility.

Nowicka-Zagrajek and Weron (2001) considered replacing the constant term in the GARCH(p, q) model by a linear function of i.i.d. stable random variables and defined the randomized GARCH, or R-GARCH(r, p, q), model,

$$\sigma_t^2 = \sum_{i^*=1}^{r} (c_{i^*} \eta_{t-i^*}) + \sum_{i=1}^{q} \left(a_i \varepsilon_{t-i}^2 \right) + \sum_{j=1}^{p} \left(b_j \sigma_{t-j}^2 \right), \qquad (2.112)$$

where $c_{i^*} \geq 0$, $i^* = 1, \ldots, r$, $a_i \geq 0$, $i = 1, \ldots, q$, $b_j \geq 0$, $j = 1, \ldots, p$, the innovations η_t are positive i.i.d. stable random variables expressed by the characteristic function in (5.7) and, $\{\eta_t\}$ and $\{z_t\}$ are independent.

The ARCH specifications presented above are based on asset returns that are observed over intervals of the same size. Müller et al. (1997) proposed a volatility specification which is based on price changes over intervals of different sizes. Their motivation derived from the heterogeneous market hypothesis (HMH) presented by Müller et al. (1993). The basic idea of the HMH is that participants in a financial market differ in their reactions to news, possession of information, perception of market risk, investment time horizon, etc. Therefore, agents with different time horizons create different types of volatility components. Based on the hypothesis that participants in a heterogeneous market make volatilities of different time resolutions behave differently, Müller et al. (1997) proposed the heterogeneous interval GARCH, or HARCH(p, n), model that takes into account the squared price changes over time intervals of different sizes:

$$\sigma_t^2 = a_0 + \sum_{i=1}^{n} \sum_{k=1}^{i} a_{ik} \left(\sum_{i^*=k}^{i} \varepsilon_{t-i^*} \right)^2 + \sum_{j=1}^{p} \left(b_j \sigma_{t-j}^2 \right), \qquad (2.113)$$

where $a_0 > 0$, $a_{ik} \geq 0$, for $i = 1, \ldots, n$, $k = 1, \ldots, i$, and $b_j \geq 0$, $j = 1, \ldots, p$. The HARCH model is able to reproduce the long-memory property of volatility. Dacorogna et al. (1998) proposed a representation of Müller et al.'s specification by keeping in the conditional variance specification only a parsimonious number of interval sizes and replacing the influence of the neighbouring interval sizes by an exponential moving average of the representative returns. They named their modification the exponential moving average HARCH, or EMA-HARCH, model.

Many financial markets impose restrictions on the maximum allowable daily change in price. As pointed out by Wei and Chiang (1997), the common practice of ignoring the problem by treating the observed censored observations as if they were actually the equilibrium prices, or by dropping the limited prices from the sample studied, leads to the underestimation of conditional volatility. Morgan and

Trevor (1999) proposed the rational expectation (RE) algorithm (which can be interpreted as an EM algorithm (Dempster et al., 1977) for censored observations in the presence of heteroscedasticity, which replaces the unobservable components of the likelihood function of the ARCH model by their rational expectations. As an alternative to the RE algorithm, Wei (2002), based on Kodres's (1993) study, proposed a censored GARCH model and developed a Bayesian estimation procedure for it. Moreover, on the basis of Kodres's (1988) research, L.F. Lee (1999), Wei (1999) and Calzorari and Fiorentini (1998) developed the class of Tobit-GARCH models.

Bollerslev and Ghysels (1996) proposed a generalization of ARCH models allowing periodicity in the conditional heteroscedasticity. Liu and Morimune (2005) modified ARCH models to capture the effect on volatilities of the consecutive number of positive or negative shocks. Tsay (1987) proposed the conditional heteroscedastic ARMA, or CHARMA(q), model of the form

$$
\begin{aligned}
y_t &= \mu_t + \sum_{i=1}^{q} c_{i,t} a_{t-i} + z_t, \\
\mu_t &= \mu(\theta | I_{t-1}), \\
z_t &\sim f(0, \sigma_z^2).
\end{aligned}
\tag{2.114}
$$

Here $c_t = (c_{1,t}, c_{2,t}, \ldots, c_{q,t})'$ is a sequence of i.i.d. distributed random vectors with mean zero and variance–covariance matrix H, or $c_t \sim f(0, H)$. The conditional variance specification of y_t is computed as

$$
\sigma_t^2 = \sigma_z^2 + (a_{t-1}, a_{t-2}, \ldots, a_{t-q}) H (a_{t-1}, a_{t-2}, \ldots, a_{t-q})'.
\tag{2.115}
$$

For example, the conditional variance of the CHARMA(2) model has the form $\sigma_t^2 = \sigma_z^2 + h_{11} a_{t-1}^2 + 2 h_{12} a_{t-1} a_{t-2} + h_{22} a_{t-2}^2$, as $H = \begin{bmatrix} h_{11} & h_{12} \\ h_{12} & h_{22} \end{bmatrix}$.

Brooks et al. (2001) reviewed the best-known software packages for estimation of ARCH models, and concluded that the estimation results differ considerably from one package to the other. Table 2.11 summarizes the ARCH models that have been presented. It should be noted that the most complex model may not necessarily be the best to use, as usually the most complex model is the most difficult model for fitting the data, in the sense that the estimation of parameters is complicated and not efficient.

2.9   Other methods of volatility modelling

Stochastic volatility models (Barndorff-Nielsen et al., 2002; Chib et al., 1998; Ghysels et al., 1996; Harvey and Shephard, 1993; Jacquier et al., 1994; Shephard, 1996; Taylor, 1994), historical volatility models (Beckers, 1983; Garman and Klass,

Table 2.11 The ARCH models presented in Chapter 2. Numbers in parentheses below each model refer to the model's defining equations given in the text

ARCH(q) Engle (1982) (2.1)	$\sigma_t^2 = a_0 + \sum_{i=1}^{q} \left(a_i \varepsilon_{t-i}^2 \right)$
GARCH(p, q) Bollerslev (1986) (2.5)	$\sigma_t^2 = a_0 + \sum_{i=1}^{q} \left(a_i \varepsilon_{t-i}^2 \right) + \sum_{j=1}^{p} \left(b_j \sigma_{t-j}^2 \right)$
IGARCH(p, q) Engle and Bollerslev (1986) (2.7)	$\sigma_t^2 = a_0 + \sum_{i=1}^{q} \left(a_i \varepsilon_{t-i}^2 \right) + \sum_{j=1}^{p} \left(b_j \sigma_{t-j}^2 \right),$ for $\sum_{i=1}^{q} a_i + \sum_{j=1}^{p} b_j = 1$
Riskmetrics™ or EWMA J.P. Morgan (1996) (2.10)	$\sigma_t^2 = 0.06 \varepsilon_{t-1}^2 + 0.94 \sigma_{t-1}^2$
ALT-GARCH(1,1) Knight and Satchell (2002a) (2.11)	$\sigma_t^2 = a_0 + a_1 z_{t-1}^2 + b_1 \sigma_{t-1}^2$
EGARCH(p, q) Nelson (1991) (2.55)	$\log \left(\sigma_t^2 \right) = a_0 + \left(1 + \sum_{i=1}^{q} a_i L^i \right) \left(1 - \sum_{j=1}^{p} b_j L^j \right)^{-1}$ $\times \left(\gamma_1 (\varepsilon_{t-1}/\sigma_{t-1}) \right.$ $\left. + \gamma_2 (|\varepsilon_{t-1}/\sigma_{t-1}| - E|\varepsilon_{t-1}/\sigma_{t-1}|) \right)$
GJR(p, q) Glosten et al. (1993) (2.57)	$\sigma_t^2 = a_0 + \sum_{i=1}^{q} \left(a_i \varepsilon_{t-i}^2 \right)$ $+ \sum_{i=1}^{q} \left(\gamma_i d(\varepsilon_{t-i} < 0) \varepsilon_{t-i}^2 \right) + \sum_{j=1}^{p} \left(b_j \sigma_{t-j}^2 \right)$
TARCH(p, q) Zakoian (1990) (2.58)	$\sigma_t = a_0 + \sum_{i=1}^{q} \left(a_i \varepsilon_{t-i}^+ \right) - \sum_{i=1}^{q} \left(\gamma_i \varepsilon_{t-i}^- \right) + \sum_{j=1}^{p} \left(b_j \sigma_{t-j} \right)$
AGARCH(p, q) Taylor (1986), Schwert (1989a, 1989b) (2.59)	$\sigma_t = a_0 + \sum_{i=1}^{q} a_i |\varepsilon_{t-i}| + \sum_{j=1}^{p} b_j \sigma_{t-j}$
Log-GARCH(p, q) Geweke (1986), Pantula (1986) (2.60)	$\log \left(\sigma_t^2 \right) = a_0 + \sum_{i=1}^{q} a_i \log \left(\varepsilon_{t-i}^2 \right) + \sum_{j=1}^{p} b_j \log \left(\sigma_{t-j}^2 \right).$

(continued)

Table 2.11   (*Continued*)

Stdev-ARCH(q)
Schwert (1990)
(2.61)

$$\sigma_t^2 = \left(a_0 + \sum_{i=1}^q a_i |\varepsilon_{t-i}|\right)^2$$

NARCH(p,q)
Higgins and Bera (1992)
(2.62)

$$\sigma_t^\delta = a_0 + \sum_{i=1}^q a_i |\varepsilon_{t-i}^2|^{\delta/2} + \sum_{j=1}^p b_j \sigma_{t-j}^\delta$$

AGARCH(p,q)
Engle (1990)
(2.64)

$$\sigma_t^2 = a_0 + \sum_{i=1}^q \left(a_i \varepsilon_{t-i}^2 + \gamma_i \varepsilon_{t-i}\right) + \sum_{j=1}^p b_j \sigma_{t-j}^2$$

NAGARCH(p,q)
Engle and Ng (1993)
(2.65)

$$\sigma_t^2 = a_0 + \sum_{i=1}^q a_i (\varepsilon_{t-i} + \gamma_i \sigma_{t-i})^2 + \sum_{j=1}^p b_j \sigma_{t-j}^2$$

VGARCH(p,q)
Engle and Ng (1993)
(2.66)

$$\sigma_t^2 = a_0 + \sum_{i=1}^q a_i (\varepsilon_{t-i}/\sigma_{t-i} + \gamma_i)^2 + \sum_{j=1}^p b_j \sigma_{t-j}^2$$

APARCH(p,q)
Ding et al. (1993)
(2.67)

$$\sigma_t^\delta = a_0 + \sum_{i=1}^q a_i (|\varepsilon_{t-1}| - \gamma_i \varepsilon_{t-i})^\delta + \sum_{j=1}^p b_j \sigma_{t-j}^\delta$$

GQARCH(p,q)
Sentana (1995)
(2.68)

$$\sigma_t^2 = a_0 + \sum_{i=1}^q a_i \varepsilon_{t-i}^2 + \sum_{i=1}^q \gamma_i \varepsilon_{t-i}$$

$$+ 2\sum_{i=1}^q \sum_{j=i+1}^q a_{ij} \varepsilon_{t-i} \varepsilon_{t-j} + \sum_{j=1}^p b_j \sigma_{t-j}^2$$

GQTARCH(p,q)
Gouriéroux and
    Monfort (1992)
(2.71)

$$\sigma_t^2 = a_0 + \sum_{i=1}^q \sum_{j=1}^J a_{ij} d_j (\varepsilon_{t-i}) + \sum_{i=1}^p b_j \sigma_{t-j}^2$$

SWARCH($\tilde{K}, q$)
Hamilton and
    Susmel (1994)
(2.95)

$$\tilde{\varepsilon}_t \equiv \sqrt{g_{s_t}} \varepsilon_t$$

$$\sigma_t^2 = a_0 + \sum_{i=1}^q \left(a_i \varepsilon_{t-i}^2\right)$$

SWARCH-L($\tilde{K}, q$)
Hamilton and
    Susmel (1994)
(2.96)

$$\tilde{\varepsilon}_t \equiv \sqrt{g_{s_t}} \varepsilon_t$$

$$\sigma_t^2 = a_0 + \gamma_1 d(\varepsilon_{t-1} < 0)\varepsilon_{t-1}^2 + \sum_{i=1}^q \left(a_i \varepsilon_{t-i}^2\right)$$

GRS
Gray (1996)
(2.97)

$$\sigma_{i,t}^2 = a_{i,0} + a_{i,1} \varepsilon_{t-1}^2 + b_{i,1} \sigma_{t-1}^2$$

$$\sigma_{t-1}^2 = p_{1,t-1}\left(\mu_{1,t-1}^2 + \sigma_{1,t-1}^2\right)$$

$$+ \left(1 - p_{1,t-1}\right)\left(\mu_{2,t-1}^2 + \sigma_{2,t-1}^2\right)$$

$$- \left(p_{1,t-1}\mu_{1,t-1} + \left(1 - p_{1,t-1}\right)\mu_{2,t-1}\right)^2$$

Table 2.11   (*Continued*)

VSARCH(p, q): Fornari and Mele (1995) (2.99)	$\sigma_t^2 = a_0 + \sum_{i=1}^{q} a_i \varepsilon_{t-i}^2 + \gamma d(\varepsilon_{t-1} > 0)\dfrac{\varepsilon_{t-1}^2}{\sigma_{t-1}^2} + \sum_{j=1}^{p} b_j \sigma_{t-j}^2$
AVSARCH(p, q) Fornari and Mele (1995) (2.100)	$\sigma_t^2 = a_0 + \sum_{i=1}^{q} a_i \varepsilon_{t-i}^2 + \sum_{i=1}^{p} b_j \sigma_{t-j}^2 + \gamma d(\varepsilon_{t-1} > 0)\varepsilon_{t-1}^2$ $\qquad + \delta\big((\varepsilon_{t-1}^2 / \sigma_{t-1}^2) - k\big) d(\varepsilon_{t-1} > 0)$
LST-GARCH(p, q) Hagerud (1996) (2.102)	$\sigma_t^2 = a_0 + \sum_{i=1}^{q} (a_i + \gamma_i F(\varepsilon_{t-i}))\varepsilon_{t-i}^2 + \sum_{j=1}^{p} b_j \sigma_{t-j}^2,$ $F(\varepsilon_{t-i}) = (1 + \exp(-\theta \varepsilon_{t-i}))^{-1} - 0.5$
EST-GARCH(p, q) Hagerud (1996) (2.103)	$\sigma_t^2 = a_0 + \sum_{i=1}^{q} (a_i + \gamma_i F(\varepsilon_{t-i}))\varepsilon_{t-i}^2 + \sum_{j=1}^{p} b_j \sigma_{t-j}^2,$ $F(\varepsilon_{t-i}) = 1 - \exp\left(-\theta \varepsilon_{t-i}^2\right)$
GLST-GARCH(p, q) Lubrano (1998) (2.105)	$\sigma_t^2 = a_0 + \sum_{i=1}^{q} (a_i + \gamma_i F(\varepsilon_{t-i}))\varepsilon_{t-i}^2 + \sum_{j=1}^{p} b_j \sigma_{t-j}^2,$ $F(\varepsilon_{t-i}) = \dfrac{1 - \exp\left(-\theta \varepsilon_{t-i}^2\right)}{1 + \exp\left(-\theta\left(\varepsilon_{t-i}^2 - c^2\right)\right)}$
GEST-GARCH(p, q) Lubrano (1998) (2.106)	$\sigma_t^2 = a_0 + \sum_{i=1}^{q} (a_i + \gamma_i F(\varepsilon_{t-i}))\varepsilon_{t-i}^2 + \sum_{j=1}^{p} b_j \sigma_{t-j}^2,$ $F(\varepsilon_{t-i}) = 1 - \exp\left(-\theta(\varepsilon_{t-i} - c)^2\right)$
CGARCH(1,1) Engle and Lee (1993) (2.108)	$\sigma_t^2 = q_t + a_1\left(\varepsilon_{t-1}^2 - q_{t-1}\right) + b_1\left(\sigma_{t-1}^2 - q_{t-1}\right)$ $q_t = a_0 + p q_{t-1} + \phi\left(\varepsilon_{t-1}^2 - \sigma_{t-1}^2\right)$
ACGARCH(1,1) Engle and Lee (1993) (2.110)	$\sigma_t^2 = q_t + a_1\left(\varepsilon_{t-1}^2 - q_{t-1}\right)$ $\qquad + \gamma_1\left(d(\varepsilon_{t-1} < 0)\varepsilon_{t-1}^2 - 0.5 q_{t-1}\right) + b_1\left(\sigma_{t-1}^2 - q_{t-1}\right)$ $q_t = a_0 + p q_{t-1} + \phi\left(\varepsilon_{t-1}^2 - \sigma_{t-1}^2\right)$ $\qquad + \gamma_2\left(d(\varepsilon_{t-1} < 0)\varepsilon_{t-1}^2 - 0.5 \sigma_{t-1}^2\right)$
N-component $\quad$ GARCH(1,1) Ding and Granger (1996) (2.111)	$\sigma_t^2 = \sum_{i=1}^{N} w_i \sigma_{i,t}^2$ $\sigma_{i,t}^2 = \sigma_2(1 - a_i - b_i) + a_i \varepsilon_{t-1}^2 + b_i \sigma_{i,t-1}^2,$
R-GARCH(r, p, q) Nowicka-Zagrajek $\quad$ and Weron (2001) (2.112)	$\sigma_t^2 = \sum_{i^*=1}^{r} (c_{i^*} \eta_{t-i^*}) + \sum_{i=1}^{q} \left(a_i \varepsilon_{t-i}^2\right) + \sum_{j=1}^{p} \left(b_j \sigma_{t-j}^2\right)$

(*continued*)

Table 2.11   (*Continued*)

HARCH(p, n)
Müller et al. (1997)
(2.113)

$$\sigma_t^2 = a_0 + \sum_{i=1}^{n} \sum_{k=1}^{i} a_{ik} \left(\sum_{i^*=k}^{i} \varepsilon_{t-i^*}^2 \right)^2 + \sum_{j=1}^{p} \left(b_j \sigma_{t-j}^2 \right)$$

FIGARCH(p, d, q)
Baillie et al. (1996a)
(3.10)

$$\sigma_t^2 = a_0 + \left(1 - B(L) - \Phi(L)(1-L)^d \right) \varepsilon_t^2 + B(L)\sigma_t^2$$

FIGARCHC(p, d, q)
Chung (1999)
(3.11)

$$\sigma_t^2 = \sigma^2(1-B) + \left(1 - B(L) - \Phi(L)(1-L)^d \right) \left(\varepsilon_t^2 - \sigma^2 \right)$$
$$+ B(L)\sigma_t^2$$

FIEGARCH(p, d, q)
Bollerslev and
  Mikkelsen (1996)
(3.12)

$$\log \left(\sigma_t^2 \right) = a_0 + \Phi(L)^{-1}(1-L)^{-d}(1+A(L))g(z_{t-1})$$

FIAPARCH(p, d, q)
Tse (1998)
(3.13)

$$\sigma_t^\delta = a_0 + \left(1 - (1-B(L))^{-1}\Phi(L)(1-L)^d \right) (|\varepsilon_t| - \gamma\varepsilon_t)^\delta$$

FIAPARCHC(p, d, q)
Chung (1999)
(3.14)

$$\sigma_t^\delta = \sigma^2(1-B) + \left(1 - B(L) - \Phi(L)(1-L)^d \right)$$
$$\times \left((|\varepsilon_t| - \gamma\varepsilon_t)^\delta - \sigma^2 \right) + B(L)\sigma_t^\delta$$

ASYMM-
  FIFGARCH($1, d, 1$)
Hwang (2001)
(3.15)

$$\sigma_t^\lambda = \frac{k}{1-\delta} + \left(1 - \frac{(1-\varphi L)(1-L)^d}{1-\delta L} \right) f^v(\varepsilon_t)\sigma_t^\lambda$$

$$f(\varepsilon_t) = \left| \frac{\varepsilon_t}{\sigma_t} - b \right| - c\left(\frac{\varepsilon_t}{\sigma_t} - b \right),$$

ASYMM-
  FIFGARCH($1, d, 1$)
modified
Ruiz and Pérez (2003)
(3.16)

$$(1-\varphi L)(1-L)^d \frac{\sigma_t^\lambda - 1}{\lambda} = \omega' + a(1+\psi L)\sigma_{t-1}^\lambda$$
$$\times (f^v(z_{t-1}) - 1)$$
$$f\left(\frac{\varepsilon_t}{\sigma_t} \right) = \left| \frac{\varepsilon_t}{\sigma_t} - b \right| - c\left(\frac{\varepsilon_t}{\sigma_t} - b \right)$$

HYGARCH(p, d, q)
Davidson (2004)
(3.17)

$$\sigma_t^2 = a_0 + \left(1 - B(L) - \Phi(L)\left(1 + a\left((1-L)^d - 1 \right) \right) \right) \varepsilon_t^2$$
$$+ B(L)\sigma_t^2$$

1980; Kunitomo, 1992; Parkinson, 1980; Rogers and Satchell, 1991), implied volatility models (Day and Lewis, 1988; Latane and Rendleman, 1976; Schmalensee and Trippi, 1978)) and realized volatility models are examples from the financial econometric literature of estimating asset returns volatility.

A typical representation of a *stochastic volatility* model can be given by

$$\varepsilon_t = z_{1,t}\sigma e^{0.5\sigma_t},$$
$$\sigma_t^2 = a\sigma_{t-1}^2 + z_{2,t},$$
$$z_{1,t} \overset{i.i.d.}{\sim} f[E(z_t) = 0, V(z_t) = 1], \quad (2.116)$$
$$z_{2,t} \overset{i.i.d.}{\sim} g\left[E(z_t) = 0, V(z_t) = \sigma_{z_2}^2\right],$$

where σ is a positive scale parameter, $|a| < 1$, and the error terms $z_{1,t}$ and $z_{2,t}$ could be contemporaneously correlated. The additional error term, $z_{2,t}$, in the conditional variance equation means that the stochastic volatility model has no closed-form solution. Hence, the estimation of the parameters is a quite difficult task. For this reason, stochastic volatility models are not as popular as ARCH processes. Jacquier et al. (1994) considered a Markov chain Monte Carlo (MCMC) framework in order to estimate stochastic volatility models, and Jacquier et al. (1999, 2004) extended the MCMC technique to allow for the leverage effect and fat-tailed conditional errors. For extensions and applications of MCMC techniques of ARCH models the interested reader is referred to Brooks et al. (1997), Dellaportas and Roberts (2003), Kaufmann and Frühwirth-Schnatter (2002) and Nakatsuma (2000). Nelson (1990b) was the first to show that the continuous time limit of an ARCH process, which is a stochastic difference equation, is a diffusion process with stochastic volatility (which is a stochastic differential equation). Duan (1996, 1997) extended Nelson's study.

The term *historical volatility models*, in a wider context, could naturally include every technique that is based on past data, i.e. the ARCH family. The term *historical volatility techniques* refers to procedures that are solely data-driven. They are not based on parameter estimation or on any theoretical properties, but they filter the historical data set in order to compute volatility. Models based on the daily open, high, low and close asset prices, and exponential smoothing methods, such as J.P. Morgan's RiskMetrics™ method, are procedures which can be included in the class of historical volatility models. Parkinson (1980) proposed the price range, which is the difference between the highest and the lowest asset log-prices,

$$Range_t = 0.361(\log(\max(P_t)) - \log(\min(P_t)))^2. \quad (2.117)$$

Alizadeh et al. (2002) and Sadorsky (2005) proposed the log-range measure of volatility defined as the logarithmic difference between the daily highest log-price and the daily lowest log-price,

$$lRange_t = \log(\log(\max(P_t)) - \log(\min(P_t))). \quad (2.118)$$

The price range contains more information for the true volatility than the squared daily returns. Consider, for example, the case where the asset prices fluctuate greatly over a trading day but this day's closing price is similar to the previous day's closing price. In such a case, squared daily returns indicate little volatility. The price range is recorded in all business newspapers and technical reports through the Japanese candlestick charting techniques. Alizadeh et al. (2002) showed that the price range is highly efficient and approximately Gaussian. It is interesting to note that Feller (1951) derived the probability distribution for the price range. Brandt and Diebold (2006) looked into price range volatility estimation in the multivariate case.

The measure of *realized volatility* is based on the idea of using high-frequency data to compute measures of volatility at a lower frequency, i.e. using hourly log-returns to generate a measure of daily volatility. We will deal with the problem of modelling realized volatility separately in Chapter 7, as it is a recently developed promising area of volatility model building.

Implied volatility is the instantaneous standard deviation of the return on the underlying asset, which would have to be input into a theoretical pricing model in order to yield a theoretical value identical to the price of the option in the marketplace, assuming all other inputs are known. Day and Lewis (1992) examined whether implied volatilities contain incremental information relative to the estimated volatility from ARCH models. Noh et al. (1994) made a comparison of the forecasting performance of ARCH and implied volatility models in the context of option pricing. In 1993, the Chicago Board of Options Exchange (CBOE) published the first implied volatility index. In 2003, a new computation of the volatility index was launched. The enhanced computation technique took into account the latest advances in financial theory, eliminating the measurement errors that had hitherto characterized the implied volatility measures. As a result, market participants consider the implied volatility index of CBOE as the world's premier barometer of investor sentiment and market volatility. An investigation on the CBOE volatility index and the interrelationship of the information provided by implied volatility and ARCH volatility is carried out in Chapter 9.

Andersen et al. (2006) provided a systematic categorization of various ways of modelling volatility, and Poon and Granger (2003) conducted a comparative review based on the forecasting performance of ARCH, implied volatility, and historical volatility models. The concept of volatility has been extended to other types of processes (e.g., Brillinger, 2007, 2008), and has been looked at from a dynamical system perspective (e.g., Tong, 2007).

2.10   Interpretation of the ARCH process

A number of studies have aimed to explain the prominence of ARCH processes in financial applications. Parke and Waters (2007) quoted Engle (2001a, p. 165): 'The goal of volatility analysis must ultimately be to explain the causes of volatility. While time series structure is valuable for forecasting, it does not satisfy our need to explain volatility. . . . Thus far, attempts to find the ultimate cause of volatility are not very satisfactory.'

Stock (1987, 1988) established the time deformation model, in which economic and calendar time proceed at different speeds, and noted the relationship between time deformation and ARCH models. Any economic variable evolves on an operational time scale, while in practice it is measured on a calendar time scale. The inappropriate use of calendar time scale leads to volatility clustering since the variable may evolve more quickly or more slowly relative to calendar time. The time deformation model for a random variable y_t has the form

$$y_t = p_t y_{t-1} + \varepsilon_t,$$
$$\varepsilon_t | I_{t-1} \sim N(0, \sigma_t^2), \qquad (2.119)$$
$$\sigma_t^2 = a_0 + a_1 \varepsilon_{t-1}^2.$$

According to Stock, when a long segment of operational time has elapsed during a unit of calendar time, p_t is small and σ_t^2 is large. In order words, the time-varying autoregressive parameter is inversely related to the conditional variance.

Mizrach (1990) developed a model that has the same notion with the 'adaptable expectations hypothesis' in macroeconomics. In Mizrach's framework, the errors made by participants in the investment market, are strongly dependent on the past errors. He therefore associated the ARCH process with the errors of the economic agents' learning processes, and inferred that the strong persistence of the errors forces the volatility of asset returns to have an ARCH-like structure.

Gallant et al. (1991), based on earlier work by Mandelbrot and Taylor (1967), Clark (1973), Westerfield (1977) and Tauchen and Pitts (1983), provided a theoretical interpretation of the ARCH effect. Let us assume that the asset returns are defined by a stochastic number of intra-period price revisions so that they can be decomposed as

$$y_t = \mu_t + \sum_{i=1}^{\omega_t} \zeta_i, \qquad (2.120)$$

where μ_t is the forecastable component, $\zeta_i \overset{i.i.d.}{\sim} N(0, s^2)$ denotes the incremental changes and ω_t is the number of times new information comes to the market at time t. The random variable ω_t is unobservable and independent of the incremental changes. In such a case, the asset returns are not normally distributed, as their distribution is a mixture of normal distributions. Rewriting (2.120) as $y_t = \mu_t + s^2 \sqrt{\omega_t} z_t$, with z_t, $t = 1, 2, \ldots$, as i.i.d. standard normal variables, the y_t conditional on any information set, ω_t and I_{t-1}, is normally distributed:

$$y_t | (\omega_t, I_{t-1}) \sim N(\mu_t, s^2 \omega_t). \qquad (2.121)$$

However, the knowledge of information that flows into the market is an unrealistic assumption. Hence, the y_t conditional on the information set available to the market participants is

$$y_t | I_{t-1} \sim N(\mu_t, s^2 E_{t-1}(\omega_t)). \qquad (2.122)$$

Note that the conditional kurtosis, $3 E_{t-1}(\omega_t^2) / E_{t-1}(\omega_t)^2$, exceeds 3, as in the ARCH process where the innovation, ε_t, always has fatter tails than its unconditional normal distribution:

$$E(\varepsilon_t^4) / E(\varepsilon_t^2)^2 \geq 3. \qquad (2.123)$$

Lamoureux and Lastrapes (1990a) assumed that the number of information arrivals is serially correlated and used the daily trading volume as a proxy variable for the daily information that flows into the stock market. Hence, ω_t can be expressed as an autoregressive process:

$$\omega_t = b_0 + \sum_{i=1}^{\kappa} b_i \omega_{t-i} + z_t,$$

$$z_t \overset{iid}{\sim} N(0, 1).$$

(2.124)

From (2.122) we know that $E((y_t - \mu_t)^2 | I_{t-1}) = s^2 \omega_t$, thus (2.124) becomes

$$E\left((y_t - \mu_t)^2 | I_{t-1}\right) = s^2 b_0 + \sum_{i=1}^{\kappa} b_i E\left((y_{t-i} - \mu_{t-i})^2 | I_{t-i-1}\right) + s^2 z_t.$$

(2.125)

The structure in (2.125) expresses the persistence in conditional variance, a characteristic that is captured by the ARCH process. Lamoureux and Lastrapes (1990a) used the trading volume as a proxy variable for ω_t. Including the daily trading volume, V_t, as an exogenous variable in the GARCH(1,1) model, they found that its coefficient was highly significant, whereas the ARCH coefficients became negligible:

$$\sigma_t^2 = a_0 + a_1 \varepsilon_t^2 + b_1 \sigma_t^2 + \delta V_t.$$

(2.126)

The heteroscedastic mixture model assumes that $\delta > 0$ and that the persistence of variance as measured by $a_1 + b_1$ should become negligible. Their work provided empirical evidence that the ARCH process is a manifestation of the time dependence on the rate of information arrival in the market.

Brailsford (1996) and Pyun et al. (2000) applied versions of the heteroscedastic mixture model and reported that the degree of persistence reduced as a proxy for information arrival enters into the variance equation. On the other hand, a number of studies (Najand and Yung, 1991; Bessembinder and Seguin, 1993; Locke and Sayers, 1993; Abhyankar, 1995; Sharma et al., 1996) tested the mixture of distributions hypothesis, for various sets of data, and found that the ARCH coefficients remain statistically significant even after a trading volume is included as an exogenous variable in the model. This contradiction forced Miyakoshi (2002) to re-examine the relationship between ARCH effects and rate of information arrival in the market. By using data from the Tokyo Stock Exchange, Miyakoshi showed that, for periods with important market announcements, the trading volume affects the return volatility and the ARCH coefficients become negligible, while for periods which lack big news the ARCH structure characterizes the conditional variance adequately. The mixture of distributions hypothesis was also re-examined by Luu and Martens (2003) in the context of realized volatility.

Engle et al. (1990a) evaluated the role of the information arrival process in the determination of volatility in a multivariate framework providing a test of two hypotheses: heat waves and meteor showers. Using meteorological analogies, they supposed that information follows a process like a heat wave so that a hot day in New York is likely to be followed by another hot day in New York but not typically by a hot day in Tokyo. On the other hand, a meteor shower in New York, which rains down on the earth as it turns, will almost surely be followed by one in Tokyo. Thus, the heat wave hypothesis is that the volatility has only country-specific autocorrelation, while the meteor shower hypothesis states that volatility in one market spills over to the next. They examined intradaily volatility in the foreign exchange markets, focusing on time periods corresponding to the business hours of different countries. Their research was based on the yen–dollar exchange rate while the Tokyo, European and New York market are open. They found that foreign news was more important than past domestic news. Thus the major effect is more like a meteor shower, i.e. Japanese news had a greater impact on the volatility of all markets except the Tokyo market. This is interpreted as evidence that volatility in part arises from trading rather than purely from news. Conrad et al. (1991), Pyun et al. (2000) and Ross (1989) examined the volatility spillover effect across large- and small-capitalization companies. The main finding is that volatility propagates asymmetrically in the sense that the effect of shocks of larger firms on the volatility of smaller companies is more significant than that from smaller firms on larger companies.

Bollerslev and Domowitz (1991) showed how the actual market mechanisms may themselves result in a very different temporal dependence in the volatility of transaction prices, with a particular automated trade execution system inducing a very high degree of persistence in the conditional variance process.

Parke and Waters (2007) proposed that time-varying volatility is a natural feature of models with forward-looking agents. They considered forward-looking mean–variance optimizing agents who differ only because they choose among three forecasting strategies that are all consistent with the notion of rational expectations. If agents are not very aggressive in pursuing the optimal forecasting strategy (existence of small martingale variance), they tend to agree numerically on the fundamentalist forecast and there is little evidence of ARCH effects. If agents are more aggressive (existence of larger martingale variance), ARCH effects appear in significant fractions of the sample runs.

Alternative expositions for theoretical evidence on the sources of the ARCH effect have been presented by Attanasio and Wadhwani (1989), Backus et al. (1989), Brock and Kleidon (1990), Diebold and Pauly (1988), Domowitz and Hakkio (1985), Engle and Susmel (1990), Giovannini and Jorion (1989), Hodrick (1989), Hong and Lee (2001), Hsieh (1988), Lai and Pauly (1988), Laux and Ng (1993), Ng (1988), Schwert (1989a), Smith (1987) and Thum (1988). Nelson (1990b) was the first to show how ARCH models can emerge from diffusion processes. The problem of estimation of discretely sampled diffusions, such as ARCH processes, and their relationship with continuous time models has also been considered in the literature – see Aït-Sahalia (2001, 2002) and the references therein.

Appendix

EViews 6

- *chapter2_GARCH.prg*

```
load chapter2_data
matrix(9,4) schwarz
'Estimate GARCH models for S&P500
equation sp500_garch01.arch(1,0,m=10000,h) sp500ret c ar(1)
  sp500_garch01.makegarch sp500_vargarch01
  sp500_garch01.makeresid sp500_unresgarch01
  sp500_garch01.makeresid(s) sp500_resgarch01
      schwarz(1,1)=sp500_garch01.@schwarz
equation sp500_garch02.arch(2,0,m=10000,h) sp500ret c ar(1)
  sp500_garch02.makegarch sp500_vargarch02
  sp500_garch02.makeresid sp500_unresgarch02
  sp500_garch02.makeresid(s) sp500_resgarch02
      schwarz(2,1)=sp500_garch02.@schwarz
equation sp500_garch03.arch(3,0,m=10000,h) sp500ret c ar(1)
  sp500_garch03.makegarch sp500_vargarch03
  sp500_garch03.makeresid sp500_unresgarch03
  sp500_garch03.makeresid(s) sp500_resgarch03
      schwarz(3,1)=sp500_garch03.@schwarz
equation sp500_garch11.arch(1,1,m=10000,h) sp500ret c ar(1)
  sp500_garch11.makegarch sp500_vargarch11
  sp500_garch11.makeresid sp500_unresgarch11
      sp500_garch11.makeresid(s) sp500_resgarch11
      schwarz(4,1)=sp500_garch11.@schwarz
equation sp500_garch12.arch(2,1,m=10000,h) sp500ret c ar(1)
  sp500_garch12.makegarch sp500_vargarch12
  sp500_garch12.makeresid sp500_unresgarch12
  sp500_garch12.makeresid(s) sp500_resgarch12
      schwarz(5,1)=sp500_garch12.@schwarz
equation sp500_garch13.arch(3,1,m=10000,h) sp500ret c ar(1)
  sp500_garch13.makegarch sp500_vargarch13
  sp500_garch13.makeresid sp500_unresgarch13
  sp500_garch13.makeresid(s) sp500_resgarch13
      schwarz(6,1)=sp500_garch13.@schwarz
equation sp500_garch21.arch(1,2,m=10000,h) sp500ret c ar(1)
  sp500_garch21.makegarch sp500_vargarch21
  sp500_garch21.makeresid sp500_unresgarch21
  sp500_garch21.makeresid(s) sp500_resgarch21
      schwarz(7,1)=sp500_garch21.@schwarz
```

```
equation sp500_garch22.arch(2,2,m=10000,h) sp500ret c ar(1)
  sp500_garch22.makegarch sp500_vargarch22
  sp500_garch22.makeresid sp500_unresgarch22
  sp500_garch22.makeresid(s) sp500_resgarch22
        schwarz(8,1)=sp500_garch22.@schwarz
equation sp500_garch23.arch(3,2,m=10000,h) sp500ret c ar(1)
  sp500_garch23.makegarch sp500_vargarch23
  sp500_garch23.makeresid sp500_unresgarch23
  sp500_garch23.makeresid(s) sp500_resgarch23
        schwarz(9,1)=sp500_garch23.@schwarz

'Estimate GARCH models for FTSE100
equation ftse100_garch01.arch(1,0,m=10000,h) ftse100ret
  c ar(1)
  ftse100_garch01.makegarch ftse100_vargarch01
  ftse100_garch01.makeresid ftse100_unresgarch01
  ftse100_garch01.makeresid(s) ftse100_resgarch01
        schwarz(1,2)=ftse100_garch01.@schwarz
equation ftse100_garch02.arch(2,0,m=10000,h) ftse100ret
  c ar(1)
  ftse100_garch02.makegarch ftse100_vargarch02
  ftse100_garch02.makeresid ftse100_unresgarch02
  ftse100_garch02.makeresid(s) ftse100_resgarch02
        schwarz(2,2)=ftse100_garch02.@schwarz
equation ftse100_garch03.arch(3,0,m=10000,h) ftse100ret
  c ar(1)
  ftse100_garch03.makegarch ftse100_vargarch03
  ftse100_garch03.makeresid ftse100_unresgarch03
  ftse100_garch03.makeresid(s) ftse100_resgarch03
        schwarz(3,2)=ftse100_garch03.@schwarz
equation ftse100_garch11.arch(1,1,m=10000,h) ftse100ret
  c ar(1)
  ftse100_garch11.makegarch ftse100_vargarch11
  ftse100_garch11.makeresid ftse100_unresgarch11
        ftse100_garch11.makeresid(s) ftse100_resgarch11
        schwarz(4,2)=ftse100_garch11.@schwarz
equation ftse100_garch12.arch(2,1,m=10000,h) ftse100ret
  c ar(1)
  ftse100_garch12.makegarch ftse100_vargarch12
  ftse100_garch12.makeresid ftse100_unresgarch12
  ftse100_garch12.makeresid(s) ftse100_resgarch12
        schwarz(5,2)=ftse100_garch12.@schwarz
equation ftse100_garch13.arch(3,1,m=10000,h) ftse100ret
  c ar(1)
  ftse100_garch13.makegarch ftse100_vargarch13
```

```
ftse100_garch13.makeresid ftse100_unresgarch13
ftse100_garch13.makeresid(s) ftse100_resgarch13
        schwarz(6,2)=ftse100_garch13.@schwarz
equation ftse100_garch21.arch(1,2,m=10000,h) ftse100ret
  c ar(1)
ftse100_garch21.makegarch ftse100_vargarch21
ftse100_garch21.makeresid ftse100_unresgarch21
ftse100_garch21.makeresid(s) ftse100_resgarch21
        schwarz(7,2)=ftse100_garch21.@schwarz
equation ftse100_garch22.arch(2,2,m=10000,h) ftse100ret
  c ar(1)
ftse100_garch22.makegarch ftse100_vargarch22
ftse100_garch22.makeresid ftse100_unresgarch22
ftse100_garch22.makeresid(s) ftse100_resgarch22
        schwarz(8,2)=ftse100_garch22.@schwarz
equation ftse100_garch23.arch(3,2,m=10000,h) ftse100ret
  c ar(1)
ftse100_garch23.makegarch ftse100_vargarch23
ftse100_garch23.makeresid ftse100_unresgarch23
ftse100_garch23.makeresid(s) ftse100_resgarch23
        schwarz(9,2)=ftse100_garch23.@schwarz

'Estimate GARCH models for $/£
equation usuk_garch01.arch(1,0,m=10000,h) usukret c ar(1)
  usuk_garch01.makegarch usuk_vargarch01
  usuk_garch01.makeresid usuk_unresgarch01
  usuk_garch01.makeresid(s) usuk_resgarch01
        schwarz(1,3)=usuk_garch01.@schwarz
equation usuk_garch02.arch(2,0,m=10000,h) usukret c ar(1)
  usuk_garch02.makegarch usuk_vargarch02
  usuk_garch02.makeresid usuk_unresgarch02
  usuk_garch02.makeresid(s) usuk_resgarch02
        schwarz(2,3)=usuk_garch02.@schwarz
equation usuk_garch03.arch(3,0,m=10000,h) usukret c ar(1)
  usuk_garch03.makegarch usuk_vargarch03
  usuk_garch03.makeresid usuk_unresgarch03
  usuk_garch03.makeresid(s) usuk_resgarch03
        schwarz(3,3)=usuk_garch03.@schwarz
equation usuk_garch11.arch(1,1,m=10000,h) usukret c ar(1)
  usuk_garch11.makegarch usuk_vargarch11
  usuk_garch11.makeresid usuk_unresgarch11
        usuk_garch11.makeresid(s) usuk_resgarch11
        schwarz(4,3)=usuk_garch11.@schwarz
equation usuk_garch12.arch(2,1,m=10000,h) usukret c ar(1)
  usuk_garch12.makegarch usuk_vargarch12
```

```
  usuk_garch12.makeresid usuk_unresgarch12
  usuk_garch12.makeresid(s) usuk_resgarch12
      schwarz(5,3)=usuk_garch12.@schwarz
equation usuk_garch13.arch(3,1,m=10000,h) usukret c ar(1)
  usuk_garch13.makegarch usuk_vargarch13
  usuk_garch13.makeresid usuk_unresgarch13
  usuk_garch13.makeresid(s) usuk_resgarch13
      schwarz(6,3)=usuk_garch13.@schwarz
equation usuk_garch21.arch(1,2,m=10000,h) usukret c ar(1)
  usuk_garch21.makegarch usuk_vargarch21
  usuk_garch21.makeresid usuk_unresgarch21
  usuk_garch21.makeresid(s) usuk_resgarch21
      schwarz(7,3)=usuk_garch21.@schwarz
equation usuk_garch22.arch(2,2,m=10000,h) usukret c ar(1)
  usuk_garch22.makegarch usuk_vargarch22
  usuk_garch22.makeresid usuk_unresgarch22
  usuk_garch22.makeresid(s) usuk_resgarch22
      schwarz(8,3)=usuk_garch22.@schwarz
equation usuk_garch23.arch(3,2,m=10000,h) usukret c ar(1)
  usuk_garch23.makegarch usuk_vargarch23
  usuk_garch23.makeresid usuk_unresgarch23
  usuk_garch23.makeresid(s) usuk_resgarch23
      schwarz(9,3)=usuk_garch23.@schwarz

'Estimate GARCH models for Gold
equation gold_garch01.arch(1,0,m=10000,h) goldret c ar(1)
  gold_garch01.makegarch gold_vargarch01
  gold_garch01.makeresid gold_unresgarch01
  gold_garch01.makeresid(s) gold_resgarch01
      schwarz(1,4)=gold_garch01.@schwarz
equation gold_garch02.arch(2,0,m=10000,h) goldret c ar(1)
  gold_garch02.makegarch gold_vargarch02
  gold_garch02.makeresid gold_unresgarch02
  gold_garch02.makeresid(s) gold_resgarch02
      schwarz(2,4)=gold_garch02.@schwarz
equation gold_garch03.arch(3,0,m=10000,h) goldret c ar(1)
  gold_garch03.makegarch gold_vargarch03
  gold_garch03.makeresid gold_unresgarch03
  gold_garch03.makeresid(s) gold_resgarch03
      schwarz(3,4)=gold_garch03.@schwarz
equation gold_garch11.arch(1,1,m=10000,h) goldret c ar(1)
  gold_garch11.makegarch gold_vargarch11
  gold_garch11.makeresid gold_unresgarch11
      gold_garch11.makeresid(s) gold_resgarch11
      schwarz(4,4)=gold_garch11.@schwarz
```

```
equation gold_garch12.arch(2,1,m=10000,h) goldret c ar(1)
  gold_garch12.makegarch gold_vargarch12
  gold_garch12.makeresid gold_unresgarch12
  gold_garch12.makeresid(s) gold_resgarch12
      schwarz(5,4)=gold_garch12.@schwarz
equation gold_garch13.arch(3,1,m=10000,h) goldret c ar(1)
  gold_garch13.makegarch gold_vargarch13
  gold_garch13.makeresid gold_unresgarch13
  gold_garch13.makeresid(s) gold_resgarch13
      schwarz(6,4)=gold_garch13.@schwarz
equation gold_garch21.arch(1,2,m=10000,h) goldret c ar(1)
  gold_garch21.makegarch gold_vargarch21
  gold_garch21.makeresid gold_unresgarch21
  gold_garch21.makeresid(s) gold_resgarch21
      schwarz(7,4)=gold_garch21.@schwarz
equation gold_garch22.arch(2,2,m=10000,h) goldret c ar(1)
  gold_garch22.makegarch gold_vargarch22
  gold_garch22.makeresid gold_unresgarch22
  gold_garch22.makeresid(s) gold_resgarch22
      schwarz(8,4)=gold_garch22.@schwarz
equation gold_garch23.arch(3,2,m=10000,h) goldret c ar(1)
  gold_garch23.makegarch gold_vargarch23
  gold_garch23.makeresid gold_unresgarch23
  gold_garch23.makeresid(s) gold_resgarch23
      schwarz(9,4)=gold_garch23.@schwarz
```

- *chapter2_MSE.prg*

```
load chapter2_data_after_chapter2_garch
matrix(9,4) mse
'Compute the in-sample mean squared error for S&P500
matrix(4437,9) squared_dist_sp500
!k=0
for !p=0 to 2
  for !q=1 to 3
    !k=!k+1
    for !i=1 to 4437
      squared_dist_sp500(!i,!k)=(sp500_vargarch{!p}{!q}
      (!i)-sp500_unresgarch{!p}{!q}(!i)^2)^2
    next
  mse(!k,1)=(@mean(@columnextract(squared_dist_sp500,
  !k)))
  next
next
```

```
'Compute the in-sample mean squared error for FTSE100
matrix(4437,9) squared_dist_ftse100
!k=0
for !p=0 to 2
  for !q=1 to 3
    !k=!k+1
    for !i=1 to 4437
      squared_dist_ftse100(!i,!k)=(ftse100_vargarch{!p}
      {!q}(!i)-ftse100_unresgarch{!p}{!q}(!i)^2)^2
    next
  mse(!k,2)=(@mean(@columnextract(squared_dist_ftse100,
  !k)))
  next
next

'Compute the in-sample mean squared error for $/£
matrix(4437,9) squared_dist_usuk
!k=0
for !p=0 to 2
  for !q=1 to 3
    !k=!k+1
    for !i=1 to 4437
      squared_dist_usuk(!i,!k)=(usuk_vargarch{!p}{!q}
      (!i)-usuk_unresgarch{!p}{!q}(!i)^2)^2
    next
  mse(!k,3)=(@mean(@columnextract(squared_dist_usuk,!k)))
  next
next

'Compute the in-sample mean squared error for Gold
matrix(4437,9) squared_dist_gold
!k=0
for !p=0 to 2
  for !q=1 to 3
    !k=!k+1
    for !i=1 to 4437
      squared_dist_gold(!i,!k)=(gold_vargarch{!p}{!q}
      (!i)-gold_unresgarch{!p}{!q}(!i)^2)^2
    next
  mse(!k,4)=(@mean(@columnextract(squared_dist_gold,!k)))
  next
next
```

- *chapter2.autocorrelation_sim.prg*

```
load chapter2.simgarch11.output
smpl @all
for !j=0.5 to 3 step .5
  !j100 = !j*10
  series d_abs_z_{!j100}
  series d_abs_e_{!j100}
  series d_abs_v_{!j100}
  series d_sqr_h_{!j100}
  d_abs_z_{!j100} = abs(z)^!j
  d_abs_e_{!j100} = abs(e)^!j
  d_abs_v_{!j100} = abs(v)^!j
  d_sqr_h_{!j100} = sqr(h)^!j
next
matrix(100,6) corr_abs_z
matrix(100,6) corr_abs_e
matrix(100,6) corr_abs_v
matrix(100,6) corr_sqr_h
for !i=1 to 100
  corr_abs_z(!i,1) = @cor(d_abs_z_5,d_abs_z_5(-!i))
  corr_abs_z(!i,2) = @cor(d_abs_z_10,d_abs_z_10(-!i))
  corr_abs_z(!i,3) = @cor(d_abs_z_15,d_abs_z_15(-!i))
  corr_abs_z(!i,4) = @cor(d_abs_z_20,d_abs_z_20(-!i))
  corr_abs_z(!i,5) = @cor(d_abs_z_25,d_abs_z_25(-!i))
  corr_abs_z(!i,6) = @cor(d_abs_z_30,d_abs_z_30(-!i))
next
for !i=1 to 100
  corr_abs_e(!i,1) = @cor(d_abs_e_5,d_abs_e_5(-!i))
  corr_abs_e(!i,2) = @cor(d_abs_e_10,d_abs_e_10(-!i))
  corr_abs_e(!i,3) = @cor(d_abs_e_15,d_abs_e_15(-!i))
  corr_abs_e(!i,4) = @cor(d_abs_e_20,d_abs_e_20(-!i))
  corr_abs_e(!i,5) = @cor(d_abs_e_25,d_abs_e_25(-!i))
  corr_abs_e(!i,6) = @cor(d_abs_e_30,d_abs_e_30(-!i))
next
for !i=1 to 100
  corr_abs_v(!i,1) = @cor(d_abs_v_5,d_abs_v_5(-!i))
  corr_abs_v(!i,2) = @cor(d_abs_v_10,d_abs_v_10(-!i))
  corr_abs_v(!i,3) = @cor(d_abs_v_15,d_abs_v_15(-!i))
  corr_abs_v(!i,4) = @cor(d_abs_v_20,d_abs_v_20(-!i))
  corr_abs_v(!i,5) = @cor(d_abs_v_25,d_abs_v_25(-!i))
  corr_abs_v(!i,6) = @cor(d_abs_v_30,d_abs_v_30(-!i))
next
for !i=1 to 100
  corr_sqr_h(!i,1) = @cor(d_sqr_h_5,d_sqr_h_5(-!i))
```

```
  corr_sqr_h(!i,2) = @cor(d_sqr_h_10,d_sqr_h_10(-!i))
  corr_sqr_h(!i,3) = @cor(d_sqr_h_15,d_sqr_h_15(-!i))
  corr_sqr_h(!i,4) = @cor(d_sqr_h_20,d_sqr_h_20(-!i))
  corr_sqr_h(!i,5) = @cor(d_sqr_h_25,d_sqr_h_25(-!i))
  corr_sqr_h(!i,6) = @cor(d_sqr_h_30,d_sqr_h_30(-!i))
next
```

- *chapter2.simgjr01.prg*

```
create chapter2.simgjr01 u 1 3000
series _temp = 1
!length = @obs(_temp)
delete _temp
  !a0 = 0.0001
  !a1 = 0.6
  !gamma1 = -0.3 ' 0.3
  !b1 = 0
series z = nrnd
series e
series h
series p
h(1) = !a0
e(1) = sqr(h(1))*z(1)
p(1) = 100
for !i = 2 to !length
  h(!i) = !a0 + (!a1*(e(!i-1)^2)) + (!gamma1*(e(!i-1)<0)*
  (e(!i-1)^2)) + (!b1*h(!i-1))
  e(!i) = sqr(h(!i))*z(!i)
  p(!i) = exp(log(p(!i-1)) + e(!i))
next
```

- *chapter2.simegarch01.prg*

```
create chapter2.simegarch01 u 1 3000
series _temp = 1
!length = @obs(_temp)
delete _temp
  !a0 = -8.95
  !gamma1 = -0.09
  !gamma2 = 0.2
  !b1 = 0
series z = nrnd
series e
series h
series p
h(1) = exp(!a0)
```

```
e(1) = sqr(h(1))*z(1)
p(1) = 100
for !i = 2 to !length
  h(!i) = exp(!a0 + !gamma1*z(!i-1) + !gamma2*abs(z(!i-1)) +
  (!b1*log(h(!i-1))))
  e(!i) = sqr(h(!i))*z(!i)
  p(!i) = exp(log(p(!i-1)) + e(!i))
next
```

G@rch 4.2

- *chapter2.sp500.ox*

```
#import <packages/Garch42/garch>
main()
{
{
decl garchobj;
garchobj = new Garch();
//*** DATA ***//
garchobj.Load("Chapter2Data.xls");
garchobj.Info();
garchobj.Select(Y_VAR, {"SP500RET",0,0});
//garchobj.Select(Z_VAR, {"NAME",0,0});//REGRESSOR IN THE
  VARIANCE
garchobj.SetSelSample(-1, 1, -1, 1);
//*** SPECIFICATIONS ***//
garchobj.CSTS(1,1);//cst in Mean (1 or 0), cst in Variance
  (1 or 0)
garchobj.DISTRI(0);//0 for Gauss, 1 for Student, 2 for GED,
  3 for Skewed-Student
garchobj.ARMA_ORDERS(1,0);//AR order (p), MA order (q).
garchobj.ARFIMA(0);//1 if Arfima wanted, 0 otherwise
garchobj.GARCH_ORDERS(1,1);//p order, q order
garchobj.ARCH_IN_MEAN(0);//ARCH-in-mean: 1 or 2 to add the
  variance or std. dev in the cond. mean

garchobj.MODEL(1);//0:RISKMETRICS,1:GARCH,2:EGARCH,
  3:GJR,4:APARCH,5:IGARCH,6:FIGARCH-BBM,7:FIGARCH-CHUNG,
  8:FIEGARCH,9:FIAPARCH-BBM,10:FIAPARCH-CHUNG,11:HYGARCH
garchobj.TRUNC(1000);//Truncation order(only F.I. models
  with BBM method)
//*** TESTS & FORECASTS ***//
garchobj.BOXPIERCE(<10;15;20>);//Lags for the Box-Pierce
  Q-statistics, <> otherwise
```

```
garchobj.ARCHLAGS(<2;5;10>);//Lags for Engle's LM ARCH
  test, <> otherwise
garchobj.NYBLOM(1);//1 to compute the Nyblom stability test,
  0 otherwise
garchobj.SBT(1);//1 to compute the Sign Bias test, 0 otherwise
garchobj.PEARSON(<40;50;60>);//Cells (<40;50;60>) for the
  adjusted Pearson Chi-square Goodness-of-fit test,
  <> otherwise //G@RCH1.12
garchobj.RBD(<10;15;20>);//Lags for the Residual-Based
  Diagnostic test of Tse, <> otherwise
garchobj.FORECAST(0,15,0);//Arg.1 : 1 to launch the
  forecasting procedure, 0 otherwise, Arg.2 : Number of
  forecasts, Arg.3 : 1 to Print the forecasts, 0 otherwise
//*** OUTPUT ***//
garchobj.MLE(2);//0:MLE (Second derivatives), 1:MLE (OPG
  Matrix), 2:QMLE
garchobj.COVAR(0);//if 1, prints variance-covariance matrix
  of the parameters.
garchobj.ITER(0);//Interval of iterations between printed
  intermediary results (if no intermediary results wanted,
  enter '0')
garchobj.TESTS(0,1);//Arg. 1 : 1 to run tests PRIOR to
  estimation, 0 otherwise // Arg. 2 : 1 to run tests AFTER
  estimation, 0 otherwise
garchobj.GRAPHS(0,0,"");//Arg.1 : if 1, displays graphics of
  the estimations (only when using GiveWin). // Arg.2 : if 1,
  saves these graphics in a EPS file (OK with all Ox versions)
  // Arg.3 : Name of the saved file.
garchobj.FOREGRAPHS(0,0,"");//Same as GRAPHS(p,s,n) but for
  the graphics of the forecasts.
//*** PARAMETERS ***//
garchobj.BOUNDS(0);// 1 if bounded parameters wanted,
  0 otherwise
garchobj.FIXPARAM(0,<0;0;0;0;0;0>);// Arg.1 : 1 to fix some
  parameters to their starting values, 0 otherwise// Arg.2 : 1
  to fix (see garchobj.DoEstimation(<>)) and 0 to estimate the
  corresponding parameter
//*** ESTIMATION ***//
garchobj.MAXSA(0,5,0.5,20,5,2,1);// Arg.1 : 1 to use the
  MaxSA algorithm of Goffe, Ferrier and Rogers (1994) and
  implemented in Ox by Charles Bos// Arg.2 : dT=initial
  temperature // Arg.3 : dRt=temperature reduction factor
  // Arg.4 : iNS=number of cycles// Arg.5 : iNT=Number of
  iterations before temperature reduction// Arg.6 : vC=step
  length adjustment// Arg.7 : vM=step length vector used in
  initial step
```

```
garchobj.Initialization(<>);//m_vPar=m_clevel | m_vbetam |
  m_dARFI | m_vAR | m_vMA | m_calpha0 | m_vgammav | m_dD |
  m_vbetav |m_valphav | m_vleverage | m_vtheta1 | m_vtheta2 |
  m_vpsy | m_ddelta | m_cA | m_cV | m_vHY | m_v_in_mean
garchobj.PrintStartValues(0);// 1: Prints the S.V. in a table
  form; 2: Individually;  3: in a Ox code to use in StartValues
garchobj.PrintBounds(1);
garchobj.DoEstimation(<>);
garchobj.Output();
garchobj.STORE(1,1,1,0,0,"sp500_GARCH11",0);
  //Arg.1,2,3,4,5:if 1 -> stored.(Res-SqRes-CondV-
  MeanFor-VarFor),Arg.6:Suffix. The name of the saved series
  will be "Res_ARG6" (or "MeanFor_ARG6", ...).Arg.7 : if 0,
  saves as an Excel spreadsheet (.xls). If 1, saves as a GiveWin
  dataset (.in7)
delete garchobj;
}
{
decl garchobj;
garchobj = new Garch();
//*** DATA ***//
garchobj.Load("Chapter2Data.xls");
garchobj.Info();
garchobj.Select(Y_VAR, {"SP500RET",0,0});
//garchobj.Select(Z_VAR, {"NAME",0,0});//REGRESSOR IN THE
  VARIANCE
garchobj.SetSelSample(-1, 1, -1, 1);
//*** SPECIFICATIONS ***//
garchobj.CSTS(1,1);//cst in Mean (1 or 0), cst in Variance
  (1 or 0)
garchobj.DISTRI(0);//0 for Gauss, 1 for Student, 2 for GED,
  3 for Skewed-Student
garchobj.ARMA_ORDERS(1,0);//AR order (p), MA order (q).
garchobj.ARFIMA(0);//1 if Arfima wanted, 0 otherwise
garchobj.GARCH_ORDERS(1,1);//p order, q order
garchobj.ARCH_IN_MEAN(0);//ARCH-in-mean: 1 or 2 to add the
  variance or std. dev in the  cond. mean

garchobj.MODEL(5);//0:RISKMETRICS,1:GARCH,2:EGARCH,
  3:GJR,4:APARCH,5:IGARCH,6:FIGARCH-BBM,7:FIGARCH-CHUNG,
  8:FIEGARCH,9:FIAPARCH-BBM,10:FIAPARCH-CHUNG,11:HYGARCH
garchobj.TRUNC(1000);//Truncation order(only F.I. models
  with BBM method)
//*** TESTS & FORECASTS ***//
garchobj.BOXPIERCE(<10;15;20>);//Lags for the Box-Pierce
  Q-statistics, <> otherwise
```

```
garchobj.ARCHLAGS(<2;5;10>);//Lags for Engle's LM ARCH test,
    <> otherwise
garchobj.NYBLOM(1);//1 to compute the Nyblom stability test,
    0 otherwise
garchobj.SBT(1);//1 to compute the Sign Bias test, 0 otherwise
garchobj.PEARSON(<40;50;60>);//Cells (<40;50;60>) for the
    adjusted Pearson Chi-square Goodness-of-fit test,
    <> otherwise //G@RCH1.12
garchobj.RBD(<10;15;20>);//Lags for the Residual-Based
    Diagnostic test of Tse, <> otherwise
garchobj.FORECAST(0,15,0);//Arg.1 : 1 to launch the
    forecasting procedure, 0 otherwise, Arg.2 : Number of
    forecasts, Arg.3 : 1 to Print the forecasts, 0 otherwise
//*** OUTPUT ***//
garchobj.MLE(2);//0:MLE (Second derivatives), 1:MLE (OPG
    Matrix), 2:QMLE
garchobj.COVAR(0);//if 1, prints variance-covariance matrix
    of the parameters.
garchobj.ITER(0);//Interval of iterations between printed
    intermediary results (if no intermediary results wanted,
    enter '0')
garchobj.TESTS(0,1);//Arg. 1 : 1 to run tests PRIOR to
    estimation, 0 otherwise // Arg. 2 : 1 to run tests AFTER
    estimation, 0 otherwise
garchobj.GRAPHS(0,0,"");//Arg.1 : if 1, displays graphics of
    the estimations (only when using GiveWin). // Arg.2 : if 1,
    saves these graphics in a EPS file (OK with all Ox versions)
    // Arg.3 : Name of the saved file.
garchobj.FOREGRAPHS(0,0,"");//Same as GRAPHS(p,s,n) but for
    the graphics of the forecasts.
//*** PARAMETERS ***//
garchobj.BOUNDS(0);// 1 if bounded parameters wanted,
    0 otherwise
garchobj.FIXPARAM(0,<0;0;0;0;0;0>);// Arg.1 : 1 to fix some
    parameters to their starting values, 0 otherwise//Arg.2 : 1 to
    fix (see garchobj.DoEstimation(<>)) and 0 to estimate the
    corresponding parameter
//*** ESTIMATION ***//
garchobj.MAXSA(0,5,0.5,20,5,2,1);// Arg.1 : 1 to use the MaxSA
    algorithm of Goffe, Ferrier and Rogers (1994) and implemented
    in Ox by Charles Bos// Arg.2 : dT=initial temperature
    // Arg.3 :dRt=temperature reduction factor // Arg.4 :
    iNS=number of cycles// Arg.5 : iNT=Number of iterations
    before temperature reduction// Arg.6 : vC=step length
    adjustment// Arg.7 : vM=step length vector used in initial
    step
```

```
garchobj.Initialization(<>);//m_vPar=m_clevel|m_vbetam|
  m_dARFI | m_vAR | m_vMA | m_calpha0 | m_vgammav | m_dD |
  m_vbetav |m_valphav | m_vleverage | m_vtheta1 | m_vtheta2 |
  m_vpsy | m_ddelta | m_cA | m_cV | m_vHY | m_v_in_mean
garchobj.PrintStartValues(0);// 1: Prints the S.V. in a table
  form; 2: Individually;  3: in a Ox code to use in StartValues
garchobj.PrintBounds(1);
garchobj.DoEstimation(<>);
garchobj.Output();
garchobj.STORE(1,1,1,0,0,"sp500_IGARCH11",0);
  //Arg.1,2,3,4,5:if 1 -> stored.(Res-SqRes-CondV-MeanFor-
  VarFor),Arg.6:Suffix. The name of the saved series will be
  "Res_ARG6" (or "MeanFor_ARG6", ...).Arg.7 : if 0, saves as an
  Excel spreadsheet (.xls). If 1, saves as a GiveWin dataset
  (.in7)
delete garchobj;
}
{
decl garchobj;
garchobj = new Garch();
//*** DATA ***//
garchobj.Load("Chapter2Data.xls");
garchobj.Info();
garchobj.Select(Y_VAR, {"SP500RET",0,0});
//garchobj.Select(Z_VAR, {"NAME",0,0});//REGRESSOR IN THE
  VARIANCE
garchobj.SetSelSample(-1, 1, -1, 1);
//*** SPECIFICATIONS ***//
garchobj.CSTS(1,1);//cst in Mean (1 or 0), cst in Variance
  (1 or 0)
garchobj.DISTRI(0);//0 for Gauss, 1 for Student, 2 for GED,
  3 for Skewed-Student
garchobj.ARMA_ORDERS(1,0);//AR order (p), MA order (q).
garchobj.ARFIMA(0);//1 if Arfima wanted, 0 otherwise
garchobj.GARCH_ORDERS(1,1);//p order, q order
garchobj.ARCH_IN_MEAN(0);//ARCH-in-mean: 1 or 2 to add the
  variance or std. dev in the  cond. mean
garchobj.MODEL(2);//0:RISKMETRICS,1:GARCH,2:EGARCH,
  3:GJR,4:APARCH,5:IGARCH,6:FIGARCH-BBM,7:FIGARCH-CHUNG,
  8:FIEGARCH,9:FIAPARCH-BBM,10:FIAPARCH-CHUNG,11:HYGARCH
garchobj.TRUNC(1000);//Truncation order(only F.I. models
  with BBM method)
//*** TESTS & FORECASTS ***//
garchobj.BOXPIERCE(<10;15;20>);//Lags for the Box-Pierce
  Q-statistics, <> otherwise
```

garchobj.ARCHLAGS(<2;5;10>);//Lags for Engle's LM ARCH test,
  <> otherwise
garchobj.NYBLOM(1);//1 to compute the Nyblom stability test,
  0 otherwise
garchobj.SBT(1);//1 to compute the Sign Bias test, 0 otherwise
garchobj.PEARSON(<40;50;60>);//Cells (<40;50;60>) for the
  adjusted Pearson Chi-square Goodness-of-fit test, <>
  otherwise //G@RCH1.12
garchobj.RBD(<10;15;20>);//Lags for the Residual-Based
  Diagnostic test of Tse, <> otherwise
garchobj.FORECAST(0,15,0);//Arg.1 : 1 to launch the
  forecasting procedure, 0 otherwise, Arg.2 : Number of
  forecasts, Arg.3 : 1 to Print the forecasts, 0 otherwise
//*** OUTPUT ***//
garchobj.MLE(2);//0:MLE (Second derivatives), 1:MLE (OPG
  Matrix), 2:QMLE
garchobj.COVAR(0);//if 1, prints variance-covariance matrix
  of the parameters.
garchobj.ITER(0);//Interval of iterations between printed
  intermediary results (if no intermediary results wanted,
  enter '0')
garchobj.TESTS(0,1);//Arg. 1 : 1 to run tests PRIOR to
  estimation, 0 otherwise// Arg. 2 : 1 to run tests AFTER
  estimation, 0 otherwise
garchobj.GRAPHS(0,0,"");//Arg.1 : if 1, displays graphics of
  the estimations (only when using GiveWin). // Arg.2 : if 1,
  saves these graphics in a EPS file (OK with all Ox versions)
  // Arg.3 : Name of the saved file.
garchobj.FOREGRAPHS(0,0,"");//Same as GRAPHS(p,s,n) but for
  the graphics of the forecasts.
//*** PARAMETERS ***//
garchobj.BOUNDS(0);// 1 if bounded parameters wanted,
  0 otherwise
garchobj.FIXPARAM(0,<0;0;0;0;0;0>);// Arg.1 : 1 to fix some
  parameters to their starting values, 0 otherwise//Arg.2 : 1 to
  fix (see garchobj.DoEstimation(<>)) and 0 to estimate the
  corresponding parameter
//*** ESTIMATION ***//
garchobj.MAXSA(0,5,0.5,20,5,2,1);// Arg.1 : 1 to use the MaxSA
  algorithm of Goffe, Ferrier and Rogers (1994) and implemented
  in Ox by Charles Bos// Arg.2 : dT=initial temperature
  // Arg.3 : dRt=temperature reduction factor // Arg.4 :
  iNS=number of cycles// Arg.5 : iNT=Number of iterations
  before temperature reduction// Arg.6 : vC=step length
  adjustment// Arg.7 : vM=step length vector used in initial

```
step
garchobj.Initialization(<>);// m_vPar = m_clevel |
  m_vbetam | m_dARFI | m_vAR | m_vMA | m_calpha0 | m_vgammav |
  m_dD | m_vbetav |m_valphav | m_vleverage | m_vtheta1 |
  m_vtheta2 | m_vpsy | m_ddelta | m_cA | m_cV | m_vHY |
  m_v_in_mean
garchobj.PrintStartValues(0);// 1: Prints the S.V. in a table
  form; 2: Individually;  3: in a Ox code to use in StartValues
garchobj.PrintBounds(1);
garchobj.DoEstimation(<>);
garchobj.Output();
garchobj.STORE(1,1,1,0,0,"sp500_EGARCH11",0);
  //Arg.1,2,3,4,5:if 1 -> stored.(Res-SqRes-CondV-MeanFor-
  VarFor),Arg.6:Suffix. The name of the saved series will be
  "Res_ARG6" (or "MeanFor_ARG6", ...).Arg.7 : if 0, saves as an
  Excel spreadsheet (.xls). If 1, saves as a GiveWin dataset
  (.in7)
delete garchobj;
}
{
decl garchobj;
garchobj = new Garch();
//*** DATA ***//
garchobj.Load("Chapter2Data.xls");
garchobj.Info();
garchobj.Select(Y_VAR, {"SP500RET",0,0});
//garchobj.Select(Z_VAR, {"NAME",0,0});//REGRESSOR IN THE
  VARIANCE
garchobj.SetSelSample(-1, 1, -1, 1);
//*** SPECIFICATIONS ***//
garchobj.CSTS(1,1);//cst in Mean (1 or 0), cst in Variance
  (1 or 0)
garchobj.DISTRI(0);//0 for Gauss, 1 for Student, 2 for GED,
  3 for Skewed-Student
garchobj.ARMA_ORDERS(1,0);//AR order (p), MA order (q).
garchobj.ARFIMA(0);//1 if Arfima wanted, 0 otherwise
garchobj.GARCH_ORDERS(1,1);//p order, q order
garchobj.ARCH_IN_MEAN(0);//ARCH-in-mean: 1 or 2 to add the
  variance or std. dev in the  cond. mean

garchobj.MODEL(3);//0:RISKMETRICS,1:GARCH,2:EGARCH,3:
  GJR,4:APARCH,5:IGARCH,6:FIGARCH-BBM,7:FIGARCH-CHUNG,8:
  FIEGARCH,9:FIAPARCH-BBM,10:FIAPARCH-CHUNG,11:HYGARCH
garchobj.TRUNC(1000);//Truncation order (only F.I. models
  with BBM method)
```

```
//*** TESTS & FORECASTS ***//
garchobj.BOXPIERCE(<10;15;20>);//Lags for the Box-Pierce
   Q-statistics, <> otherwise
garchobj.ARCHLAGS(<2;5;10>);//Lags for Engle's LM ARCH test,
   <> otherwise
garchobj.NYBLOM(1);//1 to compute the Nyblom stability test,
   0 otherwise
garchobj.SBT(1);//1 to compute the Sign Bias test, 0 otherwise
garchobj.PEARSON(<40;50;60>);//Cells (<40;50;60>) for the
   adjusted Pearson Chi-square Goodness-of-fit test, <>
   otherwise //G@RCH1.12
garchobj.RBD(<10;15;20>);//Lags for the Residual-Based
   Diagnostic test of Tse, <> otherwise
garchobj.FORECAST(0,15,0);//Arg.1 : 1 to launch the
   forecasting procedure, 0 otherwise, Arg.2 : Number of
   forecasts, Arg.3 : 1 to Print the forecasts, 0 otherwise
//*** OUTPUT ***//
garchobj.MLE(2);//0:MLE (Second derivatives), 1:MLE (OPG
   Matrix), 2:QMLE
garchobj.COVAR(0);//if 1, prints variance-covariance matrix
   of the parameters.
garchobj.ITER(0);//Interval of iterations between printed
   intermediary results (if no intermediary results wanted,
   enter '0')
garchobj.TESTS(0,1);//Arg. 1 : 1 to run tests PRIOR to
   estimation, 0 otherwise// Arg. 2 : 1 to run tests AFTER
   estimation, 0 otherwise
garchobj.GRAPHS(0,0,"");//Arg.1 : if 1, displays graphics of
   the estimations (only when using GiveWin). // Arg.2 : if 1,
   saves these graphics in a EPS file (OK with all Ox versions)
   // Arg.3 : Name of the saved file.
garchobj.FOREGRAPHS(0,0,"");//Same as GRAPHS(p,s,n) but for
   the graphics of the forecasts.
//*** PARAMETERS ***//
garchobj.BOUNDS(0);// 1 if bounded parameters wanted,
   0 otherwise
garchobj.FIXPARAM(0,<0;0;0;0;0;0>);// Arg.1 : 1 to fix some
   parameters to their starting values, 0 otherwise// Arg.2 : 1 to
   fix (see garchobj.DoEstimation(<>)) and 0 to estimate the
   corresponding parameter
//*** ESTIMATION ***//
garchobj.MAXSA(0,5,0.5,20,5,2,1);// Arg.1 : 1 to use the MaxSA
   algorithm of Goffe, Ferrier and Rogers (1994) and implemented
   in Ox by Charles Bos// Arg.2 : dT=initial temperature
   // Arg.3 : dRt=temperature reduction factor // Arg.4 :
```

```
iNS=number of cycles// Arg.5 : iNT=Number of iterations
before temperature reduction// Arg.6 : vC=step length
adjustment// Arg.7 : vM=step length vector used in initial
step
garchobj.Initialization(<>);//m_vPar=m_clevel|m_vbetam|
   m_dARFI | m_vAR | m_vMA | m_calpha0 | m_vgammav | m_dD |
   m_vbetav |m_valphav | m_vleverage | m_vtheta1 | m_vtheta2 |
   m_vpsy | m_ddelta | m_cA | m_cV | m_vHY | m_v_in_mean
garchobj.PrintStartValues(0);// 1: Prints the S.V. in a table
   form; 2: Individually; 3: in a Ox code to use in StartValues
garchobj.PrintBounds(1);
garchobj.DoEstimation(<>);
garchobj.Output();
garchobj.STORE(1,1,1,0,0,"sp500_GJR11",0);
   //Arg.1,2,3,4,5:if 1 -> stored.(Res-SqRes-CondV-MeanFor-
   VarFor),Arg.6:Suffix. The name of the saved series will be
   "Res_ARG6" (or "MeanFor_ARG6", ...).Arg.7 : if 0, saves as an
   Excel spreadsheet (.xls). If 1, saves as a GiveWin dataset
   (.in7)
delete garchobj;
}
{
decl garchobj;
garchobj = new Garch();
//*** DATA ***//
garchobj.Load("Chapter2Data.xls");
garchobj.Info();
garchobj.Select(Y_VAR, {"SP500RET",0,0});
//garchobj.Select(Z_VAR, {"NAME",0,0});//REGRESSOR IN THE
VARIANCE
garchobj.SetSelSample(-1, 1, -1, 1);
//*** SPECIFICATIONS ***//
garchobj.CSTS(1,1);//cst in Mean (1 or 0), cst in Variance
   (1 or 0)
garchobj.DISTRI(0);//0 for Gauss, 1 for Student, 2 for GED,
   3 for Skewed-Student
garchobj.ARMA_ORDERS(1,0);//AR order (p), MA order (q).
garchobj.ARFIMA(0);//1 if Arfima wanted, 0 otherwise
garchobj.GARCH_ORDERS(1,1);//p order, q order
garchobj.ARCH_IN_MEAN(0);//ARCH-in-mean: 1 or 2 to add the
   variance or std. dev in the  cond. mean

garchobj.MODEL(4);//0:RISKMETRICS,1:GARCH,2:EGARCH,
   3:GJR,4:APARCH,5:IGARCH,6:FIGARCH-BBM,7:FIGARCH-CHUNG,
   8:FIEGARCH,9:FIAPARCH-BBM,10:FIAPARCH-CHUNG,11:HYGARCH
```

```
garchobj.TRUNC(1000);//Truncation order(only F.I. models
   with BBM method)
//*** TESTS & FORECASTS ***//
garchobj.BOXPIERCE(<10;15;20>);//Lags for the Box-Pierce
   Q-statistics, <> otherwise
garchobj.ARCHLAGS(<2;5;10>);//Lags for Engle's LM ARCH test,
   <> otherwise
garchobj.NYBLOM(0);//1 to compute the Nyblom stability test,
   0 otherwise
garchobj.SBT(0);//1 to compute the Sign Bias test, 0 otherwise
garchobj.PEARSON(<40;50;60>);//Cells (<40;50;60>) for the
   adjusted Pearson Chi-square Goodness-of-fit test, <>
   otherwise //G@RCH1.12
garchobj.RBD(<10;15;20>);//Lags for the Residual-Based
   Diagnostic test of Tse, <> otherwise
garchobj.FORECAST(0,15,0);//Arg.1 : 1 to launch the
   forecasting procedure, 0 otherwise, Arg.2 : Number of
   forecasts, Arg.3 : 1 to Print the forecasts, 0 otherwise
//*** OUTPUT ***//
garchobj.MLE(2);//0:MLE (Second derivatives), 1:MLE (OPG
   Matrix), 2:QMLE
garchobj.COVAR(0);//if 1, prints variance-covariance matrix
   of the parameters.
garchobj.ITER(0);//Interval of iterations between printed
   intermediary results (if no intermediary results wanted,
   enter '0')
garchobj.TESTS(0,1);//Arg. 1 : 1 to run tests PRIOR to
   estimation, 0 otherwise// Arg. 2 : 1 to run tests AFTER
   estimation, 0 otherwise
garchobj.GRAPHS(0,0,"");//Arg.1 : if 1, displays graphics of
   the estimations (only when using GiveWin). // Arg.2 : if 1,
   saves these graphics in a EPS file (OK with all Ox versions)
   // Arg.3 : Name of the saved file.
garchobj.FOREGRAPHS(0,0,"");//Same as GRAPHS(p,s,n) but for
   the graphics of the forecasts.
//*** PARAMETERS ***//
garchobj.BOUNDS(0);// 1 if bounded parameters wanted,
   0 otherwise
garchobj.FIXPARAM(0,<0;0;0;0;0;0>);// Arg.1 : 1 to fix some
   parameters to their starting values, 0 otherwise//Arg.2 : 1 to
   fix (see garchobj.DoEstimation(<>)) and 0 to estimate the
   corresponding parameter
//*** ESTIMATION ***//
garchobj.MAXSA(0,5,0.5,20,5,2,1);// Arg.1 : 1 to use the MaxSA
   algorithm of Goffe, Ferrier and Rogers (1994) and implemented
```

```
in Ox by Charles Bos// Arg.2 : dT=initial temperature
// Arg.3 : dRt=temperature reduction factor // Arg.4 :
iNS=number of cycles// Arg.5 : iNT=Number of iterations
before temperature reduction// Arg.6 : vC=step length
adjustment// Arg.7 : vM=step length vector used in initial
step
garchobj.Initialization(<>);//m_vPar=m_clevel|m_vbetam|
  m_dARFI | m_vAR | m_vMA | m_calpha0 | m_vgammav | m_dD |
  m_vbetav |m_valphav | m_vleverage | m_vtheta1 | m_vtheta2 |
  m_vpsy | m_ddelta | m_cA | m_cV | m_vHY | m_v_in_mean
garchobj.PrintStartValues(0);// 1: Prints the S.V. in a table
  form; 2: Individually;  3: in a Ox code to use in StartValues
garchobj.PrintBounds(1);
garchobj.DoEstimation(<>);
garchobj.Output();
garchobj.STORE(1,1,1,0,0,"sp500_APARCH11",0);
  //Arg.1,2,3,4,5:if 1 -> stored.(Res-SqRes-CondV-MeanFor-
  VarFor),Arg.6:Suffix. The name of the saved series will be
  "Res_ARG6" (or "MeanFor_ARG6", ...).Arg.7: if 0, saves as an
  Excel spreadsheet (.xls). If 1, saves as a GiveWin dataset
  (.in7)
delete garchobj;
}
}
```

- *chapter2.sp500.VarCovar.ox*

```
#import <packages/Garch42/garch>
main()
{
{
decl garchobj;
garchobj = new Garch();
garchobj.Load("Chapter2Data.xls");
garchobj.Select(Y_VAR, {"SP500RET",0,0});
garchobj.SetSelSample(-1, 1, -1, 1);
garchobj.CSTS(1,1);
garchobj.DISTRI(0);
garchobj.ARMA_ORDERS(1,0);
garchobj.ARFIMA(0);
garchobj.GARCH_ORDERS(1,1);
garchobj.ARCH_IN_MEAN(0);
garchobj.MODEL(2);
garchobj.MLE(0);
garchobj.Initialization(<>);
```

```
garchobj.DoEstimation(<>);
//garchobj.Output();
decl VarCovarMatrix=garchobj.GetCovar();
print ("variance covariance matrix based on the inverse of the
  Hessian ", VarCovarMatrix);
delete garchobj;
}
{
decl garchobj;
garchobj = new Garch();
garchobj.Load("Chapter2Data.xls");
garchobj.Select(Y_VAR, {"SP500RET",0,0});
garchobj.SetSelSample(-1, 1, -1, 1);
garchobj.CSTS(1,1);
garchobj.DISTRI(0);
garchobj.ARMA_ORDERS(1,0);
garchobj.ARFIMA(0);
garchobj.GARCH_ORDERS(1,1);
garchobj.ARCH_IN_MEAN(0);
garchobj.MODEL(2);
garchobj.MLE(1);
garchobj.Initialization(<>);
garchobj.DoEstimation(<>);
//garchobj.Output();
decl VarCovarMatrix=garchobj.GetCovar();
print ("variance covariance matrix is based on the outer
  product of the gradients", VarCovarMatrix);
delete garchobj;
}
{
decl garchobj;
garchobj = new Garch();
garchobj.Load("Chapter2Data.xls");
garchobj.Select(Y_VAR, {"SP500RET",0,0});
garchobj.SetSelSample(-1, 1, -1, 1);
garchobj.CSTS(1,1);
garchobj.DISTRI(0);
garchobj.ARMA_ORDERS(1,0);
garchobj.ARFIMA(0);
garchobj.GARCH_ORDERS(1,1);
garchobj.ARCH_IN_MEAN(0);
garchobj.MODEL(2);
garchobj.MLE(2);
garchobj.Initialization(<>);
garchobj.DoEstimation(<>);
```

```
//garchobj.Output();
decl VarCovarMatrix=garchobj.GetCovar();
print ("the QMLE estimation of variance covariance matrix",
  VarCovarMatrix);
delete garchobj;
}
}
```

3

Fractionally integrated ARCH models

3.1 Fractionally integrated ARCH model specifications

Motivated by the fractionally integrated autoregressive moving average, or ARFIMA, model, Baillie et al. (1996a) proposed the fractionally integrated generalized autoregressive conditional heteroscedasticity, or FIGARCH, model. The ARFIMA (κ, d, l) model for the discrete time real-valued process $\{y_t\}$, initially developed by Granger (1980) and Granger and Joyeux (1980), is defined as

$$C(L)(1-L)^d y_t = D(L)\varepsilon_t, \tag{3.1}$$

where $C(L) = 1 - \sum_{i=1}^{\kappa} c_i L^i$, $D(L) = 1 + \sum_{i=1}^{l} d_i L^i$, and $\{\varepsilon_t\}$ is a mean-zero serially uncorrelated process. For $|d| < 0.5$, the dependent variable is covariance stationary. The fractional differencing operator, $(1-L)^d$, is usually interpreted in terms of its binomial expansion

$$(1-L)^d = \sum_{j=0}^{\infty} \pi_j L^j, \quad \text{for} \quad \pi_j = \frac{\Gamma(j-d)}{\Gamma(j+1)\Gamma(-d)} = \prod_{k=1}^{j} \frac{k-1-d}{k}, \tag{3.2}$$

where $\Gamma(.)$ denotes the gamma function.

The stationary ARMA process, defined by equation (3.1) for $d = 0$, is a short-memory process, the autocorrelations of which are geometrically bounded:

$$|Cor(y_t, y_{t+m})| \leq cr^m, \tag{3.3}$$

for $m = 1, 2, \ldots$, where $c > 0$ and $0 < r < 1$. As $m \to \infty$ the dependence, or memory, between y_t and y_{t+m} decreases rapidly. However, some observed time series appeared

ARCH Models for Financial Applications   Evdokia Xekalaki and Stavros Degiannakis
© 2010 John Wiley & Sons, Ltd

to exhibit a substantially larger degree of persistence than allowed for by stationary ARMA processes. For example, Ding et al. (1993) found that the absolute values or powers, particularly squares, of returns on the S&P500 index tend to have very slowly decaying autocorrelations. Similar evidence of this feature for other types of financial series is provided by Dacorogna et al. (1993), Mills (1996) and Taylor (1986). Such time series have autocorrelations that seem to satisfy the condition

$$Cor(y_t, y_{t+m}) \approx cm^{2d-1}, \tag{3.4}$$

as $m \to \infty$, where $c \neq 0$ and $d < 0.5$. Such processes are said to have a long memory because the autocorrelations display substantial persistence.

The concept of long memory and fractional Brownian motion was originally developed by Hurst (1951) and extended by Mandelbrot (1963, 1982) and Mandelbrot and Van Ness (1968). However, the ideas achieved prominence in the work of Granger (1980, 1981), Granger and Joyeux (1980) and Hosking (1981). Hurst was a hydrologist who worked on the Nile river dam project. He had studied an 847-year record of Nile overflows and observed that larger than average overflows were more likely to be followed by even larger overflows. Then suddenly the water flow would change to a lower than average overflow, which would be followed by lower than average overflows. Such a process could be examined neither with standard statistical correlation analysis nor by assuming that the water inflow was a random process, so that it could be analysed as a Brownian motion. In his work on Brownian motion Einstein (1905) found that the distance that a random particle covers increases with the square root of time used to measure it,

$$d \propto t^{1/2}, \tag{3.5}$$

where d is the distance covered and t is the time index. This, though, applies only to time series reprsenting Brownian motion, i.e. zero-mean and unity-variance independent processes. Hurst (1951) generalized (3.5) to account for processes other than Brownian motion in the form

$$d/s = ct^H, \tag{3.6}$$

where, for any process $\{y_t\}_{t=1}^{T}$ (e.g. asset returns) with mean $\bar{y}_T = T^{-1} \sum_{t=1}^{T} y_t$, d is given by

$$d = \max_{1 \leq k \leq T} \sum_{t=1}^{k} (y_t - \bar{y}_T) - \min_{1 \leq k \leq T} \sum_{t=1}^{k} (y_t - \bar{y}_T), \tag{3.7}$$

with s being the standard deviation of $\{y_t\}_{t=1}^{T}$ and c a constant. The ratio d/s is termed the *rescaled range* and H is the Hurst exponent. If $\{y_t\}$ is a sequence of independently and identically distributed random variables, then $H = 0.5$. Hurst's investigations for the Nile led to $H = 0.9$. Thus, the rescaled range was increasing at a faster rate than the square root of time.

The IGARCH(p, q) model in equation (2.7) could be rewritten as

$$\Phi(L)(1 - L)\varepsilon_t^2 = a_0 + (1 - B(L))v_t, \tag{3.8}$$

where $\Phi(L) \equiv (1 - A(L) - B(L))(1 - L)^{-1}$ is of order $(\max(p, q) - 1)$, or

$$\sigma_t^2 = a_0 + (1 - B(L) - \Phi(L)(1 - L))\varepsilon_t^2 + B(L)\sigma_t^2. \tag{3.9}$$

The FIGARCH(p, d, q) model is simply obtained by replacing the first difference operator in equation (3.8) by the fractional differencing operator

$$\sigma_t^2 = a_0 + (1 - B(L) - \Phi(L)(1 - L)^d)\varepsilon_t^2 + B(L)\sigma_t^2, \tag{3.10}$$

which is strictly stationary and ergodic for $0 \leq d \leq 1$ and covariance stationary for $|d| < 0.5$. The FIGARCH(p, d, q) model includes as special cases the IGARCH(p, q), for $d = 1$, and the GARCH(p, q), for $d = 0$. In contrast to the GARCH and IGARCH models, where shocks to the conditional variance either dissipate exponentially or persist indefinitely, for the case of the FIGARCH model the response of the conditional variance to past shocks decays at a slow hyperbolic rate. The sample autocorrelations of the daily absolute returns, $|y_t|$, as investigated by Ding et al. (1993) and Bollerslev and Mikkelsen (1996) among others, are outside the 95% confidence intervals for no serial dependence for more than 1000 lags. Moreover, the sample autocorrelations for the first difference of absolute returns, $(1 - L)|y_t|$, still show statistically significant long-term dependence. In contrast, the fractional difference of absolute returns, $(1 - L)^{0.5}|y_t|$, shows much less long-term dependence. Bollerslev and Mikkelsen (1996) provided evidence that illustrates the importance of using fractional integrated conditional variance models in the context of pricing options with maturity time of 1 year or longer. Note that the practical importance of the fractional integrated variance models stems from the added flexibility when modelling long-run volatility characteristics.

As Mills (1999, p. 140) noted, the implication of IGARCH models 'that shocks to the conditional variance persist indefinitely' does not reconcile with the 'persistence observed after large shocks, such as the Crash of October 1987, and ... with the perceived behaviour of agents who do not appear to frequently and radically alter the composition of their portfolios'. So the widespread observation of the IGARCH behaviour may be an artefact of a long-memory FIGARCH data-generating process. Baillie et al. (1996b) provided a simulation experiment that supports this line of argument. Beine et al. (2002) applied the FIGARCH(1,d,1) model in order to investigate the effects of official interventions on the volatility of exchange rates. One interesting remark made by them is that measuring the volatility of exchange rates through the FIGARCH model instead of a traditional ARCH model leads to different results. The GARCH and IGARCH models tend to underestimate the effect of central bank interventions on the volatility of exchange rates. Vilasuso (2002) fitted conditional volatility models to daily spot exchange rates and found that the FIGARCH(1,d,1) model generates volatility forecasts superior to those generated by a GARCH(1,1) or IGARCH(1,1) model.

Chung (1999) noted that the parallel use of a FIGARCH model for the conditional variance with an ARFIMA model for the conditional mean presents the difficulty that the fractional differencing operator does not apply to the constant term in the variance equation as in the case of the mean equation. Chung's (1999) proposal was to estimate the FIGARCH model, henceforth referred to as FIGARCHC(p, d, q), by the expression

$$\sigma_t^2 = \sigma^2(1 - B) + (1 - B(L) - \Phi(L)(1 - L)^d)(\varepsilon_t^2 - \sigma^2) + B(L)\sigma_t^2, \quad (3.11)$$

where σ^2 is the unconditional variance of ε_t.

Bollerslev and Mikkelsen (1996) extended the idea of fractional integration to the exponential GARCH model, while Tse (1998) built the fractional integration form of the APARCH model. Factorizing the autoregressive polynomial $(1 - B(L)) = \Phi(L)(1 - L)^d$, where all the roots of $\Phi(z) = 0$ lie outside the unit circle, the fractionally integrated exponential GARCH, or FIEGARCH(p, d, q), model is defined as

$$\log(\sigma_t^2) = a_0 + \Phi(L)^{-1}(1 - L)^{-d}(1 + A(L))g(z_{t-1}). \quad (3.12)$$

The fractionally integrated asymmetric power ARCH, or FIAPARCH(p, d, q), model has the form

$$\sigma_t^\delta = a_0 + (1 - (1 - B(L))^{-1}\Phi(L)(1 - L)^d)(|\varepsilon_t| - \gamma\varepsilon_t)^\delta. \quad (3.13)$$

Chung's modification can also be applied to the FIAPARCH specification. Therefore, the FIAPARCHC(p, d, q) model can also be proposed:

$$\sigma_t^\delta = \sigma^2(1 - B) + (1 - B(L) - \Phi(L)(1 - L)^d)((|\varepsilon_t| - \gamma\varepsilon_t)^\delta - \sigma^2) + B(L)\sigma_t^\delta, \quad (3.14)$$

where σ^2 is the unconditional variance of ε_t.

Hwang (2001) presented the asymmetric fractionally integrated family GARCH$(1, d, 1)$, or ASYMM-FIFGARCH$(1, d, 1)$, model, which is defined as

$$\sigma_t^\lambda = \frac{k}{1 - \delta} + \left(1 - \frac{(1 - \varphi L)(1 - L)^d}{1 - \delta L}\right) f^v(\varepsilon_t)\sigma_t^\lambda, \quad (3.15)$$

where

$$f(\varepsilon_t) = \left|\frac{\varepsilon_t}{\sigma_t} - b\right| - c\left(\frac{\varepsilon_t}{\sigma_t} - b\right), \quad \text{for} \quad |c| \le 1.$$

Hwang points out that, for different parameter values in (3.15), the following fractionally integrated ARCH models are obtained: FIEGARCH for $\lambda = 0$, $v = 1$; and FIGARCH for $\lambda = 2$, $v = 2$.

However, Ruiz and Pérez (2003) noted that Hwang's (2001) model is poorly specified and does not have the FIEGARCH model nested within it. Thus, they suggested an alternative specification, which is a direct generalization of Hentschel's model in (2.69):

$$(1 - \varphi L)(1 - L)^d \frac{\sigma_t^\lambda - 1}{\lambda} = \omega' + a(1 + \psi L)\sigma_{t-1}^\lambda (f^v(z_{t-1}) - 1), \qquad (3.16)$$

for

$$f\left(\frac{\varepsilon_t}{\sigma_t}\right) = \left|\frac{\varepsilon_t}{\sigma_t} - b\right| - c\left(\frac{\varepsilon_t}{\sigma_t} - b\right).$$

Imposing appropriate restrictions on the parameters of (3.16), a number of models are obtained as special cases, such as the FIGARCH model in (3.10), the FIEGARCH model in (3.12), and Hentschel's model in (2.69).

Davidson (2004) suggested a generalized version of the FIGARCH model, which he termed the hyperbolic GARCH, or HYGARCH(p, d, q), model:

$$\sigma_t^2 = a_0 + (1 - B(L) - \Phi(L)(1 + a((1 - L)^d - 1)))\varepsilon_t^2 + B(L)\sigma_t^2. \qquad (3.17)$$

The fractional differencing operator, $(1 - L)^d$, can also be expressed as:

$$(1 - L)^d = 1 - \sum_{j=1}^{\infty} \pi_j L^j = 1 - dL - \frac{1}{2!}d(1 - d)L^2 - \frac{1}{3!}d(1 - d)(2 - d)L^3 - \cdots.$$

$$(3.18)$$

For $d > 0$, the infinite sum $\sum_{j=1}^{\infty} \pi_j$ tends to unity. Therefore, the HYGARCH model weights the components π_j by a. Note that for $a = 1$, the HYGARCH model has the the FIGARCH model nested within it.

Bollerslev and Wright (2000), based on Parke (1999), illustrated a method for simulating an ARFIMA$(0, d, 0)$ time series as the weighted sum of i.i.d. shocks each having a random duration with a certain distribution.

Ding and Granger (1996) proposed the long-memory ARCH, or LM-ARCH(q), model to capture the long-memory property of volatility. The LM-ARCH(q) model is the generalization of the N-component GARCH(1,1) model of Ding and Granger (1996), in equation (2.111), for $N \to \infty$ and on the assumption that the parameters of the model are beta distributed. Recently, Baillie and Morana (2009) introduces the adaptive FIGARCH, or A-FIGARCH, model which is designed to account for both long memory and structural change in the conditional variance.

3.2 Estimating fractionally integrated ARCH models using G@RCH 4.2 OxMetrics: an empirical example

In the paragraphs that follow, we estimate six fractionally integrated ARCH specifications for the FTSE100 equity index: FIGARCH (Baillie et al., 1996a),

FIGARCHC (Chung, 1999), FIEGARCH (Bollerslev and Mikkelsen, 1996), FIAPARCH (Tse, 1998), FIAPARCHC (Chung, 1999) and HYGARCH (Davidson, 2004). We make use of the same data set, for the period from 4 April 1988 to 5 April 2005, as in Section 2.3, and retain the ARCH framework of (1.6) with only the constant term in the conditional mean specification:

$$
\begin{aligned}
y_t &= c_0 + \varepsilon_t, \\
\varepsilon_t &= \sigma_t z_t, \\
\sigma_t^2 &= g(\theta | I_{t-1}), \\
z_t &\overset{i.i.d.}{\sim} N(0,1).
\end{aligned}
\tag{3.19}
$$

The models are estimated using the G@RCH OxMetrics package in the following specific forms:[1]

FIGARCH$(1, d, 1)$,

$$
\sigma_t^2 = a_0 + (1 - b_1 L - (1 - a_1 L)(1 - L)^d)\varepsilon_t^2 + b_1 \sigma_{t-1}^2;
\tag{3.20}
$$

FIGARCHC$(1, d, 1)$,

$$
\sigma_t^2 = \sigma^2 (1 - b_1) + (1 - b_1 L - (1 - a_1 L)(1 - L)^d)(\varepsilon_t^2 - \sigma^2) + b_1 \sigma_{t-1}^2;
\tag{3.21}
$$

FIEGARCH $(1, d, 1)$,

$$
\begin{aligned}
\log \sigma_t^2 &= a_0 (1 - b_1) + ((1 + \alpha_1 L)(1 - L)^{-d}) \left(\gamma_1 \frac{\varepsilon_{t-1}}{\sigma_{t-1}} + \gamma_2 \left(\left| \frac{\varepsilon_{t-1}}{\sigma_{t-1}} \right| - E \left| \frac{\varepsilon_{t-1}}{\sigma_{t-1}} \right| \right) \right) \\
&\quad + b_1 \log \sigma_{t-1}^2;
\end{aligned}
\tag{3.22}
$$

FIAPARCH$(1, d, 1)$,

$$
\sigma_t^\delta = a_0 + (1 - b_1 L - (1 - a_1 L)(1 - L)^d)(|\varepsilon_t| - \gamma \varepsilon_t)^\delta + b_1 \sigma_{t-1}^\delta;
\tag{3.23}
$$

FIAPARCHC$(1, d, 1)$,

$$
\sigma_t^\delta = \sigma^2 (1 - b_1) + (1 - b_1 L - (1 - a_1 L)(1 - L)^d)\left((|\varepsilon_t| - \gamma \varepsilon_t)^\delta - \sigma^2 \right) + b_1 \sigma_{t-1}^\delta;
\tag{3.24}
$$

and HYGARCH$(1, d, 1)$,

$$
\sigma_t^2 = a_0 + (1 - b_1 L - (1 - a_1 L)(1 + a((1 - L)^d - 1)))\varepsilon_t^2 + b_1 \sigma_{t-1}^2.
\tag{3.25}
$$

Note that the FIGARCH$(1, d, 1)$ specification can be expanded as

$$
\begin{aligned}
\sigma_t^2 &= a_0 + \varepsilon_t^2 - b_1 \varepsilon_{t-1}^2 - (1 - L)^d \varepsilon_t^2 + a_1 (1 - L)^d \varepsilon_{t-1}^2 + b_1 \sigma_{t-1}^2 \\
&= a_0 - b_1 \varepsilon_{t-1}^2 + \left(dL + \frac{1}{2!} d(1 - d) L^2 + \cdots \right) \varepsilon_t^2 + a_1 \varepsilon_{t-1}^2 \\
&\quad - \left(dL + \frac{1}{2!} d(1 - d) L^2 + \cdots \right) a_1 \varepsilon_{t-1}^2 + b_1 \sigma_{t-1}^2,
\end{aligned}
$$

---

[1] In G@RCH, $\Phi(L) = 1 - \sum_{i=1}^{p} a_i L^i$.

or, more compactly,

$$\sigma_t^2 = a_0 + (a_1 - b_1)\varepsilon_{t-1}^2 + \left(dL + \frac{1}{2!}d(1-d)L^2 + \cdots\right)(\varepsilon_t^2 - a_1\varepsilon_{t-1}^2) + b_1\sigma_{t-1}^2.$$

For example, the FIGARCH$(1, 0, 0)$ corresponds to the IGARCH$(1, 1)$ as for $d = 0$ and $a_1 = 1$,

$$\sigma_t^2 = a_0 + (1 - b_1 L - (1 - L))\varepsilon_t^2 + b_1\sigma_{t-1}^2 = a_0 + (1 - b_1)\varepsilon_{t-1}^2 + b_1\sigma_{t-1}^2.$$

In a similar manner, the FIGARCHC$(1, d, 1)$ can be expanded as

$$\sigma_t^2 = \sigma^2(1 - b_1) + (1 - b_1 L - (1 - L)^d + a_1 L(1 - L)^d)(\varepsilon_t^2 - \sigma^2) + b_1\sigma_{t-1}^2$$

$$= \sigma^2(1 - b_1) + \varepsilon_t^2 - \sigma^2 - b_1(\varepsilon_{t-1}^2 - \sigma^2) - (\varepsilon_t^2 - \sigma^2)$$

$$+ \left(dL + \frac{1}{2!}d(1-d)L^2 + \cdots\right)(\varepsilon_t^2 - \sigma^2) + a_1(\varepsilon_{t-1}^2 - \sigma^2)$$

$$- \left(dL + \frac{1}{2!}d(1-d)L^2 + \cdots\right)a_1(\varepsilon_{t-1}^2 - \sigma^2) + b_1\sigma_{t-1}^2$$

$$= \sigma^2(1 - b_1) + (a_1 - b_1)\varepsilon_{t-1}^2 + \sigma^2(b_1 - 1)$$

$$+ \left(dL + \frac{1}{2!}d(1-d)L^2 + \cdots\right)(\varepsilon_t^2 - a_1\varepsilon_{t-1}^2) + b_1\sigma_{t-1}^2.$$

The FIEGARCH$(1, d, 1)$ can be represented as

$$\log\sigma_t^2 = a_0(1 - b_1) + (1 - L)^{-d}\left[\gamma_1\frac{\varepsilon_{t-1}}{\sigma_{t-1}} + \gamma_2\left(\left|\frac{\varepsilon_{t-1}}{\sigma_{t-1}}\right| - E\left|\frac{\varepsilon_{t-1}}{\sigma_{t-1}}\right|\right) + \alpha_1\gamma_1\frac{\varepsilon_{t-2}}{\sigma_{t-2}}\right.$$

$$+ \alpha_1\gamma_2\left(\left|\frac{\varepsilon_{t-2}}{\sigma_{t-2}}\right| - E\left|\frac{\varepsilon_{t-2}}{\sigma_{t-2}}\right|\right)\right] + b_1\log\sigma_{t-1}^2.$$

where, for $d > 0$,

$$(1 - L)^{-d} = 1 + \sum_{j=1}^{\infty}\left(\frac{\Gamma(j+d)}{\Gamma(d)\Gamma(j+1)}L^j\right) = 1 + \frac{1}{1!}dL + \frac{1}{2!}d(1+d)L^2 - \cdots.$$

Thus,

$$\log\sigma_t^2 = a_0(1 - b_1) + \left[\gamma_1\frac{\varepsilon_{t-1}}{\sigma_{t-1}} + \gamma_2\left(\left|\frac{\varepsilon_{t-1}}{\sigma_{t-1}}\right| - E\left|\frac{\varepsilon_{t-1}}{\sigma_{t-1}}\right|\right) + \alpha_1\gamma_1\frac{\varepsilon_{t-2}}{\sigma_{t-2}}\right.$$

$$+ \alpha_1\gamma_2\left(\left|\frac{\varepsilon_{t-2}}{\sigma_{t-2}}\right| - E\left|\frac{\varepsilon_{t-2}}{\sigma_{t-2}}\right|\right)\right] + \sum_{j=1}^{\infty}\left(\frac{\Gamma(j+d)}{\Gamma(d)\Gamma(j+1)}L^j\right)$$

$$\left[\gamma_1\frac{\varepsilon_{t-1}}{\sigma_{t-1}} + \gamma_2\left(\left|\frac{\varepsilon_{t-1}}{\sigma_{t-1}}\right| - E\left|\frac{\varepsilon_{t-1}}{\sigma_{t-1}}\right|\right) + \alpha_1\gamma_1\frac{\varepsilon_{t-2}}{\sigma_{t-2}}\right.$$

$$+ \alpha_1\gamma_2\left(\left|\frac{\varepsilon_{t-2}}{\sigma_{t-2}}\right| - E\left|\frac{\varepsilon_{t-2}}{\sigma_{t-2}}\right|\right)\right]$$

$$+ b_1\log\sigma_{t-1}^2.$$

Note also that, for $d > 0$,

$$\sum_{j=1}^{\infty}\left(\frac{\Gamma(j+d)}{\Gamma(d)\Gamma(j+1)}L^j\right) = \frac{1}{1!}dL + \frac{1}{2!}d(1+d)L^2 + \frac{1}{3!}d(1+d)(2+d)L^3 + \cdots$$

and

$$\sum_{j=1}^{\infty}d\left(\frac{\Gamma(j-d)}{\Gamma(1-d)\Gamma(j+1)}L^j\right) = (1!)^{-1}dL + (2!)^{-1}d(1-d)L^2$$

$$+ (3!)^{-1}d(1-d)(2-d)L^3 + \cdots.$$

For example, the FIEGARCH$(0,d,0)$ specification can simply be represented by

$$\log\sigma_t^2 = (1-L)^{-d}\left(\gamma_1\frac{\varepsilon_{t-1}}{\sigma_{t-1}} + \gamma_2\left(\left|\frac{\varepsilon_{t-1}}{\sigma_{t-1}}\right| - E\left|\frac{\varepsilon_{t-1}}{\sigma_{t-1}}\right|\right)\right).$$

The FIAPARCH$(1,d,1)$ model can be expanded as

$$\sigma_t^\delta = a_0 + (1 - b_1 L - (1-L)^d + \alpha_1 L(1-L)^d(|\varepsilon_t| - \gamma\varepsilon_t)^\delta + b_1\sigma_{t-1}^\delta$$

$$= a_0 + (|\varepsilon_t| - \gamma\varepsilon_t)^\delta - b_1(|\varepsilon_{t-1}| - \gamma\varepsilon_{t-1})^\delta - (1-L)^d(|\varepsilon_t| - \gamma\varepsilon_t)^\delta$$

$$+ \alpha_1(1-L)^d(|\varepsilon_{t-1}| - \gamma\varepsilon_{t-1})^\delta + b_1\sigma_{t-1}^\delta$$

$$= a_0 + (|\varepsilon_t| - \gamma\varepsilon_t)^\delta - b_1(|\varepsilon_{t-1}| - \gamma\varepsilon_{t-1})^\delta - (|\varepsilon_t| - \gamma\varepsilon_t)^\delta$$

$$+ \left(dL + \frac{1}{2!}d(1-d)L^2 + \cdots\right)(|\varepsilon_t| - \gamma\varepsilon_t)^\delta + \alpha_1(|\varepsilon_{t-1}| - \gamma\varepsilon_{t-1})^\delta$$

$$- \alpha_1\left(dL + \frac{1}{2!}d(1-d)L^2 + \cdots\right)(|\varepsilon_{t-1}| - \gamma\varepsilon_{t-1})^\delta + b_1\sigma_{t-1}^\delta$$

$$= a_0 + (\alpha_1 - b_1)(|\varepsilon_{t-1}| - \gamma\varepsilon_{t-1})^\delta + \left(dL + \frac{1}{2!}d(1-d)L^2 + \cdots\right)(|\varepsilon_t| - \gamma\varepsilon_t)^\delta$$

$$- \alpha_1\left(dL + \frac{1}{2!}d(1-d)L^2 + \cdots\right)(|\varepsilon_{t-1}| - \gamma\varepsilon_{t-1})^\delta + b_1\sigma_{t-1}^\delta.$$

The FIAPARCHC$(1,d,1)$ can be presented as

$$\sigma_t^\delta = (\alpha_1 - b_1)(|\varepsilon_{t-1}| - \gamma\varepsilon_{t-1})^\delta + \left(dL + \frac{1}{2!}d(1-d)L^2 + \cdots\right)$$

$$\left((|\varepsilon_t| - \gamma\varepsilon_t)^\delta - \alpha_1(|\varepsilon_{t-1}| - \gamma\varepsilon_{t-1})^\delta\right) + b_1\sigma_{t-1}^\delta.$$

Finally, the HYGARCH$(1,d,1)$ can be rewritten as

$$\sigma_t^2 = a_0 + (1 - b_1 L - (1 + a(1-L)^d - a) + \alpha_1 L(1 + a(1-L)^d - a))\varepsilon_t^2 + b_1\sigma_{t-1}^2$$

$$= a_0 + (-b_1 L + a + \alpha_1 L - a\alpha_1 L)\varepsilon_t^2 + (1-L)^d a(\alpha_1 L - 1)\varepsilon_t^2 + b_1\sigma_{t-1}^2$$

$$= a_0 + (-b_1 L + a + \alpha_1 L - a\alpha_1 L)\varepsilon_t^2 + a(\alpha_1 L - 1)\varepsilon_t^2$$

$$- \left(dL + \frac{1}{2}d(1-d)L^2 + \cdots \right) a(\alpha_1 L - 1)\varepsilon_t^2 + b_1 \sigma_{t-1}^2$$

$$= a_0 + (\alpha_1 - b_1)L\varepsilon_t^2 - \left(dL + \frac{1}{2}d(1-d)L^2 + \cdots \right) a(\alpha_1 L - 1)\varepsilon_t^2 + b_1 \sigma_{t-1}^2,$$

or

$$\sigma_t^2 = \alpha_0 + (\alpha_1 - b_1)L\varepsilon_t^2 + \left(dL + \frac{1}{2}d(1-d)L^2 + \cdots \right) a(1 - \alpha_1 L)\varepsilon_t^2 + b_1 \sigma_{t-1}^2.$$

It should be noted at this point that G@RCH reports values of log a instead of a.

The functional forms of the in-sample conditional standard deviation estimators, $\hat{\sigma}_{t+1}$, are:

FIGARCH$(1, d, 1)$,

$$\hat{\sigma}_{t+1} = \sqrt{a_0^{(T)} + \left(a_1^{(T)} - b_1^{(T)} \right)\hat{\varepsilon}_t^2 + \sum_{j=1}^{\infty} \left(\frac{d^{(T)}\Gamma(j - d^{(T)})}{\Gamma(1 - d^{(T)})\Gamma(j+1)} L^j \left(\hat{\varepsilon}_{t+1}^2 - a_1^{(T)}\hat{\varepsilon}_t^2 \right) \right) + b_1^{(T)}\hat{\sigma}_t^2};$$

$$(3.26)$$

FIGARCHC$(1, d, 1)$,

$$\hat{\sigma}_{t+1} = \sqrt{\left(a_1^{(T)} - b_1^{(T)} \right)\hat{\varepsilon}_t^2 + \sum_{j=1}^{\infty} \left(\frac{d^{(T)}\Gamma(j - d^{(T)})}{\Gamma(1 - d^{(T)})\Gamma(j+1)} L^j \left(\hat{\varepsilon}_{t+1}^2 - a_1^{(T)}\hat{\varepsilon}_t^2 \right) \right) + b_1^{(T)}\hat{\sigma}_t^2};$$

$$(3.27)$$

FIEGARCH$(1, d, 1)$,

$$\hat{\sigma}_{t+1} = \left(\exp\left(a_0^{(T)} \left(1 - b_1^{(T)} \right) + \left[\gamma_1^{(T)}\frac{\hat{\varepsilon}_t}{\hat{\sigma}_t} + \gamma_2^{(T)} \left(\left|\frac{\hat{\varepsilon}_t}{\hat{\sigma}_t}\right| - \sqrt{2/\pi} \right) \right] \right.\right.$$

$$+ a_1^{(T)} \left[\gamma_1^{(T)}\frac{\hat{\varepsilon}_{t-1}}{\hat{\sigma}_{t-1}} + \gamma_2^{(T)} \left(\left|\frac{\hat{\varepsilon}_{t-1}}{\hat{\sigma}_{t-1}}\right| - \sqrt{2/\pi} \right) \right] + b_1^{(T)}\log\hat{\sigma}_t^2$$

$$+ \sum_{j=1}^{\infty} \left(\frac{\Gamma(j + d^{(T)})}{\Gamma(d^{(T)})\Gamma(j+1)} \right) \left(L^j + a_1^{(T)}L^{j+1} \right)$$

$$\times \left. \left. \left(\gamma_1^{(T)}\frac{\hat{\varepsilon}_t}{\hat{\sigma}_t} + \gamma_2^{(T)} \left(\left|\frac{\hat{\varepsilon}_t}{\hat{\sigma}_t}\right| - \sqrt{2/\pi} \right) \right) \right) \right)^{1/2};$$

$$(3.28)$$

FIAPARCH$(1,d,1)$,

$$\hat{\sigma}_{t+1} = \left(a_0^{(T)} + \left(a_1^{(T)} - b_1^{(T)} \right) \left(|\hat{\varepsilon}_t| - \gamma^{(T)}\hat{\varepsilon}_t \right)^{\delta^{(T)}} + b_1^{(T)}\hat{\sigma}_t^{\delta^{(T)}} \right.$$

$$+ \sum_{j=1}^{\infty} \left(\frac{d^{(T)}\Gamma(j-d^{(T)})}{\Gamma(1-d^{(T)})\Gamma(j+1)} L^j \left(|\hat{\varepsilon}_{t+1}| - \gamma^{(T)}\hat{\varepsilon}_{t+1} \right)^{\delta^{(T)}} \right.$$

$$\left. \left. - a_1^{(T)} \left(|\hat{\varepsilon}_t| - \gamma^{(T)}\hat{\varepsilon}_t \right)^{\delta^{(T)}} \right) \right)^{1/\delta^{(T)}} ; \tag{3.29}$$

FIAPARCHC$(1,d,1)$

$$\hat{\sigma}_{t+1} = \left(\left(a_1^{(T)} - b_1^{(T)} \right) \left(|\hat{\varepsilon}_t| - \gamma^{(T)}\hat{\varepsilon}_t \right)^{\delta^{(T)}} + b_1^{(T)}\hat{\sigma}_t^{\delta^{(T)}} \right.$$

$$+ \sum_{j=1}^{\infty} \left(\frac{d^{(T)}\Gamma(j-d^{(T)})}{\Gamma(1-d^{(T)})\Gamma(j+1)} L^j \left(|\hat{\varepsilon}_{t+1}| - \gamma^{(T)}\hat{\varepsilon}_{t+1} \right)^{\delta^{(T)}} \right.$$

$$\left. \left. - a_1^{(T)} \left(|\hat{\varepsilon}_t| - \gamma^{(T)}\hat{\varepsilon}_t \right)^{\delta^{(T)}} \right) \right)^{1/\delta^{(T)}} ; \tag{3.30}$$

and HYGARCH$(1,d,1)$,

$$\hat{\sigma}_{t+1} = \left(a_0^{(T)} + \left(a_1^{(T)} - b_1^{(T)} \right)\hat{\varepsilon}_t^2 + b_1^{(T)}\hat{\sigma}_t^2 \right.$$

$$\left. + \sum_{j=1}^{\infty} \left(\frac{d^{(T)}\Gamma(j-d^{(T)})}{\Gamma(1-d^{(T)})\Gamma(j+1)} L^j a^{(T)} \left(\hat{\varepsilon}_{t+1}^2 - a_1^{(T)}\hat{\varepsilon}_t^2 \right) \right) \right)^{1/2} . \tag{3.31}$$

The autocorrelations for the absolute log-returns of the FTSE100 and S&P500 equity indices, the US dollar to British pound exchange rate and the price of a troy ounce of gold bullion are presented in Figure 3.1.[2] Parallel lines indicate the 95% confidence interval of the form $\pm 1.96/\sqrt{T} = \pm 0.0294$, for the estimated sample autocorrelation of a process with independently and identically normally distributed components. The absolute log-returns are significantly positively serially autocorrelated over long lags. The sample autocorrelations decrease at a fast rate in the first lags and at a slower rate in the higher lags, thus providing evidence for the long-memory property of volatility processes and the use of fractionally integrated volatility specifications.

---

[2] The autocorrelation functions for a lag longer than 200 can be computed with the help of the EViews program named *chapter3.autocorrelations.prg*, in the Appendix to the present chapter.

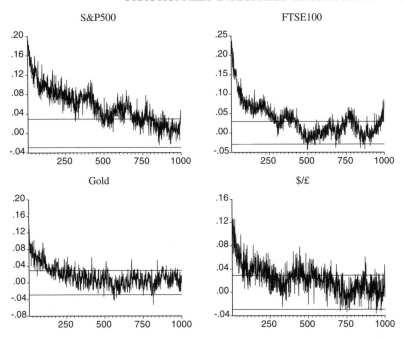

Figure 3.1   The lag $\tau = 1, \ldots, 1000$ autocorrelations for the absolute log-returns, $Cor(|y_t|, |y_{t-\tau}|)$, from 4 April 1988 to 5 April 2005.

To estimate the six fractionally integrated ARCH models for the FTSE100 index, we create the program named *chapter3.ftse100.ox*. The core G@RCH program that carries out the necessary estimations for the AR(0)-FIGARCH(1,1) model is as follows:[3]

```
#import <packages/Garch42/garch>
main()
{
   decl garchobj;
   garchobj = new Garch();
   garchobj.Load("Chapter3Data.xls");
   garchobj.Select(Y_VAR, {"FTSE100RET",0,0});
   garchobj.SetSelSample(-1, 1, -1, 1);
   garchobj.CSTS(1,1);
   garchobj.DISTRI(0);
   garchobj.ARMA_ORDERS(0,0);
   garchobj.GARCH_ORDERS(1,1);
```

---

[3] The models can be estimated either by the program presented here or by its extended version, *chapter3. ftse100.ox*, presented in the Appendix to this chapter.

```
garchobj.MODEL(6);
garchobj.MLE(2);
garchobj.MAXSA(1,5,0.5,20,5,2,1);
garchobj.Initialization(<>);
garchobj.DoEstimation(<>);
garchobj.Output();
delete garchobj;
}
```

G@RCH reads the FTSE100 log-returns from the Excel file named *chapter3Data. xls*. The command *garchobj.MODEL(.)* for values 0 to 11 estimates the FIGARCH, FIGARCHC, FIEGARCH, FIAPARCH, FIAPARCHC and HYGARCH conditional volatility specifications, respectively. The models are estimated by means of the simulated annealing algorithm, *garchobj.MAXSA(1,5,0.5,20,5,2,1)*. Note that the simulated annealing algorithm takes much longer to carry out the estimation of the six models (273 minutes) than the BFGS method takes (6 minutes and 18 seconds).[4] Table 3.1 presents the estimated parameters of the models. All the GARCH parameters are statistically significant in all cases but one, the FIEGARCH specification. The variance is characterized by a slowly mean-reverting fractionally integrated process with a degree of integration estimated to be between 0.403 and 0.579. The estimates of *d* are statistically significantly different from 0 as well as from 1, except for the case of the FIEGARCH specification, giving support to the modelling of fractionally integrated conditional heteroscedasticity. With the exception of the FIAPARCHC specification, in all cases the fractional integration parameter is not statistically different from 0.5, providing evidence for a non-covariance stationary volatility. In similar studies, *d* was estimated between 0.40 and 0.48 by Giot and Laurent (2004), less than 0.5 by Ebens (1999), between 0.25 and 0.45 by Thomakos and Wang (2003) and between 0.19 and 0.27 by Tse (1998). The asymmetry between past bad or good news and volatility is statistically significant for all models but the FIEGARCH. Moreover, the Box–Cox power transformation was correctly imposed in the ARCH specification as δ is around 1.5 and not equal to 1 or 2.

Table 3.2 depicts the *p*-values of (i) the Jarque–Bera test of normally distributed standardized residuals, (ii) Engle's test of standardized homoscedastic residuals and (iii) the Box and Pierce *Q*-test for the null hypothesis that there is no autocorrelation of up to the 10th, 15th and 20th order, respectively, on standardized and squared standardized residuals. In general lines, the residuals of the six models obey the assumption of autocorrelation and heteroscedasticity absence but the normality assumption is rejected.

Utilizing the SBC criterion for checking the models' in-sample fitting, the AR(0)-FIAPARCH(1,1) model is judged to provide the best fit on the basis of Table 3.3 as it is the one that minimizes the SBC criterion. Figure 3.2 plots the

---

[4] Both computed on the same computer, with an AMD Turion 1.8GHz processor and 1GB of RAM. Both methods, simulated annealing and BFGS, provide almost the same estimates of parameters. There are some differences in the values of parameters after the fourth decimal place. The program estimates the models via the BFGS method if *garchobj.MAXSA(1,5,0.5,20,5,2,1)* is replaced by *garchobj.MAXSA(0,5,0.5,20,5,2,1)*.

Table 3.1 FTSE100 equity index: estimation of parameters using the full set of data, from 4 April 1988 to 5 April 2005

Parameters	FIGARCH	FIGARCHC	FIEGARCH	FIAPARCH	FIAPARCHC	HYGARCH
c_0	0.037447	0.037442	0.028225	0.023049	0.025842	0.03766
	(3.154)	(3.124)	(2.147)	(1.857)	(2.075)	(3.17)
a_0	0.02462	1.214307	-0.04431	0.051112	1.385502	0.040369
	(2.623)	(3.497)	(-0.2201)	(3.842)	(4.771)	(3.149)
a_1	0.172241	0.178637	-0.27406	0.199465	0.218639	0.147567
	(3.222)	(3.072)	(-0.2154)	(5.252)	(4.676)	(3.144)
b_1	0.60469	0.580673	0.869008	0.650555	0.574548	0.647551
	(6.218)	(7.164)	(2.166)	(8.554)	(8.942)	(6.851)
d	0.48924	0.458395	0.479106	0.499577	0.403152	0.57962
	(5.617)	(8.597)	(1.213)	(6.103)	(9.213)	(5.382)
γ_1	—	—	-0.04107	0.323031	0.319042	—
			(-1.257)	(4.062)	(3.877)	
γ_2	—	—	0.104351	—	—	—
			(1.597)			
δ	—	—	—	1.447049	1.674008	—
				(9.129)	(15.7)	
$\log a$	—	—	—	—	—	-0.04544
						(-2.119)

Numbers in parentheses represent standard error ratios.

Table 3.2 FTSE100 equity index: skewness and kurtosis of standardized residuals and p-values of the Jarque–Bera test, Engle's LM test, and the Box–Pierce Q statistic

	FIGARCH	FIGARCHC	FIEGARCH	FIAPARCH	FIAPARCHC	HYGARCH
Standardized residuals						
Skewness	−0.141	−0.140	−0.125	−0.123	−0.118	−0.142
Excess kurtosis	1.032	1.034	0.961	1.027	1.061	1.014
Jarque–Bera p-value	0.000	0.000	0.000	0.000	0.000	0.000
p-value of Engle's LM test 2 lags	0.712	0.700	0.341	0.653	0.837	0.575
p-value of Engle's LM test 5 lags	0.237	0.282	0.377	0.238	0.332	0.251
p-value of Engle's LM test 10 lags	0.460	0.494	0.423	0.371	0.443	0.457
p-value of Q statistic 10 lags	0.154	0.142	0.135	0.142	0.119	0.169
p-value of Q statistic 15 lags	0.160	0.149	0.152	0.155	0.136	0.174
p-value of Q statistic 20 lags	0.067	0.062	0.072	0.076	0.065	0.075
Squared standardized residuals						
p-value of Q statistic 10 lags	0.280	0.306	0.268	0.215	0.265	0.281
p-value of Q statistic 15 lags	0.182	0.201	0.098	0.089	0.110	0.193
p-value of Q statistic 20 lags	0.096	0.103	0.049	0.031	0.033	0.102

Table 3.3   FTSE100 equity index: values of the SBC criterion for fractionally integrated models of the FTSE100 index

Model	SBC criterion
AR(0)-FIGARCH(1,1)	2.603202
AR(0)-FIGARCHC(1,1)	2.603613
AR(0)-FIEGARCH(1,1)	2.59786
AR(0)-FIAPARCH(1,1)	2.596345
AR(0)-FIAPARCHC(1,1)	2.598078
AR(0)-HYGARCH(1,1)	2.604057

in-sample annualized volatility estimated by the AR(0)-FIAPARCH(1,1) model in
comparison with the difference between volatility estimates by the FIAPARCH(1,1)
and GARCH(1,1) models. The difference ranges from −5.9 to 9.5, its mean squared
value is 2.11, and its average value of 0.1 is virtually equal to zero.

To summarize, the models presented in this chapter provide a description of the
volatility dynamics. For instance, the FIAPARCH specification takes into consideration the leverage effect, the long memory in conditional variance and the Box–Cox
power transformation of the fractionally integrated volatility. The autocorrelations for
the absolute standardized residuals of the AR(0)-FIAPARCH(1,1) model are presented in Figure 3.3.[5] About 7% of the estimated autocorrelations are outside the 95%
confidence interval defined by the endpoints $\pm 1.96/\sqrt{T}$, providing evidence that the
model adequately captures the long-memory property of the volatility process.

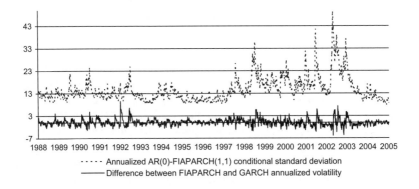

- - - - - Annualized AR(0)-FIAPARCH(1,1) conditional standard deviation
———— Difference between FIAPARCH and GARCH annualized volatility

*Figure 3.2   FTSE100 equity index: The in-sample AR(0)-FIAPARCH(1,1) annual-
ized volatility and its difference from the AR(1)-GARCH(1,1) volatility.*

[5] The autocorrelations were computed by the EViews program *chapter3.stdresiduals.ar0fiaparch11.
prg*, which is available in the Appendix to this chapter.

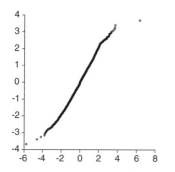 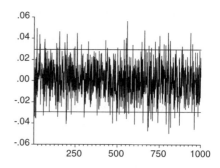

Figure 3.3   FTSE100 equity index: the quantile–quantile plot of the standardized residuals (left panel) and the lag 1 to 1000 autocorrelations for the absolute standardized residuals of the AR(0)-FIAPARCH(1,1) model, $Cor(|\hat{\varepsilon}_t \hat{\sigma}_t^{-1}|,$ $|\hat{\varepsilon}_{t-\tau} \hat{\sigma}_{t-\tau}^{-1}|)$, for $\tau = 1, \cdots, 1000$ (right panel).

However, as the Jarque–Bera test in Table 3.2 and the quantile–quantile (QQ) plot[6] in Figure 3.3 reveal, the standardized residuals still do not agree with the assumption of normality.

3.3   A more detailed investigation of the normality of the standardized residuals: goodness-of-fit tests

As mentioned in Chapter 1, the normality assumption of the standardized residuals has been investigated by the Jarque–Bera test (Jarque and Bera, 1980, 1987; see also Bera and Jarque, 1982) whose test statistic is given by

$$JB = T(Sk^2 + ((Ku - 3)^2 / 4))/6, \qquad (3.32)$$

where T denotes the number of observations, Sk the skewness and Ku the kurtosis. The JB statistic is χ^2 distributed with 2 degrees of freedom. Its main disadvantage is the over-rejection of normality in the presence of serially correlated observations (for details, see Thomakos and Wang, 2003). Jarque and Bera (1987) noted that the JB test behaves better than alternative normality tests if the alternatives to the normal distribution belong to the Pearson family. Thadewald and Büning (2007) provided simulated evidence that the JB test, in general, works best in comparison to several alternative normality tests they considered, although, for certain sample cases, other tests should be preferred. Urzúa (1996) introduced a modification of the JB test by standardizing the skewness and kurtosis.

---

[6] The QQ plot is a graphical method used to judge whether the quantiles of the empirical distribution of the standardized residuals are close to those of the assumed distribution. In the present case, the empirical distribution is close to the normal but with heavier tails. For details, see Section 3.3.3.

The JB test is certainly not the only approach to test whether the standardized residuals are normally distributed. There is a literature on *goodness-of-fit tests* with a vast number of test variations. The term 'goodness-of-fit test' refers to statistical functions that measure the distance between theoretical and empirical distribution functions. There are two main categories of such tests: tests based on the empirical distribution function (EDF tests) and tests based on the chi-square distribution (chi-square tests). The EDF tests are based on the deviation of the empirical distribution function from the hypothesized cumulative distribution function, and the distributions of their test statistics are obtained by Monte Carlo simulations. The chi-square tests are used when data are grouped into discrete classes, and observed frequencies are compared to expected frequencies based on a model.

3.3.1 EDF tests

The Kolmogorov–Smirnov (KS) test is a non-parametric procedure that compares the observed cumulative distribution function of the standardized residuals with the assumed theoretical distribution function. The KS statistic is based on the maximum absolute distance between the observed and theoretical cumulative distribution functions:

$$KS = \sqrt{T} \max_i \left| F(x_i) - \hat{F}(x_i) \right|, \tag{3.33}$$

where $F(x_i)$ and $\hat{F}(x_i)$ denote the theoretical and empirical cumulative distribution functions, respectively, and $x_1 < x_2 < \cdots < x_N$, for $N \geq T$. The KS statistic is evaluated over the set $\{x_i\}_{i=1}^{N}$ of ordered data points. If the observed value of the statistic exceeds the appropriate critical value, the null hypothesis of no significant differences between the theoretical and empirical distributions of the standardized residuals is rejected. The critical values can be simulated by a Monte Carlo technique (for details, see Dowd, 2005, p. 346). The advantage of the KS statistic is that it can be applied under any distributional assumption. Its use is not limited to cases of normality, which is very useful as we will find out in Chapter 5, where we will do away with the assumption of normally distributed standardized residuals. A main disadvantage is its sensitivity to distributional differences near the main mass of the distribution. There is a vast literature on the Kolmogorov–Smirnov type of tests. The fundamental ideas were developed by Cramér (1928), Kolmogorov (1933), von Mises (1931) and Smirnov (1937, 1939).

Kuiper (1960) defined an alternative to the KS test that is based on the sum of the maximum distances between the observed and theoretical cumulative distribution functions:

$$KS^* = \sqrt{T} \left(\max_i (F(x_i) - \hat{F}(x_i)) + \max_i (\hat{F}(x_i) - F(x_i)) \right). \tag{3.34}$$

The main difference between the KS test and Kuiper's alternative is the latter's sensitivity to distributional differences near the tails of the distribution. The critical values can be also obtained by a Monte Carlo simulation. Its small-sample distributional properties were studied by Stephens (1965).

Both the KS and KS^* statistics based tests are valid when the distributional parameters are assumed to be known. If the parameters of the distribution are unknown and the theoretical distribution is normal, then Lilliefors's (1967, 1969) approximation can be applied (see also Dallal and Wilkinson, 1986). Lilliefors (1967) obtained tables via Monte Carlo simulations for testing whether a data set comes from a normal population when the mean and the variance are not specified but must be estimated.

The Anderson–Darling test is along the same philosophical lines as the KS test, but it gives more weight to the tails than the KS test does. In particular, Anderson and Darling (1952, 1954) proposed the use of the following statistic for testing normality:

$$AD = T \int_{-\infty}^{\infty} \frac{(\hat{F}(z) - F(z))^2}{F(z)(1 - F(z))} dF(z). \tag{3.35}$$

With $z_{(1)}, z_{(2)}, \cdots, z_{(T)}$ denoting the standardized residuals arranged in ascending order, a computationally more convenient expression is

$$AD = -T - T^{-1} \sum_{t=1}^{T} (2t - 1)(\log F(\hat{z}_{(t)}) + \log(1 - F(\hat{z}_{(T+1-t)}))). \tag{3.36}$$

A similar test is the Cramér–von Mises test (see also Csörgő and Faraway, 1996) based on the statistic

$$CM = (12T)^{-1} + \sum_{t=1}^{T} \left(F(\hat{z}_{(t)}) - \frac{2t-1}{2T} \right)^2. \tag{3.37}$$

Both the AD and CM test statistics remain valid when the parameters of the distribution have to be estimated. Analytic approximations for the test statistics are provided in Stephens (1970), computational details are available in Lewis (1961), Stephens (1986) and Thomakos and Wang (2003), and theoretical background properties can be found in Abrahamson (1967), Blackman (1958), Darling (1957), Doob (1949), Feller (1948), Gnedenko (1952), Gnedenko and Rvaceva (1952), Maag (1973), Rothman and Woodroofe (1972) and Stephens (1969).

3.3.2 Chi-square tests

Chi-square tests group the standardized residuals into g discrete classes, and their observed frequencies, n_i, for $i = 1, \cdots, g$, are compared to their respective expected frequencies, E_i, based on the assumed distribution. To test the null hypothesis H_0 that the underlying distribution $F(.)$ of a population is of the form $F_0(.)$, Pearson (1900) introduced a statistic measuring the closeness of the observed frequencies (n_i) to the frequencies expected under the null hypothesis (E_i) in g discrete cells, which followed a chi-square distribution. In particular, Pearson showed that, if no parameters are estimated from the data, the statistic

$$\chi^2_{(g)} = \sum_{i=1}^{g} \frac{(n_i - E_i)^2}{E_i},$$

(3.38)

is asymptotically chi-square distributed with $g - 1$ degrees of freedom. If the value of $\chi^2_{(g)}$ is higher than some critical value chosen to be an appropriate percentile of the χ^2_{g-1} distribution, the null hypothesis $H_0 : F(.) = F_0(.)$ is rejected.

Mann and Wald (1942) proposed determining the number of cells g using the formula $\tilde{g} = 4 \sqrt[5]{2(T-1)^2 \Phi_a^{-2}}$ where Φ_a is the $(1 - a)$-quantile of the standard normal distribution. Schorr (1974) took the view that fewer cells than $\tilde{g}$ are optimal and proposed using $2T^{0.4}$ cells. Moore (1986) suggested using between $\tilde{g}/2$ and $\tilde{g}$ cells. König and Gaab (1982) argued that the value of g must follow increases in the values of T at a rate equal to $T^{0.4}$.

If a set of parameters must be estimated, as will often be the case within our framework, the usual approach is the application of the chi-square test by reducing the number of degrees of freedom by the number of estimated parameters. However, Fisher (1924) noted that the type of estimation procedure used affects the outcome of the test.[7] As first stated by Fisher, the chi-square test can be applied even when the unknown parameters are replaced by their maximum likelihood estimators. In such a case, the distribution of the $\chi^2_{(g)}$ statistic is bounded between the χ^2_{g-1} and χ^2_{g-k-1} distributions, where k is the number of estimated parameters (see also Palm and Vlaar, 1997).

Normality tests are not restricted to those presented above. Numerous modifications have been suggested; see D'Agostino and Stephens (1986), D'Agostino et al. (1990), Shapiro and Francia (1972), Shapiro and Wilk (1965) and Shapiro et al. (1968).

3.3.3 QQ plots

The QQ graph plots the empirical quantiles of the series against the quantiles of the assumed distribution.[8] A linear QQ plot indicates an observed series that conforms to the hypothesized distribution. A QQ plot that falls on a straight line in its middle part, but curves upward at its left end and downward at its right end indicates a distribution that is leptokurtic and has thicker tails than the normal distribution.

---

[7] In such cases, the appropriate parameter estimation method to use is the minimization of the chi-square statistical function with respect to the parameters under estimation (minimum chi-square, or MSC, estimators).

[8] It is a common practice to refer to the plot of the quantiles of the data against the theoretical distribution as a probability (PP) plot, and to the plot of the quantiles of the data against the quantiles of a second data set as a QQ plot.

Table 3.4   FTSE100 and CAC40 equity indices: values of the $\chi^2_{(g)}$ statistic and p-values of the chi-square distribution with $g-1$ and $g-k-1$ degrees of freedom, for $g = 20, 30, 40$ and $k = 7$ for testing the null hypothesis that the standardized residuals are normally distributed

	FTSE100*					
	Full sample			Subsample		
g	$\chi^2_{(g)}$	χ^2_{g-1}	χ^2_{g-k-1}	$\chi^2_{(g)}$	χ^2_{g-1}	χ^2_{g-k-1}
20	95.4	0.000	0.000	30.0	0.052	0.003
30	145.8	0.000	0.000	38.7	0.107	0.015
40	204.8	0.000	0.000	53.0	0.066	0.011
	CAC40**					
	Full sample			Subsample		
g	$\chi^2_{(g)}$	χ^2_{g-1}	χ^2_{g-k-1}	$\chi^2_{(g)}$	χ^2_{g-1}	χ^2_{g-k-1}
20	41.8	0.002	0.000	17.0	0.589	0.149
30	52.0	0.006	0.000	39.7	0.089	0.012
40	62.2	0.010	0.001	52.5	0.073	0.013

*The full sample spans the period from April 1988 to April 2005. The subsample spans the period from January 1995 to December 1998.
**The full sample spans the period from July 1987 to June 2003. The subsample spans the period from January 1995 to December 1998.

3.3.4   Goodness-of-fit tests using EViews and G@RCH

EViews covers a set of EDF tests. When the parameters of the distribution are (assumed to be) known, EViews allows computation of the Kolmogorov, Kuiper, Cramér–von Mises, Watson and Anderson–Darling statistics. Computation when parameters need to be estimated is also provided for the last three test statistics. For normality testing, the value of the Lilliefors test statistic is yielded. The chi-square test for the standardized residuals is available in G@RCH.

As an illustration, suppose that we wish to test the assumption of normally distributed standardized residuals for the FTSE100 and CAC40 indices. We start by estimating the AR(0)-FIAPARCH(1,1) model using G@RCH for the full data set (from April 1988 to April 2005 for the case of FTSE100 and from July 1987 to June 2003 for the case of CAC40) as well as for the subsample corresponding to the period from January 1995 to December 1998.[9] The command line *garchobj.PEARSON (<10>)* computes the chi-square test for $g = 10$.

---

[9] The estimation is carried out by the program *chapter3.TestDiffSampleSizes.ox*, which is available on the accompanying CD-ROM.

Table 3.5   FTSE100 and CAC40 equity indices: values of the Lilliefors, Cramér–von Mises, Watson, and Anderson–Darling test statistics for testing the hypothesis that the standardized residuals are normally distributed

	FTSE100*			
	Full sample		Subsample	
Test statistic	Statistic value	*p*-value	Statistic value	*p*-value
Lilliefors	0.03	0.00	0.04	0.005
Cramér–von Mises	0.54	0.00	0.19	0.007
Watson	0.51	0.00	0.17	0.007
Anderson–Darling	3.09	0.00	1.06	0.009

	CAC40**			
	Full sample		Subsample	
Test statistic	Statistic value	*p*-value	Statistic value	*p*-value
Lilliefors	0.02	0.00	0.02	>0.1
Cramér–von Mises	0.45	0.00	0.09	0.14
Watson	0.40	0.00	0.08	0.17
Anderson–Darling	2.83	0.00	0.72	0.06

*The full sample spans the period from April 1988 to April 2005. The subsample spans the period from January 1995 to December 1998.
**The full sample spans the period from July 1987 to June 2003. The subsample spans the period from January 1995 to December 1998.

In order to compute the goodness-of-fit test statistics that are available in EViews, we also need to save the residuals and the in-sample estimates of the conditional variance. For example, the command line *garchobj.STORE(1,0,1,0,0,"ch3. cac40_full",0)* saves in Excel format (because of the value 0 in the last argument) the residuals (value 1 in the first argument) and the conditional variance (value 1 in the third argument). The saved series are named *res_ch3.ftse100_full.xls* and *condv_ch3. cac40_full.xls*, because of the name *ch3.cac40_full* in the sixth argument.

Table 3.4 presents values of the $\chi^2_{(g)}$ statistic[10] and the *p*-values of the chi-square distribution with $g - 1$ and $g - k - 1$ degrees of freedom, for $g = 20(10)40$ and $k = 7$ for testing the null hypothesis that the standardized residuals are normally distributed, while Table 3.5 presents corresponding results for the Lilliefors, Cramér–von Mises, Watson and Anderson–Darling tests.[11] The EViews workfile *chapter3.testdiffsam-plesizes.wf1* contains the standardized residuals. The computation of the tests is an option available in EViews on the View button of the series, e.g. View → Descriptive Statistics & Tests → Empirical Distribution Tests. ... The null hypothesis is rejected for both samples on the basis of the majority of the tests.

---

[10] Estimated using G@RCH.
[11] Estimated using EViews.

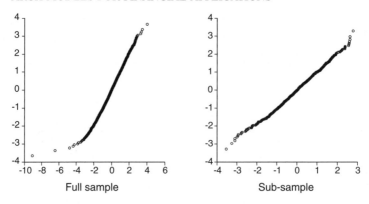

Figure 3.4   CAC40 equity index: The quantile–quantile plot of the standardized residuals of the AR(0)-FIAPARCH(1,1) model.

It should be noted at this point that the distributional form of the standardized residuals of a series may be judged by a test to conform to the normal distribution over one sample period, but not for another.[12] The data set on the standardized residuals of the CAC40 equity index represents a typical case. Two sample periods are considered (from July 1987 to January 2003 and from January 1995 to December 1998) for which the AR(0)-FIAPARCH(1,1) model is estimated. As reflected by Tables 3.4 and 3.5, the distribution of the standardized residuals is judged by the majority of the tests not to differ significantly from the normal on the basis of the subsample, but to be far from normal on the basis of the full sample. The situation is more clearly revealed by the corresponding QQ plot of the standardized residuals (Figure 3.4).

---

[12] In Section 4.1 we will see that the values of the estimated parameters may also differ as the sample changes.

Appendix

EViews 6

- *chapter3.autocorrelations.prg*

```
load chapter3.data.wf1
smpl @all
!length = 1000
series abssp500ret=@abs(sp500ret)
series absftse100ret=@abs(ftse100ret)
series absusukret=@abs(usukret)
series absgoldret=@abs(goldret)
series corr_abssp500ret
series corr_absftse100ret
series corr_absusukret
series corr_absgoldret
for !i=1 to !length
  corr_abssp500ret(!i)=@cor(abssp500ret,abssp500ret(-!i))
  corr_absftse100ret(!i)=@cor(absftse100ret,absftse100ret
  (-!i))
  corr_absusukret(!i)=@cor(absusukret,absusukret(-!i))
  corr_absgoldret(!i)=@cor(absgoldret,absgoldret(-!i))
  next
```

- *chapter3.stdresiduals.ar0fiaparch11.prg*

```
load chapter3.stdresiduals.ar0fiaparch11.wf1
smpl @all
!length = 1000
series absstdresiduals=@abs(stdresiduals)
series corr_absstdresiduals
for !i=1 to !length
  corr_absstdresiduals(!i)=@cor(absstdresiduals,
    absstdresiduals(-!i))
next
smpl 1 1000
series lower = -1.96/sqr(4437)
series upper = 1.96/sqr(4437)
group group_corr_ftse100 upper lower corr_absstdresiduals
```

G@RCH 4.2

- *chapter3.ftse100.ox*

```
#import <packages/Garch42/garch>
main()
```

```
{
{
decl garchobj;
garchobj = new Garch();
//*** DATA ***//
garchobj.Load("Chapter3Data.xls");
garchobj.Info();
garchobj.Select(Y_VAR, {"FTSE100RET",0,0});
//garchobj.Select(Z_VAR, {"NAME",0,0});//REGRESSOR IN THE
  VARIANCE
garchobj.SetSelSample(-1, 1, -1, 1);
//*** SPECIFICATIONS ***//
garchobj.CSTS(1,1);//cst in Mean (1 or 0), cst in Variance
  (1 or 0)
garchobj.DISTRI(0);//0 for Gauss, 1 for Student, 2 for GED,
  3 for Skewed-Student
garchobj.ARMA_ORDERS(0,0);//AR order (p), MA order (q).
garchobj.ARFIMA(0);//1 if Arfima wanted, 0 otherwise
garchobj.GARCH_ORDERS(1,1);//p order, q order
garchobj.ARCH_IN_MEAN(0);//ARCH-in-mean: 1 or 2 to add the
  variance or std. dev in the cond. mean

garchobj.MODEL(6);//0:RISKMETRICS,1:GARCH,2:EGARCH,
  3:GJR,4:APARCH,5:IGARCH,6:FIGARCH-BBM,7:FIGARCH-CHUNG,
  8:FIEGARCH,9:FIAPARCH-BBM,10:FIAPARCH-CHUNG,11:HYGARCH
garchobj.TRUNC(1000);//Truncation order(only F.I. models
  with BBM method)
//*** TESTS & FORECASTS ***//
garchobj.BOXPIERCE(<l10;15;20>);//Lags for the Box-Pierce
  Q-statistics, <> otherwise
garchobj.ARCHLAGS(<2;5;10>);//Lags for Engle's LM ARCH
  test, <> otherwise
garchobj.NYBLOM(1);//1 to compute the Nyblom stability test,
  0 otherwise
garchobj.SBT(1);//1 to compute the Sign Bias test, 0 otherwise
garchobj.PEARSON(<40;50;60>);//Cells (<40;50;60>) for the
  adjusted Pearson Chi-square Goodness-of-fit test, <>
  otherwise //G@RCH1.12
garchobj.RBD(<10;15;20>);//Lags for the Residual-Based
  Diagnostic test of Tse, <> otherwise
garchobj.FORECAST(0,15,0);//Arg.1 : 1 to launch the
  forecasting procedure, 0 otherwise, Arg.2 : Number of
  forecasts, Arg.3 : 1 to Print the forecasts, 0 otherwise
//*** OUTPUT ***//
garchobj.MLE(2);//0:MLE (Second derivatives), 1:MLE (OPG
```

```
   Matrix), 2:QMLE
garchobj.COVAR(0);//if 1, prints variance-covariance
   matrix of the parameters.
garchobj.ITER(0);//Interval of iterations between printed
   intermediary results (if no intermediary results wanted,
   enter '0')
garchobj.TESTS(0,1);//Arg. 1 : 1 to run tests PRIOR to
   estimation, 0 otherwise//Arg. 2 : 1 to run tests AFTER
   estimation, 0 otherwise
garchobj.GRAPHS(0,0,"");//Arg.1 : if 1, displays graphics of
   the estimations (only when using GiveWin). // Arg.2 : if 1,
   saves these graphics in a EPS file (OK with all Ox versions)
   // Arg.3 : Name of the saved file.
garchobj.FOREGRAPHS(0,0,"");//Same as GRAPHS(p,s,n) but for
   the graphics of the forecasts.
//*** PARAMETERS ***//
garchobj.BOUNDS(0);// 1 if bounded parameters wanted,
   0 otherwise
garchobj.FIXPARAM(0,<0;0;0;0;0;0>);// Arg.1 : 1 to fix some
   parameters to their starting values, 0 otherwise// Arg.2 : 1
   to fix (see garchobj.DoEstimation(<>)) and 0 to estimate the
   corresponding parameter
//*** ESTIMATION ***//
garchobj.MAXSA(1,5,0.5,20,5,2,1);// Arg.1 : 1 to use the
   MaxSA algorithm of Goffe, Ferrier and Rogers (1994) and
   implemented in Ox by Charles Bos// Arg.2 : dT=initial
   temperature // Arg.3 : dRt=temperature reduction factor
   // Arg.4 : iNS=number of cycles// Arg.5 : iNT=Number of
   iterations before temperature reduction// Arg.6 : vC=step
   length adjustment// Arg.7 : vM=step length vector used in
   initial step
garchobj.Initialization(<>);//m_vPar=m_clevel|m_vbetam|
   m_dARFI | m_vAR | m_vMA | m_calpha0 | m_vgammav | m_dD |
   m_vbetav |   m_valphav | m_vleverage | m_vtheta1 | m_vtheta2 |
   m_vpsy | m_ddelta | m_cA | m_cV | m_vHY | m_v_in_mean
garchobj.PrintStartValues(0);// 1: Prints the S.V. in a table
   form; 2: Individually;  3: in a Ox code to use in StartValues
garchobj.PrintBounds(1);
garchobj.DoEstimation(<>);
garchobj.Output();
delete garchobj;
}
{
decl garchobj;
garchobj = new Garch();
```

```
//*** DATA ***//
garchobj.Load("Chapter3Data.xls");
garchobj.Info();
garchobj.Select(Y_VAR, {"FTSE100RET",0,0});
//garchobj.Select(Z_VAR, {"NAME",0,0});//REGRESSOR IN THE
   VARIANCE
garchobj.SetSelSample(-1, 1, -1, 1);
//*** SPECIFICATIONS ***//
garchobj.CSTS(1,1);//cst in Mean (1 or 0), cst in Variance
   (1 or 0)
garchobj.DISTRI(0);//0 for Gauss, 1 for Student, 2 for GED,
   3 for Skewed-Student
garchobj.ARMA_ORDERS(0,0);//AR order (p), MA order (q).
garchobj.ARFIMA(0);//1 if Arfima wanted, 0 otherwise
garchobj.GARCH_ORDERS(1,1);//p order, q order
garchobj.ARCH_IN_MEAN(0);//ARCH-in-mean: 1 or 2 to add the
   variance or std. dev in the  cond. mean

garchobj.MODEL(7);//0:RISKMETRICS,1:GARCH,2:EGARCH,3:
   GJR,4:APARCH,5:IGARCH,6:FIGARCH-BBM,7:FIGARCH-CHUNG,8:
   FIEGARCH,9:FIAPARCH-BBM,10:FIAPARCH-CHUNG,11:HYGARCH
garchobj.TRUNC(1000);//Truncation order(only F.I. models
   with BBM method)
//*** TESTS & FORECASTS ***//
garchobj.BOXPIERCE(<10;15;20>);//Lags for the Box-Pierce
   Q-statistics, <> otherwise
garchobj.ARCHLAGS(<2;5;10>);//Lags for Engle's LM ARCH test,
   <> otherwise
garchobj.NYBLOM(1);//1 to compute the Nyblom stability test,
   0 otherwise
garchobj.SBT(1);//1 to compute the Sign Bias test, 0 otherwise
garchobj.PEARSON(<40;50;60>);//Cells (<40;50;60>) for the
   adjusted Pearson Chi-square Goodness-of-fit test, <>
   otherwise //G@RCH1.12
garchobj.RBD(<10;15;20>);//Lags for the Residual-Based
   Diagnostic test of Tse, <> otherwise
garchobj.FORECAST(0,15,0);//Arg.1 : 1 to launch the
   forecasting procedure, 0 otherwise, Arg.2 : Number of
   forecasts, Arg.3 : 1 to Print the forecasts, 0 otherwise
//*** OUTPUT ***//
garchobj.MLE(2);//0:MLE (Second derivatives), 1:MLE (OPG
   Matrix), 2:QMLE
garchobj.COVAR(0);//if 1, prints variance-covariance matrix
   of the parameters.
garchobj.ITER(0);//Interval of iterations between printed
```

```
intermediary results (if no intermediary results wanted,
enter '0')
garchobj.TESTS(0,1);//Arg. 1 : 1 to run tests PRIOR to
estimation, 0 otherwise // Arg. 2 : 1 to run tests AFTER
estimation, 0 otherwise
garchobj.GRAPHS(0,0,"");//Arg.1 : if 1, displays graphics of
the estimations (only when using GiveWin). // Arg.2 : if 1,
saves these graphics in a EPS file (OK with all Ox versions)
// Arg.3 : Name of the saved file.
garchobj.FOREGRAPHS(0,0,"");//Same as GRAPHS(p,s,n) but for
the graphics of the forecasts.
//*** PARAMETERS ***//
garchobj.BOUNDS(0);// 1 if bounded parameters wanted,
0 otherwise
garchobj.FIXPARAM(0,<0;0;0;0;0;0>);// Arg.1 : 1 to fix some
parameters to their starting values, 0 otherwise// Arg.2 : 1 to
fix (see garchobj.DoEstimation(<>)) and 0 to estimate the
corresponding parameter
//*** ESTIMATION ***//
garchobj.MAXSA(1,5,0.5,20,5,2,1);// Arg.1 : 1 to use the MaxSA
algorithm of Goffe, Ferrier and Rogers (1994) and implemented
in Ox by Charles Bos// Arg.2 : dT=initial temperature // Arg.3 :
dRt=temperature reduction factor // Arg.4 : iNS=number of
cycles// Arg.5 : iNT=Number of iterations before
temperature reduction// Arg.6 : vC=step length
adjustment// Arg.7 : vM=step length vector used in
initial step
garchobj.Initialization(<>);//m_vPar=m_clevel | m_vbetam |
m_dARFI | m_vAR | m_vMA | m_calpha0 | m_vgammav | m_dD |
m_vbetav |   m_valphav | m_vleverage | m_vtheta1 | m_vtheta2 |
m_vpsy | m_ddelta | m_cA | m_cV | m_vHY | m_v_in_mean
garchobj.PrintStartValues(0);// 1: Prints the S.V. in a table
form; 2: Individually;  3: in a Ox code to use in StartValues
garchobj.PrintBounds(1);
garchobj.DoEstimation(<>);
garchobj.Output();
delete garchobj;
}
{
decl garchobj;
garchobj = new Garch();
//*** DATA ***//
garchobj.Load("Chapter3Data.xls");
garchobj.Info();
garchobj.Select(Y_VAR, {"FTSE100RET",0,0});
```

```
//garchobj.Select(Z_VAR, {"NAME",0,0});//REGRESSOR IN THE
   VARIANCE
garchobj.SetSelSample(-1, 1, -1, 1);
//*** SPECIFICATIONS ***//
garchobj.CSTS(1,1);//cst in Mean (1 or 0), cst in Variance
   (1 or 0)
garchobj.DISTRI(0);//0 for Gauss, 1 for Student, 2 for GED,
   3 for Skewed-Student
garchobj.ARMA_ORDERS(0,0);//AR order (p), MA order (q).
garchobj.ARFIMA(0);//1 if Arfima wanted, 0 otherwise
garchobj.GARCH_ORDERS(1,1);//p order, q order
garchobj.ARCH_IN_MEAN(0);//ARCH-in-mean: 1 or 2 to add the
   variance or std. dev in the cond. mean
garchobj.MODEL(8);//0:RISKMETRICS,1:GARCH,2:EGARCH,3:
   GJR,4:APARCH,5:IGARCH,6:FIGARCH-BBM,7:FIGARCH-CHUNG,8:
   FIEGARCH,9:FIAPARCH-BBM,10:FIAPARCH-CHUNG,11:HYGARCH
garchobj.TRUNC(1000);//Truncation order(only F.I.
   models with BBM method)
//*** TESTS & FORECASTS ***//
garchobj.BOXPIERCE(<10;15;20>);//Lags for the Box-Pierce
   Q-statistics, <> otherwise
garchobj.ARCHLAGS(<2;5;10>);//Lags for Engle's LM ARCH test,
   <> otherwise
garchobj.NYBLOM(1);//1 to compute the Nyblom stability test,
   0 otherwise
garchobj.SBT(1);//1 to compute the Sign Bias test, 0 otherwise
garchobj.PEARSON(<40;50;60>);//Cells (<40;50;60>) for the
   adjusted Pearson Chi-square Goodness-of-fit test, <>
   otherwise //G@RCH1.12
garchobj.RBD(<10;15;20>);//Lags for the Residual-Based
   Diagnostic test of Tse, <> otherwise
garchobj.FORECAST(0,15,0);//Arg.1 : 1 to launch the
   forecasting procedure, 0 otherwise, Arg.2 : Number off
   forecasts, Arg.3 : 1 to Print the forecasts, 0 otherwise
//*** OUTPUT ***//
garchobj.MLE(2);//0:MLE (Second derivatives), 1:MLE
   (OPG Matrix), 2:QMLE
garchobj.COVAR(0);//if 1, prints variance-covariance
   matrix of the parameters.
garchobj.ITER(0);//Interval of iterations between printed
   intermediary results (if no intermediary results wanted,
   enter '0')
garchobj.TESTS(0,1);//Arg. 1 : 1 to run tests PRIOR to
   estimation, 0 otherwise // Arg. 2 : 1 to run tests AFTER
   estimation, 0 otherwise
```

```
garchobj.GRAPHS(0,0,"");//Arg.1 : if 1, displays graphics of
   the estimations (only when using GiveWin). // Arg.2 : if 1,
   saves these graphics in a EPS file (OK with all Ox versions)
   // Arg.3 : Name of the saved file.
garchobj.FOREGRAPHS(0,0,"");//Same as GRAPHS(p,s,n) but for
   the graphics of the forecasts.
//*** PARAMETERS ***//
garchobj.BOUNDS(0);// 1 if bounded parameters wanted,
   0 otherwise
garchobj.FIXPARAM(0,<0;0;0;0;0;0>);// Arg.1 : 1 to fix some
   parameters to their starting values, 0 otherwise//Arg.2 : 1 to
   fix (see garchobj.DoEstimation(<>)) and 0 to estimate the
   corresponding parameter
//*** ESTIMATION ***//
garchobj.MAXSA(1,5,0.5,20,5,2,1);//Arg.1 : 1 to use the MaxSA
   algorithm of Goffe, Ferrier and Rogers (1994) and implemented
   in Ox by Charles Bos// Arg.2 : dT=initial temperature
   // Arg.3 : dRt=temperature reduction factor // Arg.4 :
   iNS=number of cycles// Arg.5 : iNT=Number of iterations
   before temperature reduction// Arg.6 : vC=step length
   adjustment// Arg.7 : vM=step length vector used in
   initial step
garchobj.Initialization(<>);//m_vPar=m_clevel|m_vbetam|
   m_dARFI | m_vAR | m_vMA | m_calpha0 | m_vgammav | m_dD |
   m_vbetav | m_valphav | m_vleverage | m_vtheta1 | m_vtheta2 |
   m_vpsy | m_ddelta | m_cA | m_cV | m_vHY | m_v_in_mean
garchobj.PrintStartValues(0);// 1: Prints the S.V. in a table
   form; 2: Individually; 3: in a Ox code to use in StartValues
garchobj.PrintBounds(1);
garchobj.DoEstimation(<>);
garchobj.Output();
delete garchobj;
}
{
decl garchobj;
garchobj = new Garch();
//*** DATA ***//
garchobj.Load("Chapter3Data.xls");
garchobj.Info();
garchobj.Select(Y_VAR, {"FTSE100RET",0,0});
//garchobj.Select(Z_VAR, {"NAME",0,0});//REGRESSOR IN THE
   VARIANCE
garchobj.SetSelSample(-1, 1, -1, 1);
//*** SPECIFICATIONS ***//
garchobj.CSTS(1,1);//cst in Mean (1 or 0), cst in Variance
```

(1 or 0)

garchobj.DISTRI(0);//0 for Gauss, 1 for Student, 2 for GED,
3 for Skewed-Student

garchobj.ARMA_ORDERS(0,0);//AR order (p), MA order (q).

garchobj.ARFIMA(0);//1 if Arfima wanted, 0 otherwise

garchobj.GARCH_ORDERS(1,1);//p order, q order

garchobj.ARCH_IN_MEAN(0);//ARCH-in-mean: 1 or 2 to add the
variance or std. dev in the cond. mean

garchobj.MODEL(9);//0:RISKMETRICS,1:GARCH,2:EGARCH,3:
GJR,4:APARCH,5:IGARCH,6:FIGARCH-BBM,7:FIGARCH-CHUNG,8:
FIEGARCH,9:FIAPARCH-BBM,10:FIAPARCH-CHUNG,11:HYGARCH

garchobj.TRUNC(1000);//Truncation order(only F.I. models
with BDM method)

//*** TESTS & FORECASTS ***//

garchobj.BOXPIERCE(<10;15;20>);//Lags for the Box-Pierce
Q-statistics, <> otherwise

garchobj.ARCHLAGS(<2;5;10>);//Lags for Engle's LM ARCH test,
<> otherwise

garchobj.NYBLOM(1);//1 to compute the Nyblom stability test,
0 otherwise

garchobj.SBT(1);//1 to compute the Sign Bias test, 0 otherwise

garchobj.PEARSON(<40;50;60>);//Cells (<40;50;60>) for the
adjusted Pearson Chi-square Goodness-of-fit test, <>
otherwise //G@RCH1.12

garchobj.RBD(<10;15;20>);//Lags for the Residual-Based
Diagnostic test of Tse, <> otherwise

garchobj.FORECAST(0,15,0);//Arg.1 : 1 to launch the
forecasting procedure, 0 otherwise, Arg.2 : Number of
forecasts, Arg.3 : 1 to Print the forecasts, 0 otherwise

//*** OUTPUT ***//

garchobj.MLE(2);//0:MLE (Second derivatives), 1:MLE (OPG
Matrix), 2:QMLE

garchobj.COVAR(0);//if 1, prints variance-covariance matrix
of the parameters.

garchobj.ITER(0);//Interval of iterations between printed
intermediary results (if no intermediary results wanted,
enter '0')

garchobj.TESTS(0,1);//Arg. 1 : 1 to run tests PRIOR to
estimation, 0 otherwise // Arg. 2 : 1 to run tests AFTER
estimation, 0 otherwise

garchobj.GRAPHS(0,0,"");//Arg.1 : if 1, displays graphics of
the estimations (only when using GiveWin). // Arg.2 : if 1,
saves these graphics in a EPS file (OK with all Ox versions)

```
// Arg.3 : Name of the saved file.
garchobj.FOREGRAPHS(0,0,"");//Same as GRAPHS(p,s,n) but for
  the graphics of the forecasts.
//*** PARAMETERS ***//
garchobj.BOUNDS(0);// 1 if bounded parameters wanted,
  0 otherwise
garchobj.FIXPARAM(0,<0;0;0;0;0;0>);// Arg.1 : 1 to fix some
  parameters to their starting values, 0 otherwise//Arg.2 : 1 to
  fix (see garchobj.DoEstimation(<>)) and 0 to estimate the
  corresponding parameter
//*** ESTIMATION ***//
garchobj.MAXSA(1,5,0.5,20,5,2,1);// Arg.1 : 1 to use the MaxSA
  algorithm of Goffe, Ferrier and Rogers (1994) and implemented
  in Ox by Charles Bos// Arg.2 : dT=initial temperature
  // Arg.3 : dRt=temperature reduction factor // Arg.4 :
  iNS=number of cycles// Arg.5 : iNT=Number of iterations
  before temperature reduction// Arg.6 : vC=step length
  adjustment// Arg.7 : vM=step length vector used in
  initial step
garchobj.Initialization(<>);//m_vPar=m_clevel|m_vbetam|
  m_dARFI | m_vAR | m_vMA | m_calpha0 | m_vgammav | m_dD |
  m_vbetav |   m_valphav | m_vleverage | m_vtheta1 | m_vtheta2 |
  m_vpsy | m_ddelta | m_cA | m_cV | m_vHY | m_v_in_mean
garchobj.PrintStartValues(0);// 1: Prints the S.V. in a table
  form; 2: Individually; 3: in a Ox code to use in StartValues
garchobj.PrintBounds(1);
garchobj.DoEstimation(<>);
garchobj.Output();
delete garchobj;
}
{
decl garchobj;
garchobj = new Garch();
//*** DATA ***//
garchobj.Load("Chapter3Data.xls");
garchobj.Info();
garchobj.Select(Y_VAR, {"FTSE100RET",0,0});
//garchobj.Select(Z_VAR, {"NAME",0,0});//REGRESSOR IN THE
  VARIANCE
garchobj.SetSelSample(-1, 1, -1, 1);
//*** SPECIFICATIONS ***//
garchobj.CSTS(1,1);//cst in Mean (1 or 0), cst in Variance
  (1 or 0)
garchobj.DISTRI(0);//0 for Gauss, 1 for Student, 2 for GED,
```

```
     3 for Skewed-Student
garchobj.ARMA_ORDERS(0,0);//AR order (p), MA order (q).
garchobj.ARFIMA(0);//1 if Arfima wanted, 0 otherwise
garchobj.GARCH_ORDERS(1,1);//p order, q order
garchobj.ARCH_IN_MEAN(0);//ARCH-in-mean: 1 or 2 to add the
     variance or std. dev in the cond. mean

garchobj.MODEL(10);//0:RISKMETRICS,1:GARCH,2:EGARCH,3:
     GJR,4:APARCH,5:IGARCH,6:FIGARCH-BBM,7:FIGARCH-CHUNG,8:
     FIEGARCH,9:FIAPARCH-BBM,10:FIAPARCH-CHUNG,11:HYGARCH
garchobj.TRUNC(1000);//Truncation order(only F.I. models
     with BBM method)
//*** TESTS & FORECASTS ***//
garchobj.BOXPIERCE(<10;15;20>);//Lags for the Box-Pierce
     Q-statistics, <> otherwise
garchobj.ARCHLAGS(<2;5;10>);//Lags for Engle's LM ARCH test,
     <> otherwise
garchobj.NYBLOM(1);//1 to compute the Nyblom stability test,
     0 otherwise
garchobj.SBT(1);//1 to compute the Sign Bias test, 0 otherwise
garchobj.PEARSON(<40;50;60>);//Cells (<40;50;60>) for the
     adjusted Pearson Chi-square Goodness-of-fit test, <>
     otherwise //G@RCH1.12
garchobj.RBD(<10;15;20>);//Lags for the Residual-Based
     Diagnostic test of Tse, <> otherwise
garchobj.FORECAST(0,15,0);//Arg.1 : 1 to launch the
     forecasting procedure, 0 otherwise, Arg.2 : Number of
     forecasts, Arg.3 : 1 to Print the forecasts, 0 otherwise
//*** OUTPUT ***//
garchobj.MLE(2);//0:MLE (Second derivatives), 1:MLE (OPG
     Matrix), 2:QMLE
garchobj.COVAR(0);//if 1, prints variance-covariance matrix
     of the parameters.
garchobj.ITER(0);//Interval of iterations between printed
     intermediary results (if no intermediary results wanted,
     enter '0')
garchobj.TESTS(0,1);//Arg. 1 : 1 to run tests PRIOR to
     estimation, 0 otherwise // Arg. 2 : 1 to run tests AFTER
     estimation, 0 otherwise
garchobj.GRAPHS(0,0,"");//Arg.1 : if 1, displays graphics of
     the estimations (only when using GiveWin). // Arg.2 : if 1,
     saves these graphics in a EPS file (OK with all Ox versions)
     // Arg.3 : Name of the saved file.
garchobj.FOREGRAPHS(0,0,"");//Same as GRAPHS(p,s,n) but for
```

```
the graphics of the forecasts.
//*** PARAMETERS ***//
garchobj.BOUNDS(0);// 1 if bounded parameters wanted,
  0 otherwise
garchobj.FIXPARAM(0,<0;0;0;0;0;0>);// Arg.1 : 1 to fix some
  parameters to their starting values, 0 otherwise//Arg.2 : 1 to
  fix (see garchobj.DoEstimation(<>)) and 0 to estimate the
  corresponding parameter
//*** ESTIMATION ***//
garchobj.MAXSA(1,5,0.5,20,5,2,1);// Arg.1 : 1 to use the MaxSA
  algorithm of Goffe, Ferrier and Rogers (1994) and implemented
  in Ox by Charles Bos// Arg.2 : dT=initial temperature
  // Arg.3 : dRt=temperature reduction factor // Arg.4 :
  iNS=number of cycles// Arg.5 : iNT=Number of iterations
  before temperature reduction// Arg.6 : vC=step length
  adjustment// Arg.7 : vM=step length vector used in
  initial step
garchobj.Initialization(<>);//m_vPar = m_clevel | m_vbetam |
  m_dARFI | m_vAR | m_vMA | m_calpha0 | m_vgammav | m_dD |
  m_vbetav |  m_valphav | m_vleverage | m_vtheta1 | m_vtheta2 |
  m_vpsy | m_ddelta | m_cA | m_cV | m_vHY | m_v_in_mean
garchobj.PrintStartValues(0);// 1: Prints the S.V. in a table
  form; 2: Individually; 3: in a Ox code to use in StartValues
garchobj.PrintBounds(1);
garchobj.DoEstimation(<>);
garchobj.Output();
delete garchobj;
}
{
decl garchobj;
garchobj = new Garch();
//*** DATA ***//
garchobj.Load("Chapter3Data.xls");
garchobj.Info();
garchobj.Select(Y_VAR, {"FTSE100RET",0,0});
//garchobj.Select(Z_VAR, {"NAME",0,0});//REGRESSOR IN THE
  VARIANCE
garchobj.SetSelSample(-1, 1, -1, 1);
//*** SPECIFICATIONS ***//
garchobj.CSTS(1,1);//cst in Mean (1 or 0), cst in Variance
  (1 or 0)
garchobj.DISTRI(0);//0 for Gauss, 1 for Student, 2 for GED,
  3 for Skewed-Student
garchobj.ARMA_ORDERS(0,0);//AR order (p), MA order (q).
```

```
garchobj.ARFIMA(0);//1 if Arfima wanted, 0 otherwise
garchobj.GARCH_ORDERS(1,1);//p order, q order
garchobj.ARCH_IN_MEAN(0);//ARCH-in-mean: 1 or 2 to add the
  variance or std. dev in the  cond. mean

garchobj.MODEL(11);//0:RISKMETRICS,1:GARCH,2:EGARCH,3:
  GJR,4:APARCH,5:IGARCH,6:FIGARCH-BBM,7:FIGARCH-CHUNG,8:
  FIEGARCH,9:FIAPARCH-BBM,10:FIAPARCH-CHUNG,11:HYGARCH
garchobj.TRUNC(1000);//Truncation order(only F.I. models
  with BBM method)
//*** TESTS & FORECASTS ***//
garchobj.BOXPIERCE(<10;15;20>);//Lags for the Box-Pierce
  Q-statistics, <> otherwise
garchobj.ARCHLAGS(<2;5;10>);//Lags for Engle's LM ARCH test,
  <> otherwise
garchobj.NYBLOM(1);//1 to compute the Nyblom stability test,
  0 otherwise
garchobj.SBT(1);//1 to compute the Sign Bias test, 0 otherwise
garchobj.PEARSON(<40;50;60>);//Cells (<40;50;60>) for the
  adjusted Pearson Chi-square Goodness-of-fit test, <>
  otherwise //G@RCH1.12
garchobj.RBD(<10;15;20>);//Lags for the Residual-Based
  Diagnostic test of Tse, <> otherwise
garchobj.FORECAST(0,15,0);//Arg.1 : 1 to launch the
  forecasting procedure, 0 otherwise, Arg.2 : Number of
  forecasts, Arg.3 : 1 to Print the forecasts, 0 otherwise
//*** OUTPUT ***//
garchobj.MLE(2);//0:MLE (Second derivatives), 1:MLE (OPG
  Matrix), 2:QMLE
garchobj.COVAR(0);//if 1, prints variance-covariance
  matrix of the parameters.
garchobj.ITER(0);//Interval of iterations between printed
  intermediary results (if no intermediary results wanted,
  enter '0')
garchobj.TESTS(0,1);//Arg. 1 : 1 to run tests PRIOR to
  estimation, 0 otherwise // Arg. 2 : 1 to run tests AFTER
  estimation, 0 otherwise
garchobj.GRAPHS(0,0,"");//Arg.1 : if 1, displays graphics of
  the estimations (only when using GiveWin). // Arg.2 : if 1,
  saves these graphics in a EPS file (OK with all Ox versions)
  // Arg.3 : Name of the saved file.
garchobj.FOREGRAPHS(0,0,"");//Same as GRAPHS(p,s,n) but for
  the graphics of the forecasts.
//*** PARAMETERS ***//
garchobj.BOUNDS(0);// 1 if bounded parameters wanted,
```

```
  0 otherwise
garchobj.FIXPARAM(0,<0;0;0;0;0;0>);// Arg.1 : 1 to fix some
  parameters to their starting values, 0 otherwise// Arg.2 : 1 to
  fix (see garchobj.DoEstimation(<>)) and 0 to estimate the
  corresponding parameter
//*** ESTIMATION ***//
garchobj.MAXSA(1,5,0.5,20,5,2,1);// Arg.1 : 1 to use the MaxSA
  algorithm of Goffe, Ferrier and Rogers (1994) and implemented
  in Ox by Charles Bos// Arg.2 : dT=initial temperature
  // Arg.3 : dRt=temperature reduction factor // Arg.4 :
  iNS=number of cycles// Arg.5 : iNT=Number of iterations
  before temperature reduction// Arg.6 : vC=step length
  adjustment// Arg.7 : vM=step length vector used in
  initial step
garchobj.Initialization(<>);//m_vPar = m_clevel | m_vbetam |
  m_dARFI | m_vAR | m_vMA | m_calpha0 | m_vgammav | m_dD |
  m_vbetav |  m_valphav | m_vleverage | m_vtheta1 | m_vtheta2 |
  m_vpsy | m_ddelta | m_cA | m_cV | m_vHY | m_v_in_mean
garchobj.PrintStartValues(0);// 1: Prints the S.V. in a table
  form; 2: Individually; 3: in a Ox code to use in StartValues
garchobj.PrintBounds(1);
garchobj.DoEstimation(<>);
garchobj.Output();
delete garchobj;
}
}
```

4

Volatility forecasting: an empirical example using EViews 6

In this chapter we will obtain volatility forecasts in terms of four ARCH specifications on the basis of the data set on the S&P500 equity index for the period from 4 April 1988 to 5 April 2005 considered in Chapter 2. We will look, in particular, into generating one-step-ahead and more-than-one-step-ahead volatility forecasts using EViews. The package provides the estimation of ARCH models with GARCH(p, q), IGARCH(p, q), EGARCH(p, q), APARCH(p, q), GRJ(p, q), CGARCH(1,1) and ACGARCH(1,1) volatility specifications, for $p = 1, \ldots, 9$ and $q = 0, \ldots, 9$.

4.1 One-step-ahead volatility forecasting

Using a rolling sample of constant size $\tilde{T} = 3000$, we will generate $\tilde{\tilde{T}} = 1435$[1] one-step-ahead volatility forecasts on the basis of the above observed series of the S&P500 index in terms of four ARCH models in framework (1.6) with conditional mean $\mu_t = c_0$:

$$
\begin{aligned}
y_t &= c_0 + \varepsilon_t, \\
\varepsilon_t &= \sigma_t z_t, \\
\sigma_t^2 &= g(\theta | I_{t-1}), \\
z_t &\overset{i.i.d.}{\sim} N(0, 1),
\end{aligned}
\qquad (4.1)
$$

---

[1] $T = \tilde{T} + \tilde{\tilde{T}}$.

ARCH Models for Financial Applications   Evdokia Xekalaki and Stavros Degiannakis
© 2010 John Wiley & Sons, Ltd

and Engle's (1982) ARCH(q), Bollerslev's (1986) GARCH(p, q), Nelson's (1991) EGARCH(p, q), and Glosten's et al. (1993) GJR(p, q) conditional volatility specifications, for $p = 1, q = 1, 2$.[2] The models are estimated using the EViews package in the following forms:

ARCH(q),

$$\sigma_t^2 = a_0 + \sum_{i=1}^{q}(a_i \varepsilon_{t-i}^2); \qquad (4.2)$$

GARCH(p, q),

$$\sigma_t^2 = a_0 + \sum_{i=1}^{q}(a_i \varepsilon_{t-i}^2) + \sum_{j=1}^{p}(b_j \sigma_{t-j}^2); \qquad (4.3)$$

EGARCH(p, q),

$$\log(\sigma_t^2) = a_0 + \sum_{i=1}^{q}\left(a_i \left|\frac{\varepsilon_{t-i}}{\sigma_{t-i}}\right| + \gamma_i \left(\frac{\varepsilon_{t-i}}{\sigma_{t-i}}\right)\right) + \sum_{j=1}^{p}(b_j \log(\sigma_{t-j}^2)); \qquad (4.4)$$

and GJR(p, q),

$$\sigma_t^2 = a_0 + \sum_{i=1}^{q}(a_i \varepsilon_{t-i}^2) + \gamma_1 d(\varepsilon_{t-1} < 0)\varepsilon_{t-1}^2 + \sum_{j=1}^{p}(b_j \sigma_{t-j}^2), \qquad (4.5)$$

with $d(\varepsilon_t < 0) = 1$ if $\varepsilon_t < 0$ and $d(\varepsilon_t < 0) = 0$ otherwise.[3]

σ_{t+1}^2 is the true but unobservable value of the conditional variance at time $t + 1$. Hitherto, we have dealt with obtaining an estimate $\hat{\sigma}_{t+1}^2$ of σ_{t+1}^2, based on the entire available data set T. In what follows, we will look into obtaining an estimate of the one-step-ahead (or out-of-sample) conditional variance, $\sigma_{t+1|t}^2$, based on any information that is available up to time t. The in-sample and the out-of-sample conditional variance estimators are quite different. As an illustration of this, consider the GARCH (1,1) model. The estimator of the in-sample conditional variance in this case, given by $\hat{\sigma}_{t+1}^2 = a_0^{(T)} + a_1^{(T)}\hat{\varepsilon}_t^2 + b_1^{(T)}\hat{\sigma}_t^2$, with parameters a_0, a_1, b_1 estimated on the basis of entire data set T, depends on previous values of it and on innovations, which belong to I_t. The corresponding out-of-sample conditional variance estimator is given by $\sigma_{t+1|t}^2 = a_0^{(t)} + a_1^{(t)}\varepsilon_{t|t}^2 + b_1^{(t)}\sigma_{t|t}^2$. Not only does the conditional variance depend on values of it and on innovations that belong to I_t, but its parameters are estimated based on information that also belongs to I_t.

The one-step-ahead conditional variance of the GARCH(p, q) model is given by

$$\sigma_{t+1|t}^2 = a_0^{(t)} + \sum_{i=1}^{q}\left(a_i^{(t)}\varepsilon_{t-i+1|t}^2\right) + \sum_{j=1}^{p}(b_j^{(t)}\sigma_{t-j+1|t}^2). \qquad (4.6)$$

For $p = 0$, the ARCH(q) specification is obtained. The corresponding forecast recursion relationship associated with the EGARCH(p, q) model is

---

[2] In total, eight models are defined.

[3] In EViews 4, the parameter γ_i is fixed for $i = 1$. The addition of more lags of the asymmetric component is available in EViews 5 or later.

$$\log(\sigma^2_{t+1|t}) = a_0^{(t)} + \sum_{i=1}^{q}\left(a_i^{(t)}\left|\frac{\varepsilon_{t-i+1|t}}{\sigma_{t-i+1|t}}\right| + \gamma_i^{(t)}\left(\frac{\varepsilon_{t-i+1|t}}{\sigma_{t-i+1|t}}\right)\right) + \sum_{j=1}^{p}(b_j^{(t)}\log(\sigma^2_{t-j+1|t})),$$

$$(4.7)$$

while that associated with the GJR(p, q) model is

$$\sigma^2_{t+1|t} = a_0^{(t)} + \sum_{i=1}^{q}(a_i^{(t)}\varepsilon^2_{t-i+1|t}) + \gamma_1^{(t)}d(\varepsilon_{t|t} < 0)\varepsilon^2_{t|t} + \sum_{j=1}^{p}(b_j^{(t)}\sigma^2_{t-j+1|t}). \quad (4.8)$$

The usual approach adopted in the literature, mainly to minimize computation costs, is to calculate volatility forecasts by estimating the models once (see Hansen and Lunde, 2005a; Klaassen, 2002; Vilasuso, 2002). For greater accuracy, Billio and Pelizzon (2000) considered re-estimating the model parameters every 50 trading days. In our case, the models will be re-estimated every trading day, thus aiming at a closer to real world forecast accuracy.

The core program that carries out the necessary computations for the GARCH (1,1) model is as follows:

```
load chapter4.data.wf1
smpl @all
!ssize = 3000
!length = 4435
matrix(!length-!ssize,9) estg11_1
for !i = 1 to !length-!ssize
    statusline !i
    smpl !i !i+!ssize-1
    equation tempg11_1.arch(1,1,m=10000,h) sp500ret c
    tempg11_1.makegarch garchg11_1
    tempg11_1.makeresid resg11_1
    !ht = garchg11_1(!i+!ssize-1)
    !c1 = @coefs(1)
    !c2 = @coefs(2)
    !c3 = @coefs(3)
    !c4 = @coefs(4)
    !h_hat=!c2+(!c3*(resg11_1(!ssize+!i-1)^2))+(!c4*!ht)
    !y_hat = !c1
    !res = sp500ret(!ssize + !i) - !y_hat
    !r = (!res^2) / (!h_hat)
    estg11_1(!i,1) = !y_hat
    estg11_1(!i,2) = !h_hat
    estg11_1(!i,3) = !res
    estg11_1(!i,4) = !r
    estg11_1(!i,5) = sp500ret(!ssize + !i)
    estg11_1(!i,6) = !c1
```

```
      estg11_1(!i,7) = !c2
      estg11_1(!i,8) = !c3
      estg11_1(!i,9) = !c4
next
```

The full version of the program named *chapter4.forecast.prg* is available in the Appendix to this chapter.[4] The variable *!ssize* assigns the size of the rolling sample, ($\tilde{T} = 3000$). The difference *!length-!ssize* defines the number of one-step-ahead forecasts that will be computed ($\tilde{T} = 1435$). The matrix *estg11_1* contains 1435 rows and nine columns. The first row contains the estimates from the sample of the first 3000 observations, the second those from the sample of the next 3000 observations and so on up to the 1435th row, where the estimates from the sample of observations 1435 to 4434 are saved. The first column contains the one-step-ahead conditional mean forecast, or $y_{t+1|t} = c_0^{(t)}$, the second column the one-step-ahead conditional variance, or $\sigma_{t+1|t}^2 = a_0^{(t)} + a_1^{(t)} \varepsilon_{t|t}^2 + b_1^{(t)} \sigma_{t|t}^2$, and the third column the one-step-ahead residuals, or $\varepsilon_{t+1|t} = y_{t+1} - y_{t+1|t}$. The squared standardized residuals, $\varepsilon_{t+1|t}^2 / \sigma_{t+1|t}^2$, and the log-returns, y_{t+1}, are saved in columns four and five, respectively. Finally, in the last four columns, the values of the estimated parameters, $c_0^{(t)}, a_0^{(t)}, a_1^{(t)}, b_1^{(t)}$, are saved.

Figure 4.1 presents the one-day-ahead conditional variance forecasts of the ARCH(1), GARCH(1,1) and GJR(1,1) models. As revealed by the graphs, the ARCH(1) model yields different volatility forecasts from the other two models during the entire period under investigation. For a clearer visual comparison of the volatility forecasts of all models, Figure 4.2 provides the forecasts for a shorter period (the last six months of 2002). Similarly, the models corresponding to values of q other than 1 provide forecasts which are visually indistinguishable. Note that the conditional variance forecasts of the GARCH(1,1) and GARCH(1,2) processes are visually distinguishable in a few cases, and that the GJR(1,1) and GJR(1,2) processes have quite similar volatility forecasts. Only the EGARCH(1,1) and EGARCH(1,2) models yield clearly different volatility forecasts.

---

[4] Note a slight difference in estimating asymmetric models for $q > 1$, in EViews 6 or earlier versions. Consider, for example, the GJR(1,2) model. In EViews 6 the coefficients are saved in the order: *GARCH =* $C(2) + C(3)^*RESID(-1)^\wedge 2 + C(4)^*RESID(-1)^\wedge 2^*(RESID(-1)<0) + C(5)^*RESID(-2)^\wedge 2 + C(6)^*GARCH(-1)$, whereas in EViews 5 or in earlier versions the order is: *GARCH =* $C(2) + C(3)^*RESID(-1)^\wedge 2 + C(4)^*RESID(-2)^\wedge 2 + C(5)^*RESID(-1)^\wedge 2^*(RESID(-1)<0) + C(6)^*GARCH(-1)$. Coefficients $C(4)$ and $C(5)$ represent different parameters of the model. The same holds in the case of the EGARCH model. Consider, for example, the EGARCH(0,2) model. In EViews 6 and 5, the coefficients are saved in the order: *LOG(GARCH) =* $C(2) + C(3)^*ABS(RESID(-1)/@SQRT(GARCH(-1))) + C(4)^*ABS(RESID(-2)/@SQRT(GARCH(-1))) + C(5)^*RESID(-1)/@SQRT(GARCH(-1)) + C(6)^*RESID(-2)/@SQRT(GARCH(-2))$, whereas in EViews 4 or earlier the order is: *LOG(GARCH) =* $C(2) + C(3)^*ABS(RESID(-1)/@SQRT(GARCH(-1))) + C(4)^*RESID(-1)/@SQRT(GARCH(-1)) + C(5)^*ABS(RESID(-2)/@SQRT(GARCH(-2))) + C(6)^*RESID(-2)/@SQRT(GARCH(-2))$.

The reader who wishes to run the code in an earlier version than EViews 6 can reverse the order of the coefficients in the program *chapter4.forecast.prg* for the GJR(1,2) and EGARCH(1,2) models. For the EGARCH(1,2) and the GJR(1,2) models, for example, replace *!c4=@coefs(5) !c5=@coefs(4)* by *! c4=@coefs(4) !c5=@coefs(5)*.

In Chapter 9, the program named *chapter9.ARCH_with_vix.prg* computes the forecasts based on the command *forecast*. In such a case, one does not need to worry about the specification of the models.

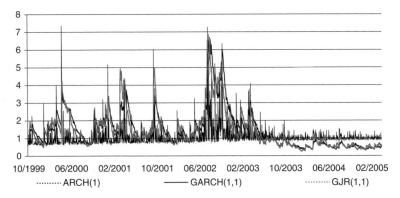

Figure 4.1   One-step-ahead conditional variance forecasts for the S&P500 equity index yielded by the ARCH(1), GARCH(1,1) and GJR(1,1) models on the basis of the observed series from 4 April 1998 to 5 April 2005.

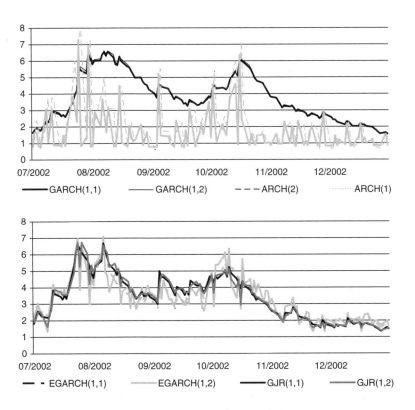

Figure 4.2   One-step-ahead conditional variance forecasts for the S&P500 equity index yielded by the eight models on the basis of the observed series over the period from 1 July 2002 to 31 December 2002.

Table 4.1   S&P500 equity index: values of the AIC, SBC, $\bar{\Psi}_{(SE)(i)}$ and $\bar{\Psi}_{(SPEC)(i)}$ used as evaluation criteria, for models $i = 1, \ldots, 8$

Model	AIC	SBC	$\bar{\Psi}_{(SE)(i)}$	$\bar{\Psi}_{(SPEC)(i)}$
ARCH(1)	2.785	2.789	9.110	1.480
ARCH(2)	2.739	2.745	8.683	1.404
GARCH(1,1)	2.604	2.610	8.125	1.030
GARCH(1,2)	2.605	2.612	8.117	1.030
EGARCH(1,1)	2.578	2.586	7.616	1.010
EGARCH(1,2)	**2.568**	**2.578**	**7.595**	**0.991**
GJR(1,1)	2.588	2.595	7.675	1.029
GJR(1,2)	2.585	2.594	7.634	1.033

Best-performing models are shown in bold face.

In Section 2.3, a first attempt was made to deal with the topic of selecting the most adequate model in terms of criteria based on the goodness of fit of the candidate models. So, if we were at time T and wished to select the most appropriate of the above eight candidate models having been estimated based on all of the available data (4435 observations), we could proceed by evaluating the ability of the models to fit the data. Then, on the basis of the values of the SBC and AIC information criteria, given in Table 4.1, it appears that one would choose the EGARCH(1,2) model. In this section, we will look into the question of model selection on the basis of the forecasting performance of the candidate models.

To this end, we consider using as criteria the average squared distance between the one-day-ahead conditional variance and the squared log-returns:[5]

$$\bar{\Psi}_{(SE)(i)} = \tilde{T}^{-1} \sum_{t=1}^{\tilde{T}} (\sigma_{t+1|t(i)}^2 - y_{t+1}^2)^2, \tag{4.9}$$

or the value of the standardized prediction error criterion (SPEC) based on the average of the squared standardized one-day-ahead residuals:[6]

$$\bar{\Psi}_{(SPEC)(i)} = \tilde{T}^{-1} \sum_{t=1}^{\tilde{T}} (\varepsilon_{t+1|t(i)} / \sigma_{t+1|t(i)})^2, \tag{4.10}$$

where $\varepsilon_{t+1|t(i)}$ and $\sigma_{t+1|t(i)}$ are the one-day-ahead residuals and one-day-ahead conditional standard deviation, respectively, associated with model i, $i = 1, \ldots, 8$. On the basis of the accuracy of the one-step predictions yielded by the models as this is reflected by the values of criteria 4.9 and 4.10, the EGARCH(1,2) model appears to remain the best-performing model among the eight competitors. Figure 4.3 presents the

---

[5] The criterion will remain qualitatively the same if, instead of log-returns, one-day-ahead residuals are used in analogy to equation (2.52). However, the result remains qualitatively the same.

[6] We will deal with the SPEC loss function (4.10) in greater detail in Chapter 10. For a given model, e.g. for the GARCH(1,1), the value of this function as given by (4.10) can be computed by clicking on matrix *estg11_1* in the View menu and then choosing Descriptive Statistics by Column. The mean in the fourth column is the average of the squared standardized one-day-ahead residuals.

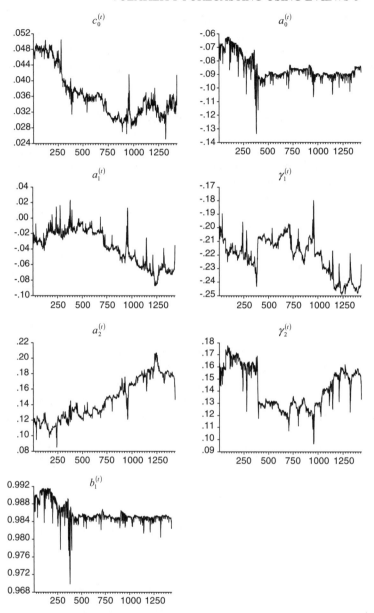

Figure 4.3   The 1435 estimates of parameters of the EGARCH(1,2) model on the basis of a rolling sample of 3000 observations on the S&P500 equity index.

1435 rolling sample based estimates of parameters of the EGARCH(1,2) model, for the period from 5 October 1999 to 4 April 2005. A detailed empirical study of rolling sample based estimates of parameters of an EGARCH model for the DAX30, FTSE100 and S&P500 financial indices has been made by Degiannakis et al. (2008). It was noted that the values of the estimated parameters change over time, reflecting the fact that structural properties and trading behaviour alter over time. It was also noted that the time-varying performance holds in the case of a generated ARCH data process, revealing that even in that case the rolling sample estimates of the parameters are time-varying.

4.2   Ten-step-ahead volatility forecasting

In this section we will obtain more-than-one-step-ahead forecasts of volatility of the S&P500 index on the basis of a parameter vector estimated at a given point in time t in terms of the same data set as in the previous section. For the sake of illustration, we will discuss the ten-day-ahead case.

More-than-one-step-ahead forecasts can be computed by repeated substitution. The forecast recursion relation of the GARCH(p, q) model is

$$
\sigma^2_{t+\tau|t} = a_0^{(t)} + \underbrace{\sum_{i=1}^{q} \left(a_i^{(t)} \sigma^2_{t-i+\tau|t} \right)}_{\text{for } i<\tau} + \underbrace{\sum_{i=\tau}^{q} \left(a_i^{(t)} \varepsilon^2_{t-i+\tau|t} \right)}_{\text{for } i\geq\tau} + \sum_{j=1}^{p} \left(b_j^{(t)} \sigma^2_{t-j+\tau|t} \right).
$$

$$(4.11)$$

For $t < \tau$, the forecast of the predictive error ε_τ conditional on information available at time t is equal to its zero expected value, $E(\varepsilon_s|I_t) = 0$. On the other hand, the estimated value of ε_τ^2 measured at time t should be equal to $\sigma^2_{\tau|t}$ for $t < \tau$. For $t \geq \tau$, the predictive error and its square are computed by the model with the available information at time t. The forecast recursion relationship associated with the EGARCH(p, q) model is

$$
\log\left(\sigma^2_{t+\tau|t} \right) = a_0^{(t)} + \underbrace{\sum_{i=\tau}^{q} \left(a_i^{(t)} \left| \frac{\varepsilon_{t-i+\tau|t}}{\sigma_{t-i+\tau|t}} \right| + \gamma_i^{(t)} \left(\frac{\varepsilon_{t-i+\tau|t}}{\sigma_{t-i+\tau|t}} \right) \right)}_{\text{for } i\geq\tau}
$$

$$
+ \sqrt{\frac{2}{\pi}} \underbrace{\sum_{i=1}^{q} \left(a_i^{(t)} \right)}_{\text{for } i<\tau} + \sum_{j=1}^{p} (b_j^{(t)} \log(\sigma^2_{t-j+\tau|t})), \qquad (4.12)
$$

where $E\left| \varepsilon_{t-i+\tau|t} \sigma^{-1}_{t-i+\tau|t} \right| = \sqrt{2/\pi}$ for $i < \tau$, as $E|z_t| = \sqrt{2/\pi}$ under the assumption that $z_t \sim N(0, 1)$. That associated with the GJR(p, q) model is given by

$$
\sigma^2_{t+\tau|t} = a_0^{(t)} + \underbrace{\sum_{i=1}^{q} \left(a_i^{(t)} \sigma^2_{t-i+\tau|t} \right)}_{\text{for } i<\tau} + \underbrace{\sum_{i=1}^{q} \left(a_i^{(t)} \varepsilon^2_{t-i+\tau|t} \right)}_{\text{for } i\geq\tau} + \gamma_1^{(t)} E(d(\varepsilon_t < 0)) \sigma^2_{t-1+\tau|t}
$$

$$
+ \sum_{j=1}^{p} \left(b_j^{(t)} \sigma^2_{t-j+\tau|t} \right). \qquad (4.13)
$$

Here, $E(d(\varepsilon_t<0))$ denotes the percentage of negative innovations out of all innovations. Under the assumption of normally distributed innovations, the expected number
of negative shocks is equal to the expected number of positive shocks, or
$E(d(\varepsilon_t<0))=0.5$.

The core of a program that carries out the necessary computations for the GJR(1,2)
process is the following:[7]

```
load chapter4.one.day.ahead.wf1
smpl @all
!ssize = 3000
!length = 4435
!models =8
!pi = @acos(-1)
!ahead2=10
'_____
matrix(!length-!ssize,!models) volatility{!ahead2}
for !ahead=!ahead2 to !length-!ssize
statusline !ahead
vector(!ahead2) h
h(1)= estt12_1(!ahead-!ahead2+1,2)
    h(2)=estt12_1(!ahead-!ahead2+1,7)+(estt12_1(!ahead-
    !ahead2+1,8)+estt12_1(!ahead-!ahead2+1,10)*estt12_1
    (!ahead-!ahead2+1,12)+estt12_1(!ahead-!ahead2+1,11))
    *estt12_1(!ahead-!ahead2+1,2)+estt12_1(!ahead-
    !ahead2+1,9)*estt12_1(!ahead-!ahead2+1,13)^2
for !r=3 to !ahead2
    h(!r)=estt12_1(!ahead-!ahead2+1,7)+(estt12_1(!ahead-
    !ahead2+1,8)+estt12_1(!ahead-!ahead2+1,10)*estt12_1
    (!ahead-!ahead2+1,12)+estt12_1(!ahead-!ahead2+1,11))
    *h(!r-1)+estt12_1(!ahead-!ahead2+1,9)*h(!r-2)
next
volatility{!ahead2}(!ahead,8)=h(!ahead2)
delete h
next
'_____
series Psi{!ahead2}
for !j=!ahead2 to !length-!ssize
Psi{!ahead2}(!j) = (volatility{!ahead2}(!j,8) - sp500ret
    (!ssize+!j)^2)^2
next
vector(1) averPsi{!ahead2}
averPsi{!ahead2}(1) = @mean(Psi{!ahead2})
```

[7] The full version of the program named *chapter4.forecast10.prg* is available in the Appendix to this
chapter.

The workfile after estimating the one-step-ahead forecasts in the previous section has been saved as *chapter4.one.day.ahead.wf1*. So, in the first line of the core program this workfile is loaded as it contains the rolling sample based estimates of parameters. The total sample size (*!length*) and size of the rolling sample (*!ssize*) are $T = 4435$ and $\tilde{T} = 3000$, respectively. The variable *!ahead2* defines the value of k for which k-step-ahead forecasts will be computed. The matrix *volatility10* has 1435 rows and eight columns.

The first nine rows remain unchanged (they were filled with zeros). The tenth row contains the ten-day-ahead volatility forecasts obtained by the eight models on the basis of the sample of the first 3000 observations. The next row contains those obtained on the basis of the next 3000 observations, and so on up to the 1435th row that contains the ten-day-ahead volatility forecasts obtained on the basis of the last 3000 sample values from the 1426th to the 4425th observation.

Therefore, the columns of the matrix *volatility10* contain the ten-day-ahead conditional variance forecasts obtained from the eight models. The eighth column, in particular, contains the ten-day-ahead conditional variance forecasts from the GJR (1,2) model. In the program presented above only the forecasts obtained by this model are computed, but in the full version of the program that is available in the Appendix to this chapter, each of the eight columns contains the conditional variance forecasts of one of the eight competing models. In each iteration of the loop *for !ahead=!ahead2 to !length-!ssize* the vector *h* is created. The vector *h* has 10 rows. In each row τ, the τ-step-ahead forecasts are saved. Thus, the first cell of *h* contains next-day's conditional variance forecast, $\sigma^2_{t+1|t} = a_0^{(t)} + (a_1^{(t)} + \gamma_1^{(t)} d(\varepsilon_{t|t} < 0)) \varepsilon^2_{t|t} + a_2^{(t)} \varepsilon^2_{t-1|t} + b_1^{(t)} \sigma^2_{t|t}$, up to the tenth cell, where the ten-day-ahead forecast $\sigma^2_{t+10|t} = a_0^{(t)} + (a_1^{(t)} + \gamma_1^{(t)} E(d(\varepsilon_{t|t} < 0)) + b_1^{(t)}) \sigma^2_{t+9|t} + a_2^{(t)} \sigma^2_{t+8|t}$ is saved. In each iteration i ($i = 1, \ldots, \tilde{T}$), $\sigma^2_{t+10|t}$ (the tenth row of the *h* vector) is saved in row i of the matrix *volatility10* and the *h* vector is deleted. Finally, in each row of series *psi10*, the value of $\Psi_{(SE)t+10} = (\sigma^2_{t+10|t} - y^2_{t+10})^2$ is kept, and *averPsi* keeps the $\bar{\Psi}_{(SE)} = 1435^{-1} \sum_{t=1}^{1435} \Psi_{(SE)t}$.

The workfile after estimating the ten-step-ahead forecasts has been saved as *chapter4.ten.days.ahead.wf1*. Figure 4.4 presents the ten-day-ahead conditional variance forecasts of the eight models for the period from 18 October 1999 to 4 April 2005. The ARCH(1) and ARCH(2) models provide different volatility forecasts than their competitors. As in the case of the one-day-ahead forecasts, similar models with different order of q provide forecasts which are visually indistinguishable.

In Table 4.2 the average squared distance between ten-day-ahead conditional variance and squared log-returns is computed:

$$\bar{\Psi}_{(SE)(i)} = \tilde{T}^{-1} \sum_{t=1}^{\tilde{T}} (\sigma^2_{t+10|t(i)} - y^2_{t+10})^2, \tag{4.14}$$

for each model i, for $i = 1, \ldots, 8$. Based on the accuracy of the ten-day-ahead predictions, the EGARCH(1,1) model appears to provide the most accurate volatility predictions.

It should be noted at this point that the evaluation was considered for overlapping forecasts as in Brooks and Persand (2003a) and Degiannakis and Xekalaki (2007a). Statistical inference is less complex for non-overlapping time periods, because of the absence of serially correlated observations. Engle et al. (1997) were the first to discuss

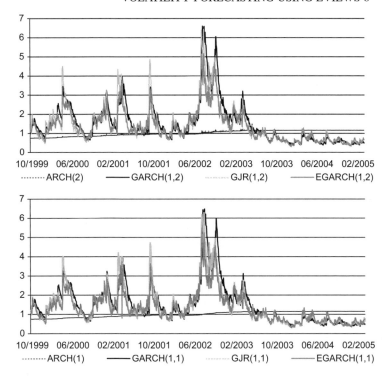

Figure 4.4   Ten-day-ahead conditional variance forecasts for the S&P500 equity index yielded by the eight models for the period from 18 October 1999 to 4 April 2005.

the evaluation of standard errors for overlapping returns. They constructed statistics that were robust to the presence of serial correlation based on the methods of Hansen and Hodrick (1980) and Richardson and Stock (1989); see also Bod et al. (2002) and Campbell et al. (1997).

Table 4.2   S&P500 equity index: average squared distance between the ten-day-ahead conditional variance yielded by the eight models and squared log-returns, $\bar{\Psi}_{(SE)(i)}$

Model	$\bar{\Psi}_{(SE)(i)}$
ARCH(1)	9.418
ARCH(2)	9.406
GARCH(1,1)	8.600
GARCH(1,2)	8.607
EGARCH(1,1)	**8.180**
EGARCH(1,2)	8.223
GJR(1,1)	8.220
GJR(1,2)	8.231

The best-performing model is shown in bold face.

Appendix

EViews 6

- *chapter4.forecast.prg*

```
load chapter4.data.wf1
smpl @all
!ssize = 3000
!length = 4435
matrix(!length-!ssize,8) estg01_1
matrix(!length-!ssize,10) estg02_1
matrix(!length-!ssize,9) estg11_1
matrix(!length-!ssize,11) estg12_1
matrix(!length-!ssize,10) este11_1
matrix(!length-!ssize,14) este12_1
matrix(!length-!ssize,11) estt11_1
matrix(!length-!ssize,13) estt12_1

for !i = 1 to !length-!ssize
    statusline !i
  smpl !i !i+!ssize-1

equation tempg01_1.arch(1,0,m=10000,h) sp500ret c
  tempg01_1.makegarch garchg01_1
  tempg01_1.makeresid resg01_1
!ht = garchg01_1(!i+!ssize-1)
!c1 = @coefs(1)
!c2 = @coefs(2)
!c3 = @coefs(3)
!h_hat = !c2 + (!c3*(resg01_1(!ssize + !i -1)^2) )
!y_hat = !c1
!res = sp500ret(!ssize + !i) - !y_hat
!r = (!res^2)/(!h_hat)
estg01_1(!i,1) = !y_hat
estg01_1(!i,2) = !h_hat
estg01_1(!i,3) = !res
estg01_1(!i,4) = !r
estg01_1(!i,5) = sp500ret(!ssize + !i)
estg01_1(!i,6) = !c1
estg01_1(!i,7) = !c2
estg01_1(!i,8) = !c3

equation tempg02_1.arch(2,0,m=10000,h) sp500ret c
  tempg02_1.makegarch garchg02_1
  tempg02_1.makeresid resg02_1
```

```
!ht = garchg02_1(!i+!ssize-1)
!c1 = @coefs(1)
!c2 = @coefs(2)
!c3 = @coefs(3)
!c4 = @coefs(4)
!h_hat = !c2 + (!c3*(resg02_1(!ssize + !i -1)^2) ) +
  (!c4*(resg02_1(!ssize + !i -2)^2) )
!y_hat = !c1
!res = sp500ret(!ssize + !i) - !y_hat
!r = (!res^2)/(!h_hat)
estg02_1(!i,1) = !y_hat
estg02_1(!i,2) = !h_hat
estg02_1(!i,3) = !res
estg02_1(!i,4) = !r
estg02_1(!i,5) = sp500ret(!ssize + !i)
estg02_1(!i,6) = !c1
estg02_1(!i,7) = !c2
estg02_1(!i,8) = !c3
estg02_1(!i,9) = !c4
estg02_1(!i,10) = resg02_1(!ssize + !i -1)

equation tempg11_1.arch(1,1,m=10000,h) sp500ret c
  tempg11_1.makegarch garchg11_1
  tempg11_1.makeresid resg11_1
!ht = garchg11_1(!i+!ssize-1)
!c1 = @coefs(1)
!c2 = @coefs(2)
!c3 = @coefs(3)
!c4 = @coefs(4)
!h_hat = !c2 + (!c3*(resg11_1(!ssize + !i -1)^2) ) + (!c4*!ht)
!y_hat = !c1
!res = sp500ret(!ssize + !i) - !y_hat
!r = (!res^2)/(!h_hat)
estg11_1(!i,1) = !y_hat
estg11_1(!i,2) = !h_hat
estg11_1(!i,3) = !res
estg11_1(!i,4) = !r
estg11_1(!i,5) = sp500ret(!ssize + !i)
estg11_1(!i,6) = !c1
estg11_1(!i,7) = !c2
estg11_1(!i,8) = !c3
estg11_1(!i,9) = !c4

equation tempg12_1.arch(2,1,m=10000,h) sp500ret c
  tempg12_1.makegarch garchg12_1
```

```
  tempg12_1.makeresid resg12_1
!ht = garchg12_1(!i+!ssize-1)
!c1 = @coefs(1)
!c2 = @coefs(2)
!c3 = @coefs(3)
!c4 = @coefs(4)
!c5 = @coefs(5)
!h_hat = !c2 + (!c3*(resg12_1(!ssize + !i -1)^2) ) +
    (!c4*(resg12_1(!ssize + !i -2)^2) ) + (!c5*!ht)
!y_hat = !c1
!res = sp500ret(!ssize + !i) - !y_hat
!r = (!res^2)/(!h_hat)
estg12_1(!i,1) = !y_hat
estg12_1(!i,2) = !h_hat
estg12_1(!i,3) = !res
estg12_1(!i,4) = !r
estg12_1(!i,5) = sp500ret(!ssize + !i)
estg12_1(!i,6) = !c1
estg12_1(!i,7) = !c2
estg12_1(!i,8) = !c3
estg12_1(!i,9) = !c4
estg12_1(!i,10) = !c5
estg12_1(!i,11) = resg12_1(!ssize + !i -1)

equation tempe11_1.arch(1,1,e,m=10000,h) sp500ret c
  tempe11_1.makegarch garche11_1
  tempe11_1.makeresid rese11_1
!ht = garche11_1(!i+!ssize-1)
!ht2 = garche11_1(!i+!ssize-2)
!c1 = @coefs(1)
!c2 = @coefs(2)
!c3 = @coefs(3)
!c4 = @coefs(4)
!c5 = @coefs(5)
!logh_hat = !c2 + (!c3*@abs(rese11_1(!ssize + !i -1)/
    (sqr(!ht)))) + (!c4*(rese11_1(!ssize + !i -1)/
    (sqr(!ht)))) + (!c5*@log(!ht))
!h_hat = @exp(!logh_hat)
!y_hat = !c1
!res = sp500ret(!ssize + !i) - !y_hat
!r = (!res^2)/(!h_hat)
este11_1(!i,1) = !y_hat
este11_1(!i,2) = !h_hat
este11_1(!i,3) = !res
este11_1(!i,4) = !r
```

```
este11_1(!i,5) = sp500ret(!ssize + !i)
este11_1(!i,6) = !c1
este11_1(!i,7) = !c2
este11_1(!i,8) = !c3
este11_1(!i,9) = !c4
este11_1(!i,10) = !c5

equation tempe12_1.arch(2,1,e,m=10000,h) sp500ret c
  tempe12_1.makegarch garche12_1
  tempe12_1.makeresid rese12_1
!ht = garche12_1(!i+!ssize-1)
!ht2 = garche12_1(!i+!ssize-2)
!c1 = @coefs(1)
!c2 = @coefs(2)
!c3 = @coefs(3)
!c4 = @coefs(5)
!c5 = @coefs(4)
!c6 = @coefs(6)
!c7 = @coefs(7)
!logh_hat = !c2 + (!c3*@abs(rese12_1(!ssize + !i -1)/
  (sqr(!ht)))) + (!c4*(rese12_1(!ssize + !i -1)/
  (sqr(!ht)))) + (!c5*@abs(rese12_1(!ssize + !i -2)/
  (sqr(!ht2)))) + (!c6*(rese12_1(!ssize + !i -2)/
  (sqr(!ht2)))) + (!c7*@log(!ht))
!h_hat = @exp(!logh_hat)
!y_hat = !c1
!res = sp500ret(!ssize + !i) - !y_hat
!r = (!res^2)/(!h_hat)
este12_1(!i,1) = !y_hat
este12_1(!i,2) = !h_hat
este12_1(!i,3) = !res
este12_1(!i,4) = !r
este12_1(!i,5) = sp500ret(!ssize + !i)
este12_1(!i,6) = !c1
este12_1(!i,7) = !c2
este12_1(!i,8) = !c3
este12_1(!i,9) = !c4
este12_1(!i,10) = !c5
este12_1(!i,11) = !c6
este12_1(!i,12) = !c7
este12_1(!i,13) = rese12_1(!ssize + !i -1)
este12_1(!i,14) = garche12_1(!ssize + !i -1)

equation tempt11_1.arch(1,1,t,m=10000,h) sp500ret c
  tempt11_1.makegarch garcht11_1
```

```
  tempt11_1.makeresid rest11_1
!ht = garcht11_1(!i+!ssize-1)
!ht2 = garcht11_1(!i+!ssize-2)
!c1 = @coefs(1)
!c2 = @coefs(2)
!c3 = @coefs(3)
!c4 = @coefs(4)
!c5 = @coefs(5)
if rest11_1(!ssize + !i -1) < 0 then
!D = 1
else
!D = 0
endif
!h_hat = !c2 + (!c3*(rest11_1(!ssize + !i -1)^2) ) +
  (!c4*!D*(rest11_1(!ssize + !i -1)^2)) + (!c5*!ht)
!y_hat = !c1
!res = sp500ret(!ssize + !i) - !y_hat
!r = (!res^2)/(!h_hat)
estt11_1(!i,1) = !y_hat
estt11_1(!i,2) = !h_hat
estt11_1(!i,3) = !res
estt11_1(!i,4) = !r
estt11_1(!i,5) = sp500ret(!ssize + !i)
estt11_1(!i,6) = !c1
estt11_1(!i,7) = !c2
estt11_1(!i,8) = !c3
estt11_1(!i,9) = !c4
estt11_1(!i,10) = !c5
series avDt11_1 = 1*(rest11_1<0)
estt11_1(!i,11) = @mean(avDt11_1)
delete avDt11_1

equation tempt12_1.arch(2,1,t,m=10000,h) sp500ret c
  tempt12_1.makegarch garcht12_1
  tempt12_1.makeresid rest12_1
!ht = garcht12_1(!i+!ssize-1)
!ht2 = garcht12_1(!i+!ssize-2)
!c1 = @coefs(1)
!c2 = @coefs(2)
!c3 = @coefs(3)
!c4 = @coefs(5)
!c5 = @coefs(4)
!c6 = @coefs(6)
if rest12_1(!ssize + !i -1) < 0 then
!D = 1
```

```
else
!D = 0
endif
!h_hat = !c2 + (!c3*(rest12_1(!ssize + !i -1)^2) ) +
   (!c4*(rest12_1(!ssize + !i -2)^2) ) + (!c5*!D*
   (rest12_1(!ssize + !i -1)^2)) + (!c6*!ht)
!y_hat = !c1
!res = sp500ret(!ssize + !i) - !y_hat
!r = (!res^2)/(!h_hat)
estt12_1(!i,1) = !y_hat
estt12_1(!i,2) = !h_hat
estt12_1(!i,3) = !res
estt12_1(!i,4) = !r
estt12_1(!i,5) = sp500ret(!ssize + !i)
estt12_1(!i,6) = !c1
estt12_1(!i,7) = !c2
estt12_1(!i,8) = !c3
estt12_1(!i,9) = !c4
estt12_1(!i,10) = !c5
estt12_1(!i,11) = !c6
series avDt12_1 = 1*(rest12_1<0)
estt12_1(!i,12) = @mean(avDt12_1)
delete avDt12_1
estt12_1(!i,13) = rest12_1(!ssize + !i -1)

next
smpl @all
```

- *chapter4.forecast10.prg*

```
load chapter4.one.day.ahead.wf1
smpl @all
!ssize = 3000
!length = 4435
!models =8
!pi = @acos(-1)
!ahead2=10
'_____
matrix(!length-!ssize,!models) volatility{!ahead2}
for !ahead=!ahead2 to !length-!ssize
  statusline !ahead

  vector(!ahead2) h
  h(1)= estg01_1(!ahead-!ahead2+1,2)
  h(2)= estg01_1(!ahead-!ahead2+1,7) + estg01_1
```

```
    (!ahead-!ahead2+1,8)*estg01_1(!ahead-!ahead2+1,2)
for !r=3 to !ahead2
h(!r) = estg01_1(!ahead-!ahead2+1,7) + estg01_1
    (!ahead-!ahead2+1,8)*h(!r-1)
next
volatility{!ahead2}(!ahead,1)=h(!ahead2)
delete h

vector(!ahead2) h
h(1)= estg02_1(!ahead-!ahead2+1,2)
h(2)= estg02_1(!ahead-!ahead2+1,7) + estg02_1(!ahead-
    !ahead2+1,8)*estg02_1(!ahead-!ahead2+1,2) + estg02_1
    (!ahead-!ahead2+1,9)*estg02_1(!ahead-!ahead2+1,10)^2
for !r=3 to !ahead2
    h(!r) = estg02_1(!ahead-!ahead2+1,7) + estg02_1
    (!ahead-!ahead2+1,8)*h(!r-1) + estg02_1(!ahead-!
    ahead2+1,9)*h(!r-2)
next
volatility{!ahead2}(!ahead,2)=h(!ahead2)
delete h

vector(!ahead2) h
h(1)= estg11_1(!ahead-!ahead2+1,2)
h(2)= estg11_1(!ahead-!ahead2+1,7) + (estg11_1(!ahead-!
    ahead2+1,8)+estg11_1(!ahead-!ahead2+1,9)) * estg11_1
    (!ahead-!ahead2+1,2)
for !r=3 to !ahead2
    h(!r) = estg11_1(!ahead-!ahead2+1,7) + (estg11_1
    (!ahead-!ahead2+1,8)+estg11_1(!ahead-!ahead2+1,9))
    * h(!r-1)
next
volatility{!ahead2}(!ahead,3)=h(!ahead2)
delete h
vector(!ahead2) h
h(1)= estg12_1(!ahead-!ahead2+1,2)
h(2)= estg12_1(!ahead-!ahead2+1,7) + (estg12_1(!ahead-!
    ahead2+1,8)+estg12_1(!ahead-!ahead2+1,10))*estg12_1
    (!ahead-!ahead2+1,2) + estg12_1(!ahead-!ahead2+1,9)
    *estg12_1(!ahead-!ahead2+1,11)^2
for !r=3 to !ahead2
    h(!r) = estg12_1(!ahead-!ahead2+1,7) + (estg12_1
    (!ahead-!ahead2+1,8)+estg12_1(!ahead-!ahead2+
    1,10))*h(!r-1) + estg12_1(!ahead-!ahead2+1,9)*h
    (!r-2)
next
```

```
volatility{!ahead2}(!ahead,4)=h(!ahead2)
delete h

vector(!ahead2) h
h(1)= este11_1(!ahead-!ahead2+1,2)
h(2)= @exp(este11_1(!ahead-!ahead2+1,7) + este11_1
  (!ahead-!ahead2+1,8)*sqr(2/!pi) + este11_1
  (!ahead-!ahead2+1,10)*log(este11_1(!ahead-!ahead2+
  1,2)))
for !r=3 to !ahead2
  h(!r) = @exp(este11_1(!ahead-!ahead2+1,7) + este11_1
    (!ahead-!ahead2+1,8)*sqr(2/!pi) + este11_1(!ahead-!
    ahead2+1,10)*log(h(!r-1)))
next
volatility{!ahead2}(!ahead,5)=h(!ahead2)
delete h

vector(!ahead2) h
h(1)= este12_1(!ahead-!ahead2+1,2)
h(2)= @exp(este12_1(!ahead-!ahead2+1,7) + este12_1
  (!ahead-!ahead2+1,8)*sqr(2/!pi) + este12_1(!ahead-
  !ahead2+1,10)*@abs(este12_1(!ahead-!ahead2+1,13)/
  sqr(este12_1(!ahead-!ahead2+1,14))) + este12_1
  (!ahead-!ahead2+1,11)*este12_1(!ahead-!ahead2+1,13)/
  sqr(este12_1(!ahead-!ahead2+1,14)) + este12_1(!ahead-
  !ahead2+1,12)*log(este12_1(!ahead-!ahead2+1,2)))
for !r=3 to !ahead2
  h(!r) = @exp(este12_1(!ahead-!ahead2+1,7) + (este12_1
    (!ahead-!ahead2+1,8)+este12_1(!ahead-!ahead2+
    1,10))*sqr(2/!pi) + este12_1(!ahead-!ahead2+1,12)*
    log(h(!r-1)))
next
volatility{!ahead2}(!ahead,6)=h(!ahead2)
delete h

vector(!ahead2) h
h(1)= estt11_1(!ahead-!ahead2+1,2)
h(2)= estt11_1(!ahead-!ahead2+1,7) + (estt11_1(!ahead-
  !ahead2+1,8) + estt11_1(!ahead-!ahead2+1,9)*estt11_1
  (!ahead-!ahead2+1,11) + estt11_1(!ahead-!ahead2+
  1,10)) * estt11_1(!ahead-!ahead2+1,2)
for !r=3 to !ahead2
  h(!r) = estt11_1(!ahead-!ahead2+1,7) + (estt11_1
    (!ahead-!ahead2+1,8) + estt11_1(!ahead-!ahead2+1,9)
    *estt11_1(!ahead-!ahead2+1,11) + estt11_1(!ahead-
```

```
        !ahead2+1,10)) * h(!r-1)
  next
  volatility{!ahead2}(!ahead,7)=h(!ahead2)
  delete h

  vector(!ahead2) h
  h(1)= estt12_1(!ahead-!ahead2+1,2)
  h(2)= estt12_1(!ahead-!ahead2+1,7) + (estt12_1(!ahead-
    !ahead2+1,8) + estt12_1(!ahead-!ahead2+1,10)*estt12_1
    (!ahead-!ahead2+1,12) + estt12_1(!ahead-!ahead2+
    1,11))*estt12_1(!ahead-!ahead2+1,2) + estt12_1
    (!ahead-!ahead2+1,9)*estt12_1(!ahead-!ahead2+1,13)^2
  for !r=3 to !ahead2
    h(!r) = estt12_1(!ahead-!ahead2+1,7) + (estt12_1
      (!ahead-!ahead2+1,8) +estt12_1(!ahead-!ahead2+1,10)
      *estt12_1(!ahead-!ahead2+1,12) + estt12_1(!ahead-
      !ahead2+1,11))*h(!r-1) + estt12_1(!ahead-!ahead2+
      1,9)*h(!r-2)
  next
  volatility{!ahead2}(!ahead,8)=h(!ahead2)
  delete h

next
'_____

  for !i=1 to !models
    series Psi{!ahead2}_{!i}
    for !j=!ahead2 to !length-!ssize
      Psi{!ahead2}_{!i}(!j) = (volatility{!ahead2}(!j,!i)
        - sp500ret(!ssize+!j)^2)^2
    next
    vector(!models) averPsi{!ahead2}
    averPsi{!ahead2}(!i) = @mean(Psi{!ahead2}_{!i})
  next
```

5

Other distributional assumptions

5.1 Non-normally distributed standardized innovations

As already mentioned, an attractive feature of the ARCH process is that even though the conditional distribution of the innovations is normal, the unconditional distribution has thicker tails than the normal distribution. However, the degree of leptokurtosis induced by the ARCH process often does not capture all of the leptokurtosis present in high-frequency speculative prices. Thus, there is a fair amount of evidence that the conditional distribution of ε_t is non-normal as well.

To circumvent this problem, Bollerslev (1987) proposed using the standardized Student t distribution with $v > 2$ degrees of freedom:

$$f_{(t)}(z_t; v) = \frac{\Gamma((v+1)/2)}{\Gamma(v/2)\sqrt{\pi(v-2)}} \left(1 + \frac{z_t^2}{v-2}\right)^{-(v+1)/2}, \quad v > 2, \qquad (5.1)$$

where $\Gamma(.)$ is the gamma function. The number v of degrees of freedom is regarded as a parameter to be estimated (i.e. here the parameter vector w reduces to the scalar v). The Student t distribution is symmetric about zero. For $v > 4$, it has conditional kurtosis $3(v-2)(v-4)^{-1}$, which is greater than 3, the corresponding value for the normal distribution; while for $v \to \infty$, it converges to (2.16), which is the density function of the standard normal distribution.

Nelson (1991) suggested the use of the generalized error distribution (GED) with density function[1]

---

[1] The GED is sometimes referred to as the exponential power distribution.

ARCH Models for Financial Applications   Evdokia Xekalaki and Stavros Degiannakis
© 2010 John Wiley & Sons, Ltd

$$f_{(GED)}(z_t; v) = \frac{v \exp(-0.5|z_t/\lambda|^v)}{\lambda 2^{(1+1/v)} \Gamma(v^{-1})}, \quad v > 0, \tag{5.2}$$

where v is the tail-thickness parameter and $\lambda \equiv \sqrt{2^{-2/v} \Gamma(v^{-1})/\Gamma(3v^{-1})}$; for more details on the GED, see Box and Tiao (1973), Harvey (1981) and Johnson et al. (1995). When $v = 2$, z_t is standard normally distributed and so (5.2) reduces to (2.16). For $v < 2$, the distribution of z_t has thicker tails than the normal distribution (e.g. for $v = 1$, z_t has a double exponential, or Laplace, distribution), while for $v > 2$, the distribution of z_t has thinner tails than the normal distribution (e.g. for $v = \infty$, z_t has a uniform distribution on the interval $(-\sqrt{3}, \sqrt{3})$). The GED was first introduced by Subbotin (1923) and was used by Box and Tiao (1962) to model prior densities in Bayesian estimation.

The densities presented above account for fat tails, but they are symmetric. Lee and Tse (1991) considered the case where the conditional distribution of innovations may be not only leptokurtotic but also asymmetric. Allowing for skewness can potentially be important in modelling interest rates as they are lower-bounded by zero and may therefore be skewed. To allow for both skewness and leptokurtosis, Lee and Tse (1991) considered using a Gram–Charlier type distribution (see Kendal and Stuart, 1969, p.157) with density function given by

$$f_{(GC)}(z_t; v, g) = f(z_t)\left(1 + \frac{v}{6}H_3(z_t) + \frac{g}{24}H_4(z_t)\right), \tag{5.3}$$

where $f(.)$ is the standard normal density function, and $H_3(z_t) \equiv z_t^3 - 3z_t$ and $H_4(z_t) \equiv z_t^4 - 6z_t^2 + 3$ are Hermite polynomials. The quantities v and g denote the measures of skewness and kurtosis, respectively. Jondeau and Rockinger (2001) examined the properties of the Gram–Charlier conditional density function and estimated ARCH models with a Gram–Charlier density function for a set of exchange rate series.

Bollerslev et al. (1994) applied the generalized t (GT) distribution of McDonald and Newey (1988),

$$f_{(GT)}(z_t; v, g) = \frac{v}{2\sigma_t b g^{1/v} B(v^{-1}, g)\left(1 + \left(|\varepsilon_t|^v / g b^v \sigma_t^v\right)\right)^{g+1/v}}, \tag{5.4}$$

for $v > 0$, $g > 0$ and $vg > 2$, where $B(v^{-1}, g) \equiv \Gamma(v^{-1})\Gamma(g)/\Gamma(v^{-1} + g)$ is the beta function and $b \equiv \sqrt{\Gamma(v^{-1})\Gamma(g)/\Gamma(3v^{-1})\Gamma(g - 2v^{-1})}$. The advantage that the generalized t distribution has is that it has both (5.1) and (5.2) nested within it. For $v = 2$, (5.4) reduces to the Student t distribution with $2g$ degrees of freedom, while for $v = \infty$, it takes the form of the GED.

Lambert and Laurent (2000, 2001) considered an extension of the skewed Student t, or skT, density proposed by Fernandez and Steel (1998) in the ARCH framework as given by

$$f_{(skT)}(z_t; v, g) = \frac{\Gamma((v+1)/2)}{\Gamma(v/2)\sqrt{\pi(v-2)}}\left(\frac{2s}{g + g^{-1}}\right)\left(1 + \frac{sz_t + m}{v-2}g^{-II_t}\right)^{-(v+1)/2}, \quad v > 2, \tag{5.5}$$

where g is the asymmetry parameter, v denotes the number of degrees of freedom of the distribution, $\Gamma(.)$ is the gamma function, $II_t = 1$ if $z_t \geq -ms^{-1}$ and $II_t = -1$ otherwise, $m = \Gamma((v-1)/2)\sqrt{(v-2)}(\Gamma(v/2)\sqrt{\pi})^{-1}(g-g^{-1})$, and $s = \sqrt{g^2 + g^{-2} - m^2 - 1}$. Angelidis and Degiannakis (2005), Degiannakis (2004) and Giot and Laurent (2003a) considered using ARCH models based on the skewed Student t distribution to fully take into account the fat left and right tails of the returns distribution.

Theodossiou (2002) derived a skewed version of the GED (SGED) to model the empirical distribution of log-returns of financial assets and to price their call options. The SGED density function can be written as

$$f_{(SGED)}(z_t; v, \theta) = \frac{v}{2\lambda}\Gamma(v^{-1})^{-1}\exp\left(-\frac{|z_t + \delta|^v}{(1-\text{sign}(z_t + \delta)\theta)^v\lambda^v}\right), \quad v > 0, \ -1 < \theta < 1,$$

(5.6)

where

$$\lambda = \left(\sqrt{1 + 3\theta^2 - 4\Gamma(2v^{-1})^2\Gamma(v^{-1})^{-1}\Gamma(3v^{-1})^{-1}\theta^2}\right)^{-1}\Gamma(v^{-1})^{1/2}\Gamma(3v^{-1})^{-1/2}$$

is the skewness parameter and

$$\delta = 2\theta\Gamma(2v^{-1})\Gamma(v^{-1})^{-1/2}\Gamma(3v^{-1})^{-1/2}$$
$$\times\left(\sqrt{1 + 3\theta^2 - 4\Gamma(2v^{-1})^2\Gamma(v^{-1})^{-1}\Gamma(3v^{-1})^{-1}\theta^2}\right)^{-1}$$

is Pearson's skewness, or alternatively $\delta = -\text{mode}(z_t)$; see Stuart and Ord (1994). The parameter v controls the height and tails of the density function, while the skewness parameter θ controls the rate of descent of the density around the mode of z_t. In the case of positive skewness ($\theta > 0$), the density function is skewed to the right as z_t is weighted by a value greater than unity for $z_t < -\delta$ and by a value less than unity for values of $z_t > -\delta$. Note that θ and δ have the same sign; thus, for $\theta > 0$, $\text{mode}(z_t)$ is less than zero. Note also that the SGED reduces to the GED for $\theta = 0$. In particular, for $\theta = 0$, (5.6) reduces to

$$f(z_t; v) = \frac{v}{2\lambda}\Gamma(v^{-1})^{-1}\exp\left(-\frac{|z_t + \delta|^v}{\lambda^v}\right).$$

Moreover, since $\lambda = \Gamma(v^{-1})^{1/2}\Gamma(3v^{-1})^{-1/2}$ and $\delta = 0$, (5.6) reduces to (5.2), or

$$f(z_t; v) = \frac{v}{\lambda 2^{(1+1/v)}\Gamma(v^{-1})}\exp\left(-\frac{|z_t|^v}{2\lambda^v}\right), \quad \text{for } \lambda \equiv \sqrt{2^{-2/v}\Gamma(v^{-1})/\Gamma(3v^{-1})}.$$

The SGED may take the form of several well-known probability distributions. For example, for $v = 1$, it leads to the skewed Laplace distribution, while for $v = 2$, it reduces to the skewed normal distribution. Further, for $\theta = 0$, the SGED reduces to the Laplace distribution when $v = 1$, to the normal distribution when $v = 2$, and to the uniform distribution when $v = \infty$. (For more details about the special cases of the SGED, the interested reader is referred to Theodossiou and Trigeorgis, 2003). Bali and Lu (2004) introduced the SGED within a discrete-time ARCH framework. Michelfelder (2005) estimated an EGARCH volatility specification with SGED distributed residuals for seven emerging markets. Note that the SGED encompasses the skewed Student t distribution of Hansen (1994) and the generalized t distribution of McDonald and Newey (1988) and can be obtained as a limiting case of Theodossiou's (1998) skewed generalized t (SGT) distribution. Bali and Theodossiou (2007) proposed a conditional technique for the estimation of value-at-risk measures based on the SGT distribution. (For more information about the SGT distribution and its nested distributions, see Allen and Bali, 2007; Ioannides et al., 2002; Hansen et al., 2003, 2007a, 2007b.)

McDonald and Xu (1995) introduced a generalized form of the beta distribution, which includes more than 30 distributions as limiting or special cases and, in particular, an exponential generalized beta (EGB) distribution leading to generalized forms of the logistics, exponential, Gompertz and Gumbell distributions, as well as the normal distribution as special cases. This general form of the beta distribution was considered for modelling purposes in various applied contexts. For example, Wang et al. (2001) applied an ARCH process associated with an EGB distribution to daily exchange rate data for six major currencies, Harris and Küçüközmen (2001) fitted the distributions of daily, weekly and monthly equity returns in the UK and US by EGB and SGT distributions, while Li (2007) proposed the EGB distribution to model Dow Jones industrial stock returns.

De Vries (1991) noted that the unconditional distribution of an ARCH process can be stable and that under suitable conditions the conditional distribution is stable as well. Stable Paretian conditional distributions have been introduced in ARCH models by Liu and Brorsen (1995), Mittnik et al. (1999) and Panorska et al. (1995). As the stable Paretian distribution does not have an analytical expression for its density function, it is expressed by its characteristic function,

$$\varphi(t, a, \beta, \sigma, \mu) = \exp\left(i\mu t - |\sigma t|^a \left(1 - i\beta \frac{t}{|t|} \omega(|t|, a) \right) \right), \qquad (5.7)$$

where $i = \sqrt{-1}$ is the imaginary unit, $0 < a \leq 2$ is the characteristic exponent, $-1 \leq \beta \leq 1$ is the skewness parameter, $\sigma > 0$ is the scale parameter, $\mu \in R$ is the location parameter, and

$$\omega(|t|, a) = \begin{cases} \tan\dfrac{\pi a}{2}, & a \neq 1 \\ -\dfrac{2}{\pi}\log|t|, & a = 1. \end{cases} \qquad (5.8)$$

The standardized innovations, z_t, are assumed as independently, identically stable Pareto distributed random variables with zero location parameter and unit scale parameter. The way GARCH models are constructed imposes limits on the heaviness of the tails of their unconditional distribution. Given that a wide range of financial data exhibit remarkably fat tails, this assumption represents a major shortcoming of GARCH models in financial time series analysis. Stable Paretian conditional distributions have been employed in a number of studies, such as Mittnik et al. (1998a,1998b) and Mittnik and Paolella (2003). Tsionas (1999) established a framework for Monte Carlo posterior inference in models with stable distributed errors by combining a Gibbs sampler with Metropolis independence chains and representing the symmetric stable variates as normal scale mixtures. Mittnik et al. (2002) and Panorska et al. (1995) derived conditions for strict stationarity of GARCH and APARCH models with stable Paretian conditional distributions. De Vries (1991) provided relationships between ARCH and stable processes. De Vries (1991), Ghose and Kroner (1995) and Groenendijk et al. (1995) demonstrated that ARCH models share many of the properties with Lévy stable distributions, but the true data generating process for an examined set of financial data is more likely ARCH than Lévy stable. Tsionas (2002) compared a stable Paretian model with ARCH errors to a stable Paretian model with stochastic volatility. The randomized GARCH model with stable Paretian innovations totally skewed to the right and with $0 < a < 1$ was studied by Nowicka-Zagrajek and Weron (2001), who obtained the unconditional distributions and analysed the dependence structure by means of the codifference. It turns out that R-GARCH models with conditional variance dependent on the past can have very heavy tails. The class is very flexible as it includes GARCH models and de Vries (1991) processes as special cases. (For more details of stable Paretian models in financial topics, see Rachev and Mittnik, 2000.) Recently, Tavares et al. (2008) estimated asymmetric ARCH models on the S&P500 and the FTSE100 index assuming stable Paretian distribution for innovations.

Hansen (1994) suggested an approach that allows not only the conditional variance, but also the higher moments of conditional distribution such as skewness and kurtosis to be time-varying. He suggested the autoregressive conditional density (ARCD) model, where the density function, $f(z_t; w)$, is presented by

$$f_{(ARCD)}(z_t; w_t | I_{t-1}).$$ (5.9)

The parameter vector of the conditional density function in (5.9) is assumed to be a function of the current information set, I_{t-1}.

Other distributions, that have been employed, include Poisson mixtures of the normal distribution (Brorsen and Yang, 1994; Drost et al., 1998; Jorion, 1988; Lin and Yeh, 2000; Vlaar and Palm, 1993), log-normal mixtures of the normal distribution (Hsieh, 1989), and mixtures of serially dependent normally distributed variables (Cai, 1994) or mixtures of serially dependent Student t distributed variables (Hamilton and Susmel, 1994).[2] Recently, Politis (2003, 2004, 2006, 2007a) developed an implicit ARCH model that gives motivation towards a more *natural* and less ad hoc distribution for the residuals. He proposed to *studentize* the ARCH residuals by dividing them by a time-localized measure of standard deviation.

---

[2] Cai (1994) and Hamilton and Susmel (1994) used such mixtures to estimate the class of regime-switching ARCH models, presented in Section 2.8.1.

5.2   Estimating ARCH models with non-normally distributed standardized innovations using G@RCH 4.2 OxMetrics: an empirical example

In this section we will deal with the estimation of three conditional volatility models that capture the main characteristics of asset returns under three distributional assumptions for the S&P500 equity index. As in the previous chapters, the data set considered refers to the period from 4 April 1988 to 5 April 2005. The conditional mean is considered as a first order autoregressive process and, on the basis of framework (1.6), we can write the general framework of our model specification as

$$
\begin{aligned}
y_t &= \mu_t + \varepsilon_t, \\
\mu_t &= c_0(1-c_1) + c_1 y_{t-1}, \\
\varepsilon_t &= \sigma_t z_t, \\
\sigma_t^2 &= g(\theta | I_{t-1}), \\
z_t &\overset{i.i.d.}{\sim} f(0,1;w),
\end{aligned}
\tag{5.10}
$$

with GARCH(1,1), APARCH(1,1) and FIAPARCH $(1, d, 1)$ conditional volatility specifications under the assumption of standardized innovations distributed according to a GED, a Student t, or a skewed Student t distribution. In total, nine models will be estimated. The models are constructed as a combination of conditional volatility specifications (2.5), (2.67) and (3.13), for $p = q = 1$, and density functions (5.1), (5.2) and (5.5). For example, the AR(1)-APARCH(1,1) with skewed Student t distributed standardized innovations model (AR(1)-APARCH(1,1)-skT) can be represented as the combination of equations (2.67), (5.5) and (5.10):

$$
\begin{aligned}
y_t &= c_0(1-c_1) + c_1 y_{t-1} + \varepsilon_t, \\
\varepsilon_t &= \sigma_t z_t, \\
\sigma_t^\delta &= a_0 + a_1 \left(|\varepsilon_{t-1}| - \gamma_1 \varepsilon_{t-1} \right)^\delta + b_1 \sigma_{t-1}^\delta, \\
z_t &\overset{i.i.d.}{\sim} skT(0,1;v,g), \\
f_{(skT)}(z_t;v,g) &= \frac{\Gamma((v+1)/2)}{\Gamma(v/2)\sqrt{\pi(v-2)}} \left(\frac{2s}{g+g^{-1}} \right) \left(1 + \frac{sz_t+m}{v-2} g^{-II_t} \right)^{-(v+1)/2}.
\end{aligned}
\tag{5.11}
$$

In this case, based on the full set of T observations, the parameter vector $\psi^{(T)} = ((\theta^{(T)}), (w^{(T)})) = ((c_0^{(T)}, c_1^{(T)}, \delta^{(T)}, a_0^{(T)}, a_1^{(T)}, \gamma_1^{(T)}, b_1^{(T)}), (v^{(T)}, g^{(T)}))$ must be estimated.

The nine models are estimated using the G@RCH OxMetrics package using the program named *chapter5.sp500.ox*, which is available on the accompanying CD-ROM. The core G@RCH program that carries out the necessary estimations for the AR(1)-APARCH(1,1)-skT model is given below:

```
#import <packages/Garch42/garch>
main()
{
```

```
    decl garchobj;
    garchobj = new Garch();
    garchobj.Load("Chapter5Data.xls");
    garchobj.Select(Y_VAR, {"SP500RET",0,0});
    garchobj.SetSelSample(-1, 1, -1, 1);
    garchobj.CSTS(1,1);
    garchobj.DISTRI(3);
    garchobj.ARMA_ORDERS(1,0);
    garchobj.GARCH_ORDERS(1,1);
    garchobj.MODEL(4);
    garchobj.MLE(2);
    garchobj.Initialization(<>);
    garchobj.DoEstimation(<>);
    garchobj.Output();
    delete garchobj;
}
```

The type of conditional volatility specification and the type of distributional assumption can be specified by specializing the parameter values given in parentheses in the lines *garchobj.MODEL* and *garchobj.DISTRI*, respectively. Model estimation is carried through quasi-maximum likelihood estimation of the variance–covariance matrix, as indicated by the line *garchobj.MLE(2)*. Table 5.1 presents the estimated parameters of the models. The autoregressive coefficient is insignificant, as in models estimated in previous chapters. In almost all the cases, the conditional volatility parameters are statistically significant, except in the AR(1)-FIAPARCH(1,1)-GED case. The Box–Cox power transformation parameter ranges from 1.223 to 1.585 and does not equal 2, and the degree of integration is estimated between 0.193 and 0.240. For all the distributional assumptions, the parameter v that accounts for leptokurtosis is statistically significant at any level of significance. In contrast, the parameter $\log(g)$, which G@RCH reports rather than g, in the skewed Student t distribution, which captures the asymmetry of the innovations, is statistically insignificant at the 5% level of significance, which indicates that the standardized residuals are symmetric.

For all the models, the standardized residuals are homoscedastic, as any test considers the models adequate, e.g. Engle's Lagrange multiplier test and the Box and Pierce Q statistic for the squared standardized residuals. Table 5.2 presents, for 10 lags, Engle's test for the homoscedasticity of the standardized residuals, the Box and Pierce Q statistic for testing whether there is no autocorrelation of up to the 10th order in squared standardized residuals and Tse's residual based diagnostic test proposed by Tse (see Section 2.7 for details on misspecification tests).

In Table 5.2, the values of the SBC and AIC criteria are also provided. Since the SBC imposes a larger penalty for additional coefficients than the AIC does, it tends to select parsimonious models in comparison to the AIC. On the basis of the AIC, the AR(1)-FIAPARCH(1,1)-GED model is picked, although almost all of its parameters are not statistically significant. Based on the SBC, the AR(1)-APARCH(1,1)-GED model appears to have the best in-sample fitting. Figure 5.1 provides plots of the

Table 5.1   S&P500 equity index: parameter estimates of the nine models on the basis of the full set of data, from 4 April 1988 to 5 April 2005

Parameters	AR(1)-GARCH(1,1)-t	AR(1)-GARCH(1,1)-GED	AR(1)-GARCH(1,1)-skT	AR(1)-APARCH(1,1)-t	AR(1)-APARCH(1,1)-GED	AR(1)-APARCH(1,1)-skT
c_0	0.055	0.042	0.050	0.040	0.031	0.033
	(5.149)	(3.437)	(4.294)	(3.901)	(2.716)	(2.840)
c_1	−0.008	−0.016	−0.010	−0.002	−0.010	−0.004
	(−0.585)	(−1.022)	(−0.721)	(−0.148)	(−0.710)	(−0.284)
a_0	0.003	0.003	0.003	0.008	0.009	0.008
	(1.787)	(1.924)	(1.813)	(2.800)	(2.880)	(2.852)
a_1	0.038	0.038	0.038	0.045	0.040	0.045
	(4.508)	(4.593)	(4.525)	(4.910)	(1.704)	(4.860)
b_1	0.961	0.960	0.960	0.954	0.954	0.953
	(108.800)	(108.700)	(107.800)	(112.800)	(87.800)	(111.600)
γ_1	—	—	—	0.794	0.859	0.797
				(5.739)	(1.539)	(5.579)
δ	—	—	—	1.223	1.304	1.239
				(7.173)	(4.880)	(7.181)
v	5.738	1.233	5.792	6.280	1.266	6.345
	(10.240)	(24.070)	(10.250)	(9.301)	(22.910)	(9.317)
$\log(g)$	—	—	−0.024	—	—	−0.033
			(−1.306)			(−1.819)

Parameters	AR(1)-FIAPARCH(1,1)-t	AR(1)-FIAPARCH(1,1)-GED	AR(1)-FIAPARCH(1,1)-skT
c_0	0.038 (3.415)	0.029 (2.265)	0.032 (2.630)
c_1	0.005 (0.355)	−0.004 (−0.171)	0.003 (0.205)
a_0	0.054 (2.265)	0.074 (0.638)	0.056 (2.372)
a_1	0.248 (2.309)	0.191 (0.322)	0.244 (2.434)
b_1	0.435 (2.778)	0.339 (0.329)	0.440 (3.023)
γ_1	0.860 (2.235)	0.932 (0.337)	0.828 (2.659)
δ	1.571 (8.611)	1.585 (2.343)	1.578 (9.480)
d	0.231 (2.677)	0.193 (0.362)	0.240 (3.065)
v	6.397 (9.267)	1.274 (23.840)	6.459 (9.313)
$\log(g)$	—	—	−0.034 (−1.847)

Values in parentheses are standard error ratios.

Table 5.2 S&P500 equity index: values of Engle's LM test statistic, the Box–Pierce Q statistic for squared residuals, Tse's $RBD_{(i)}$ test statistic, the $SBC_{(i)}$ and the $AIC_{(i)}$ information criteria for the nine models (data set: observed series from 4 April 1988 to 5 April 2005)

Model	LM test (10 lags)	Q statistic (10 lags)	$RBD_{(i)}$ (10 lags)	$SBC_{(i)}$	$AIC_{(i)}$
AR(1)-GARCH(1,1)-t	0.45 [0.922]	4.44 [0.815]	4.06 [0.945]	2.548	2.540
AR(1)-GARCH(1,1)-GED	0.43 [0.932]	4.29 [0.830]	4.03 [0.946]	2.545	2.537
AR(1)-GARCH(1,1)-skT	0.43 [0.932]	4.28 [0.831]	3.96 [0.949]	2.550	2.540
AR(1)-APARCH(1,1)-t	0.14 [0.999]	2.71 [0.951]	1.27 [0.999]	2.536	2.525
AR(1)-APARCH(1,1)-GED	0.13 [0.999]	2.14 [0.976]	1.17 [0.999]	**2.534**	2.523
AR(1)-APARCH(1,1) skT	0.14 [0.999]	2.65 [0.954]	1.19 [0.999]	2.538	2.525
AR(1)-FIAPARCH(1,1)-t	0.36 [0.963]	4.22 [0.837]	2.16 [0.995]	2.537	2.524
AR(1)-FIAPARCH(1,1)-GED	0.38 [0.954]	4.78 [0.780]	2.61 [0.989]	2.535	**2.522**
AR(1)-FIAPARCH(1,1)-skT	0.37 [0.961]	4.23 [0.835]	2.11 [0.995]	2.538	2.523

p-values are reported in brackets. Best-performing models are shown in bold face.

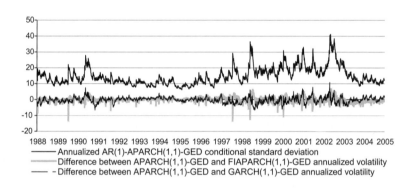

Figure 5.1   In-sample AR(1)-APARCH(1,1)-GED annualized volatility and its difference from the volatility estimated by the AR(1)-GARCH(1,1)-GED and the AR(1)-FIAPARCH(1,1)-GED models.

AR(1)-APARCH(1,1)-GED annualized volatility and its difference from the AR(1)-GARCH(1,1)-GED and of the AR(1)-FIAPARCH(1,1)-GED volatility estimates. Table 5.3 presents some information about the difference of the AR(1)-APARCH(1,1)-GED volatility estimate from the volatility estimated by each of remaining eight models. The differences range from −13.3 to 10.8, while the average squared difference ranges from 0.03 (in the case of the AR(1)-APARCH(1,1)-GED model)

Table 5.3   S&P500 equity index: descriptive statistics of the difference between the annualized volatility estimated by the AR(1)-APARCH(1,1)-GED and the volatility estimated by the other eight models (data set: observed series from 4 April 1988 to 5 April 2005)

Model	Mean difference	Mean squared difference	Minimum difference	Maximum difference
AR(1)-GARCH(1,1)-t	−0.250	3.463	−6.732	10.635
AR(1)-GARCH(1,1)-GED	−0.169	3.413	−6.593	10.823
AR(1)-GARCH(1,1)-skT	−0.230	3.444	−6.661	10.619
AR(1)-APARCH(1,1)-t	−0.054	0.039	−1.042	0.951
AR(1)-APARCH(1,1)-skT	−0.045	0.029	−0.926	0.818
AR(1)-FIAPARCH(1,1)-t	−0.021	2.167	−11.024	5.762
AR(1)-FIAPARCH(1,1)-GED	−0.001	2.690	−13.302	6.648
AR(1)- FIAPARCH(1,1)-skT	−0.014	2.146	−11.300	5.728

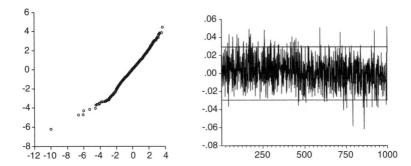

Figure 5.2   Quantile–quantile plot of the standardized residuals (left panel) and the lag 1 to 1000 autocorrelations for the absolute standardized residuals of the AR(1)-APARCH(1,1)-GED model, $Cor\left(\left|\hat{\varepsilon}_t\hat{\sigma}_t^{-1}\right|, \left|\hat{\varepsilon}_{t-\tau}\hat{\sigma}_{t-\tau}^{-1}\right|\right)$, for $\tau = 1, \ldots, 1000$ (right panel).

to 3.46 (in the case of the AR(1)-GARCH(1,1)-t model). Figure 5.2 shows the QQ plot of the standardized residuals[3] of the AR(1)-APARCH(1,1)-GED model and the autocorrelations for the absolute standardized residuals.[4] According to the QQ plot, the empirical quantiles of the standardized residuals appear not to be fully in tune with the assumption of the GED. The QQ plot falls on a straight line in the middle, but it curves upwards at the left-hand end, indicating that the distribution of the

---

[3] The series of the standardized residuals is compared to a series of random draws from the GED with 1.266 degrees of freedom. In EViews, random numbers from the GED are generated by the command @rged(1.266).

[4] The EViews program for the estimation of the autocorrelations named *chapter5.stdresiduals. ar1aparch11ged.prg* is available in the Appendix to this chapter.

standardized residuals has a thicker tail than the assumed GED. The right-hand panel of Figure 5.2 reveals that about 7% of the estimated autocorrelations are outside the 95% confidence interval, with end points given by $\pm 1.96/\sqrt{T}$. Comparing them to the absolute log-returns of S&P500 index in Figure 3.1, we see that the long-memory property of the volatility process has been captured by the model.

In summary, the models presented in this chapter describe the volatility dynamics adequately. There is no remaining heteroscedasticity in the residuals, but the standardized residuals do not agree with the assumed distribution.

5.3   Estimating ARCH models with non-normally distributed standardized innovations using EViews 6: an empirical example

EViews 6 provides the option to estimate an ARCH model assuming that the standardized residuals follow the normal, Student's t or the generalized error distribution. The volatility specifications provided by EViews are estimated for $p = q = 1$: GARCH(1,1), IGARCH(1,1), EGARCH(1,1), APARCH(1,1), GRJ(1,1), CGARCH(1,1) and ACGARCH(1,1). The conditional mean is considered as a first-order autoregressive process. Thus, seven conditional volatility models under three distributional assumptions (21 in all) are defined. The program named *chapter5. SP500.prg* carries out the estimation of the 21 models for the S&P500 equity index for the period from 4 April 1988 to 5 April 2005.

The program that executes the actual estimation of the models, saves the values of the SBC and runs Engle's Lagrange multiplier test is given below.

```
load Chapter5.SP500.wf1
smpl @all
'AR(1)-Garch(1,1)-N Model
equation model1.arch(backcast=0.7,deriv=aa)
  sp500_ret c ar(1)
'AR(1)-Iarch(1,1)-N Model
equation model2.arch(integrated,backcast=0.7,deriv=aa)
  sp500_ret c ar(1)
'AR(1)-GJR(1,1)-N Model
equation model3.arch(thrsh=1,backcast=0.7,deriv=aa)
  sp500_ret c ar(1)
'AR(1)-EGarch(1,1)-N Model
equation model4.arch(thrsh=1,egarch,backcast=0.7,
  deriv=aa) sp500_ret c ar(1)
'AR(1)-AParch(1,1)-N Model
equation model5.arch(thrsh=1,parch,backcast=0.7,
  deriv=aa) sp500_ret c ar(1)
'AR(1)-CGarch(1,1)-N Model
```

```
equation model6.arch(cgarch,backcast=0.7,deriv=aa)
  sp500_ret c ar(1)
'AR(1)-ACGarch(1,1)-N Model
equation model7.arch(thrsh=1,cgarch,backcast=0.7,
  deriv=aa) sp500_ret c ar(1)
'AR(1)-Garch(1,1)-T Model
equation model8.arch(tdist,backcast=0.7,deriv=aa)
  sp500_ret c ar(1)
'AR(1)-Iarch(1,1)-T Model
equation model9.arch(integrated,tdist,backcast=0.7,
  deriv=aa) sp500_ret c ar(1)
'AR(1)-GJR(1,1)-T Model
equation model10.arch(thrsh=1,tdist,backcast=0.7,
  deriv=aa) sp500_ret c ar(1)
'AR(1)-EGarch(1,1)-T Model
equation model11.arch(thrsh=1,tdist,egarch,backcast=0.7,
  deriv=aa) sp500_ret c ar(1)
'AR(1)-AParch(1,1)-T Model with BHHH algorithm
equation model12.arch(thrsh=1,tdist,parch,b,m=50,
  backcast=0.7,deriv=aa) sp500_ret c ar(1)
'AR(1)-CGarch(1,1)-T Model
equation model13.arch(tdist,cgarch,backcast=0.7,
  deriv=aa) sp500_ret c ar(1)
'AR(1)-ACGarch(1,1)-T Model
equation model14.arch(thrsh=1,tdist,cgarch,backcast=0.7,
deriv=aa) sp500_ret c ar(1)
'AR(1)-Garch(1,1)-ged Model
equation model15.arch(ged,backcast=0.7,
  deriv=aa) sp500_ret c ar(1)
'AR(1)-Iarch(1,1)-ged Model
equation model16.arch(integrated,ged,backcast=0.7,
  deriv=aa) sp500_ret c ar(1)
'AR(1)-GJR(1,1)-ged Model
equation model17.arch(thrsh=1,ged,backcast=0.7,deriv=aa)
  sp500_ret c ar(1)
'AR(1)-EGarch(1,1)-ged Model
equation model18.arch(thrsh=1,ged,egarch,backcast=0.7,
  deriv=aa) sp500_ret c ar(1)
'AR(1)-AParch(1,1)-ged Model
equation model19.arch(thrsh=1,ged,parch,backcast=0.7,
  deriv=aa) sp500_ret c ar(1)
'AR(1)-CGarch(1,1)-ged Model
equation model20.arch(ged,cgarch,backcast=0.7,deriv=aa)
  sp500_ret c ar(1)
'AR(1)-ACGarch(1,1)-ged Model
```

```
equation model21.arch(thrsh=1,ged,cgarch,backcast=0.7,
  deriv=aa) sp500_ret c ar(1)
vector(21) sbc
 for !i = 1 to 21
   sbc(!i) = model{!i}.@schwarz ' Schwarz Information
   Criterion
   model{!i}.archtest(10) ' Engle LM Test
 next
```

Table 5.4 provides the values of the SBC and Engle's test statistics for 10 lags. For all the models, the standardized residuals are homoscedastic. Only in the case of the AR(1)-APARCH(1,1)-t model is the p-value marginally higher than 0.05. Based on the SBC, the AR(1)-EGARCH(1,1)-GED model appears to have the best in-sample fitting. The second lowest value of the SBC corresponds to the AR(1)-APARCH(1,1)-GED model, which also had the lowest SBC value in the analysis of the previous section. The high SBC value corresponding to the AR(1)-IARCH(1,1)-GED model indicates failure of the model to fit the observed process. Table 5.5 gives the estimated parameters of the two models with the two lowest SBC values. All the parameters are statistically significant except for the autoregressive coefficient. Note that the parameters of the AR(1)-APARCH(1,1)-GED model are very close to those estimated in G@RCH.

5.4   Estimating ARCH models with non-normally distributed standardized innovations using EViews 6: the logl object

EViews offers the possibility of estimating ARCH models through the use of the log likelihood (logl) object. This is a tool for maximizing a likelihood function with respect to the parameters of the model. EViews includes example files for estimating ARCH models on the basis of the logl object. The examples include the GARCH(1,1) model with Student t distributed errors, considered by Hamilton (1994, p. 662), the GARCH model with coefficient restrictions, considered by Bollerslev et al. (1994, p. 3015), the EGARCH model with GED distributed innovations considered by Hamilton (1994, p. 668), and the multivariate BEKK-GARCH model considered by Engle and Kroner (1995).[5]

The logl object requires the specification of the log-likelihood function $l_t(y_t; \psi)$, in (2.12), for each observation y_t. Let us, for example, estimate the AR(1)-EGARCH (1,1)-GED for the S&P500 index, with the help of the logl object. The program named *chapter5.LogL_A.prg* in the Appendix to this chapter carries out the estimation on the basis of the observed series over the period from 4 April 1988 to 5 April 2005. The model is of the form

---

[5] Multivariate ARCH models are presented in Chapter 11.

Table 5.4 S&P500 equity index: values of the $SBC_{(i)}$ information criterion and Engle's LM test statistic (10 lags) for the 21 models (data set: observed series from 4 April 1988 to 5 April 2005)

Model	$SBC_{(i)}$	LM test (10 lags)	Model	$SBC_{(i)}$	LM test (10 lags)
AR(1)-GARCH(1,1)-N	2.6117	3.909 [0.95]	AR(1)-APARCH(1,1)-t	2.5754	16.94 [0.07]
AR(1)-IARCH (1,1)-N	2.6161	5.764 [0.83]	AR(1)-CGARCH(1,1)-t	2.5524	5.291 [0.87]
AR(1)-GJR(1,1)-N	2.5956	2.006 [0.99]	AR(1)-ACGARCH(1,1)-t	2.5438	2.555 [0.99]
AR(1)-EGARCH(1,1)-N	2.5866	1.119 [0.99]	AR(1)-GARCH(1,1)-GED	2.5458	3.542 [0.96]
AR(1)-APARCH(1,1)-N	2.5876	1.326 [0.99]	AR(1)-IARCH(1,1)-GED	69.778	1.005 [0.99]
AR(1)-CGARCH(1,1)-N	2.6143	5.056 [0.88]	AR(1)-GJR(1,1)-GED	2.5368	2.186 [0.99]
AR(1)-ACGARCH (1,1)-N	2.6038	2.700 [0.98]	AR(1)-EGARCH(1,1)-GED	**2.5321**	1.136 [0.99]
AR(1)-GARCH(1,1)-t	2.5491	3.653 [0.96]	AR(1)-APARCH(1,1)-GED	2.5332	1.065 [0.99]
AR(1)-IARCH(1,1)-t	2.5482	4.828 [0.90]	AR(1)-CGARCH(1,1)-GED	2.5516	4.473 [0.92]
AR(1)-GJR(1,1)-t	2.5397	2.302 [0.99]	AR(1)-ACGARCH(1,1)-GED	2.5406	2.271 [0.99]
AR(1)-EGARCH(1,1)-t	2.5353	1.235 [0.99]			

p-values are reported in brackets. Best-performing models are shown in bold face.

Table 5.5   S&P500 equity index: parameter estimates of the EGARCH(1,1) and the APARCH(1,1) models. The estimated models are in the form: $y_t = c_0(1-c_1) + c_1 y_{t-1} + \varepsilon_t$, $\varepsilon_t = \sigma_t z_t$ and $z_t \overset{i.i.d.}{\sim} GED(0, 1; v)$, with conditional volatility specifications given by $\log(\sigma_t^2) = a_0 + a_1|\varepsilon_{t-1}/\sigma_{t-1}| + \gamma_1(\varepsilon_{t-1}/\sigma_{t-1}) + b_1\log(\sigma_{t-1}^2)$ and $\sigma_t^\delta = a_0 + a_1(|\varepsilon_{t-1}| - \gamma_1\varepsilon_{t-1})^\delta + b_1\sigma_{t-1}^\delta$, respectively (data set: entire observed series from 4 April 1988 to 5 April 2005)

Parameters	AR(1)-EGARCH(1,1)-GED	AR(1)-APARCH(1,1)-GED
c_0	0.031	0.030
	(2.970)	(2.870)
c_1	−0.011	−0.011
	(−0.797)	(−0.768)
a_0	−0.073	0.011
	(−7.814)	(4.895)
a_1	0.093	0.050
	(7.595)	(6.514)
b_1	0.988	0.950
	(433)	(155)
γ_1	−0.071	0.799
	(−7.929)	(5.451)
δ	—	1.070
		(6.461)
v	1.275	1.275
	(46.7)	(46.2)

Numbers in parentheses are standard error ratios.

$$y_t = c_0 + c_1 y_{t-1} + \varepsilon_t,$$

$$\varepsilon_t = \sigma_t z_t,$$

$$z_t \overset{i.i.d.}{\sim} GED(0, 1; v),$$

$$\log(\sigma_t^2) = a_0 + \left(a_1 L\left(\left|\frac{\varepsilon_t}{\sigma_t}\right|\right) + \gamma_1 L\frac{\varepsilon_t}{\sigma_t}\right) + b_1 L \log(\sigma_t^2). \tag{5.12}$$

This is precisely the form of the AR(1)-EGARCH(1,1)-GED specification that EViews estimates via its built-in ARCH commands.[6] The estimated coefficients on the basis of the logl object and on the basis of the built-in ARCH command are provided in Table 5.6. Note that the values of the coefficients are almost identical with both methods.

The first part of *chapter5.LogL_A.prg* defines the initial values of the model. The core part of the program (part B), computes the log-likelihood function for each observation, based on the logl object:

---

[6] The built-in command could be *equation model1.arch(ged,egarch,backcast=0.7,deriv=aa) y c y(−1)*.

Table 5.6   S&P500 equity index: parameter estimates of the AR(1)-EGARCH(1,1)-GED models on the basis of the logl object. Models 1 and 2 are in the form $y_t = c_0 + c_1 y_{t-1} + \varepsilon_t$, $\varepsilon_t = \sigma_t z_t$, $z_t \overset{i.i.d.}{\sim} GED(0, 1; v)$ and $\log(\sigma_t^2) = a_0 + (a_1 L(|\varepsilon_t/\sigma_t| + \gamma_1 L\varepsilon_t/\sigma_t + b_1 L \log(\sigma_t^2)$. Model 3 is in the form $y_t = c_0^* + (c_1^* + c_2^* \exp(-\sigma_t^2/c_3^*)) y_{t-1} + c_4^* \sigma_t^2 + \varepsilon_t$, $\varepsilon_t = \sigma_t z_t$, $z_t \overset{i.i.d.}{\sim} GED(0, 1; v^*)$ and $\log(\sigma_t^2) = a_0^* + \log(1 + N_t \delta_0^*) + L(1-b_1^* L)^{-1} (a_1^*(|\varepsilon_t/\sigma_t| - E|\varepsilon_t/\sigma_t|) + \gamma_1^* \varepsilon_t/\sigma_t)$ (data set: observed series from 4 April 1988 to 5 April 2005)

Parameters	Model 1 Built in EViews	Model 2 logl Object	Parameters	Model 3 logl object
c_0	0.031	0.031	c_0^*	0.005
	(2.96)	(2.96)		(0.30)
c_1	−0.012	−0.012	c_1^*	−0.079
	(−0.84)	(−0.84)		(−2.59)
	—	—	c_2^*	0.163
				(2.54)
	—	—	c_3^*	0.991
				(—)
	—	—	c_4^*	0.034
				(1.58)
a_0	−0.071	−0.070	a_0^*	−0.266
	(−7.74)	(−7.64)		(−2.62)
δ	—	—	δ_0^*	−0.023
				(−0.82)
a_1	0.091	0.089	a_1^*	0.092
	(7.54)	(7.47)		(7.26)
γ_1	−0.070	−0.069	γ_1^*	−0.073
	(−7.84)	(−7.80)		(−7.95)
b_1	0.989	0.989	b_1^*	0.986
	(440)	(449)		(354)
v	1.274	1.271	v^*	1.273
	(47)	(47)		(46)

Numbers in parentheses are standard error ratios.

```
logl lll
lll.append @logl logl
lll.append zeta = alpha0 (1)
lll.append temp1 = alpha1 (1) *@abs (z (-1)) + gamma (1) *z (-1)
lll.append log (h) = zeta + temp1 + beta (1) *log (h (-1))
lll.append res = y -c (1) - c (2) *y (-1)
lll.append z = res/@sqrt (h)
lll.append loglam = -log (2) /nu (1) + (@gammalog (1/nu (1)) -
   @gammalog (3/nu (1))) /2
```

```
lll.append logl = log(nu(1)) - loglam - (1+1/nu(1))*log(2) -
   @gammalog(1/nu(1)) - @abs(z/exp(loglam))^nu(1)/2 -
   log(h)/2
```

Here *zeta* is the constant term of the variance specification, a_0. *Temp1* stands for the expression $(a_1L(|\varepsilon_t/\sigma_t|)+\gamma_1 L\varepsilon_t/\sigma_t)$, and *log(h)* for $a_0 + (a_1L(|\varepsilon_t/\sigma_t|)+\gamma_1 L\varepsilon_t/\sigma_t)+b_1L\log(\sigma_t^2)$. The residuals $\varepsilon_t = y_t-c_0-c_1y_{t-1}$ are defined in *res*. Then, the standardized residuals, z_t, are defined as z and the log-likelihood function for each z_t is computed as *logl*, the specification of the log-likelihood function $l_t(y_t;\psi)$, given by

$$
l_t(y_t;\psi) = \log(v)-\log(\lambda)-\left(\left(1+\frac{1}{v}\right)\log(2)\right)
$$
$$
-\log\left(\Gamma\left(\frac{1}{v}\right)\right)-\frac{1}{2}\left|\frac{z_t}{\exp(\log(\lambda))}\right|^v-\frac{1}{2}\log(\sigma_t^2),
\tag{5.13}
$$

while *loglam* is computed as $\log(\lambda) \equiv \log\left(\sqrt{2^{-2/v}\Gamma(v^{-1})/\Gamma(3v^{-1})}\right)$.

Let us now consider estimating a functional form different from those that are available in EViews. The program named *chapter5.LogL_B.prg* in the Appendix to this chapter carries out the estimation of a model of the form

$$
y_t = c_0^* + \left(c_1^* + c_2^*\exp\left(-\sigma_t^2/c_3^*\right)\right)y_{t-1} + c_4^*\sigma_t^2 + \varepsilon_t,
$$
$$
\varepsilon_t = \sigma_t z_t,
$$
$$
z_t \overset{i.i.d.}{\sim} GED(0,1;v^*),
\tag{5.14}
$$
$$
\log(\sigma_t^2) = a_0^* + \log(1+N_t\delta_0^*) + \frac{L}{(1-b_1^*L)}\left(a_1^*\left(\left|\frac{\varepsilon_t}{\sigma_t}\right|-E\left|\frac{\varepsilon_t}{\sigma_t}\right|\right)+\gamma_1^*\frac{\varepsilon_t}{\sigma_t}\right),
$$

where N_t is the number of non-trading days preceding the tth day. The conditional mean is modelled so as to capture not only the non-synchronous trading effect (parameter c_1^*), but also the relationship between investors' expected return and risk[7] (parameter c_4^*) as well as the inverse relation between volatility and serial correlation[8] (parameter c_2^*). The parameter δ_0^* allows us to explore the non-trading period effect discussed in Section 1.4. The conditional variance specification is of the EGARCH(1,0) form in the OxMetrics G@RCH package (see equation (2.75)).

The core part of the program computes the log-likelihood function of model (5.14) as follows:

```
logl lll
lll.append @logl logl
lll.append zeta = alpha0(1)+log(1+delta0(1)*ndays)
```

---

[7] It is an ARCH in mean specification as defined by equation (1.17). ARCH in mean specifications can be defined with the built-in commands of EViews, e.g. *model1.arch(ARCHM=VAR, backcast=0.7,deriv=aa)* *y c y(−1)*.

[8] See Section 1.6.2. As LeBaron stated, it is difficult to estimate c_3^* in conjunction with c_2^* when using a gradient type of algorithm. Therefore, c_3^* is set to the sample variance of the series.

```
lll.append temp1=alpha1(1)*(@abs(z(-1))-@gamma(2/nu(1))/
   (@sqr(@gamma(3/nu(1))*@gamma(1/nu(1)))))+
   gamma(1)*z(-1)
lll.append log(h) = zeta + beta(1)*(log(h(-1))-zeta(-1))+
   temp1
lll.append res = y -c(1) - (c(2)+c(3)*exp(-h/@var(y)))*
   y(-1) - c(5)*h
lll.append z = res/@sqrt(h)
lll.append loglam = -log(2)/nu(1) + (@gammalog(1/nu(1)) -
   @gammalog(3/nu(1)))/2
lll.append logl = log(nu(1)) - loglam - (1+1/nu(1))*log(2) -
   @gammalog(1/nu(1)) - @abs(z/exp(loglam))^nu(1)/2 -
   log(h)/2
```

Here *zeta* is the constant term of the variance specification given by $a_0^* + \log(1 + N_t \delta_0^*)$. *Temp1* contains the value of $(a_1^*(|\varepsilon_{t-1}/\sigma_{t-1}| - E|\varepsilon_{t-1}/\sigma_{t-1}|) + \gamma_1^* \varepsilon_{t-1}/\sigma_{t-1})$, where $E|\varepsilon_t \sigma_t^{-1}| = \Gamma(2/v)(\Gamma(1/v))^{-1/2}\Gamma(3/v)^{-1/2}$ in the case of the GED, and *log(h)* contains the value of $\log(\sigma_t^2)$. The residuals ε_t are defined in *res* as $y_t - c_0^* - (c_1^* + c_2^* \exp(-\sigma_t^2/c_3^*))y_{t-1} - c_4^*\sigma_t^2$. Finally, *logl* computes the log-likelihood function for each $z_t \equiv \varepsilon_t/\sigma_t$. Table 5.6 provides the estimated coefficients on the basis of the logl object.

Appendix

EViews 6

- *chapter5.stdresiduals.ar1aparch11ged.prg*

```
load chapter5.G@RCH.data.wf1
smpl @all
!length = 1000
series absstdres_ar1aparch11ged=@abs(stdres_
  ar1aparch11ged)
series corr_absstdres_ap11ged
for !i=1 to !length
  corr_absstdres_ap11ged(!i)=@cor(absstdres_
    ar1aparch11ged, absstdres_ar1aparch11ged(-!i))
next
smpl 1 1000
series lower = -1.96/sqr(4435)
series upper = 1.96/sqr(4435)
group group_corr_ap11ged upper lower corr_absstdres_
  ap11ged
```

- *chapter5. LogL_A.prg*

```
load chapter5_logl_a.wf1
sample s0 1 3
sample s1 4 4436
smpl s1
'_____PART A_____
' get starting values from AR(1)-EGARCH(1,1)-N
equation eq1
eq1.arch(1,1,e,m=1000) y c y(-1)
' declare and initialize parameters
'coefs on mean
c(1)=eq1.c(1)
c(2)=eq1.c(2)
' coefs on lagged variance
coef(1) beta
beta(1)=eq1.c(6)
' coefs on lagged resids
coef(1) alpha1
alpha1(1)=eq1.c(4)
' coef on asym term
coef(1) gamma
gamma(1)=eq1.c(5)
' coefs on deterministic terms
coef(1) alpha0
```

```
alpha0(1)=2*log(eq1.@se)
' 0<nu<2 is thick tails; nu>2 is thin tails
coef(1) nu=2
' set presample values of expressions in logl
smpl s0
series zeta=alpha0(1)
series h=exp(alpha0(1))
series z=0
series temp1=alpha1(1)*@abs(z(-1)) + gamma(1)*z(-1)
'_____PART B_____
' set up EGARCH likelihood
logl lll
lll.append @logl logl
lll.append zeta = alpha0(1)
lll.append temp1 = alpha1(1)*@abs(z(-1)) + gamma(1)*z(-1)
lll.append log(h) = zeta + temp1 + beta(1)*log(h(-1))
lll.append res = y -c(1) - c(2)*y(-1)
lll.append z = res/@sqrt(h)
lll.append loglam = -log(2)/nu(1) + (@gammalog(1/nu(1)) -
  @gammalog(3/nu(1)))/2
lll.append logl = log(nu(1)) - loglam - (1+1/nu(1))*log(2) -
  @gammalog(1/nu(1)) -
  @abs(z/exp(loglam))^nu(1)/2 - log(h)/2
' estimate and display output
smpl s1
lll.ml(d)
show lll.output
```

- *chapter5.LogL_B.prg*
```
load chapter5_logl_b.wf1
sample s0 1 3
sample s1 4 4436
smpl s1
'_____PART A_____
' get starting values from AR(1)-EGARCH(1,1)-N in Mean
equation eq1
eq1.arch(1,1,e,m=1000,v) y c y(-1)
' declare and initialize parameters
'coefs on mean
c(1)=eq1.c(1)
c(2)=eq1.c(2)
' coefs on lagged variance
coef(1) beta
beta(1)=eq1.c(7)
' coefs on lagged resids
```

```
coef(1) alpha1
alpha1(1)=eq1.c(5)
' coef on asym term
coef(1) gamma
gamma(1)=eq1.c(6)
' coefs on deterministic terms
coef(1) alpha0
alpha0(1)=2*log(eq1.@se)
coef(1) delta0
delta0(1)=0
' 0<nu<2 is thick tails; nu>2 is thin tails
coef(1) nu=2
!pi=@acos(-1)
' set presample values of expressions in logl
smpl s0
series zeta=alpha0(1)+log(1+delta0(1)*ndays)
series h=exp(alpha0(1))
series z=0
series temp1=alpha1(1)*(@abs(z(-1)) - @sqrt(2/!pi)) +
  gamma(1)*z(-1)
'_____PART B_____
' set up EGARCH likelihood
logl lll
lll.append @logl logl
lll.append zeta=alpha0(1)+log(1+delta0(1)*ndays)
lll.append temp1=alpha1(1)*(@abs(z(-1)) -
  @gamma(2/nu(1))/(@sqr(@gamma(3/nu(1))*
  @gamma(1/nu(1))))) +gamma(1)*z(-1)
lll.append log(h)=zeta + beta(1)*(log(h(-1))-
  zeta(-1)) + temp1
lll.append res=y -c(1)  - (c(2)+c(3)*exp(-h/@var(y)))*y
  (-1) - c(5)*h
lll.append z=res/@sqrt(h)
lll.append loglam=-log(2)/nu(1) + (@gammalog(1/nu(1)) -
  @gammalog(3/nu(1)))/2
lll.append logl=log(nu(1)) - loglam - (1+1/nu(1))*log(2) -
  @gammalog(1/nu(1)) - @abs(z/exp(loglam))^nu(1)/2 -
  log(h)/2
' estimate and display output
smpl s1
lll.ml(d)
show lll.output
```

6

Volatility forecasting: an empirical example using G@RCH Ox

In this chapter, we obtain volatility forecasts in an empirical framework using the G@RCH package. In particular, the 11 conditional volatility specifications with the four distributional assumptions that are provided for by the G@RCH package will be estimated on the basis of an observed series of values of the FTSE100 index. The daily index returns from 4 April 1988 to 5 April 2005 comprise the data set as in Chapter 2. In total, 44 models will be considered. Each model is re-estimated every trading day, for $\tilde{T} = 1435$ days, based on a rolling sample' of constant size $\breve{T} = 3000$ days.[1] The conditional mean is considered as a first-order autoregressive process and the general framework of our model specification can be described as in framework (1.6) by:

$$
\begin{aligned}
y_t &= c_0(1-c_1) + c_1 y_{t-1} + \varepsilon_t, \\
\varepsilon_t &= \sigma_t z_t, \\
\sigma_t^2 &= g(\theta | I_{t-1}), \\
z_t &\overset{i.i.d.}{\sim} f(0, 1; w).
\end{aligned}
\tag{6.1}
$$

The question of what values will be appropriate for p and q in a given situation has been of some concern in the literature. Empirical studies indicate that a model with superior in-sample performance does not necessarily yield better out-of-sample forecasts (see Pagan and Schwert, 1990; Brooks and Burke, 2003; Angelidis et al., 2004; Hansen and Lunde, 2005a). A representative example of the inability of the in-sample model selection methods to pick models with superior volatility forecasting ability is constituted by Degiannakis and Xekalaki's (2007a) findings. These reveal

---

[1] $T = \tilde{T} + \breve{T}$.

ARCH Models for Financial Applications   Evdokia Xekalaki and Stavros Degiannakis
© 2010 John Wiley & Sons, Ltd

that the commonly used in-sample methods of model selection, such as the AIC, SBC and mean squared error criteria, do not lead to the selection of a model that closely tracks future volatility. Moreover, in the majority of empirical studies, a lag of order 1 has been proven to work effectively in forecasting volatility for ARCH frameworks. Thus, for the purpose of our example, we set $p = q = 1$. An alternative approach with more accurate results would be to estimate the models for various values of lag orders p and q, but the required computation time would then increase dramatically.

Recall that the GARCH(1,1), IGARCH(1,1), EGARCH(1,1), GJR(1,1) and APARCH(1,1) conditional volatility specifications are represented by equations (2.73)–(2.77). Also, the FIGARCH(1, d, 1), FIGARCHC(1, d, 1), FIEG-ARCH(1, d, 1), FIAPARCH(1, d, 1), FIAPARCHC(1, d, 1) and HYGARCH(1, d, 1) conditional volatility frameworks are represented by equations (3.20)–(3.25). The density functions of the normal, Student t, generalized error and skewed Student t distributions are given by equations (2.16), (5.1), (5.2) and (5.5), respectively. As an example, the FIAPARCH(1, d, 1) volatility specification with a first-order autoregressive conditional mean and Student t distributed innovations (AR(1)-FIAPARCH(1, d, 1)-t model) has the form

$$y_t = c_0(1-c_1) + c_1 y_{t-1} + \varepsilon_t,$$

$$\varepsilon_t = \sigma_t z_t,$$

$$\sigma_t^\delta = a_0 + (1-b_1 L - (1-a_1 L)(1-L)^d)(|\varepsilon_t| - \gamma \varepsilon_t)^\delta + b_1 \sigma_{t-1}^\delta,$$

$$z_t \overset{i.i.d.}{\sim} t(0, 1; v),$$

$$f_{(t)}(z_t; v) = \frac{\Gamma((v+1)/2)}{\Gamma(v/2)\sqrt{\pi(v-2)}} \left(1 + \frac{z_t^2}{v-2}\right)^{-(v+1)/2},$$

(6.2)

while the parameter vector to be estimated on the basis of T observations is

$$\psi^{(T)} = \left(c_0^{(T)}, c_1^{(T)}, a_0^{(T)}, a_1^{(T)}, \delta^{(T)}, d^{(T)}, \gamma^{(T)}, b_1^{(T)}, v^{(T)}\right).$$

The in-sample conditional standard deviation, $\hat{\sigma}_{t+1}$, formulations based on the entire available data set T are given by equations (2.78)–(2.82) and (3.26)–(3.31). In the present chapter, we estimate the one-step-ahead conditional variance, $\sigma_{t+1|t}^2$, based on information that is available up to time t. The next trading day's variance forecasts are estimated using the G@RCH package in the following forms:

GARCH (1,1),

$$\sigma_{t+1|t}^2 = a_0^{(t)} + a_1^{(t)} \varepsilon_{t|t}^2 + b_1^{(t)} \sigma_{t|t}^2;$$

(6.3)

IGARCH (1,1),

$$\sigma_{t+1|t}^2 = a_0^{(t)} + a_1^{(t)} \varepsilon_{t|t}^2 + \left(1-a_1^{(t)}\right) \sigma_{t|t}^2;$$

(6.4)

EGARCH(1,1),

$$\log \sigma_{t+1|t}^2 = a_0^{(t)}\left(1-b_1^{(t)}\right) + \left(1+\alpha_1^{(t)}L\right)\left(\gamma_1^{(t)}\frac{\varepsilon_{t|t}}{\sigma_{t|t}} + \gamma_2^{(t)}\left(\left|\frac{\varepsilon_{t|t}}{\sigma_{t|t}}\right| - E\left|\frac{\varepsilon_{t|t}}{\sigma_{t|t}}\right|\right)\right) + b_1^{(t)}\log \sigma_{t|t}^2;$$

(6.5)

GJR(1,1),

$$\sigma_{t+1|t}^2 = a_0^{(t)} + \alpha_1^{(t)}\varepsilon_{t|t}^2 + \gamma_1^{(t)}d(\varepsilon_{t|t}<0)\varepsilon_{t|t}^2 + b_1^{(t)}\sigma_{t|t}^2;$$

(6.6)

APARCH(1,1).

$$\sigma_{t+1|t}^2 = \left(a_0^{(t)} + \alpha_1^{(t)}\left(\left|\varepsilon_{t|t}\right| - \gamma_1^{(t)}\varepsilon_{t|t}\right)^{\delta^{(t)}} + b_1^{(t)}\sigma_{t|t}^{\delta^{(t)}}\right)^{2/\delta^{(t)}};$$

(6.7)

FIGARCH(1, d, 1),

$$\sigma_{t+1|t} = \sqrt{a_0^{(t)} + \left(a_1^{(t)}-b_1^{(t)}\right)\varepsilon_{t|t}^2 + \sum_{j=1}^{\infty}\left(\frac{d^{(t)}\Gamma\left(j-d^{(t)}\right)}{\Gamma(1-d^{(t)})\Gamma(j+1)}L^j\left(\varepsilon_{t+1|t+j}^2 - a_1^{(t)}\varepsilon_{t|t+j}^2\right)\right) + b_1^{(t)}\sigma_{t|t}^2;}^2$$

(6.8)

FIGARCHC(1, d, 1)

$$\sigma_{t+1|t} = \sqrt{\left(a_1^{(t)}-b_1^{(t)}\right)\varepsilon_{t|t}^2 + \sum_{j=1}^{\infty}\left(\frac{d^{(t)}\Gamma\left(j-d^{(t)}\right)}{\Gamma(1-d^{(t)})\Gamma(j+1)}L^j\left(\varepsilon_{t+1|t+j}^2 - a_1^{(t)}\varepsilon_{t|t+j}^2\right)\right) + b_1^{(t)}\sigma_{t|t}^2;}$$

(6.9)

FIEGARCH(1, d, 1)

$$\sigma_{t+1|t} = \left(\exp\left(a_0^{(t)}\left(1-b_1^{(t)}\right) + \left[\gamma_1^{(t)}\frac{\varepsilon_{t|t}}{\sigma_{t|t}} + \gamma_2^{(t)}\left(\left|\frac{\varepsilon_{t|t}}{\sigma_{t|t}}\right| - \sqrt{2/\pi}\right)\right]\right.\right.$$

$$+a_1^{(t)}\left[\gamma_1^{(t)}\frac{\varepsilon_{t-1|t}}{\sigma_{t-1|t}} + \gamma_2^{(t)}\left(\left|\frac{\varepsilon_{t-1|t}}{\sigma_{t-1|t}}\right| - \sqrt{2/\pi}\right)\right] + b_1^{(t)}\log \sigma_{t|t}^2$$

$$\left.\left.+\sum_{j=1}^{\infty}\left(\frac{\Gamma(j+d^{(t)})}{\Gamma(d^{(t)})\Gamma(j+1)}\right)\left(L^j + a_1^{(t)}L^{j+1}\right)\left(\gamma_1^{(t)}\frac{\varepsilon_{t|t+j}}{\sigma_{t|t+j}} + \gamma_2^{(t)}\left(\left|\frac{\varepsilon_{t|t+j}}{\sigma_{t|t+j}}\right| - \sqrt{2/\pi}\right)\right)\right)\right)\right)^{1/2};$$

(6.10)

---

[2] Note that $L^j\varepsilon_{t+1|t+j}^2 = \varepsilon_{t+1-j|t+j-j}^2 = \varepsilon_{t+1-j|t}^2.$

FIAPARCH$(1,d,1)$,

$$
\begin{aligned}
\sigma_{t+1|t} = \Bigg(& a_0^{(t)} + \left(a_1^{(t)} - b_1^{(t)} \right) \left(|\varepsilon_{t|t}| - \gamma^{(t)}\varepsilon_{t|t} \right)^{\delta^{(t)}} + b_1^{(t)}\sigma_{t|t}^{\delta^{(t)}} \\
& + \sum_{j=1}^{\infty} \left(\frac{d^{(t)}\Gamma(j-d^{(t)})}{\Gamma(1-d^{(t)})\Gamma(j+1)} L^j \left(|\varepsilon_{t+1|t+j}| - \gamma^{(t)}\varepsilon_{t+1|t+j} \right)^{\delta^{(t)}} \right. \\
& \left. - a_1^{(t)} \left(|\varepsilon_{t|t+j}| - \gamma^{(t)}\varepsilon_{t|t+j} \right)^{\delta^{(t)}} \right) \Bigg)^{1/\delta^{(t)}};
\end{aligned}
\tag{6.11}
$$

FIAPARCHC$(1,d,1)$,

$$
\begin{aligned}
\sigma_{t+1|t} = \Bigg(& \left(a_1^{(t)} - b_1^{(t)} \right) \left(|\varepsilon_{t|t}| - \gamma^{(t)}\varepsilon_{t|t} \right)^{\delta^{(t)}} + b_1^{(t)}\sigma_{t|t}^{\delta^{(t)}} \\
& + \sum_{j=1}^{\infty} \left(\frac{d^{(t)}\Gamma(j-d^{(t)})}{\Gamma(1-d^{(t)})\Gamma(j+1)} L^j \left(|\varepsilon_{t+1|t+j}| - \gamma^{(t)}\varepsilon_{t+1|t+j} \right)^{\delta^{(t)}} \right. \\
& \left. - a_1^{(t)} \left(|\varepsilon_{t|t+j}| - \gamma^{(t)}\varepsilon_{t|t+j} \right)^{\delta^{(t)}} \right) \Bigg)^{1/\delta^{(t)}};
\end{aligned}
\tag{6.12}
$$

and HYGARCH$(1,d,1)$,

$$
\begin{aligned}
\sigma_{t+1|t} = \Bigg(& a_0^{(t)} + \left(a_1^{(t)} - b_1^{(t)} \right)\varepsilon_{t|t}^2 + b_1^{(t)}\sigma_{t|t}^2 \\
& + \sum_{j=1}^{\infty} \left(\frac{d^{(t)}\Gamma(j-d^{(t)})}{\Gamma(1-d^{(t)})\Gamma(j+1)} L^j a^{(t)} \left(\varepsilon_{t+1|t+j}^2 - a_1^{(t)}\varepsilon_{t|t+j}^2 \right) \right) \Bigg)^{1/2}.
\end{aligned}
\tag{6.13}
$$

Note that

$$
E\left| \frac{\varepsilon_{t|t}}{\sigma_{t|t}} \right| = \sqrt{\frac{2}{\pi}}
$$

for normally distributed residuals,

$$
E\left| \frac{\varepsilon_{t|t}}{\sigma_{t|t}} \right| = \Gamma\left(\frac{v+1}{2} \right) \frac{2\sqrt{v-2}}{\sqrt{\pi}(v-1)\Gamma(v/2)}
$$

for Student t distributed residuals,

$$
E\left| \frac{\varepsilon_{t|t}}{\sigma_{t|t}} \right| = \lambda 2^{v^{-1}} \frac{\Gamma(2v^{-1})}{\Gamma(v^{-1})}
$$

for residuals having the generalized error distribution, and

$$E\left|\frac{\varepsilon_{t|t}}{\sigma_{t|t}}\right| = \frac{4g^2}{g+g^{-1}} \frac{\Gamma((v+1)/2)\sqrt{v-2}}{\sqrt{\pi}(v-1)\Gamma(v/2)}$$

for residuals following the skewed Student t distribution.

The program named *chapter6.ftse100.OutSample.ox* available on the accompanying CD-ROM carries out the estimation of the 44 model specifications for the FTSE100 index. The core program that executes the necessary computations for the AR(1)-GARCH(1,1)-N model is as follows:

```
#import <packages/Garch42/garch>
main ()
{
decl ii;
decl rolling_sample=2999; //It's the constant rolling sample
   minus 1
decl matrixsize=1435; //It's the size of the output matrix or
   the number of rolling forecasts
decl ftse100_output_GARCH_n;
ftse100_output_GARCH_n = zeros (matrixsize,7);
for (ii=1; ii<matrixsize+1;++ii)
{
decl i,j,k,l,garchobj;
garchobj = new Garch ();
garchobj.Load ("chapter6Data.xls");
garchobj.Select (Y_VAR, {"ftse100",0,0} );
garchobj.SetSelSample (ii, 1, ii+rolling_sample, 1);
garchobj.CSTS (1,1);
garchobj.DISTRI (0);
garchobj.ARMA_ORDERS (1,0);
garchobj.ARFIMA (0);
garchobj.GARCH_ORDERS (1,1);
garchobj.MODEL (1);
garchobj.TRUNC (400);
garchobj.BOUNDS (1);
garchobj.MLE (2);
garchobj.FORECAST (1,1,0);
garchobj.Initialization (<>);
garchobj.DoEstimation (<>);
garchobj.Output ();
garchobj.STORE (0,0,0,1,1,"ftse100_GARCH_n",0);
decl ppp1;
decl ppp2;
ppp1 =zeros (1,1);
```

```
ppp2 =zeros(1,1);
ppp1 = loadmat ("VarFor_ftse100_GARCH_n.xls");
ppp2 = loadmat ("MeanFor_ftse100_GARCH_n.xls");
decl prmter=garchobj.GetPar();
garchobj.Append(prmter,"prmter",1);
print ("sample number ", ii);
ftse100_output_GARCH_n[ii-1][0]=ppp1[0][0];
ftse100_output_GARCH_n[ii-1][1]=ppp2[0][0];
ftse100_output_GARCH_n[ii-1][2]=prmter[0][0];
ftse100_output_GARCH_n[ii-1][3]=prmter[1][0];
ftse100_output_GARCH_n[ii-1][4]=prmter[2][0];
ftse100_output_GARCH_n[ii-1][5]=prmter[3][0];
ftse100_output_GARCH_n[ii-1][6]=prmter[4][0];
delete garchobj;
}
savemat("ftse100_GARCH_n.xls",ftse100_output_GARCH_n);
}
```

G@RCH reads the FTSE100 log-returns from the Excel file named *chapter6-Data.xls* (*Load("chapter6Data.xls")*). The variable *rolling_sample* (in line 5) assigns the size of the rolling sample, $\tilde{T} = 3000$, and the variable *matrixsize* (in the line 6) defines the $\tilde{T} = 1435$ one-step-ahead volatility forecasts that are computed. The matrix *ftse100_output_GARCH_n* contains 1435 rows and seven columns. The first two columns contain the values of the one-step-ahead conditional variance, $\sigma^2_{t+1|t}$, and of the one-step-ahead conditional mean, $y_{t+1|t}$, forecasts. The next five columns list the values of the estimated parameters, $c_0^{(t)}, c_1^{(t)}, a_0^{(t)}, a_1^{(t)}, b_1^{(t)}$.[3] The command *STORE(0,0,0,1,1,"ftse100_GARCH_n",0)* saves the one-step-ahead conditional mean and conditional variance forecasts in the Excel files, *meanFor_ftse100_GARCH_n.xls* and *varFor_ftse100_GARCH_n.xls*, respectively. Then the command *ftse100_output_GARCH_n[ii-1][0]=ppp1[0][0]* loads the saved values from file *varFor_ftse100_GARCH_n.xls* to the cell corresponding to the *ii*th row and first column of the matrix *ftse100_output_GARCH_n*. The commands *decl prmter=garchobj.GetPar()* and *garchobj.Append(prmter,"prmter",1)* store the estimated parameters, $c_0^{(t)}, c_1^{(t)}, a_0^{(t)}, a_1^{(t)}, b_1^{(t)}$. Then the commands *ftse100_output_GARCH_n[ii-1][j]=prmter[j-2][0]*, for $j = 2,\ldots,6$, load the values of the estimated parameters at the cell corresponding to the *ii*th row and $(j+1)$th column of matrix *ftse100_output_GARCH_n*. Finally, matrix *ftse100_output_GARCH_n* is saved in Excel format with the name *ftse100_GARCH_n.xls* (command *savemat ("ftse100_GARCH_n.xls",ftse100_output_GARCH_n)*).

Models whose estimation procedure failed even once were excluded from further analysis. Since the algorithm for the numerical maximization of the log-likelihood

---

[3] Older versions of the G@RCH package, such as version 2.3, save the values of the parameters $a_1^{(t)}, b_1^{(t)}$ in reverse order. Consider, for example, the AR(1)-GARCH(1,1) model. Version 2.3 would save $b_1^{(t)}$ in the fourth column and $a_1^{(t)}$ in the fifth column. The files saved in the CD-ROM were created in G@RCH 2.3.

Table 6.1   FTSE100 equity index: out-of-sample mean squared error (6.14), for models $i = 1, \ldots, 41$

Rank	$\bar{\Psi}_{(SE)}$	Model	Rank	$\bar{\Psi}_{(SE)}$	Model
1	8.3553	AR(1)-FIAPARCH(1,1)-t	23	8.7186	AR(1)-FIGARCH(1,1)-GED
2	8.3568	AR(1)-FIAPARCH(1,1)-skT	24	8.7207	AR(1)-FIGARCHC(1,1)-GED
3	8.3574	AR(1)-FIAPARCHC(1,1)-t	25	8.7218	AR(1)-FIGARCH(1,1)-N
4	8.3574	AR(1)-FIAPARCHC(1,1)-skT	26	8.7223	AR(1)-FIGARCH(1,1)-t
5	8.3641	AR(1)-FIAPARCH(1,1)-GED	27	8.7230	AR(1)-FIGARCHC(1,1)-skT
6	8.3657	AR(1)-FIAPARCHC(1,1)-GED	28	8.7237	AR(1)-FIGARCH(1,1)-skT
7	8.3912	AR(1)-FIAPARCH(1,1)-N	29	8.7242	AR(1)-FIGARCHC(1,1)-N
8	8.3928	AR(1)-FIAPARCHC(1,1)-N	30	8.7288	AR(1)-HYGARCH(1,1)-GED
9	8.4698	AR(1)-GJR(1,1)-t*	31	8.7300	AR(1)-HYGARCH(1,1)-t
10	8.4733	AR(1)-GJR(1,1)-skT	32	8.7312	AR(1)-HYGARCH(1,1)-skT
11	8.4810	AR(1)-APARCH(1,1)-t**	33	8.7358	AR(1)-HYGARCH(1,1)-N
12	8.4813	AR(1)-GJR(1,1)-GED	34	8.7918	AR(1)-GARCH(1,1)-N
13	8.4915	AR(1)-APARCH(1,1)-GED	35	8.8025	AR(1)-GARCH(1,1)-GED
14	8.5052	AR(1)-FIEGARCH(1,1)-skT	36	8.8143	AR(1)-GARCH(1,1)-t
15	8.5070	AR(1)-GJR(1,1)-N	37	8.8156	AR(1)-GARCH(1,1)-skT
16	8.5136	AR(1)-EGARCH(1,1)-skT	38	8.8815	AR(1)-IGARCH(1,1)-N
17	8.5277	AR(1)-EGARCH(1,1)-GED	39	8.8818	AR(1)-IGARCH(1,1)-GED
18	8.5346	AR(1)-APARCH(1,1)-N	40	8.8861	AR(1)-IGARCH(1,1)-t
19	8.5568	AR(1)-FIEGARCH(1,1)-GED	41	8.8865	AR(1)-IGARCH(1,1)-skT
20	8.5929	AR(1)-EGARCH(1,1)-N	42	—	AR(1)-FIGARCHC(1,1)-t
21	8.6041	AR(1)-FIEGARCH(1,1)-N	43	—	AR(1)-FIEGARCH(1,1)-t
22	8.6846	AR(1)-EGARCH(1,1)-t	44	—	AR(1)-APARCH(1,1)-skT

*The model that minimizes the SBC information criterion.
**The model that minimizes the AIC information criterion.

function for the AR(1)-FIGARCHC$(1, d, 1)$-t, AR(1)-FIEGARCH$(1, d, 1)$-t and AR(1)-APARCH(1,1)-skT models failed to converge, these models were excluded. An alternative approach for points in time when a model fails to be estimated is to compute the mean and variance forecasts of the current trading day based on the parameter estimates of the previous trading day (Angelidis et al., 2004).

To judge the ability of the models to forecast the next trading day's variance, the volatility forecasting performance of the 41 competing processes will be evaluated. As in the case of the S&P500 index dealt with in Chapter 4, the mean squared distance between the conditional variance forecast and squared log-returns is computed,

$$\bar{\Psi}_{(SE)(i)} = \tilde{T}^{-1} \sum_{t=1}^{\tilde{T}} \left(\sigma_{t+1|t(i)}^2 - y_{t+1}^2 \right)^2, \qquad (6.14)$$

for each model specification $i = 1, \ldots, 41$. Table 6.1 lists the out-of-sample mean squared errors. Minimization of the mean squared error loss function is achieved by the AR(1)-FIAPARCH$(1, d, 1)$-t model. Figure 6.1 depicts the one-day-ahead conditional variance forecasts of the AR(1)-FIAPARCH$(1, d, 1)$-t model and Figure 6.2 plots the rolling sample based estimates of its parameters. The plots of the estimated parameters display a sort of dependence on time as was the case with the S&P500 index in Chapter 4. Similar findings have also been noted by Degiannakis et al. (2008).

The selection procedure presented above was based on a comparative evaluation of the forecasting performance of the candidate models. This differs from the procedure one would adopt on the basis of information on the ability of the candidate models to fit data. The model selection procedure would then be based on in-sample information criteria that allow comparisons of the descriptive abilities of the models. The program named *chapter6.ftse100.InSample.ox*, in the Appendix to this chapter, carries out the estimation of the 44 models based on the full data set. The SBC and AIC information criteria are minimized by models AR(1)-GJR (1,1)-t and AR(1)-APARCH(1,1)-t, respectively. On the basis of Table 6.1, these

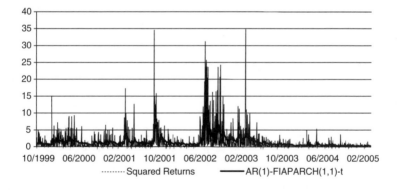

Figure 6.1   FTSE100 equity index: the next day's conditional variance forecasts of the AR(1)-FIAPARCH(1,1)-t model and the squared log-returns (from October 1999 to April 2005).

Figure 6.2   FTSE100 equity index: the rolling sample based estimates of parameters of the AR(1)-FIAPARCH(1, d, 1)-t model (6.2) (from 5 October 1999 to 4 April 2005).

models are respectively ranked ninth and eleventh with respect to the loss function $\bar{\Psi}_{(SE)}$. The results are in line with the comments of Pagan and Schwert (1990), Brooks and Burke (2003) and Hansen and Lunde (2005a) that a model which accommodates a significant in-sample relation need not result in better out-of-sample performance.[4] However, one cannot tell whether models AR(1)-GJR(1,1)-t and AR(1)-APARCH(1,1)-t are of statistically equivalent forecasting ability to model AR(1)-FIAPARCH(1, d, 1)-t. The statistical comparison of this ability will be discussed in Chapter 10.

---

[4] Of course, there are cases, as in Section 4.1, where in-sample and out-of-sample evaluation criteria result in picking the same model.

Appendix

G@RCH 4.2

- *chapter6.ftse100.InSample.ox*

```
#import <packages/Garch42/garch>
main ()
{
{
    decl i,j,k,l,garchobj;
    garchobj = new Garch();
  garchobj.Load("Chapter6Data.xls");
  garchobj.Select(Y_VAR, {"ftse100",0,0} );
    garchobj.SetSelSample(-1, 1, -1, 1);
    garchobj.CSTS(1,1);
    garchobj.DISTRI(0);
    garchobj.ARMA_ORDERS(1,0);
    garchobj.ARFIMA(0);
    garchobj.GARCH_ORDERS(1,1);
    garchobj.MODEL(1);
    garchobj.TRUNC(400);
    garchobj.BOUNDS(1);
    garchobj.MLE(2);
    garchobj.TESTS(0,1);
    garchobj.Initialization(<>);
    garchobj.DoEstimation(<>);
    garchobj.Output();
    delete garchobj;
}
{
    decl i,j,k,l,garchobj;
    garchobj = new Garch();
  garchobj.Load("Chapter6Data.xls");
  garchobj.Select(Y_VAR, {"ftse100",0,0} );
    garchobj.SetSelSample(-1, 1, -1, 1);
    garchobj.CSTS(1,1);
    garchobj.DISTRI(0);
    garchobj.ARMA_ORDERS(1,0);
    garchobj.ARFIMA(0);
    garchobj.GARCH_ORDERS(1,1);
    garchobj.MODEL(2);
    garchobj.TRUNC(400);
    garchobj.BOUNDS(1);
    garchobj.MLE(2);
    garchobj.TESTS(0,1);
```

```
    garchobj.Initialization(<>);
    garchobj.DoEstimation(<>);
    garchobj.Output();
    delete garchobj;
}
{
    decl i,j,k,l,garchobj;
    garchobj = new Garch();
  garchobj.Load("Chapter6Data.xls");
  garchobj.Select(Y_VAR, {"ftse100",0,0} );
    garchobj.SetSelSample(-1, 1, -1, 1);
    garchobj.CSTS(1,1);
    garchobj.DISTRI(0);
    garchobj.ARMA_ORDERS(1,0);
    garchobj.ARFIMA(0);
    garchobj.GARCH_ORDERS(1,1);
    garchobj.MODEL(3);
    garchobj.TRUNC(400);
    garchobj.BOUNDS(1);
    garchobj.MLE(2);
    garchobj.TESTS(0,1);
    garchobj.Initialization(<>);
    garchobj.DoEstimation(<>);
    garchobj.Output();
    delete garchobj;
}
{
    decl i,j,k,l,garchobj;
    garchobj = new Garch();
  garchobj.Load("Chapter6Data.xls");
  garchobj.Select(Y_VAR, {"ftse100",0,0} );
    garchobj.SetSelSample(-1, 1, -1, 1);
    garchobj.CSTS(1,1);
    garchobj.DISTRI(0);
    garchobj.ARMA_ORDERS(1,0);
    garchobj.ARFIMA(0);
    garchobj.GARCH_ORDERS(1,1);
    garchobj.MODEL(4);
    garchobj.TRUNC(400);
    garchobj.BOUNDS(1);
    garchobj.MLE(2);
    garchobj.TESTS(0,1);
    garchobj.Initialization(<>);
    garchobj.DoEstimation(<>);
    garchobj.Output();
```

```
   delete garchobj;
}
{

   decl i,j,k,l,garchobj;
   garchobj = new Garch();
 garchobj.Load("Chapter6Data.xls");
 garchobj.Select(Y_VAR, {"ftse100",0,0} );
   garchobj.SetSelSample(-1, 1, -1, 1);
   garchobj.CSTS(1,1);
   garchobj.DISTRI(0);
   garchobj.ARMA_ORDERS(1,0);
   garchobj.ARFIMA(0);
   garchobj.GARCH_ORDERS(1,1);
   garchobj.MODEL(5);
   garchobj.TRUNC(400);
   garchobj.BOUNDS(1);
   garchobj.MLE(2);
   garchobj.TESTS(0,1);
   garchobj.Initialization(<>);
   garchobj.DoEstimation(<>);
   garchobj.Output();
   delete garchobj;
}
{

   decl i,j,k,l,garchobj;
   garchobj = new Garch();
 garchobj.Load("Chapter6Data.xls");
 garchobj.Select(Y_VAR, {"ftse100",0,0} );
   garchobj.SetSelSample(-1, 1, -1, 1);
   garchobj.CSTS(1,1);
   garchobj.DISTRI(0);
   garchobj.ARMA_ORDERS(1,0);
   garchobj.ARFIMA(0);
   garchobj.GARCH_ORDERS(1,1);
   garchobj.MODEL(6);
   garchobj.TRUNC(400);
   garchobj.BOUNDS(1);
   garchobj.MLE(2);
   garchobj.TESTS(0,1);
   garchobj.Initialization(<>);
   garchobj.DoEstimation(<>);
   garchobj.Output();
   delete garchobj;
}
{
```

```
    decl i,j,k,l,garchobj;
    garchobj = new Garch();
  garchobj.Load("Chapter6Data.xls");
  garchobj.Select(Y_VAR, {"ftse100",0,0} );
    garchobj.SetSelSample(-1, 1, -1, 1);
    garchobj.CSTS(1,1);
    garchobj.DISTRI(0);
    garchobj.ARMA_ORDERS(1,0);
    garchobj.ARFIMA(0);
    garchobj.GARCH_ORDERS(1,1);
    garchobj.MODEL(7);
    garchobj.TRUNC(400);
    garchobj.BOUNDS(1);
    garchobj.MLE(2);
    garchobj.TESTS(0,1);
    garchobj.Initialization(<>);
    garchobj.DoEstimation(<>);
    garchobj.Output();
    delete garchobj;
}
{
    decl i,j,k,l,garchobj;
    garchobj = new Garch();
  garchobj.Load("Chapter6Data.xls");
  garchobj.Select(Y_VAR, {"ftse100",0,0} );
    garchobj.SetSelSample(-1, 1, -1, 1);
    garchobj.CSTS(1,1);
    garchobj.DISTRI(0);
    garchobj.ARMA_ORDERS(1,0);
    garchobj.ARFIMA(0);
    garchobj.GARCH_ORDERS(1,1);
    garchobj.MODEL(8);
    garchobj.TRUNC(400);
    garchobj.BOUNDS(1);
    garchobj.MLE(2);
    garchobj.TESTS(0,1);
    garchobj.Initialization(<>);
    garchobj.DoEstimation(<>);
    garchobj.Output();
    delete garchobj;
}
{
    decl i,j,k,l,garchobj;
    garchobj = new Garch();
  garchobj.Load("Chapter6Data.xls");
```

```
  garchobj.Select(Y_VAR, {"ftse100",0,0} );
    garchobj.SetSelSample(-1, 1, -1, 1);
    garchobj.CSTS(1,1);
    garchobj.DISTRI(0);
    garchobj.ARMA_ORDERS(1,0);
    garchobj.ARFIMA(0);
    garchobj.GARCH_ORDERS(1,1);
    garchobj.MODEL(9);
    garchobj.TRUNC(400);
    garchobj.BOUNDS(1);
    garchobj.MLE(2);
    garchobj.TESTS(0,1);
    garchobj.Initialization(<>);
    garchobj.DoEstimation(<>);
    garchobj.Output();
    delete garchobj;
}
{
    decl i,j,k,l,garchobj;
    garchobj = new Garch();
  garchobj.Load("Chapter6Data.xls");
  garchobj.Select(Y_VAR, {"ftse100",0,0} );
    garchobj.SetSelSample(-1, 1, -1, 1);
    garchobj.CSTS(1,1);
    garchobj.DISTRI(0);
    garchobj.ARMA_ORDERS(1,0);
    garchobj.ARFIMA(0);
    garchobj.GARCH_ORDERS(1,1);
    garchobj.MODEL(10);
    garchobj.TRUNC(400);
    garchobj.BOUNDS(1);
    garchobj.MLE(2);
    garchobj.TESTS(0,1);
    garchobj.Initialization(<>);
    garchobj.DoEstimation(<>);
    garchobj.Output();
    delete garchobj;
}
{
    decl i,j,k,l,garchobj;
    garchobj = new Garch();
  garchobj.Load("Chapter6Data.xls");
  garchobj.Select(Y_VAR, {"ftse100",0,0} );
    garchobj.SetSelSample(-1, 1, -1, 1);
    garchobj.CSTS(1,1);
```

```
    garchobj.DISTRI(0);
    garchobj.ARMA_ORDERS(1,0);
    garchobj.ARFIMA(0);
    garchobj.GARCH_ORDERS(1,1);
    garchobj.MODEL(11);
    garchobj.TRUNC(400);
    garchobj.BOUNDS(1);
    garchobj.MLE(2);
    garchobj.TESTS(0,1);
    garchobj.Initialization(<>);
    garchobj.DoEstimation(<>);
    garchobj.Output();
    delete garchobj;
}
{
    decl i,j,k,l,garchobj;
    garchobj = new Garch();
  garchobj.Load("Chapter6Data.xls");
  garchobj.Select(Y_VAR, {"ftse100",0,0} );
    garchobj.SetSelSample(-1, 1, -1, 1);
    garchobj.CSTS(1,1);
    garchobj.DISTRI(1);
    garchobj.ARMA_ORDERS(1,0);
    garchobj.ARFIMA(0);
    garchobj.GARCH_ORDERS(1,1);
    garchobj.MODEL(1);
    garchobj.TRUNC(400);
    garchobj.BOUNDS(1);
    garchobj.MLE(2);
    garchobj.TESTS(0,1);
    garchobj.Initialization(<>);
    garchobj.DoEstimation(<>);
    garchobj.Output();
    delete garchobj;
}
{
    decl i,j,k,l,garchobj;
    garchobj = new Garch();
  garchobj.Load("Chapter6Data.xls");
  garchobj.Select(Y_VAR, {"ftse100",0,0} );
    garchobj.SetSelSample(-1, 1, -1, 1);
    garchobj.CSTS(1,1);
    garchobj.DISTRI(1);
    garchobj.ARMA_ORDERS(1,0);
    garchobj.ARFIMA(0);
```

```
      garchobj.GARCH_ORDERS(1,1);
      garchobj.MODEL(2);
      garchobj.TRUNC(400);
      garchobj.BOUNDS(1);
      garchobj.MLE(2);
      garchobj.TESTS(0,1);
      garchobj.Initialization(<>);
      garchobj.DoEstimation(<>);
      garchobj.Output();
      delete garchobj;
}
{
      decl i,j,k,l,garchobj;
      garchobj = new Garch();
   garchobj.Load("Chapter6Data.xls");
   garchobj.Select(Y_VAR, {"ftse100",0,0} );
      garchobj.SetSelSample(-1, 1, -1, 1);
      garchobj.CSTS(1,1);
      garchobj.DISTRI(1);
      garchobj.ARMA_ORDERS(1,0);
      garchobj.ARFIMA(0);
      garchobj.GARCH_ORDERS(1,1);
      garchobj.MODEL(3);
      garchobj.TRUNC(400);
      garchobj.BOUNDS(1);
      garchobj.MLE(2);
      garchobj.TESTS(0,1);
      garchobj.Initialization(<>);
      garchobj.DoEstimation(<>);
      garchobj.Output();
      delete garchobj;
}
{
      decl i,j,k,l,garchobj;
      garchobj = new Garch();
   garchobj.Load("Chapter6Data.xls");
   garchobj.Select(Y_VAR, {"ftse100",0,0} );
      garchobj.SetSelSample(-1, 1, -1, 1);
      garchobj.CSTS(1,1);
      garchobj.DISTRI(1);
      garchobj.ARMA_ORDERS(1,0);
      garchobj.ARFIMA(0);
      garchobj.GARCH_ORDERS(1,1);
      garchobj.MODEL(4);
      garchobj.TRUNC(400);
```

```
    garchobj.BOUNDS(1);
    garchobj.MLE(2);
    garchobj.TESTS(0,1);
    garchobj.Initialization(<>);
    garchobj.DoEstimation(<>);
    garchobj.Output();
    delete garchobj;
}
{
    decl i,j,k,l,garchobj;
    garchobj = new Garch();
  garchobj.Load("Chapter6Data.xls");
  garchobj.Select(Y_VAR, {"ftse100",0,0} );
    garchobj.SetSelSample(-1, 1, -1, 1);
    garchobj.CSTS(1,1);
    garchobj.DISTRI(1);
    garchobj.ARMA_ORDERS(1,0);
    garchobj.ARFIMA(0);
    garchobj.GARCH_ORDERS(1,1);
    garchobj.MODEL(5);
    garchobj.TRUNC(400);
    garchobj.BOUNDS(1);
    garchobj.MLE(2);
    garchobj.TESTS(0,1);
    garchobj.Initialization(<>);
    garchobj.DoEstimation(<>);
    garchobj.Output();
    delete garchobj;
}
{
    decl i,j,k,l,garchobj;
    garchobj = new Garch();
  garchobj.Load("Chapter6Data.xls");
  garchobj.Select(Y_VAR, {"ftse100",0,0} );
    garchobj.SetSelSample(-1, 1, -1, 1);
    garchobj.CSTS(1,1);
    garchobj.DISTRI(1);
    garchobj.ARMA_ORDERS(1,0);
    garchobj.ARFIMA(0);
    garchobj.GARCH_ORDERS(1,1);
    garchobj.MODEL(6);
    garchobj.TRUNC(400);
    garchobj.BOUNDS(1);
    garchobj.MLE(2);
    garchobj.TESTS(0,1);
```

```
     garchobj.Initialization(<>);
     garchobj.DoEstimation(<>);
     garchobj.Output();
     delete garchobj;
}
{
     decl i,j,k,l,garchobj;
     garchobj = new Garch();
   garchobj.Load("Chapter6Data.xls");
   garchobj.Select(Y_VAR, {"ftse100",0,0} );
     garchobj.SetSelSample(-1, 1, -1, 1);
     garchobj.CSTS(1,1);
     garchobj.DISTRI(1);
     garchobj.ARMA_ORDERS(1,0);
     garchobj.ARFIMA(0);
     garchobj.GARCH_ORDERS(1,1);
     garchobj.MODEL(7);
     garchobj.TRUNC(400);
     garchobj.BOUNDS(1);
     garchobj.MLE(2);
     garchobj.TESTS(0,1);
     garchobj.Initialization(<>);
     garchobj.DoEstimation(<>);
     garchobj.Output();
     delete garchobj;
}
{
     decl i,j,k,l,garchobj;
     garchobj = new Garch();
   garchobj.Load("Chapter6Data.xls");
   garchobj.Select(Y_VAR, {"ftse100",0,0} );
     garchobj.SetSelSample(-1, 1, -1, 1);
     garchobj.CSTS(1,1);
     garchobj.DISTRI(1);
     garchobj.ARMA_ORDERS(1,0);
     garchobj.ARFIMA(0);
     garchobj.GARCH_ORDERS(1,1);
     garchobj.MODEL(8);
     garchobj.TRUNC(400);
     garchobj.BOUNDS(1);
     garchobj.MLE(2);
     garchobj.TESTS(0,1);
     garchobj.Initialization(<>);
     garchobj.DoEstimation(<>);
     garchobj.Output();
```

```
    delete garchobj;
}
{

   decl i,j,k,l,garchobj;
   garchobj = new Garch();
  garchobj.Load("Chapter6Data.xls");
  garchobj.Select(Y_VAR, {"ftse100",0,0} );
   garchobj.SetSelSample(-1, 1, -1, 1);
   garchobj.CSTS(1,1);
   garchobj.DISTRI(1);
   garchobj.ARMA_ORDERS(1,0);
   garchobj.ARFIMA(0);
   garchobj.GARCH_ORDERS(1,1);
   garchobj.MODEL(9);
   garchobj.TRUNC(400);
   garchobj.BOUNDS(1);
   garchobj.MLE(2);
   garchobj.TESTS(0,1);
   garchobj.Initialization(<>);
   garchobj.DoEstimation(<>);
   garchobj.Output();
   delete garchobj;
}
{

   decl i,j,k,l,garchobj;
   garchobj = new Garch();
  garchobj.Load("Chapter6Data.xls");
  garchobj.Select(Y_VAR, {"ftse100",0,0} );
   garchobj.SetSelSample(-1, 1, -1, 1);
   garchobj.CSTS(1,1);
   garchobj.DISTRI(1);
   garchobj.ARMA_ORDERS(1,0);
   garchobj.ARFIMA(0);
   garchobj.GARCH_ORDERS(1,1);
   garchobj.MODEL(10);
   garchobj.TRUNC(400);
   garchobj.BOUNDS(1);
   garchobj.MLE(2);
   garchobj.TESTS(0,1);
   garchobj.Initialization(<>);
   garchobj.DoEstimation(<>);
   garchobj.Output();
   delete garchobj;
}
{
```

```
    decl i,j,k,l,garchobj;
    garchobj = new Garch();
  garchobj.Load("Chapter6Data.xls");
  garchobj.Select(Y_VAR, {"ftse100",0,0} );
    garchobj.SetSelSample(-1, 1, -1, 1);
    garchobj.CSTS(1,1);
    garchobj.DISTRI(1);
    garchobj.ARMA_ORDERS(1,0);
    garchobj.ARFIMA(0);
    garchobj.GARCH_ORDERS(1,1);
    garchobj.MODEL(11);
    garchobj.TRUNC(400);
    garchobj.BOUNDS(1);
    garchobj.MLE(2);
    garchobj.TESTS(0,1);
    garchobj.Initialization(<>);
    garchobj.DoEstimation(<>);
    garchobj.Output();
    delete garchobj;
}
{

    decl i,j,k,l,garchobj;
    garchobj = new Garch();
  garchobj.Load("Chapter6Data.xls");
  garchobj.Select(Y_VAR, {"ftse100",0,0} );
    garchobj.SetSelSample(-1, 1, -1, 1);
    garchobj.CSTS(1,1);
    garchobj.DISTRI(3);
    garchobj.ARMA_ORDERS(1,0);
    garchobj.ARFIMA(0);
    garchobj.GARCH_ORDERS(1,1);
    garchobj.MODEL(1);
    garchobj.TRUNC(400);
    garchobj.BOUNDS(1);
    garchobj.MLE(2);
    garchobj.TESTS(0,1);
    garchobj.Initialization(<>);
    garchobj.DoEstimation(<>);
    garchobj.Output();
    delete garchobj;
}
{

    decl i,j,k,l,garchobj;
    garchobj = new Garch();
  garchobj.Load("Chapter6Data.xls");
```

```
  garchobj.Select(Y_VAR, {"ftse100",0,0} );
    garchobj.SetSelSample(-1, 1, -1, 1);
    garchobj.CSTS(1,1);
    garchobj.DISTRI(3);
    garchobj.ARMA_ORDERS(1,0);
    garchobj.ARFIMA(0);
    garchobj.GARCH_ORDERS(1,1);
    garchobj.MODEL(2);
    garchobj.TRUNC(400);
    garchobj.BOUNDS(1);
    garchobj.MLE(2);
    garchobj.TESTS(0,1);
    garchobj.Initialization(<>);
    garchobj.DoEstimation(<>);
    garchobj.Output();
    delete garchobj;
}
{
    decl i,j,k,l,garchobj;
    garchobj = new Garch();
  garchobj.Load("Chapter6Data.xls");
  garchobj.Select(Y_VAR, {"ftse100",0,0} );
    garchobj.SetSelSample(-1, 1, -1, 1);
    garchobj.CSTS(1,1);
    garchobj.DISTRI(3);
    garchobj.ARMA_ORDERS(1,0);
    garchobj.ARFIMA(0);
    garchobj.GARCH_ORDERS(1,1);
    garchobj.MODEL(3);
    garchobj.TRUNC(400);
    garchobj.BOUNDS(1);
    garchobj.MLE(2);
    garchobj.TESTS(0,1);
    garchobj.Initialization(<>);
    garchobj.DoEstimation(<>);
    garchobj.Output();
    delete garchobj;
}
{
    decl i,j,k,l,garchobj;
    garchobj = new Garch();
  garchobj.Load("Chapter6Data.xls");
  garchobj.Select(Y_VAR, {"ftse100",0,0} );
    garchobj.SetSelSample(-1, 1, -1, 1);
    garchobj.CSTS(1,1);
```

```
    garchobj.DISTRI(3);
    garchobj.ARMA_ORDERS(1,0);
    garchobj.ARFIMA(0);
    garchobj.GARCH_ORDERS(1,1);
    garchobj.MODEL(4);
    garchobj.TRUNC(400);
    garchobj.BOUNDS(1);
    garchobj.MLE(2);
    garchobj.TESTS(0,1);
    garchobj.Initialization(<>);
    garchobj.DoEstimation(<>);
    garchobj.Output();
    delete garchobj;
}
{
    decl i,j,k,l,garchobj;
    garchobj = new Garch();
  garchobj.Load("Chapter6Data.xls");
  garchobj.Select(Y_VAR, {"ftse100",0,0} );
    garchobj.SetSelSample(-1, 1, -1, 1);
    garchobj.CSTS(1,1);
    garchobj.DISTRI(3);
    garchobj.ARMA_ORDERS(1,0);
    garchobj.ARFIMA(0);
    garchobj.GARCH_ORDERS(1,1);
    garchobj.MODEL(5);
    garchobj.TRUNC(400);
    garchobj.BOUNDS(1);
    garchobj.MLE(2);
    garchobj.TESTS(0,1);
    garchobj.Initialization(<>);
    garchobj.DoEstimation(<>);
    garchobj.Output();
    delete garchobj;
}
{
    decl i,j,k,l,garchobj;
    garchobj = new Garch();
  garchobj.Load("Chapter6Data.xls");
  garchobj.Select(Y_VAR, {"ftse100",0,0} );
    garchobj.SetSelSample(-1, 1, -1, 1);
    garchobj.CSTS(1,1);
    garchobj.DISTRI(3);
    garchobj.ARMA_ORDERS(1,0);
    garchobj.ARFIMA(0);
```

```
    garchobj.GARCH_ORDERS(1,1);
    garchobj.MODEL(6);
    garchobj.TRUNC(400);
    garchobj.BOUNDS(1);
    garchobj.MLE(2);
    garchobj.TESTS(0,1);
    garchobj.Initialization(<>);
    garchobj.DoEstimation(<>);
    garchobj.Output();
    delete garchobj;
}
{
    decl i,j,k,l,garchobj;
    garchobj = new Garch();
  garchobj.Load("Chapter6Data.xls");
  garchobj.Select(Y_VAR, {"ftse100",0,0} );
    garchobj.SetSelSample(-1, 1, -1, 1);
    garchobj.CSTS(1,1);
    garchobj.DISTRI(3);
    garchobj.ARMA_ORDERS(1,0);
    garchobj.ARFIMA(0);
    garchobj.GARCH_ORDERS(1,1);
    garchobj.MODEL(7);
    garchobj.TRUNC(400);
    garchobj.BOUNDS(1);
    garchobj.MLE(2);
    garchobj.TESTS(0,1);
    garchobj.Initialization(<>);
    garchobj.DoEstimation(<>);
    garchobj.Output();
    delete garchobj;
}
{
    decl i,j,k,l,garchobj;
    garchobj = new Garch();
  garchobj.Load("Chapter6Data.xls");
  garchobj.Select(Y_VAR, {"ftse100",0,0} );
    garchobj.SetSelSample(-1, 1, -1, 1);
    garchobj.CSTS(1,1);
    garchobj.DISTRI(3);
    garchobj.ARMA_ORDERS(1,0);
    garchobj.ARFIMA(0);
    garchobj.GARCH_ORDERS(1,1);
    garchobj.MODEL(8);
    garchobj.TRUNC(400);
```

```
   garchobj.BOUNDS(1);
   garchobj.MLE(2);
   garchobj.TESTS(0,1);
   garchobj.Initialization(<>);
   garchobj.DoEstimation(<>);
   garchobj.Output();
   delete garchobj;
}
{
   decl i,j,k,l,garchobj;
   garchobj = new Garch();
  garchobj.Load("Chapter6Data.xls");
  garchobj.Select(Y_VAR, {"ftse100",0,0} );
   garchobj.SetSelSample(-1, 1, -1, 1);
   garchobj.CSTS(1,1);
   garchobj.DISTRI(3);
   garchobj.ARMA_ORDERS(1,0);
   garchobj.ARFIMA(0);
   garchobj.GARCH_ORDERS(1,1);
   garchobj.MODEL(9);
   garchobj.TRUNC(400);
   garchobj.BOUNDS(1);
   garchobj.MLE(2);
   garchobj.TESTS(0,1);
   garchobj.Initialization(<>);
   garchobj.DoEstimation(<>);
   garchobj.Output();
   delete garchobj;
}
{
   decl i,j,k,l,garchobj;
   garchobj = new Garch();
   garchobj.Load("Chapter6Data.xls");
  garchobj.Select(Y_VAR, {"ftse100",0,0} );
  garchobj.SetSelSample(-1, 1, -1, 1);
   garchobj.CSTS(1,1);
   garchobj.DISTRI(3);
   garchobj.ARMA_ORDERS(1,0);
   garchobj.ARFIMA(0);
   garchobj.GARCH_ORDERS(1,1);
   garchobj.MODEL(10);
   garchobj.TRUNC(400);
   garchobj.BOUNDS(1);
   garchobj.MLE(2);
   garchobj.TESTS(0,1);
```

```
    garchobj.Initialization(<>);
    garchobj.DoEstimation(<>);
    garchobj.Output();
    delete garchobj;
}
{
    decl i,j,k,l,garchobj;
    garchobj = new Garch();
  garchobj.Load("Chapter6Data.xls");
  garchobj.Select(Y_VAR, {"ftse100",0,0} );
    garchobj.SetSelSample(-1, 1, -1, 1);
    garchobj.CSTS(1,1);
    garchobj.DISTRI(3);
    garchobj.ARMA_ORDERS(1,0);
    garchobj.ARFIMA(0);
    garchobj.GARCH_ORDERS(1,1);
    garchobj.MODEL(11);
    garchobj.TRUNC(400);
    garchobj.BOUNDS(1);
    garchobj.MLE(2);
    garchobj.TESTS(0,1);
    garchobj.Initialization(<>);
    garchobj.DoEstimation(<>);
    garchobj.Output();
    delete garchobj;
}
{
    decl i,j,k,l,garchobj;
    garchobj = new Garch();
  garchobj.Load("Chapter6Data.xls");
  garchobj.Select(Y_VAR, {"ftse100",0,0} );
    garchobj.SetSelSample(-1, 1, -1, 1);
    garchobj.CSTS(1,1);
    garchobj.DISTRI(2);
    garchobj.ARMA_ORDERS(1,0);
    garchobj.ARFIMA(0);
    garchobj.GARCH_ORDERS(1,1);
    garchobj.MODEL(1);
    garchobj.TRUNC(400);
    garchobj.BOUNDS(1);
    garchobj.MLE(2);
    garchobj.TESTS(0,1);
    garchobj.Initialization(<>);
    garchobj.DoEstimation(<>);
    garchobj.Output();
```

```
      delete garchobj;
}
{
    decl i,j,k,l,garchobj;
    garchobj = new Garch();
  garchobj.Load("Chapter6Data.xls");
  garchobj.Select(Y_VAR, {"ftse100",0,0} );
    garchobj.SetSelSample(-1, 1, -1, 1);
    garchobj.CSTS(1,1);
    garchobj.DISTRI(2);
    garchobj.ARMA_ORDERS(1,0);
    garchobj.ARFIMA(0);
    garchobj.GARCH_ORDERS(1,1);
    garchobj.MODEL(2);
    garchobj.TRUNC(400);
    garchobj.BOUNDS(1);
    garchobj.MLE(2);
    garchobj.TESTS(0,1);
    garchobj.Initialization(<>);
    garchobj.DoEstimation(<>);
    garchobj.Output();
    delete garchobj;
}
{
    decl i,j,k,l,garchobj;
    garchobj = new Garch();
  garchobj.Load("Chapter6Data.xls");
  garchobj.Select(Y_VAR, {"ftse100",0,0} );
    garchobj.SetSelSample(-1, 1, -1, 1);
    garchobj.CSTS(1,1);
    garchobj.DISTRI(2);
    garchobj.ARMA_ORDERS(1,0);
    garchobj.ARFIMA(0);
    garchobj.GARCH_ORDERS(1,1);
    garchobj.MODEL(3);
    garchobj.TRUNC(400);
    garchobj.BOUNDS(1);
    garchobj.MLE(2);
    garchobj.TESTS(0,1);
    garchobj.Initialization(<>);
    garchobj.DoEstimation(<>);
    garchobj.Output();
    delete garchobj;
}
{
```

```
    decl i,j,k,l,garchobj;
    garchobj = new Garch();
  garchobj.Load("Chapter6Data.xls");
  garchobj.Select(Y_VAR, {"ftse100",0,0} );
    garchobj.SetSelSample(-1, 1, -1, 1);
    garchobj.CSTS(1,1);
    garchobj.DISTRI(2);
    garchobj.ARMA_ORDERS(1,0);
    garchobj.ARFIMA(0);
    garchobj.GARCH_ORDERS(1,1);
    garchobj.MODEL(4);
    garchobj.TRUNC(400);
    garchobj.BOUNDS(1);
    garchobj.MLE(2);
    garchobj.TESTS(0,1);
    garchobj.Initialization(<>);
    garchobj.DoEstimation(<>);
    garchobj.Output();
    delete garchobj;
}
{
    decl i,j,k,l,garchobj;
    garchobj = new Garch();
  garchobj.Load("Chapter6Data.xls");
  garchobj.Select(Y_VAR, {"ftse100",0,0} );
    garchobj.SetSelSample(-1, 1, -1, 1);
    garchobj.CSTS(1,1);
    garchobj.DISTRI(2);
    garchobj.ARMA_ORDERS(1,0);
    garchobj.ARFIMA(0);
    garchobj.GARCH_ORDERS(1,1);
    garchobj.MODEL(5);
    garchobj.TRUNC(400);
    garchobj.BOUNDS(1);
    garchobj.MLE(2);
    garchobj.TESTS(0,1);
    garchobj.Initialization(<>);
    garchobj.DoEstimation(<>);
    garchobj.Output();
    delete garchobj;
}
{
    decl i,j,k,l,garchobj;
    garchobj = new Garch();
  garchobj.Load("Chapter6Data.xls");
```

```
  garchobj.Select(Y_VAR, {"ftse100",0,0} );
    garchobj.SetSelSample(-1, 1, -1, 1);
    garchobj.CSTS(1,1);
    garchobj.DISTRI(2);
    garchobj.ARMA_ORDERS(1,0);
    garchobj.ARFIMA(0);
    garchobj.GARCH_ORDERS(1,1);
    garchobj.MODEL(6);
    garchobj.TRUNC(400);
    garchobj.BOUNDS(1);
    garchobj.MLE(2);
    garchobj.TESTS(0,1);
    garchobj.Initialization(<>);
    garchobj.DoEstimation(<>);
    garchobj.Output();
    delete garchobj;
}
{
    decl i,j,k,l,garchobj;
    garchobj = new Garch();
  garchobj.Load("Chapter6Data.xls");
  garchobj.Select(Y_VAR, {"ftse100",0,0} );
    garchobj.SetSelSample(-1, 1, -1, 1);
    garchobj.CSTS(1,1);
    garchobj.DISTRI(2);
    garchobj.ARMA_ORDERS(1,0);
    garchobj.ARFIMA(0);
    garchobj.GARCH_ORDERS(1,1);
    garchobj.MODEL(7);
    garchobj.TRUNC(400);
    garchobj.BOUNDS(1);
    garchobj.MLE(2);
    garchobj.TESTS(0,1);
    garchobj.Initialization(<>);
    garchobj.DoEstimation(<>);
    garchobj.Output();
    delete garchobj;
}
{
    decl i,j,k,l,garchobj;
    garchobj = new Garch();
  garchobj.Load("Chapter6Data.xls");
  garchobj.Select(Y_VAR, {"ftse100",0,0} );
    garchobj.SetSelSample(-1, 1, -1, 1);
    garchobj.CSTS(1,1);
```

```
    garchobj.DISTRI(2);
    garchobj.ARMA_ORDERS(1,0);
    garchobj.ARFIMA(0);
    garchobj.GARCH_ORDERS(1,1);
    garchobj.MODEL(8);
    garchobj.TRUNC(400);
    garchobj.BOUNDS(1);
    garchobj.MLE(2);
    garchobj.TESTS(0,1);
    garchobj.Initialization(<>);
    garchobj.DoEstimation(<>);
    garchobj.Output();
    delete garchobj;
}
{
    decl i,j,k,l,garchobj;
    garchobj = new Garch();
  garchobj.Load("Chapter6Data.xls");
  garchobj.Select(Y_VAR, {"ftse100",0,0} );
    garchobj.SetSelSample(-1, 1, -1, 1);
    garchobj.CSTS(1,1);
    garchobj.DISTRI(2);
    garchobj.ARMA_ORDERS(1,0);
    garchobj.ARFIMA(0);
    garchobj.GARCH_ORDERS(1,1);
    garchobj.MODEL(9);
    garchobj.TRUNC(400);
    garchobj.BOUNDS(1);
    garchobj.MLE(2);
    garchobj.TESTS(0,1);
    garchobj.Initialization(<>);
    garchobj.DoEstimation(<>);
    garchobj.Output();
    delete garchobj;
}
{
    decl i,j,k,l,garchobj;
    garchobj = new Garch();
  garchobj.Load("Chapter6Data.xls");
  garchobj.Select(Y_VAR, {"ftse100",0,0} );
    garchobj.SetSelSample(-1, 1, -1, 1);
    garchobj.CSTS(1,1);
    garchobj.DISTRI(2);
    garchobj.ARMA_ORDERS(1,0);
    garchobj.ARFIMA(0);
```

```
   garchobj.GARCH_ORDERS(1,1);
   garchobj.MODEL(10);
   garchobj.TRUNC(400);
   garchobj.BOUNDS(1);
   garchobj.MLE(2);
   garchobj.TESTS(0,1);
   garchobj.Initialization(<>);
   garchobj.DoEstimation(<>);
   garchobj.Output();
   delete garchobj;
}
{
   decl i,j,k,l,garchobj;
   garchobj = new Garch();
  garchobj.Load("Chapter6Data.xls");
  garchobj.Select(Y_VAR, {"ftse100",0,0} );
   garchobj.SetSelSample(-1, 1, -1, 1);
   garchobj.CSTS(1,1);
   garchobj.DISTRI(2);
   garchobj.ARMA_ORDERS(1,0);
   garchobj.ARFIMA(0);
   garchobj.GARCH_ORDERS(1,1);
   garchobj.MODEL(11);
   garchobj.TRUNC(400);
   garchobj.BOUNDS(1);
   garchobj.MLE(2);
   garchobj.TESTS(0,1);
   garchobj.Initialization(<>);
   garchobj.DoEstimation(<>);
   garchobj.Output();
   delete garchobj;
}
}
```

7

Intraday realized volatility models

7.1 Realized volatility

Andersen and Bollerslev (1998a) introduced an alternative volatility measure, the *realized volatility*. The modelling of realized volatility is based on using the sum of squared intraday returns to generate more accurate daily volatility measures. However, the idea of using high-frequency data to compute measures of volatility at a lower frequency was first introduced by Merton (1980).

Let us suppose that the instantaneous logarithmic price, $\log(P(t))$, of a financial asset conforms with a simple diffusion process,

$$d \log(P(t)) = \sigma(t)dW(t), \tag{7.1}$$

where $\sigma(t)$ is the volatility of the instantaneous returns process, $W(t)$ is the standard Wiener process, and $\sigma(t)$ and $W(t)$ are independent. The concept of realized volatility is based on that of *integrated volatility*, which is central to the stochastic volatility option pricing scheme considered by Hull and White (1987). The *integrated volatility*, $\sigma_t^{2(IV)}$, aggregated over the time interval $(t-1, t)$ is defined by

$$\sigma_t^{2(IV)} = \int_{t-1}^{t} \sigma^2(x)dx. \tag{7.2}$$

It is a latent variable and hence not observable. However, according to the theory of quadratic variation of semi-martingales, the integrated volatility can be consistently estimated by the realized volatility,[1] $\sigma_t^{2(RV)}$, which is defined as the sum of squared returns observed over very small time intervals,

---

[1] For technical details the reader is referred to Andersen et al. (2001b) and Barndorff-Nielsen and Shephard (2001).

$$\sigma_t^{2(RV)} = \sum_{j=1}^{m-1} \left(\log(P_{((j+1)/m),t}) - \log(P_{(j/m),t})\right)^2, \qquad (7.3)$$

where $P_{(m),t}$ are the financial asset prices during period t with sampling frequency m. For example, for daily time intervals and m observations per day, $\sigma_t^{2(RV)}$ denotes the realized volatility of trading day t, and

$$\sigma_t^{2(RV)(\tau)} = \sum_{i=0}^{\tau-1} \sum_{j=1}^{m-1} \left(\log(P_{((j+1)/m),t-i}) - \log(P_{(j/m),t-i})\right)^2 \qquad (7.4)$$

denotes the realized volatility for a period of τ trading days. The realized volatility converges in probability to the integrated volatility as the sampling frequency m increases, i.e.,

$$\plim_{m \to \infty} \left(\sum_{j=1}^{m-1} \left(\log(P_{((j+1)/m),t}) - \log(P_{(j/m),t})\right)^2 \right) = \sigma_t^{2(IV)}. \qquad (7.5)$$

Moreover, $\sigma_t^{2(RV)}$ is asymptotically normal and, in particular, as $m \to \infty$,

$$\frac{\sqrt{m}\left(\sigma_t^{2(RV)} - \int_{t-1}^{t} \sigma^2(x)dx\right)}{\sqrt{\int_{t-1}^{t} 2\sigma^4(x)dx}} \xrightarrow{d} N(0,1). \qquad (7.6)$$

For more details about the asymptotic distribution of $\sigma_t^{2(RV)}$ and the asymptotic volatility of volatility, $\sigma_t^{2(IQ)} = \int_{t-1}^{t} 2\sigma^4(x)dx$, termed *integrated quarticity*, see Barndorff-Nielsen and Shephard (2002a, 2002b, 2003, 2004a, 2004b, 2005, 2006).

The sampling frequency, m, is selected on a trade-off basis between accuracy and potential biases due to market microstructure frictions.[2] On the one hand, the accuracy improves as the sampling frequency increases $(m \to \infty)$, while on the other hand, at a high sampling frequency the bid–ask spread, the unavailability of the transaction at each point in time and the other market frictions are a source of additional noise in the estimator of volatility.

The sampling frequency, m, should be sufficiently high that the market microstructure[3] features do not induce bias in the volatility estimator. In the case of discrete time with a sample frequency of $m = 1$, $y_{(1),t}^2 = (\log(P_{(1),t}) - \log(P_{(1),t-1}))^2$ is an unbiased estimator of $\sigma_t^{2(IV)}$. As noted by Andersen and Bollerslev (1998a) and Ebens (1999), the discretely sampled daily returns, for $m = 1$, constitute a noisy estimator,

---

[2] The term 'market friction' refers to any factor that interferes with trade, such as transparency of transactions, transaction costs, taxes, regulatory costs and bid–ask spreads, i.e. the amount by which the ask price (the price at which a trader is offering to sell the asset) exceeds the bid price (the price that a trader is offering to pay for the asset).

[3] Madhavan (2000, p. 206) described the market microstructure theory as 'the area of finance that examines the process by which investors' latent demands are ultimately translated into transactions'. Alexander (2008, p. 180) described market microstructure as 'the study of price formation and how this is related to trading protocols and trading volume'.

but the accuracy improves as the sampling frequency increases $(m \to \infty)$. However, the observed tick-by-tick asset prices are available only at discrete points in time and asset returns are characterized by the effect of non-synchronous trading.

In the majority of studies, a sampling frequency of 5 or 30 minutes is used, in order to avoid market microstructure frictions without lessening the accuracy of the continuous record asymptotics. For example, Andersen and Bollerslev (1998a), Andersen et al. (1999a, 2000a, 2001b) and Kayahan et al. (2002) used a sampling frequency of 5 minutes for heavily traded assets. Andersen et al. (1999b, 2006) proposed the construction of the volatility signature plot, which provides a graphical representation of the average realized volatility against the sampling frequency. As the sampling frequency increases, the bias induced by microstructure frictions also increases. Thus, in the signature plot one should look for the highest frequency where the average realized volatility appears to stabilize. Corsi et al. (2001), Oomen (2001), Martens (2002), Areal and Taylor (2002), Barndorff-Nielsen et al. (2004), Aït-Sahalia et al. (2005), Engle and Sun (2005), Hansen and Lunde (2005b), Zhang et al. (2005) and Bandi and Russell (2005, 2006) deal with the effects of market microstructure and data adjustments on the construction of the realized volatility measure. Andersen et al. (2007), based on findings of Barndorff-Nielsen and Shephard (2004b, 2006), provided a non-parametric framework for measuring jumps in financial asset prices. An interesting finding of their study is that separately modelling the integrated volatility and jump components of the variation process is likely to result in important improvements in derivatives and other pricing decisions. Andersen et al. (1999b, 2000b, 2001a, 2003) were the first to explore the distributional properties of realized volatility. The main findings were that (i) although the distribution of asset returns is non-normal (with high skew and kurtosis), the distribution of returns scaled by the realized standard deviation is approximately Gaussian, and (ii) the realized logarithmic standard deviation is also nearly Gaussian.

It should be noted at this point that the inter-day variance is decomposed into the intraday variance, $\sigma_t^{2(RV)}$, and the intraday autocovariances, $y_{((j+1)/m),t}y_{((j-i+1)/m),t}$, as follows:[4]

---

[4] Note that $E(y_{((j+1)/m),t}) = 0$ and

$$y_t^2 = \left(\sum_{j=1}^{m-1} y_{((j+1)/m),t}\right)^2 = \left(\sum_{j=1}^{m-1} \left(\log\left(P_{((j+1)/m),t}\right) - \log\left(P_{(j/m),t}\right)\right)\right)^2$$

$$= \sum_{j=1}^{m-1} \left(\log\left(P_{((j+1)/m),t}/P_{(j/m),t}\right)\right)^2 + \sum_{j=1}^{m-1}\sum_{\substack{i=1 \\ i \neq j}}^{m-1} \log\left(P_{((j+1)/m),t}/P_{(j/m),t}\right)\log\left(P_{((i+1)/m),t}/P_{(i/m),t}\right)$$

$$= \sum_{j=1}^{m-1} \left(\log\left(P_{((j+1/m),t}/P_{(j/m),t}\right)\right)^2 + 2\sum_{j=1}^{m-1}\sum_{i=j+1}^{m-1} \log\left(P_{((j+1)/m),t}/P_{(j/m),t}\right)\log\left(P_{((j-i+1)/m),t}/P_{((j-i)/m),t}\right)$$

$$= \sum_{j=1}^{m-1} y_{((j+1/m),t}^2 + 2\sum_{j=1}^{m-1}\sum_{i=j+1}^{m-1} y_{((j+1)/m),t}y_{((j-i+1)/m),t} = \sigma_t^{2(RV)} + 2\sum_{j=1}^{m-1}\sum_{i=j+1}^{m-1} y_{((j+1)/m),t}y_{((j-i+1)/m),t}.$$

$$y_t^2 = \sigma_t^{2(RV)} + 2 \sum_{j=1}^{m-1} \sum_{i=j+1}^{m-1} y_{((j+1)/m),t} y_{((j-i+1)/m),t}. \tag{7.7}$$

The intra-day autocovariances are measurement errors, i.e. $E\big(y_{((j+1)/m),t}$ $y_{((j-i+1)/m),t}\big) = 0$, and $E(y_t^2) = \sigma_t^{2(RV)}$. Oomen (2001) suggested plotting the auto-covariance bias factor against the sampling frequency. The optimal sampling frequency can be chosen as the highest frequency for which the autocovariance bias term disappears.

Martens (2002) proposed accounting for changes in the asset prices during the hours that the stock market is closed without inserting the *noisy* effect of daily returns:

$$\sigma_t^{2(RV)} = \frac{\hat{\sigma}_{oc}^2 + \hat{\sigma}_{co}^2}{\hat{\sigma}_{oc}^2} \sum_{j=1}^{m-1} \big(\log(P_{((j+1)/m),t}) - \log(P_{(j/m),t})\big)^2, \tag{7.8}$$

where

$$\hat{\sigma}_{oc}^2 = T^{-1} \sum_{t=1}^{T} \big(\log(P_{(1),t}) - \log(P_{(1/m),t})\big)^2$$

is the *open-to-closed sample variance* and

$$\hat{\sigma}_{co}^2 = T^{-1} \sum_{t=1}^{T} \big(\log(P_{(1/m),t}) - \log(P_{(1),t-1})\big)^2$$

is the *closed-to-open sample variance*.

Engle and Sun (2005) proposed an econometric model for the joint distribution of tick-by-tick return and duration, with microstructure noise explicitly filtered out, and used the model to obtain an estimate of daily realized volatility. Zhang et al. (2005) noted that the usual realized volatility mainly estimates the magnitude of the noise term rather than anything to do with volatility. They showed that instead of sparsely sampling the tick-by-tick data, one should separate the observations into multiple *grids* and combine the usual single grid realized volatility with the multiple grid based device. Martens and van Dijk (2007) introduced the realized range as the summation of high–low ranges for intraday intervals. Although, the realized range is theoretically a more efficient estimator than the realized variance, in practice it is highly affected by the microstructure frictions. However, Martens and van Dijk (2007) accounted for market biasedness by scaling the realized range with the ratio of the average level of the daily range and the average level of the realized range. Their Monte Carlo simulations provided evidence that the scaled realized range outperforms various realized variance estimators that are adjusted for microstructure noise.

7.2 Intraday volatility models

A model that has been considered to capture the long-memory property of the realized volatility is the fractionally integrated ARMAX, or ARFIMAX$(\kappa, \tilde{d}, l)$, specification represented by

$$(1-c(L))(1-L)^{\tilde{d}}\left(\log\left(\sigma_t^{2(RV)}\right)-x'_t\beta\right) = (1+d(L))\varepsilon_t, \quad \varepsilon_t \overset{i.i.d.}{\sim} N\left(0,\sigma_\varepsilon^2\right), \qquad (7.9)$$

where $c(L) = \sum_{i=1}^{\kappa} c_i L^i$, $d(L) = \sum_{i=1}^{l} d_i L^i$, x_t is a $k \times 1$ vector of explanatory variables and β is a $k \times 1$ vector of unknown parameters. Applications of ARFIMAX models to the realized volatility and their properties have been considered by Taylor and Xu (1997), Andersen and Bollerslev (1997, 1998b), Ebens (1999), Andersen (2000), Oomen (2001), Bollerslev and Wright (2001), Andersen et al. (2003, 2005a), Thomakos and Wang (2003), Giot and Laurent (2004), Koopman et al. (2005), Angelidis and Degiannakis (2007) and references therein.

However, the volatility of volatility also exhibits time variation and volatility clustering. A representative example is the realized volatility of the S&P500 index futures analysed by Corsi et al. (2005). Therefore, an ARFIMAX$(\kappa, \tilde{d}, l)$-GARCH (p, q) model for the realized volatility can be applied:

$$(1-c(L))(1-L)\tilde{d}\left(\log\left(\sigma_t^{2(RV)}\right)-x'_t\beta\right) = (1+d(L))\varepsilon_t,$$
$$\varepsilon_t = h_t z_t,$$
$$h_t^2 = a_0 + \sum_{i=1}^{q} a_i \varepsilon_{t-i}^2 + \sum_{i=1}^{p} b_i h_{t-i}^2, \qquad (7.10)$$

where $z_t \sim N(0,1)$. The ARFIMAX$(\kappa, \tilde{d}, l)$-GARCH(p, q) model was first proposed by Baillie et al. (1996b) to analyse monthly consumer price index inflation data. Naturally, the model can be straightforwardly extended to take account of any ARCH volatility specification and distributional assumption existing in the ARCH literature. Degiannakis (2008a) showed that the ARFIMAX$(\kappa, \tilde{d}, l)$-TARCH(p, q) model is able to provide statistically superior one-trading-day-ahead realized volatility forecasts.

Corsi (2004) suggested the heterogeneous autoregressive model for the realized volatility (HAR-RV),

$$\sigma_t^{(RV)} = w_0 + w_1 \sigma_{t-1}^{(RV)} + w_2 \left(\frac{1}{5}\sum_{j=1}^{5}\sigma_{t-j}^{(RV)}\right) + w_3 \left(\frac{1}{22}\sum_{j=1}^{22}\sigma_{t-j}^{(RV)}\right) + \varepsilon_t,$$
$$\varepsilon_t \overset{i.i.d.}{\sim} N\left(0,\sigma_\varepsilon^2\right), \qquad (7.11)$$

with the last day's realized volatility explained by the daily, weekly and monthly realized volatilities. The HAR-RV model is an autoregressive structure of the volatilities realized over different interval sizes. Its economic interpretation stems from the heterogeneous market hypothesis presented by Müller et al. (1993). The basic idea is that market participants have a different perspective of their investment horizon. Thus, $\sigma_{t-1}^{(RV)}$ accounts for the volatility of inter-day or intraday trading strategies, $\frac{1}{5}\sum_{j=1}^{5}\sigma_{t-j}^{(RV)}$ accounts for medium-term trading and $\frac{1}{22}\sum_{j=1}^{22}\sigma_{t-j}^{(RV)}$ captures investment strategies of time horizons of one month or longer. The heterogeneity, which originates from the difference in the time horizon, creates volatility. The HAR-RV model can alternatively be represented in terms of the square root of the sum of the realized variances:

$$\sigma_t^{(RV)} = w_0 + w_1 \sigma_{t-1}^{(RV)} + w_2 \sqrt{\frac{1}{5} \sum_{j=1}^{5} \sigma_{t-j}^{2(RV)}} + w_3 \sqrt{\frac{1}{22} \sum_{j=1}^{22} \sigma_{t-j}^{2(RV)}} + \varepsilon_t, \qquad (7.12)$$

$$\varepsilon_t \overset{i.i.d.}{\sim} N(0, \sigma_\varepsilon^2),$$

where $\sqrt{\tau^{-1} \sum_{j=1}^{\tau} \sigma_{t-j}^{2(RV)}}$ defines the average τ-period realized standard deviation.

Corsi et al. (2005) extended the HAR model by including a GARCH component to account for the volatility clustering of the realized volatility. They termed the new specification the HAR-GARCH(p, q) model:

$$\sigma_t^{(RV)} = w_0 + w_1 \sigma_{t-1}^{(RV)} + w_2 \left(\frac{1}{5} \sum_{j=1}^{5} \sigma_{t-j}^{(RV)} \right) + w_3 \left(\frac{1}{22} \sum_{j=1}^{22} \sigma_{t-j}^{(RV)} \right) + \varepsilon_t,$$

$$\varepsilon_t = h_t z_t, \qquad (7.13)$$

$$h_t^2 = a_0 + \sum_{i=1}^{q} a_i \varepsilon_{t-i}^2 + \sum_{i=1}^{p} b_i \sigma_{t-i}^2,$$

where $z_t \sim N(0, 1)$. Of course, in this case too it is possible to extend the model to take account of any ARCH volatility specification and distributional assumption existing in the ARCH literature.

Another approach in modelling intraday quantities is in terms of the modelling of durations between trades. Engle and Russell (1998) proposed the modelling of the intervals between two trades in a manner similar to an ARCH process. Let $dt_i \equiv t_i - t_{i-1}$ be the interval between two transactions. This is referred to as *duration*. The autoregressive conditional duration (ACD) class of models is defined as

$$dt_i = E(dt_i | I_{i-1}) z_i,$$

$$z_i \overset{i.i.d.}{\sim} f(E(z_i) = 1; w), \qquad (7.14)$$

where z_i is a sequence of independent and identically distributed non-negative random variables and $E(dt_i | I_{i-1})$ is the conditional expectation of the duration. In the ACD(p, q) model, the conditional expectation of the duration is specified as a linear function of past durations and conditional durations:

$$E(dt_i | I_{i-1}) = a_0 + \sum_{j=1}^{p} (a_j dt_{i-j}) + \sum_{j=1}^{q} \left(b_j E(dt_{i-j} | I_{i-j-1}) \right). \qquad (7.15)$$

Engle and Russell (1998) assumed for the distribution of z_i the standard exponential and the Weibull distribution, whereas Grammig and Maurer (2000) suggested the Burr distribution, and Lunde (1999) and Zhang et al. (2001) proposed the standardized generalized gamma distribution. Zhang et al. (2001), based on the notion of threshold autoregressive models, proposed the threshold ACD, or TACD(p, q), model, in order to capture the non-linear relation between the duration and predetermined variables and to account for regime switches. Tsay (2002, p. 199) provides simulated examples for the ACD models. Bauwens et al. (2004) classify the financial duration models into five classes: ACD, logarithmic ACD (proposed by Bauwens and Giot, 2000), TACD, stochastic conditional duration and stochastic volatility duration models. Meitz and Teräsvirta (2006) introduced the smooth transition ACD, or STACD(p, q), model[5] and the time-varying ACD, or TVACD (p, q), model.

Engle (2002a) proposed the multiplicative error model (MEM) for positive-valued time series – realized volatility, duration, trading volume, etc. The MEM(p, q) model has the same structure as the ACD model, for any non-negative variable, $y_t \geq 0$ for all t:

$$
\begin{aligned}
y_t &= E(y_t|I_{t-1})z_t, \\
z_t &\overset{i.i.d.}{\sim} f(E(z_t) = 1; w), \\
E(y_t|I_{t-1}) &= a_0 + \sum_{i=1}^{q}(a_i y_{t-i}) + \sum_{j=1}^{p}(b_j E(y_t|I_{t-1})).
\end{aligned}
\tag{7.16}
$$

Lanne (2006) introduced an MEM model for realized volatility with time-varying parameters and an error term following a mixture of gamma distributions. The model is fitted to the daily realized volatility series of Deutschemark–dollar and yen–dollar returns and is shown to capture the slowly decaying autocorrelation function of the observed realized volatility better than the ARFIMAX model.

7.3   Intraday realized volatility and ARFIMAX models in G@RCH 4.2 OxMetrics: an empirical example

In this section we investigate some properties of the realized volatility for the CAC40, DAX30, FTSE100 and S&P500 indices and estimate the ARFIMAX $(\kappa, \tilde{d}, l)$ specification using the logarithm of the realized variance as the dependent variable. The data set was obtained from Olsen and Associates and consists of 5-minute linearly interpolated prices of the CAC40 from January 1995 to December 2003 (2257 trading days in all), the DAX30 from July 1995 to December 2003 (2137 trading days in all), the FTSE100 from January 1998 to December 2003 (1486 trading days in all), and the S&P500 from January 1997 to December 2003 (1735 trading days in all).

7.3.1 Descriptive statistics

Table 7.1 provides descriptive statistics for the daily log-returns. As has been the case with the data sets of previous sections, non-zero skewness and excess kurtosis are observed. Under the assumption that the log-returns are independently and identically normally distributed, the sample skewness, $\hat{s}$, and kurtosis, $\hat{\kappa}$, are distributed normally with variances $V(\hat{s}) = 6/T$ and $V(\hat{\kappa}) = 24/T$, respectively. Only the skewness parameter of the CAC40 index is within the 95% confidence interval thus indicating that only the distribution of this index may be symmetric.

To avoid market microstructure frictions without lessening the accuracy of the continuous record asymptotics, we use 5-minute linearly interpolated prices. We follow

---

[5] Inspired by the smooth transition GARCH(p, q) model.

Table 7.1 Descriptive statistics of the daily log-returns, $y_t = 100(\log(P_t) - \log(P_{t-1}))$, and the annualized intraday standard deviation, $\sqrt{252}\sigma_t^{(RV)}$, for the CAC40 (January 1995 to December 2003), DAX30 (December 2003), DAX30 (July 1995 to November 2003), FTSE100 (January 1998 to December 2003) and S&P500 (January 1997 to December 2003) indices

	Daily log-returns				Annualized realized volatility (standard deviation)			
	CAC40	DAX30	FTSE100	S&P500	CAC40	DAX30	FTSE100	S&P500
Mean	0.02787	0.02995	−0.0093	0.022	24.014	23.610	16.993	15.805
Median	0.03949	0.10805	−0.0046	0.029	19.914	21.080	15.274	14.530
Maximum	8.88540	7.44950	5.91238	5.576	152.33	120.00	104.61	57.870
Minimum	−7.6930	−8.8768	−6.3508	−8.762	6.6150	3.2511	5.3032	4.590
Std. dev.	1.57241	1.70246	1.31767	1.300	15.579	13.313	7.3257	6.520
Skewness	0.09383	−0.1752	−0.1399	−0.125	3.7115	1.7641	2.8113	1.882
Kurtosis	5.64017	5.18385	4.75622	5.670	21.900	9.2065	21.239	8.778
Jarque–Bera	658.833	435.601	195.819	520.05	38775	4538	22555	3437
Probability	0.00	0.00	0.00	0.00	0.00	0.00	0.00	0.00

Martens (2002) and compute the realized volatility as in (7.8) for scaling the intraday returns to account for overnight returns without inserting the 'noisy' effect of daily returns.

Table 7.1 also gives descriptive statistics for the the annualized volatility, $\sqrt{252\sigma_t^{2(RV)}}$. The most volatile indices appear to be the CAC40 and the DAX30, while the safest market is that of the UK. Ebens (1999), Andersen et al. (2001a), Thomakos and Wang (2003) and Giot and Laurent (2004), among others, reached similar findings, noting that the risk investors face is not normally distributed as it exhibits positive skewness and excess kurtosis.

Figure 7.1 is a plot of the annualized realized standard deviation, $\sqrt{252}\sigma_t^{(RV)}$. The daily closing prices are also plotted in order to make feasible the tendency of stock

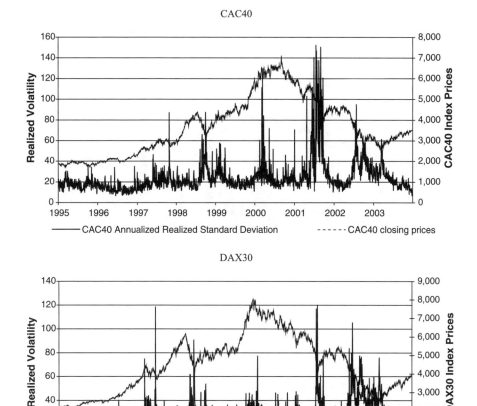

Figure 7.1   Annualized realized standard deviation and daily index prices for the CAC40, DAX30, FTSE100 and S&P500 indices.

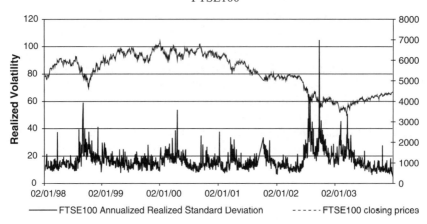

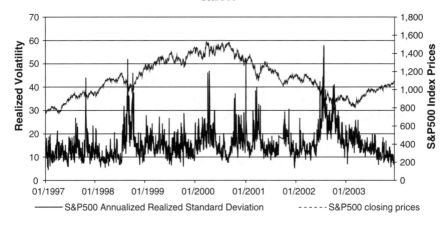

Figure 7.1   (Continued)

returns to be negatively correlated with changes in returns volatility. Both ARCH
volatility and realized volatility have an asymmetric relationship with asset returns
(compare Figures 2.19–2.22 with Figure 7.1).

Figures 7.2–7.5 plot the distributions of daily log-returns, intraday standard
deviation and logarithmic variance. The density estimates are calculated as in Ebens
(1999) based on the normal kernel.[6] The intraday volatility is leptokurtic and skewed
to the right, whereas the density of logarithmic variance is close to the normal
distribution but statistically distinguishable from it. A series of normality test
statistics (Jarque–Bera, Anderson–Darling and Cramér–von Mises) are also provided

---

[6] See equation 3.31 of Silverman (1986).

Density estimate of log-returns

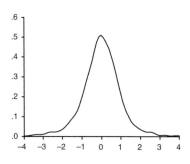

Density estimate of intraday standard deviation

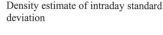

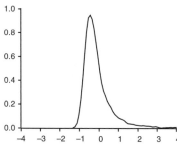

Sample skewness, $\hat{s}$, and kurtosis, $\hat{\kappa}$

Density estimate of intraday logarithmic variance

	$\hat{s}$	$\hat{\kappa}$
Log-returns	0.093838	5.640179
Variance	7.139802	65.90414
Standard deviation	3.711525	21.90015
Logarithmic variance	1.030770	5.027758
Jarque–Bera	786.3531	[0.00]
Anderson–Darling	26.54339	[0.00]
Cramér–von Mises	4.398854	[0.00]

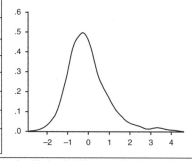

The density estimates are based on the normal kernel with bandwidths calculated according to equation 3.31 of Silverman (1986). The series are normalized to zero mean and unit variance. The bottom left-hand panel shows the estimated skewness and kurtosis of daily log-returns, and intraday variance, standard deviation and logarithmic variance. The Jarque–Bera, Anderson–Darling and Cramér–von Mises statistics test the null hypothesis that the logarithmic variance is normally distributed; p-values are displayed in square brackets.

Figure 7.2   CAC40 index: distribution of daily log-returns, intraday standard deviation and logarithmic intraday variance.

in Figures 7.2–7.5. Clearly, the assumption of normality is rejected at any level of significance.

Andersen et al. (2000b, 2000c, 2001a) point out that the distribution of $\{y_t/\sigma_t^{(RV)}\}_{t=1}^{T}$, the standardized daily returns divided by the intraday standard deviation, is very close to the normal. According to Figure 7.6 the graphs of the density of $\{y_t/\sigma_t^{(RV)}\}_{t=1}^{T}$ are indeed very close to that of the normal distribution. Moreover, Table 7.2, which lists the Anderson–Darling and Cramér–von Mises statistics, reveals that the normality assumption associated with the $\{y_t/\sigma_t^{(RV)}\}_{t=1}^{T}$ series is not rejected in the case of the DAX30 and FTSE100 indices.

Density estimate of log-returns

Density estimate of intraday standard deviation

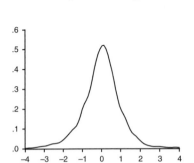
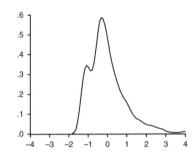

Sample skewness, $\hat{s}$, and kurtosis, $\hat{\kappa}$

Density estimate of intraday logarithmic variance

	$\hat{s}$	$\hat{\kappa}$
Log-returns	−0.17528	5.183852
Variance	5.670969	56.16863
Standard deviation	1.764126	9.206527
Logarithmic variance	−0.295738	3.077907
Jarque–Bera	31.69122	[0.00]
Anderson–Darling	11.83784	[0.00]
Cramér–von Misses	2.149275	[0.00]

See footnote to Figure 7.2.

Figure 7.3   DAX30 index: distribution of daily log-returns, intraday standard deviation and logarithmic intraday variance.

7.3.2 In-sample analysis

The in-sample analysis would require estimation of an ARFIMAX model. In particular, consider the case of the CAC40 index. Denoting the realized variance of the index by

$$\sigma_t^{2(RV)} = \frac{\hat{\sigma}_{oc}^2 + \hat{\sigma}_{co}^2}{\hat{\sigma}_{oc}^2} \sum_{j=1}^{m-1} \left(100(\log(CAC40_{((j+1)/m),t}) - \log(CAC40_{(j/m),t}))\right)^2,$$

the ARFIMAX$(1, \tilde{d}, 1)$ model represented by

$$(1 - c_1 L)(1 - L)\tilde{d}\left(\log\left(\sigma_t^{2(RV)}\right) - w_0 - w_1 y_{t-1} - w_2 d_{t-1} y_{t-1}\right) = (1 + d_1 L)\varepsilon_t, \quad (7.17)$$

should be estimated, where $y_t = 100 \log(CAC40_t / CAC40_{t-1})$ denotes the CAC40 daily log-returns, $d_t = 1$ when $y_t > 0$ and $d_t = 0$ otherwise, and $\varepsilon_t \sim N(0, \sigma_\varepsilon^2)$. The parameter w_2 refers to the asymmetric relationship between volatility and past returns.

Density estimate of log-returns

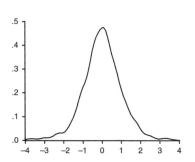

Density estimate of intraday standard deviation

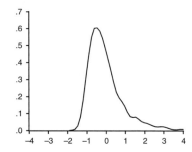

Sample skewness, $\hat{s}$, and kurtosis, $\hat{\kappa}$

Density estimate of intraday logarithmic variance

	$\hat{s}$	$\hat{\kappa}$
Log-returns	-0.13993	4.756223
Variance	11.36949	235.7506
Standard deviation	2.811338	21.23928
Logarithmic variance	0.628339	3.802112
Jarque–Bera	137.6174	[0.00]
Anderson–Darling	6.695296	[0.00]
Cramér–von Mises	1.049088	[0.00]

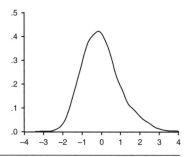

See footnote to Figure 7.2.

Figure 7.4   FTSE100 index: distribution of daily log-returns, intraday standard deviation and logarithmic intraday variance.

Since $\varepsilon_t \sim N(0, \sigma_\varepsilon^2)$, the variable $\exp(\varepsilon_t)$ is log-normally distributed, and hence the unbiased in-sample realized volatility is estimated as by

$$\hat{\sigma}_{(un),t}^{2(RV)} = \exp\left(\log\hat{\sigma}_t^{2(RV)} + 0.5\sigma_\varepsilon^2\right). \quad (7.18)$$

The model is estimated by the ARFIMA 1.04 (Doornik and Ooms (2006)) package of Ox.[7] The program named *chapter7.FullSample.cac40.ox*, given below, carries out the estimation of the ARFIMAX$(1, \tilde{d}, 1)$ model for the CAC40 index.[8]

```
#include <oxstd.h>
#include <oxfloat.h>
#import <packages/arfima104/arfima>
```

---

[7] ARFIMA 1.04 is compatible with Ox version 4 or later. An earlier version, ARFIMA 1.01 (Doornik and Ooms, 2001), is compatible with earlier versions of Ox.

[8] The programs for the rest data set model estimation are available in the CD-ROM.

Density estimate of log-returns

Density estimate of intraday standard deviation

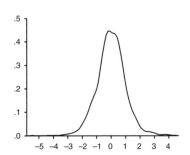

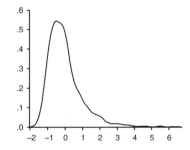

Sample skewness, $\hat{s}$, and kurtosis, $\hat{\kappa}$

Density estimate of intraday logarithmic variance

	$\hat{s}$	$\hat{\kappa}$
Log-returns	0.02023	4.72845
Variance	4.1229	28.9487
Standard deviation	1.84977	8.62704
Logarithmic variance	0.39006	3.40753
Jarque–Bera	56.4236	[0.00]
Anderson–Darling	3.94075	[0.00]
Cramér–von Mises	0.63344	[0.00]

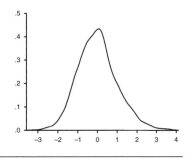

See footnote to Figure 7.2.

Figure 7.5   S&P500 index: distribution of daily log-returns, intraday standard deviation and logarithmic intraday variance.

```
main ()
{
decl rolling_sample=2257-1; // for cac40
decl jj;
decl arfima;
arfima = new Arfima ();
arfima.Load("Chapter7.cac40.xls");
arfima.Select(Y_VAR, {"lnrv",0,0 } );
arfima.Deterministic(FALSE);
arfima.Select(X_VAR, {"Constant", 0, 0 } );
arfima.Select(X_VAR, {"daily_return_lag", 0, 0 } );
arfima.Select(X_VAR, {"daily_return_lag_minus", 0, 0 } );
arfima.ARMA(1,1);
arfima.SetMethod(M_MAXLIK); //use the exact
  maximum likelihood
```

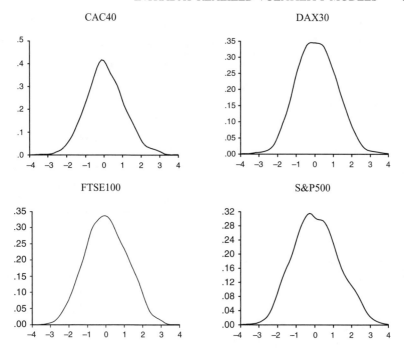

Figure 7.6   Distribution of daily log-returns standardized by the intraday standard deviation.

```
arfima.SetSelSample(-1, 1, -1, 1);
arfima.Estimate();
decl E_RVbiased=arfima.GetGroupLag(Y_VAR, 0, 0)
  -arfima.GetResiduals();
decl E_RV=exp(E_RVbiased+0.5*arfima.GetSigma2());
decl sfile;
decl cac40_ERV = zeros(rolling_sample+1,2);
decl cac40mat;
cac40mat = loadmat("Chapter7.cac40.xls");
cac40_ERV[][0] = E_RV[]; //unbiased in-sample RV[t|t-1]
for (jj=1; jj<rolling_sample+1;++jj)
{
cac40_ERV[jj-1][1] = cac40mat[jj][6];//daily return
}
sfile = sprint("Chapter7.FullSample.cac40.xls");
savemat(sfile,cac40_ERV);
delete arfima;
}
```

Table 7.2   Descriptive statistics of the daily log-returns standardized by the intraday standard deviation, $\{y_t/\sigma_t^{(RV)}\}_{t=1}^{T}$, and statistics for testing the normality assumption

	CAC40	DAX30	FTSE100	S&P500
Mean	0.083	0.110	0.032	0.090
Standard deviation	1.035	1.077	1.122	1.220
Skewness	0.346	0.135	0.099	0.210
Kurtosis	4.378	3.054	2.651	2.801
Lilliefors test	0.026	0.020	0.022	0.030
[p-value]	[0.002]	[0.042]	[>0.1]	[0.001]
Cramér–von Mises test	0.335	0.155	0.170	0.278
[p-value]	[0.000]	[0.021]	[0.013]	[0.001]

ARFIMA 1.04 reads the variables $\sigma_t^{2(RV)}$, y_{t-1} and $d_{t-1}y_{t-1}$ from the Excel file named *chapter7.cac40.xls* to estimate the ARFIMAX$(\kappa, \tilde{d}, l)$ specification. The command *arfima.Select* defines $\sigma_t^{2(RV)}$ as the dependent variable (*Y_VAR*), and y_{t-1} and $d_{t-1}y_{t-1}$ as the exogenous variables (*X_VAR*) in the ARFIMAX model. The command *arfima.ARMA(1,1)* defines the orders ($\kappa = 1$ and $l = 1$). *E_RV* declares the unbiased in-sample realized variance estimator, $\hat{\sigma}_{(un),t}^{2(RV)}$. For each trading day, the value of $\hat{\sigma}_{(un),t}^{2(RV)}$ and the log-returns are saved in the Excel file *chapter7. FullSample.cac40.xls*.

Table 7.3 contains the estimated parameters of the ARFIMAX$(1, \tilde{d}, 1)$ model. The asymmetry between past bad or good news and volatility is statistically significant for all models and indices. It should be noted at this point that the leverage effect was also found to be statistically significant in Ebens (1999) for the Dow Jones Industrial Average index and in Giot and Laurent (2004) for the CAC40 index, the S&P500 futures contracts, the yen–dollar and Deutschemark–dollar exchange rates. On the other hand, Andersen et al. (1999a) did not find evidence supporting this relationship in the Deutschemark–dollar exchange rate. Note that the variance is characterized by a slowly mean-reverting fractionally integrated process with the degree of integration estimated between 0.492 and 0.497. In all cases, the fractional integration parameter is lower than 0.5, but statistically not different from 0.5. In similar studies, $\tilde{d}$, was estimated between 0.40 and 0.48 by Giot and Laurent (2004), less than 0.5 by Ebens (1999) and between 0.25 and 0.45 by Thomakos and Wang (2003).

7.3.3 Out-of-sample analysis

For an out-of-sample analysis in terms of the model considered in the previous section, one may proceed as follows. Based on a rolling sample of $\tilde{T} = 1000$ trading days, the parameter vector $(\tilde{d}^{(t)}, w_0^{(t)}, w_1^{(t)}, w_2^{(t)}, c_1^{(t)}, d_1^{(t)}, \sigma_\varepsilon^{2(t)})'$ is re-estimated and $\tilde{T}$ one-day-ahead volatility forecasts are computed. The total sample size is

Table 7.3   ARFIMAX$(1, \tilde{d}, 1)$ estimated model parameters (equation (7.17))

Parameters	CAC40		DAX30	
w_0	0.3658	(2.8740)	0.198675	(3.006)
w_1	0.02454	(0.0118)	0.035161	(0.011)
w_2	−0.11839	(0.0205)	−0.12860	(0.0188)
c_1	0.132769	(0.0847)	0.030521	(0.1437)
d_1	−0.313349	(0.0798)	−0.16822	(0.1413)
$\tilde{d}$	0.496994	(0.0042)	0.49718	(0.0039)
σ_ε^2	0.259613		0.22598	
Parameters	FTSE100		S&P500	
w_0	−0.199619	(1.854)	−0.3452	(1.778)
w_1	0.0151035	(0.0166)	−0.02525	(0.0148)
w_2	−0.105202	(0.0282)	−0.09695	(0.0255)
c_1	0.106464	(0.1216)	−0.1584	(0.1262)
d_1	−0.296288	(0.1170)	−0.00004	(0.1295)
$\tilde{d}$	0.494346	(0.0078)	0.4925	(0.0099)
σ_ε^2	0.211864		0.2182	

Standard errors are reported in parentheses. Data set comprises 5-minute intra-day log-returns for CAC40 (January 1995 to December 2003), DAX30 (July 1995 to November 2003), FTSE100 (January 1998 to December 2003) and S&P500 (January 1997 to December 2003) indices.

$T = \tilde{T} + \breve{T}$. The one-day-ahead conditional realized standard deviation forecasts are computed as

$$\sigma_{(un),t+1|t}^{(RV)} = \sqrt{\exp\left(\log\left(\sigma_{t+1|t}^{2(RV)}\right) + 0.5\sigma_\varepsilon^{2(t)}\right)}, \qquad (7.19)$$

where

$$\log\left(\sigma_{t+1|t}^{2(RV)}\right) = d_1^{(t)}(1-L)^{-\tilde{d}^{(t)}}\varepsilon_{t|t} + w_0^{(t)} + w_1^{(t)}y_t + w_2^{(t)}d_t y_t$$
$$+ c_1^{(t)}\left(\log\left(\sigma_{t|t}^{2(RV)}\right) - w_0^{(t)} - w_1^{(t)}y_{t-1} - w_2^{(t)}d_{t-1}y_{t-1}\right) \qquad (7.20)$$

and

$$(1-L)^{-\tilde{d}^{(t)}} = 1 + \sum_{i=1}^{\infty}\left(\frac{\Gamma\left(i + \tilde{d}^{(t)}\right)}{\Gamma\left(\tilde{d}^{(t)}\right)\Gamma(i+1)}L^i\right)$$
$$= 1 + \frac{1}{1!}\tilde{d}^{(t)}L + \frac{1}{2!}\tilde{d}^{(t)}\left(1 + \tilde{d}^{(t)}\right)L^2 - \ldots, \qquad (7.21)$$

for $\tilde{d}^{(t)} > 0$, and $\Gamma(.)$ is the gamma function.

The program named *chapter7.OutSample.cac40.ox*, given below, based on a rolling sample of constant size of $\tilde{T} = 1000$, computes $\tilde{T} = 1257$ one-day-ahead CAC40 volatility forecast estimates from the ARFIMAX$(1, \tilde{d}, 1)$ model:

```
#include <oxstd.h>
#include <oxfloat.h>
#import <packages/arfima104/arfima>
main ()
{
decl rolling_sample=999; //It's the constant rolling
  sample minus 1
decl ii;
decl cac40_output;
cac40_output = zeros(2257,12);
for (ii=2; ii<1259;++ii)
{
decl arfima;
arfima = new Arfima();
arfima.Load("Chapter7.cac40.xls");
arfima.Select(Y_VAR, {"lnrv",0,0 } );
arfima.Deterministic(FALSE);
arfima.Select(X_VAR, {"Constant", 0, 0 } );
arfima.Select(X_VAR, {"daily_return_lag", 0, 0 } );
arfima.Select(X_VAR, {"daily_return_lag_minus", 0, 0 } );
arfima.ARMA(1,1);
arfima.SetMethod(M_MAXLIK);
arfima.SetSelSample(ii, 1, ii+rolling_sample, 1);
arfima.Estimate();
decl prmter=arfima.GetPar();
decl lnRV=arfima.GetGroupLag(Y_VAR, 0, 0);
decl resids=arfima.GetResiduals();
decl E_RVbiased=arfima.GetGroupLag(Y_VAR, 0, 0)
  -arfima.GetResiduals();
decl E_RV=exp(E_RVbiased+0.5*arfima.GetSigma2());
decl fore=arfima.Forecast(1);
decl unbRVfore=exp(fore+0.5*arfima.GetSigma2());
cac40_output[ii-1][0]=prmter[0][0];
cac40_output[ii-1][1]=prmter[1][0];
cac40_output[ii-1][2]=prmter[2][0];
cac40_output[ii-1][3]=prmter[3][0];
cac40_output[ii-1][4]=prmter[4][0];
cac40_output[ii-1][5]=prmter[5][0];
cac40_output[ii-1][6]=lnRV[rolling_sample][0];
  //dependent variable
cac40_output[ii-1][7]=resids[rolling_sample][0];
```

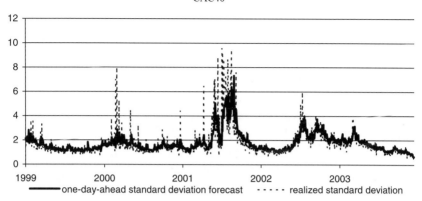

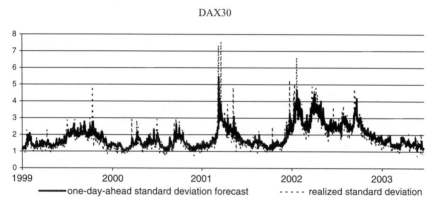

Figure 7.7   Realized intraday standard deviation and its forecast yielded by the ARFIMAX$(1, \tilde{d}, 1)$ model for the CAC40, DAX30, FTSE100 and S&P500 indices.

```
//residuals
cac40_output[ii-1][8]=E_RVbiased[rolling_sample][0];
  //biased in-sample lnRV[t|t-1]
cac40_output[ii-1][9]=E_RV[rolling_sample][0];
  //unbiased in-sample RV[t|t-1]
cac40_output[ii-1][10]=fore[0][0];//biased out-of-sample
  of lnRV[t+1|t]
cac40_output[ii-1][11]=unbRVfore[0][0];//unbiased
  out-of-sample of RV[t+1|t]
delete arfima;
}
savemat("Chapter7.OutSample.cac40.xls",cac40_output);
}
```

FTSE100

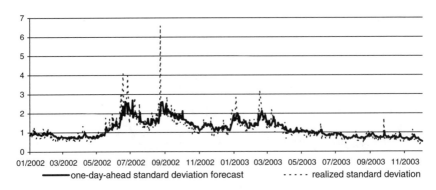

—— one-day-ahead standard deviation forecast   - - - - - realized standard deviation

S&P500

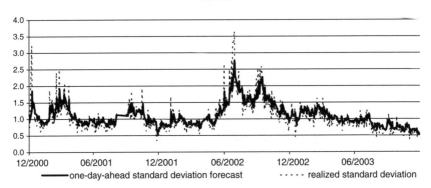

—— one-day-ahead standard deviation forecast   - - - - - realized standard deviation

Figure 7.7   (Continued)

The programs for the DAX30, FTSE100 and S&P500 data sets are available on the accompanying CD-ROM. In the tth row of the matrix *cac40_output*, the parameter vector $(\tilde{d}^{(t)}, w_0^{(t)}, w_1^{(t)}, w_2^{(t)}, c_1^{(t)}, d_1^{(t)})$ is saved. The ARFIMAX $(1, \tilde{d}, 1)$ model is estimated $\tilde{T} = 1257$ times. The values of the in-sample realized volatility estimators[9] $\hat{\sigma}_{(un),t}^{2(RV)}$ are saved in the tenth column, whereas the values of the out-of-sample realized volatility estimators $\sigma_{(un),t+1|t}^{(RV)}$ are saved in the last (twelfth) column. The matrix *cac40_output* is saved as an Excel file with the name *chapter7.OutSample.cac40.xls*.

Figure 7.7 plots the realized intraday volatility and the corresponding forecasts. It appears that the ARFIMAX$(1, \tilde{d}, 1)$ model produces quite accurate one-day-ahead variability forecasts, although it fails to capture sudden upward movements. Figure 7.8 shows the distribution of returns scaled by the realized standard deviation forecasts,

---

[9] The program in Section 7.3.2 estimates the in-sample realized volatility based on the entire available data set of T observations. The program in the present section estimates the in-sample realized volatility based on the last $\breve{T} = 1000$ observations.

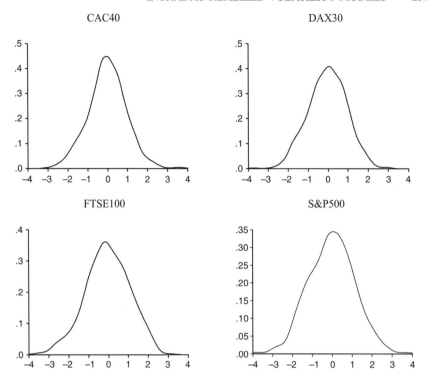

Figure 7.8   Distribution of daily log-returns standardized by the one-day-ahead intraday standard deviation.

Table 7.4   Descriptive statistics of the daily log-returns standardized by the out-of-sample realized standard deviation, $\{y_t/\sigma^{(RV)}_{(un),t|t-1}\}^T_{t=1}$, and statistics for testing the normality assumption

	CAC40	DAX30	FTSE100	S&P500
Mean	0.040	−0.033	−0.041	−0.042
Standard deviation	0.934	0.011	1.106	1.182
Skewness	0.514	−0.101	−0.185	−0.381
Kurtosis	4.756	2.913	2.923	5.244
Lilliefors test	0.033	0.021	0.024	0.025
[*p*-value]	[0.004]	[>0.1]	[>0.1]	[>0.1]
Cramér–von Mises test	0.346	0.092	0.033	0.065
[*p*-value]	[0.0001]	[0.145]	[0.809]	[0.321]

(i.e. of $\{y_t/\sigma_{(un),t|t-1}^{(RV)}\}_{t=1}^{T}$). Table 7.4 lists the descriptive statistics associated with $\{y_t/\sigma_{(un),t|t-1}^{(RV)}\}_{t=1}^{T}$. Giot and Laurent (2004) noted that although the returns scaled by the realized standard deviation are normally distributed, the daily returns standardized by the one-day-ahead intraday standard deviation forecast, $\{y_t/\sigma_{(un),t|t-1}^{(RV)}\}_{t=1}^{T}$, may not be normally distributed. On the basis of our data set, for the DAX30, FTSE100 and S&P500 indices, the series $\{y_t/\sigma_{(un),t|t-1}^{(RV)}\}_{t=1}^{T}$ appears to be normally distributed. Degiannakis (2008a) provides comparisons of the ARFIMAX$(\kappa, \tilde{d}, l)$ and ARFIMAX$(\kappa, \tilde{d}, l)$-TARCH(p, q) models in terms of their performance in forecasting the intraday realized volatility of the CAC40 and DAX30 indices and notes that, in the case of the DAX30 index, the ARFIMAX$(\kappa, \tilde{d}, l)$-TARCH(p, q) appears to provide more accurate one-day-ahead realized volatility forecasts.

8

Applications in value-at-risk, expected shortfall and options pricing

8.1   One-day-ahead value-at-risk forecasting

The need of major financial institutions to measure their exposure to risk began in 1970s after an increase in financial instability. *Value-at-risk* (VaR) is the main tool for reporting to bank regulators the risk that financial institutions face. It is a statistic of the dispersion of a distribution and, for a given portfolio, refers to the worst outcome likely to occur over a predetermined period and a given confidence level. According to the Basle Committee, VaR is used by financial institutions to calculate capital charges with respect to their financial risk. In addition, the Securities and Exchange Commission allows the use of VaR in order for institutions to report their market risk exposure.

Several methods have been proposed to estimate the risk faced by that financial institutions. These methods are divided into three main categories: (i) parametric methods that model the entire distribution and the volatility dynamics; (ii) non-parametric methods; and (iii) semi-parametric methods that combine the previous two types. For extended discussions on VaR, the interested reader is referred to Best (1999), Duffie and Pan (1997), Dowd (2005), Holton (2003), Jorion (2006) and the website http://www.gloriamundi.org.

8.1.1   Value-at-risk

At a given probability level $1-p$, VaR is the predicted amount of financial loss of a portfolio over a given time horizon. This is formally defined as follows. Let P_t be the observed value of a portfolio at time t, and let $y_t = \log(P_t/P_{t-1})$ denote the log-returns for the period from $t-1$ to t. For a long trading position[1] and under the assumption of standard normally distributed log-returns, VaR is defined to be the value $VaR_t^{(1-p)}$ satisfying the condition[2]

$$p = P\left(y_t \le VaR_t^{(1-p)}\right) = \int_{-\infty}^{VaR_t^{(1-p)}} \frac{1}{\sqrt{2\pi}} \exp\left(-\frac{1}{2}y_t^2\right) dy_t. \tag{8.1}$$

This implies that

$$VaR_t^{(1-p)} = \zeta_p,$$

where ζ_p is the $100p$th percentile of the standard normal distribution. In the more general case where $\zeta_t \sim N\left(\mu_t, \sigma_t^2\right)$, we have

$$p = P\left(\zeta_t \le VaR_t^{(1-p)}\right) = \frac{1}{\sigma_t\sqrt{2\pi}} \int_{-\infty}^{VaR_t^{(1-p)}} \exp\left(-\frac{1}{2}\left(\frac{\zeta_t-\mu_t}{\sigma_t}\right)^2\right) d\zeta_t,$$

whence

$$VaR_t^{(1-p)} = \mu_t + \zeta_p\sigma_t.$$

Hence, under the assumption that $y_t \sim N(0, 1)$, the probability of a loss less than $VaR_t^{(0.95)} \equiv VaR_t^{(1-0.05)} = -1.645$ is equal to $p = 5\%$.[3] The value -1.645 is the value of risk at a 95% level of confidence, or, in other words, for a capital of €10 million, the 95% VaR equals €164 500. Thus, if a risk manager states that the daily VaR of a portfolio is €164 500 at a 95% confidence level, it means that there are five chances in a 100 for a loss greater than €164 500 to incur. Figure 8.1 offers a visual representation of the 95% VaR for $y_t \sim N(0, 1)$ and may generally be conveniently used for graphical determination of the value of $VaR_t^{(1-p)}$ for various other values of p.

8.1.2 Parametric value-at-risk modelling

The most successful technique to model the VaR is through ARCH modelling. Consider the ARCH framework of (1.6):

$$\begin{aligned}
y_t &= \mu_t + \varepsilon_t, \\
\mu_t &= \mu(\theta|I_{t-1}), \\
\varepsilon_t &= \sigma_t z_t, \\
\sigma_t &= g(\theta|I_{t-1}), \\
z_t &\overset{i.i.d.}{\sim} f(0, 1; w).
\end{aligned} \tag{8.2}$$

---

[1] The state of owning a security is called a long position. The sale of a borrowed security with the expectation that the asset will fall in value is called a short position.

[2] Baumol (1963) was the first to attempt to estimate the risk that financial institutions face. He proposed a measure that is essentially the same as the widely known VaR.

[3] Accordingly, for a short trading position, $VaR_t^{(95\%)}$ is obtained through the condition $5\% = P\left(y_t \ge VaR_t^{(95\%)}\right) = \int_{VaR_t^{(95\%)}}^{+\infty} (2\pi)^{-1/2}\exp(-y_t^2/2)dy_t$, i.e. $VaR_t^{(95\%)} = \zeta_{0.95}$.

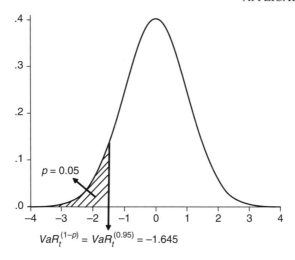

Figure 8.1 Visual representation of the value of $VaR_t^{(1-p)}$ and $p = P\left(y_t \leq VaR_t^{(1-p)}\right)$ in the case $y_t \sim N(0, 1)$ and $p = 0.05$.

where, as usual, z_t denotes a random variable having a density function $f(0, 1; w)$ of mean zero and variance one with w being some parameter vector to be estimated. The VaR value in this case is given by

$$VaR_t^{(1-p)} = \mu_t + f_a(z_t; w)\sigma_t, \qquad (8.3)$$

where $f_a(z_t; w)$ the a-quantile[4] of the distribution of z_t defined by $f(0, 1; w)$, μ_t is the conditional mean and σ_t is the conditional standard deviation at time t. Based on a sample size of T observations, the in-sample VaR at time t, for $t = 1, \ldots, T$, can be estimated by

$$\widehat{VaR}_t(1-p) = \hat{\mu}_t + f_a\left(z_t; w^{(T)}\right)\hat{\sigma}_t, \qquad (8.4)$$

where $\hat{\mu}_t$ and $\hat{\sigma}_t$ respectively denote estimates of the in-sample conditional mean and conditional standard deviation at time t, and $w^{(T)}$ denotes the estimated parameter vector of $f(0, 1; w)$ based on the basis of all of the T available observations. (Recall that the parameter vector $\psi^{(T)} = \left(\theta^{(T)}, w^{(T)}\right)$ is estimated on the basis of all of the T available observations.)

The one-step-ahead VaR forecast is given by

$$VaR_{t+1|t}^{(1-p)} = \mu_{t+1|t} + f_a\left(z_t; w^{(t)}\right)\sigma_{t+1|t}, \qquad (8.5)$$

where $f_a\left(z_t; w^{(t)}\right)$ is the a-quantile of the assumed distribution, computed on the basis of the information set available at time t, and $\mu_{t+1|t}$ and $\sigma_{t+1|t}$ are respectively the

---

[4] For long trading positions $a = p$, whereas for short trading positions $a = 1-p$.

conditional mean and conditional standard deviation forecasts at time $t + 1$ given the information at time t. The parameter vector $\psi^{(t)} = \left(\theta^{(t)}, w^{(t)}\right)$ is re-estimated at each point in time t.

ARCH models have been extensively applied in the estimation of VaR. Alexander and Leigh (1997) note that the ARCH model is preferable to the EWMA model (2.10). Giot and Laurent (2003a, 2003b) find that the APARCH-skT model provides better estimates of the VaR for equities and commodities. Huang and Lin (2004) suggest the use of the APARCH model with normally or Student t distributed innovations for the estimation of VaR for the Taiwan Stock Index futures. Degiannakis (2004) provides an extension of the family of the ARCH models considered for VaR forecasting, by including fractionally integrated volatility processes, and notes that the FIAPARCH model with skewed Student t conditionally distributed innovations provides more accurate one-day-ahead VaR forecasts in the case of the CAC40, DAX30 and FTSE100 indices. Although the APARCH volatility specification comprises several volatility specifications, its superiority has been contested by some researchers. According to Angelidis et al. (2004), the ARCH structure that produces the most accurate VaR forecasts can be different for every portfolio. Angelidis and Degiannakis (2005) point out that that a risk manager may have to employ different volatility techniques in order to obtain accurate VaR forecasts for long and short trading positions. Mittnik et al. (2000) recommend more general structures for both the volatility process and the distribution of z_t in order to improve VaR forecasts. Mittnik and Paolella (2000) deduce that general ARCH structures and non-normal innovation distributions are more adequate for modelling VaR in the case of the highly volatile exchange rates on East Asian currencies against the US dollar, while So and Yu (2006) argue that it is more important to model the underlying fat-tailed innovation distribution than the fractionally integrated structure of the volatility process.

The general line taken by the aforementioned studies is that the most flexible models generate the most accurate VaR forecasts. In contrast, Brooks and Persand (2003b) point out that the simplest volatility specifications, such as the historical average of the variance or the autoregressive volatility model, produce appropriate out-of-sample VaR forecasts, while Bams et al. (2005) note that complex (simple) tail models often lead to overestimation (underestimation) of the VaR.

Guermat and Harris (2002) apply the exponentially weighted likelihood (EWML) model in equity portfolios in the US, the UK and Japan and note its superiority to the GARCH(1,1) model under the normal or the Student t distribution. Li and Lin (2004) propose a regime-switching specification as an alternative risk management technique, while Billio and Pelizzon (2000) consider a multivariate switching regime model in order to calculate the VaR for ten Italian stocks.

8.1.3 Intraday data and value-at-risk modelling

Giot and Laurent (2004) propose a two-step framework for computing one-day-ahead VaR forecasts based on information extracted by the intra-day returns. In the first step, an ARFIMAX $(\kappa, \tilde{d}, l)$ model, of the specification (7.9), is considered to model the long-memory property of the realized volatility. In the second step, an ARCH specification, where the conditional volatility for the daily returns is proportional to the conditional realized volatility, is estimated. In particular, in the first step, the ARFIMAX $(0, d', 1)$ model for the logarithmic realized variance, $\log\left(\sigma_t^{2(RV)}\right)$, is estimated by

$$(1-L)^{\tilde{d}}\left(\log\left(\sigma_t^{2(RV)}\right) - w_0 - w_1 y_{t-1} - w_2 d_{t-1} y_{t-1}\right) = (1 + d_1 L)\varepsilon_t, \quad \varepsilon_t \sim N\left(0, \sigma_\varepsilon^2\right),$$

$$(8.6)$$

where y_t is the daily log-return on day t, $d_t = 1$ when $y_t > 0$ and $d_t = 0$ otherwise. The one-day-ahead realized variance is $\sigma_{(un),t|t-1}^{2(RV)} = \exp\left(\log\left(\sigma_{t|t-1}^{2(RV)}\right) + 0.5\sigma_\varepsilon^{2(t)}\right)$ as in equation (7.18). In the second step, the daily log-returns are modelled as an AR (0)-GARCH(0,0)-skT specification with $\sigma_{(un),t|t-1}^{2(RV)}$ regarded as an exogenous variable in the conditional variance:

$$
\begin{aligned}
y_t &= c_0 + \varepsilon_t, \\
\varepsilon_t &= \sigma_t z_t, \\
\sigma_t^2 &= \tilde{\sigma}^2 \sigma_{(un),t|t-1}^{2(RV)}, \\
z_t &\sim skT(0, 1; v, g).
\end{aligned}
\qquad (8.7)
$$

The one-day-ahead variance for day $t + 1$, given the information available on day t, is then computed as $\sigma_{t+1|t}^2 = \tilde{\sigma}^{2(t)}\sigma_{(un),t+1|t}^{2(RV)}$. The one-step-ahead $VaR_{t+1|t}^{(1-p)}$ is as given by (8.5), where now $f_a\left(z_t; w^{(t)}\right)$ is the a-quantile of the skewed Student t distribution, $w^{(t)} = \left(v^{(t)}, g^{(t)}\right)'$, $\mu_{t+1|t} = c_0^{(t)}$ and $\sigma_{t+1|t} = \tilde{\sigma}^{(t)}\sigma_{(un),t+1|t}^{(RV)}$. The parameter vector $\psi^{(t)} = \left(c_0^{(t)}, \tilde{\sigma}^{(t)}, v^{(t)}, g^{(t)}\right)'$ is computed on the basis of the information set available on day t.

Giot and Laurent (2004) compare the APARCH-skT model with the ARFIMAX specification described above in their attempt to forecast the VaR for the CAC40 stock index, the S&P500 futures contracts, and the yen–dollar and Deutschemark–dollar exchange rates. They note that the use of intraday data does not appear to improve the performance of inter-day VaR modelling. Angelidis and Degiannakis (2008) consider the ARFIMAX $(1, \tilde{d}, 1)$ and the AR(1)-GARCH(0,0)-skT specifications for the first and the second steps, respectively. They note that although the ARFIMAX model produces statistically more accurate forecasts of the realized volatility than an ARCH specification, it appears not to provide any added value in forecasting the next day's VaR. Giot (2005) estimate the VaR at intraday time horizons of 15 and 30 minutes and notice that the GARCH model with Student t distributed innovations appears to have the best overall performance, and that there are no significant differences between daily and intraday VaR models once the intraday seasonality in the volatility is taken into account.

8.1.4 Non-parametric and semi-parametric value-at-risk modelling

The problem of VaR estimation has also been widely examined in non-parametric as well as semi-parametric modelling set-ups. Among the first, *historical simulation* (HS) modelling has been most commonly used on account of its apparently better performance; the interested reader may be referred, for example, to Jackson et al.'s (1998) empirical results revealing that historical simulation based methods work better than others. In this set-up, assuming that the dynamic properties of the log-returns are well described by the sample period chosen, the value of VaR is determined by the corresponding a-quantile of the past $\breve{T}$ log-returns,

$$VaR_{t+1|t}^{(1-p)} = f_a\left(\{y_{t+1-\tau}\}_{\tau=1}^{\breve{T}}\right), \tag{8.8}$$

where, as noted above, for long trading positions $a = p$, whereas for short trading positions $a = 1-p$. At each point in time t, $VaR_{t+1|t}^{(1-p)}$ is computed based on the past $\breve{T}$ log-returns, i.e. $y_{t+1-\tau}$, for $t+1-\tau = (t, t-1, t-2, \ldots, t-\breve{T}+1)$. HS modelling has received much attention because of its simplicity and its relatively low reliance on theory as no distributional assumption about the underlying process of the returns need be made. The main advantage of the HS approach is that it can be implemented even in a spreadsheet, as it is assumed that the distribution of the portfolio returns is constant and, therefore, there is no need to model the time-varying variance of log-returns. The disadvantages are that (a) the length of $\breve{T}$ strongly affects the estimation of VaR (Van den Goorbergh and Vlaar, 1999) and (b) if $\breve{T}$ is too large, the most recent observations, which probably describe the future distribution better, carry the same weight as the earlier ones. Boudoukh et al. (1998) and Christoffersen (2003) managed to lessen the effect of the sample size choice on VaR forecasts with the introduction of an exponential weighting scheme.

Lambadiaris et al. (2003), performing historical and Monte Carlo simulations on Greek stocks and bonds, noted that the Monte Carlo method appears to be more appropriate for the stock market, while for bonds the results are mixed. Cabedo and Moya (2003) developed an ARMA historical simulation method which improves the simple historical VaR estimation.

Semi-parametric modelling methods are typically based on what are termed in the literature *filtered historical simulation* (FHS) and *extreme value theory* (EVT) set-ups. The FHS scheme is a combination of the above-mentioned methods and was proposed by Hull and White (1998) and Barone-Adesi et al. (1999). In particular, it amounts to employing a parametric model for the series of volatility innovations and using the standardized residuals of the model to obtain the value of VaR in a manner similar to that of the HS procedure. Within this framework, VaR is determined by

$$VaR_{t+1|t}^{(1-p)} = \mu_{t+1|t} + f_a\left(\{\hat{\varepsilon}_{t+1-i|t}/\hat{\sigma}_{t+1-i|t}\}_{i=1}^{\breve{T}}\right)\sigma_{t+1|t}. \tag{8.9}$$

Empirical studies have revealed that FHS based methods exhibit better performance than other parametric or semi-parametric methods (Angelidis and Benos, 2008;

Angelidis et al., 2007; Barone-Adesi and Giannopoulos, 2001), and appear to produce robust VaR estimates.

EVT techniques focus on modelling the tails of the returns distribution. They have largely been considered in the context of VaR estimation (McNeil, 1997, 1998, 1999; Embrechts, 2000; Ho et al., 2000; Longin, 2000; McNeil and Frey, 2000; Poon et al., 2001; Jondeau and Rockinger, 2003; Seymour and Polakow, 2003; Byström, 2004; Gençay and Selçuk, 2004; Brooks et al., 2005; Chan and Gray, 2006; Harmantzis et al., 2006; Kuester et al., 2006). One such technique is the so-called *peaks over threshold technique*. This is mainly used for obtaining one-step-ahead VaR forecasts in terms of the tail index of the generalized Pareto distribution. In particular,

$$VaR^{(1-p)}_{t+1|t} = \sigma_{t+1|t} u \left(\frac{p}{T_u/T} \right)^{-\tau_1}, \tag{8.10}$$

where $\tau_1 = T_u^{-1} \sum_{t=1}^{T_u} \log(y_t/u)$ is the Hill estimator of the tail index τ_1, and T_u is the number of observations above some threshold value u. The generalized Pareto distribution has the density function

$$f(x; \tau_1, \tau_2) = \begin{cases} 1 - \left(1 + \frac{\tau_1 x}{\tau_2}\right)^{-1/\tau_1} & \text{for} \quad \tau_1 \neq 0, \tau_2 > 0, \\ 1 - \exp\left(\frac{-x}{\tau_2}\right) & \text{for} \quad \tau_1 = 0, \tau_2 > 0, \end{cases} \tag{8.11}$$

with $x \geq u$ if $\tau_1 \geq 0$ and $u \leq x \leq u - \tau_2/\tau_1$ if $\tau_1 < 0$. The probability that the standardized returns, y_t^* exceed u is given by

$$P(y_t^* - u \leq x | y_t^* > u) = \frac{F(x+u) - F(u)}{1 - F(u)}, \tag{8.12}$$

where $x > u$ and $F(.)$ is the distribution function of the standardized returns.

8.1.5 Back-testing value-at-risk

How accurately VaR is estimated is very important to financial institutions. In practice, if the VaR is overestimated, regulators charge a higher amount of capital than really necessary, worsening their performance. If the VaR is underestimated, the regulatory capital set aside may not be enough to cover the risk that financial institutions face. The simplest method used by them to evaluate the accuracy of the VaR forecasts is to count how many times the losses exceed the value of VaR. If this count does not differ substantially from what is expected, then the VaR forecasts are adequately computed.

However, statistical inference for the forecasting performance of risk models would be more appropriate. Among the available statistical methods for evaluating VaR models the most commonly used are those based on back-testing measures,

e.g. Kupiec (1995) and Christoffersen (1998, 2003). Back-testing statistical inferences typically focus on testing hypotheses about percentages of times a portfolio loss
has exceeded the estimate of VaR (violation percentages).[5] These are tested on the
basis of the corresponding observed violation or exception rates.

8.1.5.1 Back-testing value-at-risk in terms of unconditional coverage

Kupiec (1995) proposed a back-testing procedure in terms of a hypothesis test for the
expected percentage of VaR violations, $p = P\big(y_{t+1} \leq VaR_{t+1|t}^{(1-p)}\big)$, i.e. for the per-
centage of times when the portfolio loss exceeds the VaR estimate. Kupiec (1995)
employs the observed rate at which the portfolio loss has been exceeded by the
estimate of VaR to test whether the true coverage percentage p^* differs statistically
significantly from the desired level p. Let $N = \sum_{t=1}^{\tilde{T}} \tilde{I}_t$ be the number of trading days
over the $\tilde{T}$ period that the portfolio loss exceeds the VaR estimate,[6] where

$$\tilde{I}_{t+1} = \begin{cases} 1, & \text{if} \quad y_{t+1} < VaR_{t+1|t}^{(1-p)}, \\ 0, & \text{if} \quad y_{t+1} \geq VaR_{t+1|t}^{(1-p)}. \end{cases} \tag{8.13}$$

The observed coverage ratio is then $N/\tilde{T}$. The hypotheses that need to be tested are

$$H_0 : p^* = p, \quad H_1 : p^* \neq p. \tag{8.14}$$

By definition, N is a binomial variable, which under the null hypothesis has a binomial
distribution with parameters $\tilde{T}$ and p, i.e. $N \sim B(\tilde{T}, p)$. Therefore, the likelihood ratio
statistic under the null hypothesis is

$$LR_{un} = 2\log\left(\left(1 - \frac{N}{\tilde{T}}\right)^{\tilde{T}-N}\left(\frac{N}{\tilde{T}}\right)^{N}\right) - 2\log\left((1-p)^{\tilde{T}-N}p^{N}\right). \tag{8.15}$$

This is asymptotically χ^2 distributed with one degree of freedom. The test is known as
the *unconditional coverage test*. It leads to the rejection of a model for both a
statistically significantly high or low failure rate. However, as noted by Kupiec (1995)
and Berkowitz (2001), its power is generally poor. Although it results in rejection of a
model over- or underestimating the *true* VaR, it does not address the question of
whether the violations are randomly distributed.

8.1.5.2 Back-testing value-at-risk in terms of conditional coverage

Randomness in the occurrence of the event $\left\{y_{t+1} < VaR_{t+1|t}^{(1-p)}\right\}$ on the basis of its past
occurrences cannot be tested with tests such as those considered by Cowles and Jones
(1937), Mood (1940) and Ljung and Box (1978), as the conditional probabilities given a
VaR violation are not serially uncorrelated. Christoffersen (1998) considered a testing

---

[5] A violation occurs if the estimated VaR is not able to cover the realized loss.
[6] We assume a long trading position. If $y_{t+1} < VaR_{t+1|t}^{(1-p)}$ a violation occurs. For short trading positions
$\tilde{I}_{t+1} = 1$ if $y_{t+1} > VaR_{t+1|t}^{(1-p)}$ and $\tilde{I}_{t+1} = 0$ if $y_{t+1} \leq VaR_{t+1|t}^{(1-p)}$.

procedure based on the expected percentage of occurrence of a VaR violation event conditional on the numbers of its past occurrences in a first-order Markov set-up.

In particular, for $\{\tilde{I}_t\}_{t=1}^{\infty}$ defined as in (8.13), Christoffersen (1998) proposed testing whether

$$\pi_{ij} = P(\tilde{I}_t = i | \tilde{I}_{t-1} = j) = P(\tilde{I}_t = i), \quad i, j = 0, 1,$$

i.e., whether the VaR violations are independent events or not. Under such a hypothesis, the probability of observing a violation in two consecutive periods must be equal to the desired value p and, hence, the hypotheses to be tested are

$$H_0 : \pi_{01} = \pi_{11} = p, \quad H_1 : \pi_{01} \neq \pi_{11}. \tag{8.16}$$

On the basis of an observed segment $\{\tilde{I}_t\}_{t=1}^{\tilde{T}}$ of the series of VaR violation occurrences $\tilde{I}_1, \tilde{I}_2, \ldots$, the hypotheses were tested utilizing a LR_{in} likelihood ratio statistic that is asymptotically χ^2 distributed with one degree of freedom:

$$LR_{in} = 2 \left(\log\left((1-\hat{\pi}_{01})^{n_{00}} \hat{\pi}_{01}{}^{n_{01}} (1-\hat{\pi}_{11})^{n_{10}} \hat{\pi}_{11}^{n_{11}} \right) \right.$$
$$\left. -\log\left(\left(1-\frac{N}{\tilde{T}}\right)^{n_{00}+n_{10}} \left(\frac{N}{\tilde{T}}\right)^{n_{01}+n_{11}} \right) \right) \sim \chi_1^2, \tag{8.17}$$

where $\hat{\pi}_{ij} = n_{ij} / \sum_j n_{ij}$ is the sample estimate of π_{ij} with n_{ij} denoting the number of trading days with value i followed by value j, for $i, j = 0, 1$.

Christoffersen (1998) further proposed the simultaneous testing of whether (i) the true percentage p^* of failures is equal to the desired percentage and (ii) the VaR failure process is independently distributed, i.e., testing the hypotheses

$$\begin{aligned} H_0 &: p^* = p \text{ and } \pi_{01} = \pi_{11} = p, \\ H_1 &: p^* \neq p \text{ or } \pi_{01} \neq \pi_{11} \neq p. \end{aligned} \tag{8.18}$$

The LR_{cc} statistic is χ^2 distributed with two degrees of freedom:

$$LR_{cc} = -2\log\left((1-p)^{\tilde{T}-N} p^N \right) + 2\log\left((1-\hat{\pi}_{01})^{n_{00}} \hat{\pi}_{01}^{n_{01}} (1-\hat{\pi}_{11})^{n_{10}} \hat{\pi}_{11}^{n_{11}} \right) \sim \chi_2^2, \tag{8.19}$$

Contrary to Kupiec's (1995) test, the conditional coverage procedure rejects a risk model that generates either too many or too few *clustered* violations.

8.1.5.3 A generalization of the conditional coverage test

Engle and Manganelli (2004) suggested examining whether the variable $Hit_t = \tilde{I}_t - p$ is uncorrelated with any information available in the information set I_{t-1}, i.e., whether Hit_t is uncorrelated with its lagged values, and with the VaR forecast and its lagged values. If it is, the risk model is deemed adequate. A risk manager may use the regression model

$$Hit_t = \delta_0 + \sum_{i=1}^{i'} \delta_i Hit_{t-i} + \sum_{j=1}^{j'} \zeta_j VaR_{t-j+1|t-j}^{(1-p)} + \varepsilon_t, \quad \varepsilon_t \sim N(0, 1), \tag{8.20}$$

and examine the hypothesis $\delta_0 = \delta_1 = \ldots = \delta_{i'} = \zeta_1 = \ldots = \zeta_{j'} = 0$.

8.1.6 Value-at-risk loss functions

Lopez (1999) suggested the development of a loss function for back-testing different models and proposed to measure the accuracy of the VaR forecasts on the basis of the distance between the observed returns and the forecasted VaR values if a violation occurs. For a long trading position, Lopez defined the penalty variable

$$\Psi_{(VaR)t+1} = \begin{cases} 1 + (y_{t+1} - VaR_{t+1|t}^{(1-p)})^2, & \text{if} \quad y_{t+1} < VaR_{t+1|t}^{(1-p)}, \\ 0, & \text{if} \quad y_{t+1} \geq VaR_{t+1|t}^{(1-p)}. \end{cases} \qquad (8.21)$$

A VaR model is penalized when an exception takes place. Hence, the model is preferred to other candidate models if it yields a lower total loss value. This is defined as the sum of these penalty scores, $\sum_{t=1}^{T} \Psi_{(VaR)t}$. This function incorporates both the cumulative number of exceptions and their magnitude. However, a model that does not generate any violation is deemed the most adequate as $\sum_{t=1}^{T} \Psi_{(VaR)t} = 0$. Thus, the risk models must be first filtered by the aforementioned back-testing measures.

Sarma et al. (2003), combining the advantages of a loss function with those of back-testing measures, suggested a two-stage back-testing procedure. When multiple risk models meet the back-testing statistical criteria of VaR evaluation, a loss function is brought into play to judge statistically the differences among VaR forecasts. In the first stage, the statistical accuracy of the models is tested by examining whether the mean number of violations is not statistically significantly different from that expected and whether these violations are independently distributed. In the second stage, they propose use of what they term the *firm's loss function*, i.e., penalizing failures but also imposing a penalty reflecting the cost of capital suffered on other days:

$$\Psi_{(VaR)t+1} = \begin{cases} (y_{t+1} - VaR_{t+1|t}^{(1-p)})^2, & \text{if} \quad y_{t+1} < VaR_{t+1|t}^{(1-p)}, \\ -\alpha_c VaR_{t+1|t}^{(1-p)}, & \text{if} \quad y_{t+1} \geq VaR_{t+1|t}^{(1-p)}, \end{cases} \qquad (8.22)$$

where α_c is a measure of cost of capital opportunity. Following this procedure, the risk manager ensures that the models that have not been rejected in the first stage forecast VaR satisfactorily. Note that, in contrast to Lopez's (1999) loss function, in the case of Sarma et al.'s (2003) loss function a score of 1 is not added when a violation occurs. The reason for this is that, for all the models, the expected exception proportion does not differ statistically significantly from its desired level, since they implement a two-stage back-testing procedure.

8.2   One-day-ahead expected shortfall forecasting

Some recent studies provide evidence that the VaR may provide an inadequate picture of risk. Artzner et al. (1997, 1999) show that the VaR of a portfolio may be greater than the sum of individual VaRs, and therefore that managing risk by using it may fail to automatically stimulate diversification – that is, VaR does not satisfy the subadditivity property. Taleb (1997a, 1997b) and Hoppe (1998, 1999) argue that the underlying statistical assumptions of VaR modelling are violated in practice, while Beder (1995) notes that different risk management techniques produce different VaR forecasts and therefore risk estimates might be imprecise. Marshall and Siegel (1997) compare the results of 11 software vendors that used a common covariance matrix, and implement the RiskMetrics™ VaR measure. They demonstrate that there may be significant differences for identical portfolios and market parameters and, hence, that banking internal models are exposed to *implementation* risk.

Moreover, VaR does not measure the size of the potential loss, given that this loss exceeds the estimate of VaR. The reason is that it indicates the potential loss at a specific confidence level, but it does not say anything about the expected loss. The magnitude of the expected loss should be the main concern of the risk manager.

To overcome such shortcomings of the VaR, Delbaen (2002) and Artzner et al. (1997) introduce the *expected shortfall* (ES) risk measure, which expresses the expected value of the loss, given that a VaR violation has occurred. In particular, ES is a probability-weighted average of tail loss and the one-step-ahead ES forecast. For a long trading position, it is defined by

$$ES_{t+1|t}^{(1-p)} = E\left(y_{t+1} \middle| \left(y_{t+1} \leq VaR_{t+1|t}^{(1-p)}\right)\right), \tag{8.23}$$

or

$$ES_{t+1|t}^{(1-p)} = E\left(VaR_{t+1|t}^{(1-\tilde{p})}\right), \quad \forall 0 < \tilde{p} < p. \tag{8.24}$$

Given the fact that there are no analytical solutions for all distributional assumptions, in order to calculate ES, Dowd (2005) suggested dividing (slicing) the tail into a large number $\tilde{k}$ of segments of equal probability mass, estimating the VaR associated with each segment and calculating the ES as the average of these VaR estimates, i.e., using

$$ES_{t+1|t}^{(1-p)} = \tilde{k}^{-1} \sum_{i=1}^{\tilde{k}} \left(VaR_{t+1|t}^{\left(1-p+ip(\tilde{k}+1)^{-1}\right)}\right). \tag{8.25}$$

Suppose that we wish to estimate $ES_{t+1|t}^{(1-5\%)}$, for $\tilde{k} = 5000$. Then we have to compute

$$ES_{t+1|t}^{(1-5\%)} = \frac{1}{5000} \sum_{i=1}^{5000} \left(VaR_{t+1|t}^{1-5\%+i5\%/5000}\right), \quad \text{for } i = 1, 2, \ldots, 5000,$$

or, equivalently,

$$ES_{t+1|t}^{(1-5\%)} = \frac{1}{5000} \sum_{i=1}^{5000} \left(VaR_{t+1|t}^{(1-\tilde{p})}\right), \quad \tilde{p} = 4.999\%, 4.998\%, \ldots, 0.01\%,$$

where $1-\tilde{p} = (1-p) + ip(\tilde{k}+1)^{-1}$. ES informs the risk manager about the expected loss if an extreme event occurs. Assuming normality, for $p = 5\%$ and a long trading

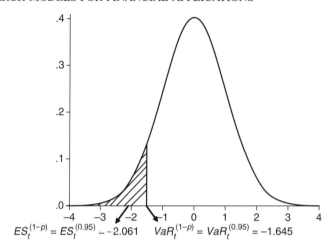

$ES_t^{(1-p)} = ES_t^{(0.95)} = -2.061 \qquad VaR_t^{(1-p)} = VaR_t^{(0.95)} = -1.645$

Figure 8.2   Visual representation of the value of $ES_t^{(1-p)}$ and $VaR_t^{(1-p)}$ in the case $y_t \sim N(0, 1)$ and $p = 0.05$.

position, the values of VaR and ES measures are $VaR_t^{(0.95)} = -1.645$ and $ES_t^{(0.95)} = -2.061$, respectively. In other words, for a capital of €10 000, at probability level 95%, the predicted amount of financial loss is equal to €164.50, and the average loss given a VaR violation equals €206.10. Figure 8.2 offers visual representations of the 95% VaR and ES for $y_t \sim N(0, 1)$ and may generally be conveniently used for graphically determining $VaR_t^{(1-p)}$ and $ES_t^{(1-p)}$ for various other values of p.

ES has in common with VaR the features of measuring the risk across positions and summarizing the risk by just one value. It is worth noting that it is a coherent risk measure[7] that satisfies the properties of (i) subadditivity, (ii) homogeneity, (iii) monotonicity, and (iv) translational invariance. Moreover, ES based risk management leads to lower expected losses than VaR based risk management. Basak and Shapiro (2001) provide theoretical evidence that under a VaR risk management the expected losses are higher than under an ES risk management. Yamai and Yoshiba (2005) compare the VaR and ES measures and note that VaR may not be reliable during market turmoil as it can mislead rational investors, whereas ES may be a better choice overall.[8]

---

[7] The concept of a coherent risk measure is defined by Artzner et al. (1997). Let $\rho(.)$ be a measure of risk and x, y represent any two random losses. The properties of a coherent risk measure are described as: (i) subadditivity, $\rho(x) + \rho(y) \geq \rho(x+y)$; (ii) positive homogeneity, $\rho(tx) = t\rho(x)$ for $t > 0$; (iii) monotonicity,$\rho(x) \geq \rho(y)$ if $x \leq y$; and (iv) translational invariance, $\rho(x+n) = \rho(x) - n$.

[8] However, Yamai and Yoshiba (2005) advised risk managers to combine the two measures for best results. Cuoco et al. (2001) also note that if VaR and ES are re-evaluated periodically, they both lead to equally accurate measures of risk.

8.2.1 Historical simulation and filtered historical simulation for expected shortfall

The slicing technique in (8.25) for calculating ES as the average of VaRs can also be applied in non-parametric and semi-parametric modelling frameworks. In equation (8.8), the one-step-ahead $VaR_{t+1|t}^{(1-p)}$, at a given probability level $1-p$, was computed according to the HS method. The HS estimation of ES is calculated as the average $VaR_{t+1|t}^{(1-\tilde{p})}$, for $0 < \tilde{p} < p$. This is the case for the FHS method too. The FHS estimate of ES is calculated as the average $VaR_{t+1|t}^{(1-\tilde{p})}$, for $0 < \tilde{p} < p$, where $VaR_{t+1|t}^{(1-p)}$ is computed according to equation (8.9).

8.2.2 Loss functions for expected shortfall

Angelidis and Degiannakis (2007) propose utilizing loss functions that are based on ES as VaR does not give any indication about the size of the expected loss:[9]

$$
\Psi_{(ES)t+1} = \begin{cases} (y_{t+1} - ES_{t+1|t}^{(1-p)})^2, & \text{if } y_{t+1} < VaR_{t+1|t}^{(1-p)}, \\ 0, & \text{if } y_{t+1} \geq VaR_{t+1|t}^{(1-p)}. \end{cases} \tag{8.26}
$$

Along the same lines as Sarma et al. (2003), Angelidis and Degiannakis suggest a two-stage back-testing procedure. In the first stage, the VaR forecasting accuracy of the candidate models is investigated by the likelihood ratio statistics given by (8.15) and (8.17). In the second stage, the average of (8.26) is calculated only for the models that are judged to forecast the VaR adequately in the first stage. Therefore, an ES based back-testing procedure picks risk models that calculate the VaR accurately, as the prerequisite of correct unconditional and conditional coverage is satisfied, and provide more precise loss forecasts, as they minimize the $\bar{\Psi}_{(ES)} = \tilde{T}^{-1} \sum_{t=0}^{\tilde{T}-1} \Psi_{(ES)t+1}$ of (8.26).

ES has been studied by Acerbi et al. (2001), Acerbi (2002) and Inui and Kijima (2005), among others. Bali and Theodossiou (2007) indicated that the AR(1)-AGARCH(1,1) and AR(1)-EGARCH(1,1) models with skewed generalized t (SGT) conditional distributed innovations exhibit the best overall in-sample and out-of-sample performance in VaR and ES forecasting for the S&P500 index in the period from January 1950 to December 2000. Angelidis and Degiannakis (2007) investigate the 44 models presented in Chapter 6 using the ES based two-stage back-testing procedure for the S&P500 equity index, the gold price and the $/£ exchange rate for the period from April 1988 to April 2005. They note the best-performing models appear to be the AR(1)-FIEGARCH-N for S&P500, the AR(1)-GARCH(1,1)-GED and AR(1)-IGARCH(1,1)-GED for gold, and the AR(1)-EGARCH(1,1)-N for the $/£

---

[9] Various modifications of (8.26) can be considered. For example, the squared distance $\left(y_{t+1} - ES_{t+1|t}^{(1-p)}\right)^2$ can be replaced with the absolute distance, $\left|y_{t+1} - ES_{t+1|t}^{(1-p)}\right|$.

exchange rate. These share some features such as asymmetry in volatility specification, and overestimation of the *true* VaR in the case of Student t and skewed Student t
innovation distributions.

8.3   FTSE100 index: one-step-ahead value-at-risk and expected shortfall forecasting

In this section we obtain the one-day-ahead 95% VaR and 95% ES values for the
FTSE100 index. The data set considered is again the series of observed daily returns of
the FTSE100 index for the period from 4 April 1988 to 5 April 2005. The AR(1)-
GARCH(1,1), AR(1)-GJR(1,1) and AR(1)-FIAPARCH $(1, d, 1)$ volatility specifications under the four distributional assumptions[10] provided by the G@RCH package
were estimated in Chapter 6. Each model was estimated for $\tilde{T} = 1435$ trading days
based on a rolling sample of $\tilde{T} = 3000$ days. For example, the AR(1)-FIAPARCH
$(1, d, 1)$-GED model was considered in the form:

$$
\begin{aligned}
y_t &= c_0(1-c_1) + c_1 y_{t-1} + \varepsilon_t, \\
\varepsilon_t &= \sigma_t z_t, \\
\sigma_t^\delta &= a_0 + \left(1 - b_1 L - (1-a_1 L)(1-L)^d\right)(|\varepsilon_t| - \gamma \varepsilon_t)^\delta + b_1 \sigma_{t-1}^\delta, \\
z_t &\overset{i.i.d.}{\sim} GED(0, 1; v), \\
f_{(GED)}(z_t; v) &= \frac{v \exp(-0.5|z_t/\lambda|^v)}{\lambda 2^{(1+1/v)} \Gamma(v^{-1})},
\end{aligned}
\tag{8.27}
$$

where $\lambda \equiv \sqrt{2^{-2/v} \Gamma(v^{-1}) / \Gamma(3v^{-1})}$.

The program named *chapter8.var.es.ftse100.prg* (in the Appendix to this chapter)
computes the one-day-ahead 95% VaR and 95% ES for long trading positions,
Kupiec's and Christoffersen's coverage test statistics and the loss function defined
by (8.26) for all 12 models. An excerpt from the code referring to the computation of
the one-day-ahead 95% VaR and 95% ES estimates for the AR(1)-FIAPARCH
$(1, d, 1)$-GED model is given below as an illustration.

```
'AR(1)-FIAPARCH(1,d,1)-ged
read(t=xls) ftse100_FIAPARCHBBM_ged.xls 11
for !i = 1 to !sample
    ftse100VaR95L(!i,place)=@sqrt(var1(!i))*@qged
        (0.05,var11(!i))+var2(!i)
    vector(!ES_repeat) ES = na
    !step1 = (1-0.95)/(!ES_repeat+1)
    !step2 = 0.05
```

---

[10] The normal, Student t, generalized error distribution and skewed Student t density functions are
respectively defined by equations (2.16), (5.1), (5.2) and (5.5).

```
   for !kk = 1 to !ES_repeat
   !step2 = !step2-!step1
   ES(!kk) = @sqrt(var1(!i))*@qged(!step2,var11(!i))
      +var2(!i)
 next
 ftse100ES951(!i,place) = @mean(ES)
next
```

EViews reads the Excel file named *ftse100_FIAPARCHBBM_ged.xls*, where the one-day-ahead conditional variance, $\sigma^2_{t+1|t}$, the one-day-ahead conditional mean, $y_{t+1|t}$, and the estimated parameter vector $\psi^{(t)} = (c_0^{(t)}, c_1^{(t)}, a_0^{(t)}, a_1^{(t)}, \delta^{(t)}, d^{(t)}, \gamma^{(t)}, b_1^{(t)}, v^{(t)})$ have been stored by the G@RCH code used in the application of Chapter 6 (*read(t=xls) ftse100_FIAPARCHBBM_ged.xls 11*). In the matrix *ftse100VaR95L(!i,place)* the values of the one-day-ahead 95% VaR are saved for each trading day ($!i = 1,\ldots, 1435$) and each model (*place* = 1,...,12). The first column of *ftse100VaR95L* contains the estimate $VaR^{(95\%)}_{t+1|t} = \mu_{t+1|t} + f_{5\%}(z_t; v^{(t)})\sigma_{t+1|t}$ from equation (8.5) for the AR(1)-FIAPARCH $(1, d, 1)$-GED model,

$$\mu_{t+1|t} = c_0^{(t)}\left(1 - c_1^{(t)}\right) + c_1^{(t)} y_t \tag{8.28}$$

$$\sigma_{t+1|t} = \left(\left(a_1^{(t)} - b_1^{(t)}\right)\left(|\varepsilon_{t|t}| - \gamma^{(t)}\varepsilon_{t|t}\right)^{\delta^{(t)}} + b_1^{(t)}\sigma_{t|t}^{\delta^{(t)}}\right.$$
$$+ \sum_{i=1}^{\infty}\left(\frac{d^{(t)}\Gamma(i - d^{(t)})}{\Gamma(1 - d^{(t)})\Gamma(i+1)}L^i\left(|\varepsilon_{t+1|t+i}| - \gamma^{(t)}\varepsilon_{t+1|t+i}\right)^{\delta^{(t)}}\right.$$
$$\left.\left.- a_1^{(t)}\left(|\varepsilon_{t|t+i}| - \gamma^{(t)}\varepsilon_{t|t+i}\right)^{\delta^{(t)}}\right)\right)^{1/\delta^{(t)}}, \tag{8.29}$$

with $f_{5\%}(z_t; v^{(t)})$ denoting the 5% quantile of the generalized error distribution with tail-thickness parameter v. The vector *ftse100ES95L(!i,place)* contains the one-day-ahead 95% ES estimates for each trading day ($!i = 1,\ldots, 1435$) and each model (*place* = 1,...,12). On each trading day *!i*, the value of the one-day-ahead 95% ES is computed as the average of $\tilde{k} = !ES\_repeat = 5000$ VaR estimates by

$$ES^{(95\%)}_{t+1|t} = \frac{1}{5000}\sum_{i=1}^{5000}\left(VaR^{(95\% + i5\%(5000+1)^{-1})}_{t+1|t}\right). \tag{8.30}$$

The matrix *ftse100result95L(!i,place)* is a matrix of zeros and ones containing the values of the variable $\tilde{I}_{t+1}$ in equation (8.13). If the one-day-ahead 95% VaR estimate by model *place* for trading day *i* does not cover the realized loss, then a one is saved in row *i* and column *place* of the matrix *ftse100result95L*. Otherwise, a zero is saved:

```
for !i = 1 to !sample
  if ftse100(!i)<ftse100VaR95L(!i,place) then
      ftse100result95L(!i,place) = 1
  else
      ftse100result95L(!i,place) = 0
  endif
next
```

The following part of the code deals with the computation of Kupiec's (1995) test statistic. The variables *kupiecresult1* and *kupiecresult2* contain the values of $-2\log\left((1-p)^{T-N}p^{N}\right)$ and $2\log\left((1-N/\tilde{T})^{T-N}(N/\tilde{T})^{N}\right)$, respectively. Note that both are multiplied by 10^{150} as they contain values that in some cases are very close to zero. The vector *LRunLong* contains the likelihood ratio statistic given in (8.15) and the first cell of *Kupiec95L* contains the p-value of LR_{un} which is χ^2 distributed with one degree of freedom (p-value $= 1$-@$cchisq(LRunLong(1),1)$).

```
'Create the vector for the probabilities of default
Vector(1) probab
probab(1)=0.05
'How many forecasts have you made?
vector(1) T
T(1)=!sample
'What is the average (expected)?
vector(1) average
average(1)= probab(1)*T(1)
'The number of brakes for long
vector(1) NLong = na
NLong(1)=@round(average95l(1,place)*!sample)
'Vectors for LR tests
vector(1) LRunLong = na
series kupiecresult1=-2*log(((1-probab(1))^(t(1)-
NLong(1)))*(probab(1)^NLong(1))*1e+150)
series kupiecresult2=2*log(((1-NLong(1)/t(1))^(t(1)-
NLong(1)))*((NLong(1)/t(1))^NLong(1))*1e+150)
LRunLong(1)=kupiecresult1(1)+kupiecresult2(1)
Kupiec95L(1,place) = 1-@cchisq(LRunLong(1),1)
```

The lines of the code that follow carry out the computation of Christoffersen's (1998) test statistic. The variables *LRin1* and *LRin2* contain the values of $(1-\pi_{01})^{n_{00}}\pi_{01}^{n_{01}}(1-\pi_{11})^{n_{10}}\pi_{11}^{n_{11}}$ and $(1-N/\tilde{T})^{n_{00}+n_{10}}(N/\tilde{T})^{n_{01}+n_{11}}$, respectively. As in the case of *kupiecresult1* and *kupiecresult2*, *LRin1* and *LRin2* are multiplied by 10^{150}. The vector *LRcc* contains the likelihood ratio statistic of (8.17). The p-value of LR_{in}, which is χ^2 distributed with one degree of freedom, is stored in the first cell of *Christ95L* (p-value $= 1 - @cchisq(LRcc(1),1)$).

```
'Christoffersen test
smpl @all
'Create the vector for the conditional coverage test
vector(1) LRcc=na
'Create the coef for the independence
'The number of observations with value 0 followed by 0
coef(1) n00=0
'The number of observations with value 0 followed by 1
coef(1) n01=0
'The number of observations with value 1 followed by 1
coef(1) n11=0
'The number of observations with value 1 followed by 0
coef(1) n10=0
'Create the probability to go to 0 given that we are at 0
coef(1) pr00(1)=0
'Create the probability to go to 1 given that we are at 0
coef(1) pr01(1)=0
'Create the probability to go to 0 given that we are at 1
coef(1) pr10(1)=0
'Create the probability to go to 1 given that we are at 1
coef(1) pr11(1)=0
'Create the series for the independence coverage
series LRin
'Estimate the nij
for !i=1 to !sample-1
  if ftse100result95l(!i,place)=0 and ftse100result95L
      (!i+1,place)=0 then
    n00(1)=n00(1)+1
  else
    n00(1)=n00(1)
  endif
  if ftse100result95l(!i,place)=0 and ftse100result95L
      (!i+1,place) =1 then
    n01(1)=n01(1)+1
  else
    n01(1)=n01(1)
  endif
  if ftse100result95l(!i,place)=1 and ftse100result95L
      (!i+1,place)=1 then
    n11(1)=n11(1)+1
  else
```

```
    n11(1)=n11(1)
    endif
    if ftse100result95l(!i,place)=1 and ftse100result95L
      (!i+1,place)=0 then
        n10(1)=n10(1)+1
    else
        n10(1)=n10(1)
    endif
next
pr01(1)=n01(1)/(n00(1)+n01(1))
pr11(1)=n11(1)/(n10(1)+n11(1))
series LRin1=(((1-pr01(1))^n00(1))*(pr01(1)^n01(1))*((1-
    pr11(1))^n10(1))*pr11(1)^n11(1))*1e+150
coef(1) p2=(n01(1)|n11(1))/(n01(1)+n00(1)+n10(1)+n11(1))
series LRin2=(((1-p2(1))^(n00(1)+n10(1)))*(p2(1)^(n01(1)
    +n11(1))))*1e+150
LRcc(1)=-2*(log(lrin2(1)/lrin1(1)))
Christ95L(1,place)=1-@cchisq(LRcc(1),1)
```

The following part of the code computes of the value of the average expected shortfall loss function of (8.26). If the one-day-ahead 95% VaR of model *place* for trading day i does not cover the realized loss, or if $y_{t+1} < VaR_{t+1|t}^{(95\%)}$, then the value of $\left(y_{t+1} - ES_{t+1|t}^{(95\%)}\right)^2$ is stored in row i and column *place* of the matrix *ftse100loss95l*. Otherwise, a zero is saved:

```
for !i = 1 to !sample
  if ftse100result95l(!i,place) = 1 then
    ftse100loss95l(!i,place) = (ftse100(!i)-ftse100es
      95l(!i,place))^2
  else
    ftse100loss95l(!i,place) = 0
  endif
next
```

Table 8.1 presents, for the 12 models, the average values for the one-day-ahead 95% VaR and 95% ES forecasts, the percentage of violations, Kupiec's and Christoffersen's p-values and the average value of the expected shortfall loss function of (8.26). The back-testing tests indicate a lower performance for the AR(1)-GARCH(1,1)-N and AR(1)-GARCH(1,1)-GED models in forecasting the next day's VaR. No model rejections are indicated by Christoffersen's LR_{in} test, while, only two rejections are signified by Kupiec's LR_{un} test. Note, however, that, by a two-stage back-testing procedure, both of these models would be excluded from further comparative evaluation on the basis of the expected shortfall loss function, $\tilde{T}^{-1} \sum_{t=0}^{\tilde{T}-1} \Psi_{(ES)t+1}$. Using this loss function as a criterion, the AR(1)-

Table 8.1 FTSE100 equity index: average values of the one-day-ahead 95% VaR and 95% ES forecasts, percentages of violations $(N/\tilde{T})$, Kupiec's (1995) and Christoffersen's (1998) p-values for LR_{in}, and values of the expected shortfall loss function $(\tilde{T}^{-1}\sum_{t=0}^{\tilde{T}-1}\Psi_{(ES)t+1})$. (Data set from 4 April 1988 to 5 April 2005.)

Model	Average 95% Var	Average 95% ES	$N/\tilde{T}$	Kupiec's LR_{un} p-value	Christoffersen's LR_{in} p-value	Loss function $\tilde{T}^{-1}\sum_{t=0}^{\tilde{T}-1}\Psi_{(ES)t+1}$	Loss function ranking
AR(1)-FIAPARCH(1,d,1)-t	−1.996	−2.627	4.6%	48.04%	57.94%	0.0149	1
AR(1)-GJR(1,1)-t	−1.980	−2.612	4.7%	56.09%	93.80%	0.0193	2
AR(1)-FIAPARCH(1,d,1)-GED	−1.851	−2.411	5.5%	38.72%	74.76%	0.0195	3
AR(1)-FIAPARCH(1,d,1)-N	−1.842	−2.314	5.6%	27.19%	83.55%	0.0210	4
AR(1)-GJR(1,1)-GED	−1.832	−2.390	5.6%	27.19%	77.13%	0.0223	5
AR(1)-GARCH(1,1)-t	−1.934	−2.559	5.2%	78.62%	27.33%	0.0230	6
AR(1)-GJR(1,1)-N	−1.822	−2.290	5.8%	18.32%	68.94%	0.0233	7
AR(1)-GARCH(1,1)-GED	−1.785	−2.337	6.4%	1.85%	37.82%	0.0264	8
AR(1)-GARCH(1,1)-N	−1.776	−2.236	6.4%	1.85%	37.82%	0.0274	9
AR(1)-FIAPARCH(1,d,1)-skT	−2.031	−2.836	4.5%	33.93%	50.20%	0.0617	10
AR(1)-GJR(1,1)-skT	−2.019	−2.870	4.7%	56.09%	93.80%	0.0624	11
AR(1)-GARCH(1,1)-skT	−1.968	−2.816	4.8%	73.75%	36.42%	0.0634	12

$\Psi_{(ES)t+1}$ is defined by (8.26).

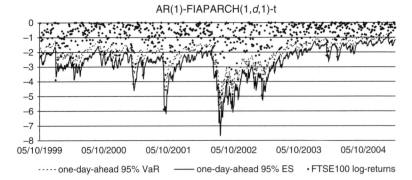

Figure 8.3   One-day-ahead 95% VaR and 95% ES forecasts obtained by the AR(1)-FIAPARCH(1,d,1)-t model and the corresponding actual FTSE100 index losses from 5 October 1999 to 4 April 2005.

FIAPARCH(1,d,1)-t model appears to exhibit the poorest performance. Therefore, the AR(1)-FIAPARCH(1,d,1)-t model appears to performing best (given that the coverage criteria are also satisfied). Figure 8.3 plots its one-day-ahead 95% VaR and 95% ES forecasts and the actual FTSE100 losses (i.e. daily log-returns). There appear to be 66 trading days (4.6% of the 1435 out-of-sample trading days) on which the one-day-ahead 95% VaR has not been able to cover the realized loss.

8.4   Multi-period value-at-risk and expected shortfall forecasting

The formulas for obtaining VaR forecasts given in the preceding sections cannot be directly applied for obtaining multi-period VaR forecasts. They can only be used when the log-returns are independently and identically normally distributed. In this case, the τ-day-ahead VaR is given by

$$VaR_{t+\tau|t}^{(1-p)} = \sqrt{\tau}VaR_{t+1|t}^{(1-p)}. \tag{8.31}$$

Diebold et al. (1996) showed that converting the daily VaR to a 10-day-ahead VaR by using the $\sqrt{\tau}$ rule for $\tau = 10$ generates inadequate forecasts. The Basle Committee proposal (1995a, 1995b, 1998) requires financial institutions to set the holding period equal to 10 trading days. However, as noted by Danielsson and Zigrand (2004), the square-root rule appears to lead to an underestimation of risk and, hence, the objective of the Basle Committee proposal is not addressed satisfactorily.

Since analytical expressions for the multi-period density are not available, Christoffersen (2003) suggested the use of numerical techniques to obtain estimates for multi-period VaR. Consider, for example, the AR(0)-GARCH(1,1) model with normally distributed conditional innovations:

$$
\begin{aligned}
y_{t+1} &= \sigma_{t+1} z_{t+1}, \\
\sigma_{t+1}^2 &= \alpha_0 + \alpha_1 y_t^2 + b_1 \sigma_t^2, \\
z_{t+1} &\sim N(0, 1).
\end{aligned}
\tag{8.32}
$$

The estimate of the τ-day-ahead VaR is obtained as follows.

- Step 1.1. Generate random numbers $\{\breve{z}_{i,1}\}_{i=1}^{MC}$ from the standard normal distribution, where MC denotes the number of draws.[11]

- Step 1.2. Create the hypothetical returns for time $t+1$: $\breve{y}_{i,t+1} = \sigma_{t+1|t} \breve{z}_{i,1}$, for $i = 1, \ldots, MC$.

- Step 1.3. Create the forecast variance for time $t+2$: $\breve{\sigma}_{i,t+2}^2 = \alpha_0^{(t)} + \alpha_1^{(t)} \breve{y}_{i,t+1}^2 + b_1^{(t)} \sigma_{t+1|t}^2$.

- Step 2.1. Generate further random numbers, $\{\breve{z}_{i,2}\}_{i=1}^{MC}$, $\breve{z}_{i,2} \sim N(0,1)$

- Step 2.2. Calculate the hypothetical returns for time $t+2$: $\breve{y}_{i,t+2} = \breve{\sigma}_{i,t+2} \breve{z}_{i,2}$, for $i = 1, \ldots, MC$.

- Step 2.3. Create the forecast variance for time $t+3$: $\breve{\sigma}_{i,t+3}^2 = \alpha_0^{(t)} + \alpha_1^{(t)} \breve{y}_{i,t+2}^2 + b_1^{(t)} \breve{\sigma}_{i,t+2}$

...

- Step τ.1. Generate further random numbers, $\{\breve{z}_{i,\tau}\}_{i=1}^{MC}$, $\breve{z}_{i,\tau} \sim N(0,1)$.

- Step τ.2. Calculate the hypothetical returns for time $t+\tau$: $\breve{y}_{i,t+\tau} = \breve{\sigma}_{i,t+\tau} \breve{z}_{i,\tau}$.

- Step τ.3. Calculate the τ-day VaR: $VaR_{t+\tau|t}^{(1-p)} = f_a\left(\{\breve{y}_{i,t+\tau}\}_{i=1}^{MC}\right)$.

This technique can be generalized for any volatility specification and distributional assumption. Brooks and Persand (2003b) and Dowd et al. (2004) have also dealt with the computation of multi-period VaR forecasts.

By analogy with (8.23) and (8.24) for the case $\tau = 1$, the τ-day-ahead expected value of the loss, given that the return at time $t + \tau$ falls below the corresponding value of the VaR forecast or the τ-day-ahead ES forecast, for a long trading position, is defined as

$$
ES_{t+\tau|t}^{(1-p)} = E\left(y_{t+\tau} \mid \left(y_{t+\tau} \leq VaR_{t+\tau|t}^{(1-p)}\right)\right),
\tag{8.33}
$$

---

[11] The MC notation is due to Christoffersen (2003).

while the value of the τ-day-ahead ES measure is given by

$$ES_{t+\tau|t}^{(1-p)} = E\left(VaR_{t+\tau|t}^{(1-\tilde{p})}\right), \quad \forall 0 < \tilde{p} < p. \tag{8.34}$$

Hence, by dividing the tail into a large number $\tilde{k}$ of slices, we can estimate the τ-day-ahead VaR associated with each slice and then take the τ-day-ahead ES as the average of these VaRs using

$$ES_{t+\tau|t}^{(1-p)} = \tilde{k}^{-1} \sum_{i=1}^{\tilde{k}} \left(VaR_{t+\tau|t}^{(1-p+ip(\tilde{k}+1)^{-1})}\right). \tag{8.35}$$

8.5 ARCH volatility forecasts in Black–Scholes option pricing

The common way to measure the performance of volatility forecasting models is by assessing their ability to predict future volatility. Another way to judge the forecasting accuracy is to construct a measure of the usefulness of the forecasts. Noh et al. (1994) considered assessing model performance by computing option prices based on the volatility forecasts of the underlying asset returns, devising trading rules to trade options on a daily basis and comparing the resulting profits.

The day-by-day rates of return are reflective of the corresponding predictive performances of the models. Comparing the results provides an indirect comparative assessment of a trading strategy based on volatility forecasts provided by a set of competing ARCH models on the basis of the option price forecast.

Within this framework, the present section examines the performance of a number of ARCH models in predicting volatility in pricing options. The comparative evaluation is carried out using options data on the basis of the cumulative profits of traders always using variance forecasts obtained by a single model. Noh et al. (1994) considered the problem of evaluating the performance of two model based methods for volatility forecasting, i.e. the ARCH modelling based method and the implied volatility regression method, within an option trading framework. The ARCH models provide one common conditional volatility estimate for both call and put option prices, while the implied-volatility forecasting method provides different volatility estimates for call and put option prices (the terms *call* and *put* are defined in Section 8.5.1). Over the April 1986 to December 1991 period, for S&P500 index options, the ARCH model based forecasting method led to a greater profit than the rule based on the implied volatility regression model. In particular, by the trading strategy based on the ARCH model a daily profit of 0.89% was earned, while by the implied volatility method a daily loss of 1.26% was made.

A comparative evaluation is carried out by comparing volatility forecasts of these models in terms of the profits of traders pricing derivatives in a real market based on these forecasts. Forecasts of option prices used in the comparative evaluation are calculated using the Black–Scholes (BS) option pricing formula.

8.5.1 Options

An option is a security that gives its owner the right, but not the obligation, to buy or sell an asset at a fixed price (the exercise price) within a specified period of time, subject to certain conditions. There are two main types of options: *calls* and *puts*. A call option is the right to buy a number of shares, of the underlying asset, at a fixed price on or before the maturity day. A put option is a right to sell a number of shares, of the underlying asset, at a fixed price on or before maturity. A *straddle* option is the purchase (or sale) of both a call and a put option, of the underlying asset, with the same expiration day. The maturity day is the last date that the option can be exercised. If the option can be exercised only at maturity, it is termed a European option, whereas an American option can be exercised on or before the expiration day.

The purchaser of a call (put) option acquires the right to buy (sell) a share of a stock for a given price on or before time T and pays for the right at the time of purchase. On the other hand, the writer of this call (put) collects both the option price on the particular day and the obligation to deliver (buy) one share of stock in the future for the exercise price, if the purchaser of the call (put) demands.

8.5.1.1 The stock and exercise price relationship

The exercise price of an *at-the-money* option is equal to the price of the underlying asset. The exercise price of a *near-the-money* option is approximately the same as the price of the underlying asset. A call (put) option is said to be *in-the-money* if its exercise price is less (greater) than the current price of the underlying asset. A call (put) option is said to be *out-of-the-money* if its exercise price is greater (less) than the price of the underlying asset.

8.5.1.2 The Black–Scholes option pricing formula

The pricing of options is a cornerstone of the financial literature. The BS option pricing model is very important and useful in estimating the fair value of an option. Based on the law of one price or no-arbitrage condition, the option pricing models of Black and Scholes (1973) and Merton (1973) gained almost immediate acceptance among academics and investment professionals.[12] Their approach can be used to price any security whose payoffs depend on the prices of other securities. The main idea is to create a costless self-financing portfolio strategy, whereby long positions are completely financed by short positions, which can replicate the payoff of the

[12] The 1997 Nobel Prize in Economics was awarded to Robert C. Merton and Myron S. Scholes for their work, in collaboration with Fischer Black, in developing the Black–Scholes option pricing model. Black, who died in 1995, would undoubtedly have shared in the prize had he still been alive. According to Daigler (1994, p. 128): 'An early version of Black and Scholes (1973) was submitted in the summer of 1970, but both the Journal of Political Economy and the Review of Economics and Statistics rejected the paper – perhaps because the ideas were so new and/or because Black was not an academic. After revising the approach and receiving encouragement from the University of Chicago professors Merton Miller and Eugene Fama, an article testing the model empirically was published in 1972 in the Journal of Finance (Black and Scholes 1972). The proof of the model was published in 1973 in the Journal of Political Economy, published by the University of Chicago.'

derivative. Under the no-arbitrage condition, the dynamic strategy reduces to a partial differential equation subject to a set of boundary conditions that are determined by the specific terms of the derivative security.

The pricing of index options is formulated on the basis of the BS option pricing model. In particular, the forecast price of a call and a put option at time $t + 1$ given the information available at time t, with τ days to maturity, denoted respectively by $C^{(\tau)}_{t+1|t}$ and $P^{(\tau)}_{t+1|t}$, is given by

$$C^{(\tau)}_{t+1|t} = S_t e^{-\gamma_t \tau} N(d_1) - K e^{-rf_t \tau} N(d_2), \qquad (8.36)$$

$$P^{(\tau)}_{t+1|t} = -S_t e^{-\gamma_t \tau} N(-d_1) + K e^{-rf_t \tau} N(-d_2), \qquad (8.37)$$

where

$$d_1 = \frac{\log(S_t/K) + \left(rf_t - \gamma_t + \frac{1}{2} \left(\bar{\sigma}^{(\tau)}_{t+2|t} \right)^2 \right) \tau}{\bar{\sigma}^{(\tau)}_{t+2|t} \sqrt{\tau}}, \qquad d_2 = d_1 - \bar{\sigma}^{(\tau)}_{t+2|t} \sqrt{\tau}.$$

Here S_t is the market closing price of the underlying asset at time t (which is used as a forecast for S_{t+1}), rf_t is the daily continuously compounded risk-free interest rate, γ_t is the daily continuously compounded dividend yield, K is the exercise (or strike) price at maturity and $N(.)$ is the cumulative standard normal distribution function. $\bar{\sigma}^{(\tau)}_{t+2|t}$ denotes the average standard deviation of the rate of return during the life of the option, from $t + 2$ until maturity, given the information available at time t. A graphical representation of the relationship between option prices and variables involved in the BS formula as their values evolve temporally is provided by Figure 8.4.

8.5.1.3 Computing theoretical option prices: an example

Consider a trader interested in evaluating the BS theoretical price of a European call and put option with 3 months to expiry. The stock price is €60 and the strike price is €65. Usually, the input variables, rf, γ and σ, are measured on an annual basis. In such a case, the risk free rate is 8% per annum (the annualized log-return of the 3-month treasury bills), the annual dividend yield is 5% and the volatility is 30% per annum. Thus, $S_t = 60$, $K = 65$, $\tau = 0.25$, $rf_t = 0.08$, $\gamma_t = 0.05$ and $\sigma_t = 0.3$. We have

$$\begin{aligned} d_1 &= -0.409, & N(d_1) &= 0.341, & N(-d_1) &= 0.659, \\ d_2 &= -0.559, & N(d_2) &= 0.288, & N(-d_2) &= 0.712, \end{aligned}$$

so that the price estimate of the call option is

$$C_t = 60 e^{-0.05*0.25} 0.341 - 65 e^{-0.08*0.25} 0.288 = €1.8674$$

and that of the put option is

$$P_t = -60 e^{-0.05*0.25} 0.659 + 65 e^{-0.08*0.25} 0.712 = €6.3256.$$

If rf, γ and σ were considered on a daily frequency basis, we would assume that options expire in $365 \times 3/12$ calendar days and $252 \times 3/12$ trading days. The daily

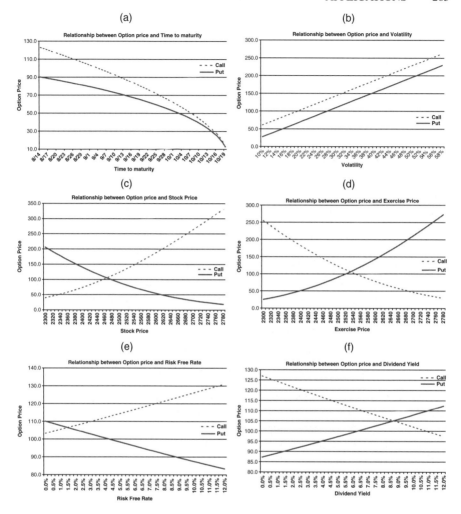

Figure 8.4   Relationship between the values of option prices and variables involved in the BS formula as they evolve temporally.

risk-free rate would be (8%)/365, the daily dividend yield (5%)/252 and the daily volatility (30%)/252. For rf_t we assume 365 calendar days per annum, whereas for $\sigma_{t+1|t}^{(\tau)}$ and y_t we assume 252 trading days per annum. So $\tau = 252 \times 3/12$, $rf_t = 0.08/365 = 0.0219\%$, $\gamma_t = 0.05/252 = 0.0198\%$, $\sigma_t = 0.3/\sqrt{252} = 1.89\%$, and therefore

$$d_1 = \frac{\log\left(\frac{60}{65}\right) + \left(\frac{0.08}{365}365\frac{3}{12} - \frac{0.05}{252}252\frac{3}{12} + \frac{1}{2}\left(\frac{0.3}{\sqrt{252}}\right)^2 252\frac{3}{12}\right)}{\frac{0.3}{\sqrt{252}}\sqrt{252\frac{3}{12}}} = -0.409$$

and

$$d_2 = d_1 - \frac{0.3}{\sqrt{252}}\sqrt{252\frac{3}{12}} = -0.559.$$

Thus, we reach the same price estimates for the call and put options,

$$C_t = 60e^{-(0.05/252)(252\times3/12)}0.341 - 65e^{-(0.08/365)(365\times3/12)}0.288 = €1.8674$$

and

$$P_t = -60e^{-(0.05/252)(252\times3/12)}0.659 + 65e^{-(0.08/365)(365\times3/12)}0.712 = €6.3256.$$

8.5.1.4 Construction of the Black–Scholes option pricing formula

Suppose we have an option, either a call or a put, whose value, $C(S, t)$, depends only on the stock price, S, and time, t. Let us create a riskless hedge portfolio, consisting of a long position in the stock (buy the stock) and short position in the option (sell the option) under the assumption that investors have full access to information, are borrowing and lending at the continuously compounded risk-free interest rate and are trading continuously in a frictionless capital market with no transaction costs, no taxes, and no short sales constraints. Moreover, we assume that the stock price exhibits a geometric Brownian motion,

$$dS(t) = \mu S(t)dt + \sigma S(t)dB(t), \tag{8.38}$$

where μ is the instantaneous expected rate of return on the underlying asset, σ is the instantaneous variance of the rate of return and $B(t)$ is a standard Brownian motion. If we denote by Q_S the number of stocks and by Q_C the number of options, the value V_H of this riskless hedge portfolio and its changes dV_H in short intervals will be determined by

$$\begin{aligned} V_H &= Q_S S + Q_C C, \\ dV_H &= Q_S dS + Q_C dC, \end{aligned} \tag{8.39}$$

respectively. If we assume that the short position is changed continuously, we can use Ito's lemma to expand $dC = C(S + dS, t + dt) - C(S, t)$ as

$$dC(S, t) = \frac{\partial C}{\partial S}dS + \frac{\partial C}{\partial t}dt + \frac{1}{2}\frac{\partial^2 C}{\partial S^2}\sigma^2 S^2 dt. \tag{8.40}$$

We then proceed with determining Q_S and Q_C so that the risk factor is eliminated, i.e.

$$Q_S dS + Q_C \frac{\partial C}{\partial S}dS = 0. \tag{8.41}$$

This leads us to a ratio of stocks to options instantaneously adjusted at the rate of $-\partial C/\partial S$,

$$\frac{Q_C}{Q_S} = -\frac{\partial S}{\partial C}. \tag{8.42}$$

Normalizing by setting $Q_S = 1$, the above equation shows that for each stock purchased we have to write $\partial S/\partial C$ options on it. Moreover, since the return on the equity on the hedge portfolio is certain, it is equal to the value of the risk-free rate, rf:

$$\frac{dV_H}{V_H} = (rf)dt. \tag{8.43}$$

Thus,

$$dV_H = \frac{-\partial S}{\partial C}\left(\frac{\partial C}{\partial t}dt + \frac{1}{2}\frac{\partial^2 C}{\partial S^2}\sigma^2 S^2 dt\right),$$

and solving for $\partial C/\partial t$ we obtain

$$\frac{\partial C}{\partial t} = (rf)V_H\left(-\frac{\partial C}{\partial S}\right) - \frac{1}{2}\frac{\partial^2 C}{\partial S^2}\sigma^2 S^2. \tag{8.44}$$

Combining $V_H = S - C(1/(\partial C/\partial S))$ with the above equation and rearranging, we reach the following partial differential equation for the value of the option:

$$\frac{1}{2}\sigma^2 S^2 \frac{\partial^2 C}{\partial S^2} + (rf)S\frac{\partial C}{\partial S} + \frac{\partial C}{\partial T} - (rf)C = 0. \tag{8.45}$$

Having derived the equation for the valuation of an option, we must next consider the boundary conditions yielding a unique solution to the partial differential equation. First, we are dealing with pricing a European call, $C(S, t)$, with exercise price K and expiry date T. At maturity, the value of the call is known with certainty and is the payoff, $C(S(T), T) = \max(0, S(T) - K)$. Moreover, if $S = 0$, the call option is worthless even if there is a long time to expiry, $C(0, t) = 0$. Finally, as the stock price increases without bound, it becomes even more likely that the option will be exercised and the magnitude of the exercise price becomes less and less important. Thus, as $S \to \infty$, the value of the option becomes that of the asset, $C(S, t) \approx S$ as $S \to \infty$. On the other hand, the boundary conditions to price a European put, denoted by $P(S, t)$, with exercise price K and expiry date T, imply that at maturity the value of the put is known with certainty and equals the payoff, $P(S(T), T) = \max(0, K - S(T))$. If the stock price is zero, the put price is the present value of the exercise price received at maturity, $P(0, t) = \exp(-rf(T - t))K$. Finally, as the asset price increases without bound, the option is unlikely to be exercised, i.e., $P(S, t) \to 0$ as $S \to \infty$.

Therefore, the unique solution of the partial differential equation given by (8.45), subject to the above boundary conditions, is obtained through the system of equations

$$\begin{aligned}
C(S(t), t) &= S(t)N(d_1) - Ke^{-rf(T-t)}N(d_2), \\
P(S(t), t) &= Ke^{-rf(T-t)}N(-d_2) - S(t)N(-d_1), \\
d_1 &= \frac{\log(S(t)/K) + (rf + \frac{1}{2}\sigma^2)(T-t)}{\sigma\sqrt{T-t}}, \\
d_2 &= d_1 - \sigma\sqrt{T-t},
\end{aligned} \tag{8.46}$$

where $(T - t) = \tau$ is the time to maturity of the option.

Merton (1973) extended the BS formula to allow for dividend yield. The model can be used to price European call and put options on a stock or stock index paying a known dividend yield equal to γ. Suppose that, within time dt, the underlying asset pays a dividend equal to $\gamma S dt$. The asset price should fall by the amount of the dividend payment. Thus, the asset price distribution satisfies

$$dS(t) = (\mu - \gamma)S(t)dt + \sigma S(t)dB(t). \qquad (8.47)$$

Proceeding exactly as before, we reach a partial differential equation of the form

$$\frac{1}{2}\sigma^2 S^2 \frac{\partial^2 C}{\partial S^2} + (rf - \gamma)S \frac{\partial C}{\partial S} + \frac{\partial C}{\partial T} - (rf)C = 0, \qquad (8.48)$$

which is uniquely solved subject to the same set of boundary conditions, except for the value of the option when the asset price increases without bound. As $S \to \infty$, the value of the call equals the price of the asset without its dividend: $C(S, t) \approx S \exp(-\gamma(T - t))$. Adding a constant dividend yield, the option pricing formula becomes

$$
\begin{aligned}
C(S(t), t) &= S(t)e^{-\gamma(T-t)}N(d_1) - Ke^{-rf(T-t)}N(d_2), \\
P(S(t), t) &= Ke^{-rf(T-t)}N(-d_2) - S(t)e^{-\gamma(T-t)}N(-d_1), \\
d_1 &= \frac{\log(S(t)/K) + (rf - \gamma + \frac{1}{2}\sigma^2)(T-t)}{\sigma\sqrt{T-t}}, \\
d_2 &= d_1 - \sigma\sqrt{T-t}.
\end{aligned}
\qquad (8.49)
$$

Note that in order to derive the option pricing formula, we make no assumptions about investors' preferences. Both an economy consisting of risk-neutral investors and one consisting of risk-averse investors must yield the same price for the derivative security. Cox and Ross (1976) assume a risk-neutral economy and define the price of the option as the expected value of its payoff discounted at the risk-free rate:

$$C(\tilde{S}(T), t) = e^{-rf(T-t)}E\big(\max(0, \tilde{S}(T) - K)|I_t\big), \qquad (8.50)$$

$$P(\tilde{S}(T), t) = e^{-rf(T-t)}E\big(\max(0, K - \tilde{S}(T))|I_t\big). \qquad (8.51)$$

The expectation is evaluated conditional on the information available at time t. $\tilde{S}(T)$ denotes the terminal stock price adjusted for risk neutrality. The procedure applied to solve the conditional expectation is called the risk-neutral pricing method and the solution yields the BS formula.

8.5.1.5 Option strategies and cash flows

Suppose that the price of stock at time t is S_t and that call and put options, with expiration date T and exercise price K, are priced at C_t and P_t, respectively. In terms of cash flows, the purchaser of an option (a long option position) always has an initial negative cash flow, the price of the option, and a future cash flow that is at worst zero. The writer of the option (a short option position) has an initial positive cash flow followed by a terminal cash flow that is at best zero. At the expiration date, T, the call

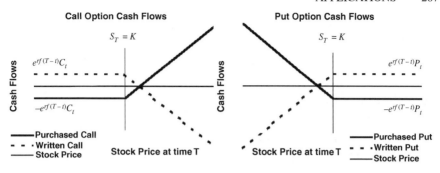

Figure 8.5   The cash flows of taking long and short positions in call and put options.

option is exercised only if $S_T > K$. Thus, the cash flow, at time T, of the call purchaser is[13]

$$\max(0, S_T - K) - e^{rf_t(T-t)}C_t = \begin{cases} -e^{rf_t(T-t)}C_t & \text{if } S_T \leq K, \\ S_T - K - e^{rf_t(T-t)}C_t & \text{if } S_T > K. \end{cases} \tag{8.52}$$

The cash flow of the call writer is the opposite of that of the call purchaser:

$$e^{rf_t(T-t)}C_t - \max(0, S_T - K) = \begin{cases} e^{rf_t(T-t)}C_t & \text{if } S_T \leq K, \\ e^{rf_t(T-t)}C_t + K - S_T & \text{if } S_T > K. \end{cases} \tag{8.53}$$

Moreover, the put is exercised only if $S_T < K$. Thus, at the maturity date, the cash flow of the put purchaser is:

$$\max(0, K - S_T) - e^{rf_t(T-t)}P_t = \begin{cases} -e^{rf_t(T-t)}P_t & \text{if } S_T \geq K, \\ K - S_T - e^{rf_t(T-t)}P_t & \text{if } S_T < K, \end{cases} \tag{8.54}$$

and the cash flow of the put writer is

$$e^{rf_t(T-t)}P_t - \max(0, K - S_T) = \begin{cases} e^{rf_t(T-t)}P_t & \text{if } S_T \geq K \\ e^{rf_t(T-t)}P_t + S_T - K & \text{if } S_T < K. \end{cases} \tag{8.55}$$

Figure 8.5 depicts the temporal trajectories of profits and losses in buying and writing options over the period considered.

A long straddle position is an option strategy whereby a call and a put of the same exercise price, maturity and underlying terms are purchased. This position is called a straddle since it will lead to a profit from a substantial change in the stock price in either direction. Traders purchase a straddle under one of two circumstances. The first is when a large change in the stock price is expected, but the direction of the change is unknown. Examples include an upcoming announcement of earnings, uncertain takeover or merger speculation, a court case for damages, a new product announcement,

---

[13] It is assumed that investors, at time t, are borrowing and lending at the same risk-free rate, rf_t. Thus, during the period from t to T the money is invested with a daily return of $(1 + rf_t) \approx \exp(rf_t)$.

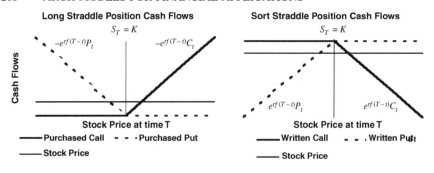

Figure 8.6   *The cash flows of taking long and short straddle positions.*

or an uncertain economic announcement such as inflation figures or a change in the prime interest rate. A straddle appears to be a risk-free trading strategy when a large change in the price of a stock is expected.

However, in the real world this is not necessarily the case. If the general view of the market is that there will be a big jump in the stock price soon, the option prices should reflect the increase in the potential volatility of the stock. A trader will find options on the stock to be significantly more expensive than options on a similar stock for which no jump is expected. For a straddle to be an effective strategy, the trader must believe that big movements in the stock price are likely, and this belief must be different from that of most of the other market participants.

The second circumstance under which straddles are purchased occurs when the trader estimates that the true future volatility of the stock will be greater than the volatility that is currently impounded in the option prices. Note that, although the long straddle has theoretically unlimited potential profit and limited risk, it should not be viewed as a low-risk strategy. Options can lose their value very quickly, and in the case of a straddle there is twice the amount of erosion of time value compared to the purchase of a call or put. The opposite of a long straddle strategy is a short straddle position. This strategy has unlimited risk and limited profit potential, and is therefore only appropriate for experienced investors with a high tolerance for risk. The short straddle will profit from limited stock movement and will suffer losses if the underlying asset moves substantially in either direction. Figure 8.6 plots the observed time series of the payoffs of taking long and short straddle positions. At expiration date, T, the cash flows of taking a long and a short straddle position are

$$|S_T - K| - e^{rf_t(T-t)}(C_t + P_t) \quad \text{and} \quad e^{rf_t(T-t)}(C_t + P_t) - |S_T - K|, \qquad (8.56)$$

respectively.

8.5.1.6 Straddle trading: an example

Consider a trader who feels that the price of a certain stock, currently valued at $54 by the market, will move significantly in the next three months. The trader could create a straddle by buying both a put and a call with a strike price of $55 and an expiration date in three months. Suppose that the call and the put costs are \$5 and \$4, respectively. The most that can be lost is the amount paid, \$9, if the stock price moves to \$55. If the stock price moves above \$63 or below \$45, the long position earns a profit. In the case of taking a short straddle position, the maximum profit is the premium received, or \$9. The maximum loss is unlimited, and the short position will lose if the stock price moves above \$63 or below \$45.

8.5.2 Assessing the performance of volatility forecasting methods

Noh et al. (1994) devised rules for trading near-the-money straddles. If the straddle price forecast is greater than the market straddle price, the straddle is bought. If the straddle price forecast is less than the market straddle price, the straddle is sold:

$$\text{if } C_{t+1|t}^{(\tau)} + P_{t+1|t}^{(\tau)} > P_t^{(\tau)} + C_t^{(\tau)} \Rightarrow \text{ the straddle is bought at time } t; \qquad (8.57)$$

$$\text{if } C_{t+1|t}^{(\tau)} + P_{t+1|t}^{(\tau)} < P_t^{(\tau)} + C_t^{(\tau)} \Rightarrow \text{ the straddle is sold at time } t. \qquad (8.58)$$

The strategy can be understood with the help of the following example. On Monday (time t), after the stock market closes,[14] Tuesday's (time $t + 1$) price for an option that expires on Friday (time $t + 4$) is estimated. The remaining life of the option is 3 days, from Tuesday to Friday. If the predicted option price on Tuesday is higher than the observed option price on Monday, the option is bought in order to be sold on Tuesday. If the predicted option price on Tuesday is lower than the observed option price on Monday, the option is short-sold in order to be bought on Tuesday.

The rate of return from trading an option is

$$RT_{t+1} = \frac{C_{t+1} + P_{t+1} - C_t - P_t}{C_t + P_t}, \text{ on buying a straddle}, \qquad (8.59)$$

$$RT_{t+1} = \frac{-C_{t+1} - P_{t+1} + C_t + P_t}{C_t + P_t}, \text{ on short-selling a straddle}. \qquad (8.60)$$

Note that the transaction costs, X, should be taken into account. If this is the case, the net rate of return from trading an option is given by

$$NRT_{t+1} = RT_{t+1} - \frac{X}{C_t + P_t}. \qquad (8.61)$$

Moreover, a filter can be applied in the trading strategy, so that traders will trade straddles only when profits are predicted to exceed transaction costs. So, straddles are traded only when the absolute difference between the forecast and today's option price exceeds the amount of the filter, F_{il}, yielding a net rate of return of:

---

[14] The trading strategy assumes that there is enough time to forecast the option prices given all the information at time t (closing prices of stocks) so that the trader can decide on the trading of an option at time t (before the option market closes). For example, the Chicago Stock Market closes at 3:00 pm local time and the Chicago Board of Option Exchange closes as 3:15 pm local time.

$$NRT_{t+1} = \begin{cases} RT_{t+1} - \dfrac{X}{C_t + P_t}, & \text{if } \left| C^{(\tau)}_{t+1|t} + P^{(\tau)}_{t+1|t} - C_t - P_t \right| > F_{il}, \\ 0, & \text{otherwise.} \end{cases} \tag{8.62}$$

Noh et al. (1994) applied the AR(1)-GARCH(1,1)-N model in order to forecast future volatility. Forecasts of option prices, on the next trading day, are calculated using the BS option pricing formula and conditional volatility forecasts. The average volatility during the life of the option is computed as the square root of the average forecast conditional variance,

$$\bar{\sigma}^{(\tau)}_{t+2|t} = \left(\tau^{-1} \sum_{i=2}^{\tau+1} \sigma^2_{t+i|t} \right)^{1/2}, \tag{8.63}$$

where $\sigma^2_{t+i|t}$ denotes the prediction of the conditional variance at time $t + i$ given the information set available at time t.

Noh et al. (1994) assessed the performance of the AR(1)-GARCH(1,1)-N model for straddles written on the S&P500 index over the period from April 1986 to December 1991 and found that the model earns a profit of \$0.885 per straddle in excess of a \$0.25 transaction cost and applying a \$0.50 filter. González-Rivera et al. (2004) also evaluated the ability of various volatility models in predicting one-period-ahead call options on the S&P500 index, with expiration dates ranging from January 2000 through November 2000, and found that simple models like the EWMA of J.P. Morgan's RiskMetrics™ (1996) performed as well as sophisticated ARCH specifications.

8.5.3 Black–Scholes option pricing using a set of ARCH processes

The GARCH(1,1) model is the one most commonly used in financial applications. The question that arises at this point is why one should use the simple GARCH(1,1) model instead of a higher-order of GARCH(p,q) model, an asymmetric ARCH model, or even a more complicated form of an ARCH process. There are a vast number of ARCH models. Which one is the most appropriate? The volatility prediction model, which gives the highest rate of return in trading options, should be preferred Moreover, under the assumption that the BS formula perfectly describes the dynamics of the market that affects the price of the option, the model that gives the most precise prediction of conditional volatility should be the model that yields the highest rate of return. Unfortunately, we come up against an important limitation. Even if one could find a model which predicts the volatility precisely, it is well known that the BS formula does not describe the dynamics pricing the options perfectly.[15] Moreover, the validity of the variance forecasts depends on which option pricing formula is used. Engle et al. (1997) used Hull and White's (1987) modification to the BS formula to price straddles on a simulated options market. As we will see in Section 8.6, a series of studies such as those by Duan (1995), Engle and Mustafa (1992) and Heston and

---

[15] The biasedness of BS formula has been widely examined. For details, see Daigler (1994) and Natenberg (1994).

Nandi (2000) derived ARCH based option pricing models assuming that a specific ARCH process generates the variance of the asset. However, despite its limitations, the BS pricing formula has found wide acceptance among floor traders on options exchanges.

Since the ARCH-based option pricing models considered in the literature for the various models being compared are different (or no ARCH based pricing formula exists for some of them), the BS option pricing model is adopted.

8.5.4 Trading straddles based on a set of ARCH processes

In the paragraphs that follow we create a straddle trading game based on a set of nine ARCH models.

Let us assume that there are nine traders and that each trader employs an ARCH model to forecast future volatility and straddle prices. Each trading day, if the straddle price forecast is greater than the market straddle price, the straddle is bought, otherwise the straddle is sold. The scheme described below is based on the study by Degiannakis and Xekalaki (2008) which proposed a straddle trading game driven by volatility forecasts produced by various methods of ARCH model selection.[16] The S&P500 daily log-returns over the period from 14 March 1996 to 2 June 2000 are considered, a total of 1064 daily returns. The models are estimated using the EViews 6 package[17] in the form

$$y_t = c_1 y_{t-1} + z_t \sigma_t,$$
$$z_t \overset{i.i.d.}{\sim} N(0, 1). \tag{8.64}$$

The GARCH$(p, 1)$, EGARCH$(p, 1)$ and GJR$(p, 1)$ conditional volatility specifications, given by (2.73), (2.75) and (2.76), are considered. The AR(1)-GARCH$(p, 1)$-N, AR(1)-EGARCH$(p, 1)$-N and AR(1)-GJR$(p, 1)$-N models are thus applied for $p = 0, 1, 2$, yielding a total of nine models. The conditional variance for the GARCH(p, q) process may be rewritten as

$$\sigma_t^2 = (u_t', \eta_t', w_t')(v, \zeta, \omega)', \tag{8.65}$$

where $u_t' = \left(1, \varepsilon_{t-1}^2, \ldots, \varepsilon_{t-q}^2\right)$, $\eta_t' = 0$, $w_t' = \left(\sigma_{t-1}^2, \ldots, \sigma_{t-p}^2\right)$, $v' = (a_0, a_1, \ldots, a_q)$, $\zeta' = 0$ and $\omega' = (b_1, \ldots, b_p)$. For the EGARCH$(p, q)$ process, the conditional variance can be expressed as

$$\log \sigma_t^2 = (u_t', \eta_t', w_t')(v, \zeta, \omega)', \tag{8.66}$$

where $u_t' = \left(1, |\varepsilon_{t-1}/\sigma_{t-1}|, \ldots, |\varepsilon_{t-q}/\sigma_{t-q}|\right)$, $\eta_t' = (\varepsilon_{t-1}/\sigma_{t-1}, \ldots, \varepsilon_{t-q}/\sigma_{t-q})$, $w_t' = \left(\log \sigma_{t-1}^2, \ldots, \log \sigma_{t-p}^2\right)$, $v' = (a_0, a_1, \ldots, a_q)$, $\zeta' = (\gamma_1, \ldots, \gamma_q)$, $\omega' = (b_1, \ldots, b_p)$.

---

[16] Methods of ARCH model selection are discussed in Chapter 10.

[17] There are minor differences in model estimations among various versions of EViews. For example, EViews 3.1, 4.1 and 6 provide slightly different parameters estimates in some models and a set of trading days.

For the GJR(p, q) process, the conditional variance may take the form

$$\sigma_t^2 = \left(u_t', \eta_t', w_t'\right)(v, \zeta, \omega)', \tag{8.67}$$

where $u_t' = \left(1, \varepsilon_{t-1}^2, \ldots, \varepsilon_{t-q}^2\right)$, $\eta_t' = \left(d(\varepsilon_{t-1} < 0)\varepsilon_{t-1}^2\right)$, $w_t' = \left(\sigma_{t-1}^2, \ldots, \sigma_{t-p}^2\right)$, $v' = (a_0, a_1, \ldots, a_q)$, $\zeta' = (\gamma_1)$, $\omega' = (b_1, \ldots, b_p)$, $d(\varepsilon_t < 0) = 1$ if $\varepsilon_t < 0$, and $d(\varepsilon_t < 0) = 0$ otherwise. In general, the conditional variance forecast recursion relations of (8.65)–(8.67) could be presented as

$$\begin{aligned}
\sigma_{t+\tau|t}^2 \equiv E\left(\sigma_{t+\tau}^2|I_t\right) &= E\left(u_{t+\tau}', \eta_{t+\tau}', w_{t+\tau}'|I_t\right)\left(v^{(t)}, \zeta^{(t)}, \omega^{(t)}\right)' \\
&= \left(u_{t+\tau|t}', \eta_{t+\tau|t}', w_{t+\tau|t}'\right)\left(v^{(t)}, \zeta^{(t)}, \omega^{(t)}\right)'.
\end{aligned} \tag{8.68}$$

The algorithm described below illustrates how the straddle trading scheme considered works in practice for each of the nine models.

- For model i, $i = 1, 2, \ldots, 9$ and for each point in time t, $t = \tilde{T}$, $\tilde{T} + 1, \ldots, \tilde{T} + \breve{T} - 1$, the vector of coefficients is estimated using a rolling sample of $\breve{T} = 500$ observations:

$$\theta_{(i)}^{(t)} \equiv \left(c_{1(i)}^{(t)}, v_{(i)}^{(t)}, \zeta_{(i)}^{(t)}, \omega_{(i)}^{(t)}\right)'. \tag{8.69}$$

- Using the vector of coefficients $\theta_{(i)}^{(t)}$, estimate the vector

$$\left(y_{t+1|t(i)}, \sigma_{t+1|t(i)}^2\right). \tag{8.70}$$

- Compute the conditional variance forecasts from time $t+2$ to $t+\tau+1$:

$$\left(\sigma_{t+2|t(i)}^2, \ldots, \sigma_{t+\tau+1|t(i)}^2\right). \tag{8.71}$$

Each model has been estimated for $\tilde{T} = 564$ trading days based on a rolling sample of $\breve{T} = 500$ days. The first one-step-ahead volatility prediction, $\sigma_{t+1|t}^2$, is available at time $t = 500$, or on 11 March 1998.

When estimating volatility, and in particular when using different ARCH models, the size of the sample is very important. One might say that there are two opposite criteria for selecting the size of the sample. On the one hand, larger data sets are often used for the estimation of ARCH models. In several studies, results indicate that for good parameter estimates you need thousands of observations. The main reason is that when modelling volatility, one tries to give the observations in the tail of the error distribution a parametric structure. Thus, if the sample is too small, not enough observations will end up in the tail, resulting in poor parameter estimates. On the other hand, as already noted in the Chapter 4, the use of a restricted sample size incorporates changes in trading behaviour more efficiently. Among others, Angelidis et al. (2004), Engle et al. (1993) and Frey and Michaud (1997) supported the use of restricted samples and provided empirical evidence that they capture changes in market activity better. Hoppe (1998), investigating the issue of the sample size in the context of VaR, argues that a smaller sample could lead to more accurate estimates than a larger one. In our case, a sample of 500 observations might be considered far too small. Based on a larger sample, the reader can re-estimate the models and investigate potential differences.

The S&P500 index options[18] data were obtained from the Datastream database for the period from 11 March 1998 to 2 June 2000 (564 trading days in all). Unfortunately, the data record is not adequate for all the trading days. Proper data are available for 456 trading days and contain information on the closing price of the call and put options, exercise price, expiration date, and the number of contracts traded. In total, 49 500 call and put prices were collected. However, in order to minimize the biasedness of the BS formula, only the straddle options with exercise prices closest to the index level, maturity longer than 10 trading days and trading volume greater than 100 were considered from the entire data set for each trading day. The choice of these data points was based on considerations for the optimal performance of the option pricing model. Also under such considerations, Sabbatini and Linton (1998), employing Duan's (1995) option pricing model in estimating the volatility of the Swiss market index, used a two-year period of daily closing prices of near-the-money options and with a period to maturity of at least 15 days. In our case, near-the-money trading is considered since practice has shown that the BS pricing model tends to misprice deep out-of-the-money and deep in-the-money options (see Black, 1975; Gultekin et al., 1982; Merton, 1976; MacBeth and Merville, 1979, 1980), while it works better for near-the-money options (see Daigler, 1994, p. 153).

Further, a maturity period of length no shorter than 10 trading days is considered to prevent mispricings attributable to causes of practical as well as of theoretical nature. In particular, experience has shown that traders pay less and less attention to the values generated by the pricing model as expiration approaches (see Natenberg, 1994, p. 398). In the Spanish market, Aragó and Fernández (2002) noted that during the week of expiration the conditional variances of spot and futures markets show a differential behaviour. Moreover, Whaley (1982) noted that the BS formula underprices short-maturity options. From the theoretical point of view, there is often a departure from the BS model's assumption that stock prices are realizations of a continuous diffusion process, as in most markets the underlying contracts conform to a combination of both a diffusion process and a jump process.[19] According to Dumas et al. (1998), the volatility estimation of options close to expiration is extremely sensitive to possible measurement errors. Most of the time, asset prices change smoothly and continuously with no gaps. However, every now and then a gap will occur, instantaneously sending

---

[18] S&P500 index options are traded on the Chicago Board Options Exchange (CBOE).

[19] A variation of the BS model, which assumes that the underlying contract follows a jump diffusion process, has been developed (see Merton, 1976; Beckers, 1981). Unfortunately, the model is considerably more complex mathematically than the traditional BS model. Moreover, in addition to the five customary inputs, the model requires two new inputs: the average size of a jump in the underlying market and the frequency with which such jumps are likely to occur. Unless the trader can adequately estimate these new inputs, the values generated by a jump diffusion model may be no better, and might be worse, than those generated by the traditional model. Most traders take the view that whatever weaknesses are encountered in a traditional model they can best be offset through intelligent decision-making based on actual trading experience rather than through the use of a more complex jump diffusion model.

the price to a new level. These prices will again be followed by a smooth diffusion process until another gap occurs. Furthermore, as Natenberg (1994, p. 397) commented: 'since a gap in the market will have its greatest effect on at-the-money options close to expiration,[20], it is these options that are likely to be mispriced by the traditional BS pricing model with its continuous diffusion process'.

---

[20] A gap in the market has its greatest effect on a high-gamma option, and at-the-money options close to expiration have the highest gamma. Delta, lambda, gamma, theta, vega and rho comprise the pricing sensitivities and represent the key relationships between the individual characteristics of the option and the option price. Delta is the change in the option price for a given change in the stock price, that is, the hedge ratio:

$$\Delta_{CALL} = \frac{\partial C}{\partial S} = e^{-\gamma\tau}N(d_1) > 0,$$

$$\Delta_{PUT} = \frac{\partial P}{\partial S} = e^{-\gamma\tau}(N(d_1)-1) < 0.$$

Lambda measures the percentage change in the option price for a given percentage change in the stock price:

$$\Lambda_{CALL} = \frac{\partial C}{\partial S}\frac{S}{C} = e^{-\gamma\tau}N(d_1)\frac{S}{C} > 1,$$

$$\Lambda_{PUT} = \frac{\partial P}{\partial S}\frac{S}{P} = e^{-\gamma\tau}(N(d_1)-1)\frac{S}{P} < 0.$$

Rho measures the change in the option price for a given change in the risk-free interest rate:

$$P_{CALL} = -\frac{\partial C}{\partial rf} = \tau Ke^{-(rf)\tau}N(d_2) > 0,$$

$$P_{PUT} = -\frac{\partial P}{\partial rf} = -\tau Ke^{-(rf)\tau}N(-d_2) < 0.$$

Theta is the change in the option price for a given a change in the time, until option expiration. As time to maturity decreases, it is normal to express theta as the negative partial derivative with respect to time:

$$\Theta_{CALL} = -\frac{\partial C}{\partial \tau} = -\frac{Se^{-(\gamma\tau + d_1^2/2)}\sigma}{2\sqrt{2\pi\tau}} + \gamma Se^{-\gamma\tau}N(d_1) - r_f Ke^{-(rf)\tau}N(d_2),$$

$$\Theta_{PUT} = -\frac{\partial P}{\partial \tau} = -\frac{Se^{-(\gamma\tau + d_1^2/2)}\sigma}{2\sqrt{2\pi\tau}} - \gamma Se^{-\gamma\tau}N(-d_1) + r_f Ke^{-(rf)\tau}N(-d_2).$$

Gamma measures the change in the delta for a given change in the stock price. Gamma is identical for call and put options:

$$\Gamma_{CALL,PUT} = \frac{\partial^2 C}{\partial S^2} = \frac{\partial^2 P}{\partial S^2} = \frac{e^{-(d_1^2/2 + \gamma\tau)}}{S\sigma\sqrt{2\pi\tau}} > 0.$$

Vega is the change in the option price for a given change in the volatility of the stock. Vega is equal for call and put options:

$$V_{CALL,PUT} = \frac{\partial C}{\partial \sigma} = \frac{\partial P}{\partial \sigma} = \frac{Se^{-(\gamma\tau + d_1^2/2)}\sqrt{\tau}}{\sqrt{2\pi}} > 0.$$

Consider the example of computing the theoretical option prices in Section 8.5.1. The option sensitivities for the call and put options are as follows:

	Price	Delta	Lambda	Rho	Theta	Gamma	Vega
Call	1.867	0.337	10.833	4.591	−6.981	0.040	10.873
Put	6.326	−0.650	−6.169	−11.337	−4.847	0.040	10.873

The necessary code is divided into three (sub)programs. The program *chapter8_ sp500_options.partA.prg* carries out the estimation of the nine models for the S&P500 log-returns for $\tilde{T} = 564$ trading days based on a rolling sample of $\tilde{T} = 500$ days. The core program that executes the necessary computations for the AR(1)-ARCH(1)-N model is as follows:

```
for !i = 1 to !length-!ssize
statusline !i
  smpl @first+!i-1 @first+!i+!ssize-2
  equation tempg01.arch(1,0,m=10000,h, deriv=f) sp500
    sp500(-1)
  tempg01.makegarch garchg01
  tempg01.makeresid resg01
!ht = garchg01(!i+!ssize-1)
!c1 = @coefs(1)
!c2 = @coefs(2)
!c3 = @coefs(3)
!h_hat = !c2 + (!c3*(resg01(!ssize + !i -1)^2) )
!y_hat = (!c1*sp500(!ssize + !i -1))
!res = sp500(!ssize + !i) - !y_hat
estg01(!i,1) = !y_hat
if !h_hat>=0 then
  estg01(!i,2) = !h_hat
else
  estg01(!i,2) = estg01(!i-1,2)
endif
estg01(!i,3) = !res
estg01(!i,4) = (!res^2) / estg01(!i,2)
estg01(!i,5) = estg01(!i,4) + !g01
!g01 = estg01(!i,5)
estg01(!i,6) = !c1
estg01(!i,7) = !c2
estg01(!i,8) = !c3
```

EViews loads the workfile named *chapter8_sp500_options.wf1* that contains the necessary data for the computation of options prices. The series *sp500* contains the S&P500 log-returns for 1661 trading days. The rows numbered 582–1645 contain the log-returns from 14 March 1996 to 2 June 2000. The series *call_obs* and *next_call* contain the values of C_t and C_{t+1}, respectively. The series *put_obs* and *next_put* contain the values of P_t and P_{t+1}, and the series *straddle_obs* and *next_straddle* store the straddle prices at times t and $t+1$, respectively. *ahead* denotes the number of trading days to maturity, τ, S is the market closing price, S_t, *rf* is the daily continuously compounded risk-free interest rate, rf_t, k is the exercise price at the maturity date and q is the dividend yield, γ_t. The series *t365* contains coefficients computed as the number of calendar days to maturity divided by 365 and is used to annualize the risk-free interest rate. Likewise, the series *t252* contains coefficients computed as the number of trading days to maturity divided by 252 and is used to annualize volatility and dividend yield.

The equation object named *tempG01* estimates the AR(1)-ARCH(1)-N model by numerical maximization of the log-likelihood function using the Marquardt algorithm[21] and the quasi-maximum likelihood estimation method.[22] Option *derive=f* corresponds to fast numerical derivatives and *m=10000* sets the maximum number of iterations to 10 000.

The matrix *estg01* contains 1161 rows and eight columns. The first two columns list the one-step-ahead conditional means, $\mu_{t+1|t}$, and conditional variances, $\sigma^2_{t+1|t}$. The third column lists the one-step-ahead residuals, $y_{t+1}-\mu_{t+1|t}$. Columns 4 and 5 store the values of the quantities $(y_{t+1}-\mu_{t+1|t})^2/\sigma^2_{t+1|t}$ and $\sum_{i=1}^{t}(y_{i+1}-\mu_{i+1|i})^2/\sigma^2_{i+1|i}$, respectively. Note that these values will be used in Chapter 10. The last three columns contain the values of the estimated parameters, $c_1^{(t)}, a_0^{(t)}, a_1^{(t)}$.

The program *chapter8_sp500_options.partB.prg* in the Appendix to the chapter provides volatility forecasts during the life of the option, from time $t+1$ until the maturity date, $\sigma_{t+1|t}^{(\tau)}$, and computes the call and put prices, $C_{t+1|t}^{(\tau)}$ and $P_{t+1|t}^{(\tau)}$, respectively. The computation of $\bar{\sigma}_{t+2|t}^{(\tau)}$ for the AR(1)-ARCH(1)-N model is carried out by the following segment of the code.

```
for !ahead=1 to !day
  if ahead(!ahead)>0 then
    vector(ahead(!ahead)) h
    h(1)= estg01(!ahead-!ssize,7) + estg01(!ahead-
      !ssize,8)*estg01(!ahead-!ssize,2)
    for !r=2 to ahead(!ahead)
    h(!r) = estg01(!ahead-!ssize,7) + estg01
      (!ahead-!ssize,8)*h(!r-1)
    next
    volatility(!ahead,1)=sqr(@sum(h)*252/ahead(!ahead))
    delete h
  endif
next
```

Each trading day $t=!ahead$, if the series *ahead* denotes a positive value, a vector named h is created. The vector h has as many rows as the number of trading days to maturity [*vector(ahead(!ahead)) h*]. Its *i*th component, $h(i)$, contains the estimated value of $\sigma^2_{t+i+1|t} = a_0^{(t)} + a_1^{(t)}\sigma^2_{t+i|t}, i = 1,2,\ldots,\tau$. The first column of the volatility matrix contains the annualized volatility forecasts during the life of the option computed as

---

[21] If the optional parameter b is added then the BHHH algorithm would be applied. With no optional parameter the Marquardt algorithm is applied as the default method.

[22] Optional parameter h. See also Sections 2.2.2 and 2.2.3.

$$\bar{\sigma}_{t+2|t}^{(\tau)} = \sqrt{252\tau^{-1} \sum_{i=2}^{\tau+1} \sigma_{t+i|t}^2},$$

for each trading day $t=!ahead$.

On each trading day and for each of the nine ARCH models, the call and put option prices are forecasted. The core program that carries out the necessary computations for forecasting options prices is:

```
for !j=1 to !models
 for !i=1 to !day
 if K(!i)>0 then
  !d1 = ( log(S(!i)*exp(-q(!i)*t252(!i))/K(!i))+ (rf(!i)
     *t365(!i)) + (Volatility(!i,!j)^2*t252(!i)/2) ) /
  (Volatility(!i,!j)*sqr(t252(!i)))
   !d2 = !d1 - (Volatility(!i,!j)*sqr(t252(!i)))
  Calls(!i,!j) = S(!i)*exp(-q(!i)*t252(!i))*@cnorm(!d1) - K
     (!i)*exp(-rf(!i)*t365(!i))*@cnorm(!d2)
  Puts(!i,!j) = K(!i)*exp(-rf(!i)*t365(!i))*@cnorm(-!d2) -
     S(!i)*exp(-q(!i)*t252(!i))*@cnorm(-!d1)
  Straddles(!i,!j) = Calls(!i,!j) + Puts(!i,!j)
  else
  Calls(!i,!j) = na
  Puts(!i,!j) = na
  Straddles(!i,!j) = na
  endif
 next
next
```

Whenever an option's price must be computed [if $K(!i)>0$ then] for each model and trading day, the matrices *Calls*, *Puts* and *Straddles* store the forecasts of calls, puts and straddles, respectively. For example, the 1082nd row and first column of the *Calls* matrix stores the call price based on the volatility forecast obtained by the AR (1)-ARCH(1)-N model for the 1082nd trading day, 11 March 1998. Once the call price is estimated, the put price can also be computed according to the put–call parity:

$$P(S(t),t) = C(S(t),t) - S(t)e^{-\gamma(T-t)} + Ke^{-rf(T-t)}. \tag{8.72}$$

The put–call parity is model-free and holds for any option pricing model. It is based on the principle that there must not be arbitrage opportunities, i.e. the dividend adjusted price of the asset plus the current value of a put option on that asset (with strike price K) must be equal to the current value of a call option on that asset (with strike price K) plus the present value of a risk-free zero coupon bond with value K at maturity, i.e. $S(t)e^{-\gamma(T-t)} + P(S(t),t) = C(S(t),t) + Ke^{-rf(T-t)}$.

The program named *chapter8_sp500_options.partC.prg* in the Appendix to the chapter computes. for each of the nine traders, the daily rate of return from trading straddles for 456 days according to equation (8.62):

```
for !x=1 to 3
  for !p=1 to 2
    for !j=1 to !models
      for !i=1 to !day
    if Straddles(!i,!j)<> na then
        if Straddles(!i,!j)>Straddle_obs(!i)+filter(!x) then
          Str_Ret_F{!x}C{!p}(!i,!j)=(next_Straddle(!i)-
            Straddle_obs(!i)-
Cost(!p))/Straddle_obs(!i)
        else
            if Straddles(!i,!j)<Straddle_obs(!i)-filter(!x)
              then
              Str_Ret_F{!x}C{!p}(!i,!j) = (-next_Straddle
              (!i)+Straddle_obs(!i)-
Cost(!p))/Straddle_obs(!i)
          else
          Str_Ret_F{!x}C{!p}(!i,!j) = na
        endif
      endif
    else
        Str_Ret_F{!x}C{!p}(!i,!j) = na
    endif
        next
      next
    next
next
```

For example, the matrix *str_Ret_F3C2* stores in row *!i* and column *!j* the rate of return from trading straddles according to the volatility forecast produced by model *!j* on the *!i*th trading day, taking into consideration a filter of $F_{il} = \$3.50$ and a transaction cost of $X = \$2$. Table 8.2 presents the means and standard deviations of the actual S&P500 option prices and their estimated values based on three ARCH volatility forecasts. The ARCH forecasts for both call and put options appear to be lower than the actual option prices, as in Noh et al.'s (1994) study.

For each trader, the average daily rate of return from trading straddles for 456 days is given in Table 8.3. According to the *t*-ratios (computed as ratios of the mean to the standard deviation divided by the square root of the number of trading days) all the traders appear to achieve profits significantly greater than zero. However, the trader who employs the AR(1)-GARCH(2,1)-N model appears to make the highest profit, 4.0% per day trading for 456 days, with a *t*-ratio of 4.84. Figure 8.7 depicts the cumulative returns of this agent from trading straddles on a daily basis. However, each

Table 8.2   Mean and standard deviation of the actual S&P500 option prices and their estimates obtained by three ARCH volatility forecasts, for each of the 456 trading days considered in the application of Section 8.5.4

ARCH model	Call option		Put option		Straddle option	
	Mean	Std. dev.	Mean	Std. dev.	Mean	Std. dev.
Next day's observed prices	33.6	13.4	27.4	14.1	61.0	18.5
Prices forecast by						
AR(1)-GARCH(1,1)	30.9	10.1	26.9	9.3	57.9	18.2
AR(1)-GJR(1,1)	30.2	9.8	26.2	8.9	56.4	17.5
AR(1)-EGARCH(1,1)	28.3	8.8	24.3	7.3	52.5	14.7

time an agent trades a contract a transaction cost has to be paid. Taking into consideration a transaction cost of $2, which reflects the bid–ask spread,[23] the rate of return would naturally be lower. Table 8.3 also provides for each trader the net rate of return after a trading cost of $2 and filters of $2 and $3.50. Although the *before transaction cost* profits are noticeably greater than zero, applying a $2 transaction cost and a $2 filter, the profits are not appreciably greater than zero for any of the agents. For a larger filter, i.e. for $X = \$2.00$ and $F_{il} = \$3.50$, the trader using the AR (1)-GJR(0,1)-N model for 310 trading days makes a daily profit of 1.8% with a t-ratio of 1.64.

8.5.5 Discussion

Selecting a model that produces accurate volatility predictions for pricing the next day's options is an intriguing problem. In the preceding subsection, using a set on the S&P500 index options data over the period from March 1998 to June 2000, we obtained and compared option prices forecasts. We did this by feeding the volatility estimated by nine ARCH models into the BS option pricing model, which is commonly used in option exchanges worldwide despite the fact that the BS formula assumes a constant variance.

However, profits noticeably greater than zero were achieved only under a perfect framework of no commissions. Only the bid–ask spread was taken into account.

---

[23] The bid price is the price that a trader is offering to pay for the option. The ask price is the price that a trader is offering to sell the option. The ask price is higher than the bid price, and the difference between them is referred to as the bid–ask spread. The exchange sets the upper limits for the bid–ask spread. For example, according to the CBOE rules, the width is supposed to be $1 for contracts above $20. However, Exchange rules allow for doubling and even tripling the width depending upon the market conditions. For a retail investor, the cost is higher and varies significantly from broker to broker. The actual amount charged is usually calculated as a fixed cost plus a proportion of the dollar amount of the trade, i.e. from a discount broker the purchase of contracts of $10 000 would cost $145 in commissions. Retail commissions from full service brokers are higher.

Table 8.3   Daily rates of return from trading straddles on the S&P500 index based on the volatility forecasts produced by the nine ARCH models of Section 8.5.4 (11 March 1998 to 2 June 2000)

Model	$0 transaction cost $0.00 filter				$2 transaction cost $2.00 filter				$2 transaction cost $3.50 filter			
	Mean	Std. dev.	t ratio	Days	Mean	Std. dev.	t ratio	Days	Mean	Std. dev.	t ratio	Days
1. AR(1)-GARCH(0,1)	3.4%	18.0%	4.10	456	0.6%	18.6%	0.63	374	1.2%	18.6%	1.14	327
2. AR(1)-GARCH(1,1)	3.9%	17.9%	4.61	456	0.7%	18.9%	0.76	371	1.6%	19.3%	1.44	308
3. AR(1)-GARCH(2,1)	4.0%	17.8%	4.84	456	0.9%	18.7%	0.91	372	1.5%	19.1%	1.42	317
4. AR(1)-GJR (0,1)	3.8%	17.9%	4.53	456	0.7%	18.8%	0.70	369	1.8%	19.5%	1.64	310
5. AR(1)-GJR (1,1)	3.9%	17.9%	4.64	456	0.8%	18.0%	0.88	396	1.3%	18.5%	1.27	350
6. AR(1)-GJR (2,1)	4.0%	17.9%	4.77	456	0.8%	17.8%	0.93	396	0.7%	18.0%	0.74	356
7. AR(1)-EGARCH(0,1)	3.4%	18.0%	4.00	456	0.3%	18.2%	0.33	375	1.1%	18.4%	1.11	330
8. AR(1)-EGARCH(1,1)	3.4%	18.0%	4.06	456	0.4%	17.2%	0.45	402	0.5%	17.7%	0.59	362
9. AR(1)-EGARCH(2,1)	3.1%	18.0%	3.67	456	0.0%	17.6%	-0.02	403	0.7%	17.9%	0.70	358

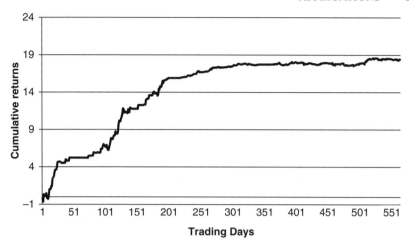

Figure 8.7   Cumulative rates of return under the AR(1)-GARCH(2,1)-N model from trading straddles on the S&P500 index (11 March 1998 to 2 June 2000).

Transaction charges for a trader and market impact costs diminish the daily profits. If a trader could really make a daily profit of 1.8%, the results presented would make a good case for market inefficiency, or at least for a huge temporary inefficiency.

8.6   ARCH option pricing formulas

The method illustrated in Section 8.5.4 is an arbitrary method of plugging volatility forecasts into an option pricing formula. As seen in Section 8.5.1, the construction of the BS option pricing formula is based on the assumption that the underlying stock price conforms to a geometric Brownian motion with constant variance for the rate of returns. On the other hand, the ARCH process assumes a discrete time framework with time-varying variance for the rate of returns. However, at least for near-the-money trading, the computation of options prices by plugging accurate volatility estimators (mainly from ARCH-type models) into the BS formula is a widely used strategy among market participants. Knight and Satchell (2002b) noted that financial practitioners use ARCH models to predict volatility, but use the BS formula coupled with ARCH to price options, although this hybrid procedure lacks theoretical rigour.
Engle (2001b) noted:

> Two different strands of financial econometrics have approached the option pricing problem. The first observes options data and infers the risk neutral conditional density. The second specifies the pricing kernel, interpreted as price per unit probability, estimates the empirical

conditional density, and computes the option price. .... An important class of models assumes that the true data generating process is a continuous-time diffusion process. In this case, both the empirical and risk neutral density can be computed.

As the risk-neutral density is estimated at each point in time, there is no time series behaviour of any kind. Therefore, the first approach has no implications for risk management. The second approach typically has some appeal for the BS continuous hedging argument.

As already noted in Section 2.9, Nelson (1990b) showed that the continuous time limit of an ARCH process defines a diffusion process with stochastic volatility. Duan (1997) generalized Nelson's finding and showed that the bivariate diffusion models, which are widely used in modelling stochastic volatility in the finance literature, can be represented as limits of specific ARCH processes (see also Nelson and Foster, 1994).

Based on certain assumptions concerning the trader's utility function, Duan (1995) developed an option pricing model for the underlying asset, S_t, whose one-period rate of return, $y_t = \log(S_t/S_{t-1})$, conforms with the ARCH in mean model with a NAGARCH(1,1) volatility specification:

$$
\begin{aligned}
y_t &= c_0 + c_1\sigma_t - 0.5\sigma_t^2 + \varepsilon_t, \\
\varepsilon_t &= z_t\sigma_t, \\
z_t &\sim N(0,1), \\
\sigma_t^2 &= a_0 + a_1(\varepsilon_{t-1} + \gamma_1\sigma_{t-1})^2 + b_1\sigma_{t-1}^2.
\end{aligned}
\tag{8.73}
$$

The one-period expected return of the underlying asset conditional on the information set I_{t-1} is given by $E(y_t|I_{t-1}) = c_0 + c_1\sigma_t - 0.5\sigma_t^2$. Noting that $E(S_t/S_{t-1}|I_{t-1}) = E(1 + y_t|I_{t-1}) = e^{c_0 + c_1\sigma_t}$, it follows that, under (8.73), c_0 can be regarded as the one-period continuously compounded risk-free interest rate and c_1 as expressing the constant unit risk premium.

Under the risk-neutral valuation relationship of (8.50), the price of the option is the expected payoff discounted at the risk-free rate. Duan defined the locally risk-neutral valuation relationship (LRNVR), under which the one-period expected rate of return on the underlying asset must equal the risk-free rate and the one-period conditional variance must be invariant with respect to a change in the risk neutralized pricing measure. Consider, for example, the expectation $E^*\big(\max(0, \tilde{S}(T) - K)|I_t\big)$. This is the expected payoff under the risk-neutral distribution. Under a pricing measure that satisfies the LRNVR we must have

$$
E^*(S_t/S_{t-1}|I_{t-1}) = \exp(c_0)
\tag{8.74}
$$

and

$$
V^*(\log(S_t/S_{t-1})|I_{t-1}) = V(\log(S_t/S_{t-1})|I_{t-1}) = \sigma_t^2.
\tag{8.75}
$$

Let us consider the standard normal distribution $N(0,1)$ as the risk-neutral distribution and let $z_t^* \sim N(0,1)$. Replace z_t in (8.73) by $z_t^* - c_1$. Then, under a pricing measure that satisfies the LRNVR, we obtain

$$y_t = c_0 - 0.5\sigma_t^2 + \sigma_t z_t^*,$$
$$\sigma_t^2 = a_0 + a_1 \left(z_{t-1}^* \sigma_{t-1} - c_1 \sigma_{t-1} + \gamma_1 \sigma_{t-1} \right)^2 + b_1 \sigma_{t-1}^2. \qquad (8.76)$$

It can be shown that for a process defined by (8.76), conditions (8.74) and (8.75) are satisfied (for a proof, see Christoffersen, 2003, p. 135).

Duan (1995) did not provide a closed-form solution for his model, but resorted to a control-variate Monte Carlo simulation method for computing option prices.

Under the risk-neutral valuation relationship, the price of the call option is given by

$$C(\tilde{S}(T), t) = e^{-rf(T-t)} E^* (\max(0, \tilde{S}(T) - K) | I_t). \qquad (8.77)$$

An algorithm similar in nature to that used in Section 8.4 can be employed to obtain an estimate of the price of the underlying asset at maturity T under risk neutrality $\tilde{S}^*(T)$:

- Step 1.1. Generate random numbers $\{\tilde{z}_{i,1}\}_{i=1}^{MC}$ from the standard normal distribution, where MC denotes the number of draws.

- Step 1.2. Create the hypothetical risk-neutral returns for time $t+1$: $\tilde{y}_{i,t+1}^* = c_0^{(t)} - 0.5\sigma_{t+1|t}^2 + \sigma_{t+1|t}\tilde{z}_{i,1}$, for $i = 1,\ldots,MC$.

- Step 1.3. Create the forecast variance at time $t+2$:

$$\tilde{\sigma}_{i,t+2}^2 = a_0^{(t)} + a_1^{(t)} \left(\tilde{z}_{i,1}\sigma_{t+1|t} - \left(c_1^{(t)} - \gamma_1^{(t)} \right)\sigma_{t+1|t} \right)^2 + b_1^{(t)}\sigma_{t+1|t}^2$$

- Step 2.1. Generate further random numbers, $\{\tilde{z}_{i,2}\}_{i=1}^{MC}$, $\tilde{z}_{i,2} \sim N(0,1)$.

- Step 2.2. Calculate the hypothetical risk-neutral returns for time $t+2$: $\tilde{y}_{i,t+2}^* = c_0^{(t)} - 0.5\tilde{\sigma}_{t+2}^2 + \tilde{\sigma}_{t+2}\tilde{z}_{i,2}$, for $i = 1,\ldots,MC$.

- Step 2.3. Create the forecast variance at time $t+3$:

$$\tilde{\sigma}_{i,t+3}^2 = a_0^{(t)} + a_1^{(t)} \left(\tilde{z}_{i,2}\tilde{\sigma}_{t+2} - \left(c_1^{(t)} - \gamma_1^{(t)} \right)\tilde{\sigma}_{t+2} \right)^2 + b_1^{(t)}\tilde{\sigma}_{t+2}^2$$

...

- Step $(T-1).1$. Generate further random numbers, $\{\tilde{z}_{i,T-1}\}_{i=1}^{MC}$, $\tilde{z}_{i,T} \sim N(0,1)$.

- Step $(T-1).2$. Calculate the hypothetical risk-neutral returns for time $T-1$: $\tilde{y}_{i,T-1}^* = c_0^{(t)} - 0.5\tilde{\sigma}_{T-1}^2 + \tilde{\sigma}_{T-1}\tilde{z}_{i,T-1}$, for $i = 1,\ldots,MC$.

- Step $(T-1).3$. Create the forecast variance at time $T-1$:

$$\tilde{\sigma}_{i,T-1}^2 = a_0^{(t)} + a_1^{(t)} \left(\tilde{z}_{i,T}\tilde{\sigma}_{T-2} - \left(c_1^{(t)} - \gamma_1^{(t)} \right)\tilde{\sigma}_{T-2} \right)^2 + b_1^{(t)}\tilde{\sigma}_{T-2}^2$$

- Step $T.1$. Generate further random numbers, $\{\breve{z}_{i,T}\}_{i=1}^{MC}$, $\breve{z}_{i,T} \sim N(0,1)$.

- Step $T.2$. Calculate the hypothetical risk-neutral returns for time T:
$\breve{y}_{i,T}^{*} = c_{0}^{(t)} - 0.5\,\breve{\sigma}_{T}^{2} + \breve{\sigma}_{T}\breve{z}_{i,T}$, for $i = 1, \ldots, MC$.

At each simulation path $i, i = 1, \ldots, MC$, the price of the underlying asset at maturity day T under risk neutrality can then be estimated as

$$\tilde{S}_{i}^{*}(T) = \tilde{S}(t)\exp\left(\sum_{j=t}^{T-1}\breve{y}_{i,j+1}^{*}\right). \qquad (8.78)$$

Therefore, the risk-neutral option pricing formula given by (8.77) is computed in practice as:

$$C(\tilde{S}(T), t) = e^{-rf(T-t)}(MC)^{-1}\sum_{i=1}^{MC}\max\left(0, \tilde{S}_{i}^{*}(T) - K\right). \qquad (8.79)$$

Christoffersen (2003) provided an Excel worksheet for simulating Duan's (1995) ARCH option pricing method.

The parameters of the ARCH process in (8.76) can be computed based either on information provided solely by the underlying asset returns or on information provided by the options prices. In the former case, the parameters of the ARCH model are estimated under the physical probability measure using the time series of log-returns as demonstrated in the preceding examples, and the ARCH option prices are computed using (8.79). In the latter case, the unknown parameters are estimated under the risk-neutral probability measure by numerically minimizing the squared distance between actual and fitted option prices,

$$\min_{a_{0}, a_{1}, c_{1}', \gamma_{1}, b_{1}, \sigma_{t+1|t}}\left(N^{-1}\sum_{i=1}^{N}\left(C_{i}(\tilde{S}(T), t) - C_{i}\right)^{2}\right), \qquad (8.80)$$

where N is the sample size of the actual option prices, C_{i} is the actual price of the ith option and $c_{1}' = c_{1} - \gamma_{1}$. Note that $\sigma_{t+1|t}$ must also be considered as an unknown parameter. Christoffersen and Jacobs (2004) provided evidence against fitting option prices based solely on time series of the underlying asset returns. They observed that 'it is dangerous to make inference about the potential for volatility models to fit option prices based only on statistical analysis on the underlying asset returns'. Note that estimating the parameters of the ARCH model under the physical probability measure does not imply information from option prices. Option prices contain forward-looking information about the expectations of the market participants for the future movement of the underlying asset price. Moreover, there is not a direct relationship of maximizing the likelihood function of log-returns by minimizing an economically motivated objective function such as that given in (8.80).

Engle and Mustafa (1992) were the first to consider the risk-neutral pricing method and to estimate the parameters of an ARCH process by equating theoretical and observed option prices. Based on the observation that Monte Carlo simulated sample paths for the underlying asset prices fail to exhibit the martingale property, Duan and Simonato (1998) introduced an empirical martingale simulation which ensures that the simulated prices satisfy the rational option pricing bounds. Duan et al. (1999b) proposed a marriage of the empirical martingale simulation and the quasi-Monte Carlo simulation in order to accelerate Monte Carlo convergence. Ritchken and Trevor (1999), Duan and Simonato (2001) and Duan et al. (2003) proposed numerical methods for pricing American options under the GARCH option pricing model. Duan et al. (1999a) provided an analytical approximation for Duan's (1995) option pricing model using the Edgeworth expansion of the risk-neutral density function, whereas Duan et al. (2006a) defined an option pricing framework under the GJR(p, q) and EGARCH(p, q) processes for the volatility of the underlying asset. Kallsen and Taqqu (1998) treat the ARCH process in a continuous time framework by letting the process evolve as a geometric Brownian motion between any two discrete ARCH times and reached the same option pricing formula as Duan (1995).

Heston and Nandi (2000) developed a closed-form solution for European options under an ARCH in mean process with a mixed volatility specification of NAGARCH (1,1) and VGARCH(1,1) models:[24]

$$
\begin{aligned}
y_t &= c_0 + c_1 \sigma_t + \varepsilon_t, \\
\varepsilon_t &= z_t \sigma_t, \\
z_t &\sim N(0, 1), \\
\sigma_t^2 &= a_0 + a_1 (z_{t-1} + \gamma_1 \sigma_{t-1})^2 + b_1 \sigma_{t-1}^2.
\end{aligned}
\tag{8.81}
$$

Hsieh and Ritchken (2005) compared the in-sample and out-of-sample performance of Duan's (1995) and Heston and Nandi's (2000) ARCH option pricing models. They recommended Duan's specification as exhibiting superior performance for all moneyness and maturity categories.

Hauksson and Rachev (2001), under the LRNVR condition, developed an option pricing model based on the GARCH(1,1) process with a symmetric standard stable Paretian conditional distribution:

$$
\begin{aligned}
y_t &= c_0 + c_1 \sigma_t - 0.5 \sigma_t^2 + \varepsilon_t, \\
\varepsilon_t &= z_t \sigma_t, \\
\sigma_t^2 &= a_0 + a_1 \varepsilon_{t-1}^2 + b_1 \sigma_{t-1}^2,
\end{aligned}
\tag{8.82}
$$

where the characteristic function of z_t is as given by (5.7) for $\varphi(t, a, \beta = 0, \sigma = 1, \mu = 0)$.

---

[24] From equations (2.66) and (2.65), the VGARCH(1,1) model is $\sigma_t^2 = a_0 + a_1 (z_{t-1} + \gamma_1)^2 + b_1 \sigma_{t-1}^2$, and the NAGARCH(1,1) model is $\sigma_t^2 = a_0 + a_1 (\varepsilon_{t-1} + \gamma_1 \sigma_{t-1})^2 + b_1 \sigma_{t-1}^2$.

According to Christoffersen and Jacobs (2004), for the purpose of option valuation, it appears that one should not look beyond a simple ARCH model that allows for volatility clustering and a leverage effect.[25] On the other hand, in recent studies there have been attempts to capture the conditional leptokurtosis and skewness of the underlying asset log-returns in ARCH option pricing. Menn and Rachev (2005), for example, propose an ARCH option pricing model with innovations distributed according to the smoothly truncated stable distribution. Kim et al. (2007) develop an ARCH option pricing framework with modified tempered stable innovations. Furthermore, Duan et al. (2006b) consider the pricing of options when there are jumps in asset returns and volatilities. Sabbatini and Linton (1998), Duan and Wei (1999), Härdle and Hafner (2000), Fofana and Brorsen (2001) and Barone-Adesi et al. (2004), among others, have also considered ARCH-based option pricing models. Chen and Leung (2005), applying option pricing econometric techniques, note that semi-parametric and non-parametric statistical projection models appear to offer better pricing performance than ARCH option pricing models.

8.6.1 Computation of Duan's ARCH option prices: an example

The program *chapter8_duan's_option_pricing_formula.prg* in the Appendix to the chapter carries out the estimation of formula (8.79) for the ARCH process

$$
\begin{aligned}
y_t &= 0.00011 - 0.5\sigma_t^2 + \sigma_t z_t^*, \\
\sigma_t^2 &= 0.00001 + 0.2\left(z_{t-1}^*\sigma_{t-1} - 0.01\sigma_{t-1}\right)^2 + 0.71\sigma_{t-1}^2, \\
z_t^* &\sim N(0,1),
\end{aligned}
\tag{8.83}
$$

under the LRNVR condition for $MC = 20\,000$ replications, for $T = 30$ trading days. The stock price is €60, the strike price is €65, the daily risk-free rate is $c_0 = 0.00011$ and today's daily variance is $\sigma_{t|t}^2 = 0.00027$.

The matrix named *random_normal* stores the necessary generated standard normal random variables $\left\{\breve{z}_{i,t}\right\}_{i=1}^{MC}$, for $i = 1, \ldots, 20\,000$ and $t = 1, \ldots, 30$. The matrix *random_variance*, of $MC\_total = 20\,000$ rows and $day = 30$ columns, contains the variance forecasts $\breve{\sigma}_{i,\,t+1}^2 = a_0^{(t)} + a_1^{(t)}\left(\breve{z}_{i,t}\breve{\sigma}_t - c_1'^{(t)}\breve{\sigma}_t\right)^2 + b_1^{(t)}\breve{\sigma}_t^2$ obtained for each day t ($t = 1, \ldots, 30$), and each simulated path i ($i = 1, \ldots, 20\,000$). The matrix named *random_returns* contains the risk-neutral returns $\breve{y}_{i,t}^* = c_0^{(t)} - 0.5\,\breve{\sigma}_t^{\,2} + \breve{\sigma}_t \breve{z}_{i,t}$. The vector *price_at_maturity* is a vector with components the 20 000 simulated prices of the underlying asset at maturity day T given by $\tilde{S}_i^*(T) = \tilde{S}(t)\exp\left(\sum_{j=t}^{T-1}\breve{y}_{i,j+1}^*\right)$,

---

[25] This is in contrast to the estimation of ARCH processes from the time series of the underlying asset log-returns, where the extended models generate more accurate volatility forecasts. Under different evaluation frameworks, Angelidis and Degiannakis (2008) and González-Rivera et al. (2004) also noted that for option pricing, simple models perform as well as more sophisticated specifications.

and the vector *payoff_LRNVR* is a vector with components the 20 000 expected payoffs given by $\max\left(0, \tilde{S}_i^*(T)-65\right)$. The estimated call option price, computed as

$$C(\tilde{S}(T), t) = e^{-0.00011*30} \frac{1}{20\ 000} \sum_{i=1}^{20\ 000} \max\left(0, \tilde{S}_i^*(T)-65\right),$$

is stored in *ARCH_call_price*. Finally, the put option price computed by the put–call parity as $P(\tilde{S}(T), t) = C(\tilde{S}(T), t) - S(T) + Ke^{-rf(T-t)}$ is contained in *ARCH_put_price*.

Appendix

EViews 6

- *chapter8.var.es.ftse100.prg*

```
load chapter8.var.es.ftse100.wf1
!sample = 1435
!models = 12
!model = 1
!ES_repeat = 5000
matrix(!sample,!models) ftse100VaR95L = NA
matrix(!sample,!models) ftse100ES95L = NA
matrix(!sample,!models) ftse100Result95L = NA
matrix(!sample,!models) ftse100Loss95L = NA
matrix(1,!models) average95L = NA
matrix(1,!models) Kupiec95L = NA
matrix(1,!models) Christ95L = NA
scalar place = 1
'********************************************************
'AR(1)-FIAPARCH(1,d,1)-ged
'********************************************************
read(t=xls) ftse100_FIAPARCHBBM_ged.xls 11
for !i = 1 to !sample
    ftse100VaR95L(!i,place) = @sqrt(var1(!i))*@qged(0.05,
      var11(!i))+var2(!i)
    vector(!ES_repeat) ES = na
    !step1 = (1-0.95)/(!ES_repeat+1)
    !step2 = 0.05
    for !kk = 1 to !ES_repeat
      !step2 = !step2-!step1
      ES(!kk) = @sqrt(var1(!i))*@qged(!step2,var11(!i))
        +var2(!i)
    next
    ftse100ES95l(!i,place) = @mean(ES)
Statusline Sample ftse100 !model !i
next
!model = !model +1
for !i = 1 to !sample
    if    ftse100(!i)<ftse100VaR95L(!i,place) then
          ftse100result95L(!i,place) = 1
    else
          ftse100result95L(!i,place) = 0
    endif
next
```

```
freeze(test) ftse100result95l.stats
average95L(1,place) = test(5,place+1)
delete test
'_____
'Kupiec test
smpl @all
'Create the vector for the probabilities of default
Vector(1) probab
probab(1)=0.05
'How many forecasts have you made?
vector(1) T
T(1)=!sample
'What is the average (expected)?
vector(1) average
average(1)= probab(1)*T(1)
'The number of brakes for long
vector(1) NLong = na
NLong(1)=@round(average95l(1,place)*!sample)
'Vectors for LR tests
vector(1) LRunLong = na
series kupiecresult1=-2*log(((1-probab(1))^(t(1)-NLong
  (1)))*(probab(1)^NLong(1))*1e+150)
series kupiecresult2=2*log(((1-NLong(1)/t(1))^(t(1)-
  NLong(1)))*((NLong(1)/t(1))^NLong(1))*1e+150)
LRunLong(1)=kupiecresult1(1)+kupiecresult2(1)

Kupiec95L(1,place) = 1-@cchisq(LRunLong(1),1)

delete average
delete NLong
delete LRunLong
delete kupiecresult1
delete kupiecresult2
'_____
'Christoffersen test
smpl @all
'Create the vector for the conditional coverage test
vector(1) LRcc=na
'Create the coef for the independence
'The number of observations with value 0 followed by 0
coef(1) n00=0
'The number of observations with value 0 followed by 1
coef(1) n01=0
'The number of observations with value 1 followed by 1
```

```
coef(1) n11=0
'The number of observations with value 1 followed by 0
coef(1) n10=0
'Create the probability to go to 0 given that we are at 0
coef(1) pr00(1)=0
'Create the probability to go to 1 given that we are at 0
coef(1) pr01(1)=0
'Create the probability to go to 0 given that we are at 1
coef(1) pr10(1)=0
'Create the probability to go to 1 given that we are at 1
coef(1) pr11(1)=0
'Create the series for the independence coverage
series LRin
'Estimate the nij
for !i=1 to !sample-1
    if ftse100result95l(!i,place)=0 and ftse100result95L
       (!i+1,place)=0 then
         n00(1)=n00(1)+1
    else
         n00(1)=n00(1)
    endif
    if ftse100result95l(!i,place)=0 and ftse100result95L
       (!i+1,place) =1 then
         n01(1)=n01(1)+1
    else
         n01(1)=n01(1)
    endif
    if ftse100result95l(!i,place)=1 and ftse100result95L
       (!i+1,place)=1 then
         n11(1)=n11(1)+1
    else
         n11(1)=n11(1)
    endif
    if ftse100result95l(!i,place)=1 and ftse100result95L
        (!i+1,place)=0 then
      n10(1)=n10(1)+1
    else
      n10(1)=n10(1)
    endif
next
pr01(1)=n01(1)/(n00(1)+n01(1))
pr11(1)=n11(1)/(n10(1)+n11(1))
series LRin1=(((1-pr01(1))^n00(1))*(pr01(1)^n01(1))*
  ((1-pr11(1))^n10(1))*pr11(1)^n11(1))*1e+150
```

```
coef(1) p2 = (n01(1)+n11(1))/(n01(1)+n00(1)+n10(1)
  +n11(1))
series LRin2 = (((1-p2(1))^(n00(1)+n10(1)))*(p2(1)^(n01
  (1)+n11(1))))*1e+150
LRcc(1) = -2*(log(lrin2(1)/lrin1(1)))
Christ95L(1,place) = 1-@cchisq(LRcc(1),1)
'_____
'MSE Loss function
for !i = 1 to !sample
   if ftse100result951(!i,place) = 1 then
     ftse100loss951(!i,place) = (ftse100(!i)-ftse100es951
        (!i,place))^2
   else
     ftse100loss951(!i,place) = 0
   endif
next
place = place + 1
for !i = 1 to 11
delete var!i
next
'*********************************************************
'AR(1)-FIAPARCH(1,d,1)-n
'*********************************************************
read(t=xls) ftse100_FIAPARCHBBM_n.xls 10
for !i = 1 to !sample
   ftse100VaR95L(!i,place) = @sqrt(var1(!i))*@qnorm
     (0.05)+var2(!i)
   vector(!ES_repeat) ES = na
   !step1 = (1-0.95)/(!ES_repeat+1)
   !step2 = 0.05
   for !kk = 1 to !ES_repeat
     !step2 = !step2-!step1
     ES(!kk) = @sqrt(var1(!i))*@qnorm(!step2)+var2(!i)
   next
   ftse100ES951(!i,place) = @mean(ES)
Statusline Sample ftse100 !model !i
next
!model = !model +1
for !i = 1 to !sample
   if    ftse100(!i)<ftse100VaR95L(!i,place) then
           ftse100result95L(!i,place) = 1
   else
           ftse100result95L(!i,place) = 0
   endif
```

```
next

freeze(test) ftse100result95l.stats
average95L(1,place) = test(5,place+1)
delete test
'_____
'Kupiec test
smpl @all
'Create the vector for the probabilities of default
Vector(1) probab
probab(1)=0.05
'How many forecasts have you made?
vector(1) T
T(1)=!sample
'What is the average (expected)?
vector(1) average
average(1)= probab(1)*T(1)
'The number of brakes for long
vector(1) NLong = na
NLong(1)=@round(average95l(1,place)*!sample)
'Vectors for LR tests
vector(1) LRunLong = na
series kupiecresult1=-2*log(((1-probab(1))^(t(1)-NLong
  (1)))*(probab(1)^NLong(1))*1e+150)
series kupiecresult2=2*log(((1-NLong(1)/t(1))^(t(1)-
  NLong(1)))*((NLong(1)/t(1))^NLong(1))*1e+150)
LRunLong(1)=kupiecresult1(1)+kupiecresult2(1)
Kupiec95L(1,place) = 1-@cchisq(LRunLong(1),1)
delete average
delete NLong
delete LRunLong
delete kupiecresult1
delete kupiecresult2
'_____
'Christoffersen test
smpl @all
'Create the vector for the conditional coverage test
vector(1) LRcc=na
'Create the coef for the independence
'The number of observations with value 0 followed by 0
coef(1) n00=0
'The number of observations with value 0 followed by 1
coef(1) n01=0
'The number of observations with value 1 followed by 1
coef(1) n11=0
```

```
'The number of observations with value 1 followed by 0
coef(1) n10=0
'Create the probability to go to 0 given that we are at 0
coef(1) pr00(1)=0
'Create the probability to go to 1 given that we are at 0
coef(1) pr01(1)=0
'Create the probability to go to 0 given that we are at 1
coef(1) pr10(1)=0
'Create the probability to go to 1 given that we are at 1
coef(1) pr11(1)=0
'Create the series for the independence coverage
series LRin
'Estimate the nij
for !i=1 to !sample-1
    if ftse100result95l(!i,place)=0 and ftse100result95L
      (!i+1,place)=0 then
      n00(1)=n00(1)+1
    else
      n00(1)=n00(1)
    endif
    if ftse100result95l(!i,place)=0 and ftse100result95L
      (!i+1,place) =1 then
      n01(1)=n01(1)+1
    else
      n01(1)=n01(1)
    endif
    if ftse100result95l(!i,place)=1 and ftse100result95L
      (!i+1,place)=1 then
      n11(1)=n11(1)+1
    else
      n11(1)=n11(1)
    endif
    if ftse100result95l(!i,place)=1 and ftse100result95L
      (!i+1,place)=0 then
      n10(1)=n10(1)+1
    else
      n10(1)=n10(1)
    endif
next
pr01(1)=n01(1)/(n00(1)+n01(1))
pr11(1)=n11(1)/(n10(1)+n11(1))
series LRin1=(((1-pr01(1))^n00(1))*(pr01(1)^n01(1))*
  ((1-pr11(1))^n10(1))*pr11(1)^n11(1))*1e+150
coef(1) p2 = (n01(1)+n11(1))/(n01(1)+n00(1)+n10(1)
  +n11(1))
```

```
series LRin2 = (((1-p2(1))^(n00(1)+n10(1)))*(p2(1)^
  (n01(1)+n11(1))))*1e+150
LRcc(1) = -2*(log(lrin2(1)/lrin1(1)))
Christ95L(1,place) = 1-@cchisq(LRcc(1),1)
'_____
'MSE Loss function
for !i = 1 to !sample
    if ftse100result95l(!i,place) = 1 then
        ftse100loss95l(!i,place) = (ftse100(!i)-ftse100es95l
            (!i,place))^2
    else
        ftse100loss95l(!i,place) = 0
    endif
next
place = place + 1
for !i = 1 to 10
delete var!i
next
'***********************************************************
'AR(1)-FIAPARCH(1,d,1)-skT
'***********************************************************
read(t=xls) ftse100_FIAPARCHBBM_skt.xls 12
!pi = @acos(-1)
for !i = 1 to !sample
    coef(1) asym = @exp(var11(!i))
    coef(1)  tdf1 = var12(!i)
    coef(1) m = ((asym(1)) - (1/(asym(1))))*@gamma
        ((tdf1(1)-1)/2)*@sqrt(tdf1(1)-2)/(@sqrt(!pi)*
        @gamma(tdf1(1)/2))
    coef(1) s = @sqrt(((asym(1))^2 + (1/(asym(1))^2) - 1 - m
        (1)^2))
    scalar p_value = 0.05
    scalar freedom = tdf1(1)
        if p_value < (1/(1+asym(1)^2)) then
        scalar quant_95 = ((1/asym(1))*(@qtdist((p_value/2)*
            (1+asym(1)^2),freedom))-m(1))/s(1)
    else
        scalar quant_95 = (-asym(1)*@qtdist(((1-p_value)/2)*
            (1+asym(1)^(-2)),freedom)-m(1))/s(1)

    endif
    ftse100VaR95L(!i,place) = @sqrt(var1(!i))*quant_95
        +var2(!i)
    vector(!ES_repeat) ES = na
    !step1 = (1-0.95)/(!ES_repeat+1)
```

```
  !step2 = 0.05
  for !kk = 1 to !ES_repeat
    !step2 = !step2-!step1
  if !step2 < (1/(1+asym(1)^2)) then
    scalar quant = ((1/asym(1))*(@qtdist((!step2/2)*
      (1+asym(1)^2),freedom))-m(1))/s(1)
    ES(!kk) = quant*var1(1) + var2(!i)
  else
    scalar quant = (-asym(1)*@qtdist(((1-!step2)/2)*
      (1+asym(1)^(-2)),freedom)-m(1))/s(1)
    ES(!kk) = quant*var1(1) + var2(!i)
  endif
 next
 ftse100ES95l(!i,place) = @mean(ES)
 Statusline Sample ftse100 !model !i
next
!model = !model +1
for !i = 1 to !sample
  if    ftse100(!i)<ftse100VaR95L(!i,place) then
                ftse100result95L(!i,place) = 1
  else
                ftse100result95L(!i,place) = 0
  endif
next
freeze(test) ftse100result95l.stats
average95L(1,place) = test(5,place+1)
delete test
'_____
'Kupiec test
smpl @all
'Create the vector for the probabilities of default
vector(1) probab
probab(1)=0.05
'How many forecasts have you made?
vector(1) T
T(1)=!sample
'What is the average (expected)?
vector(1) average
average(1)= probab(1)*T(1)
'The number of brakes for long
vector(1) NLong = na
NLong(1)=@round(average95l(1,place)*!sample)
'Vectors for LR tests
vector(1) LRunLong = na
series kupiecresult1=-2*log(((1-probab(1))^(t(1)-NLong
```

```
   (1)))*(probab(1)^NLong(1))*1e+150)
series kupiecresult2=2*log(((1-NLong(1)/t(1))^(t(1)-
   NLong(1)))*((NLong(1)/t(1))^NLong(1))*1e+150)
LRunLong(1)=kupiecresult1(1)+kupiecresult2(1)
Kupiec95L(1,place) = 1-@cchisq(LRunLong(1),1)
delete average
delete NLong
delete LRunLong
delete kupiecresult1
delete kupiecresult2
'_____
'Christoffersen test
smpl @all
'Create the vector for the conditional coverage test
vector(1) LRcc=na
'Create the coef for the independence
'The number of observations with value 0 followed by 0
coef(1) n00=0
'The number of observations with value 0 followed by 1
coef(1) n01=0
'The number of observations with value 1 followed by 1
coef(1) n11=0
'The number of observations with value 1 followed by 0
coef(1) n10=0
'Create the probability to go to 0 given that we are at 0
coef(1) pr00(1)=0
'Create the probability to go to 1 given that we are at 0
coef(1) pr01(1)=0
'Create the probability to go to 0 given that we are at 1
coef(1) pr10(1)=0
'Create the probability to go to 1 given that we are at 1
coef(1) pr11(1)=0
'Create the series for the independence coverage
series LRin
'Estimate the nij
for !i=1 to !sample-1
  if ftse100result95l(!i,place)=0 and ftse100result95L
    (!i+1,place)=0 then
    n00(1)=n00(1)+1
  else
    n00(1)=n00(1)
  endif
  if ftse100result95l(!i,place)=0 and ftse100result95L
    (!i+1,place) =1 then
    n01(1)=n01(1)+1
```

```
  else
    n01(1)=n01(1)
  endif
  if ftse100result95l(!i,place)=1 and ftse100result95L
    (!i+1,place)=1 then
    n11(1)=n11(1)+1
  else
    n11(1)=n11(1)
  endif
  if ftse100result95l(!i,place)=1 and ftse100result95L
    (!i+1,place)=0 then
    n10(1)=n10(1)+1
  else
    n10(1)=n10(1)
  endif
next
pr01(1)=n01(1)/(n00(1)+n01(1))
pr11(1)=n11(1)/(n10(1)+n11(1))
series LRin1=(((1-pr01(1))^n00(1))*(pr01(1)^n01(1))*
  ((1-pr11(1))^n10(1))*pr11(1)^n11(1))*1e+150
coef(1) p2 = (n01(1)+n11(1))/(n01(1)+n00(1)+n10(1)
  +n11(1))
series LRin2 = (((1-p2(1))^(n00(1)+n10(1)))*(p2(1)^(n01
  (1)+n11(1))))*1e+150
LRcc(1) = -2*(log(lrin2(1)/lrin1(1)))
Christ95L(1,place) = 1-@cchisq(LRcc(1),1)
'_____
'MSE Loss function
for !i = 1 to !sample
  if ftse100result95l(!i,place) = 1 then
    ftse100loss95l(!i,place) = (ftse100(!i)-ftse100es95l
      (!i,place))^2
  else
    ftse100loss95l(!i,place) = 0
  endif
next
place = place + 1
for !i = 1 to 12
delete var!i
next
'*******************************************************
'AR(1)-FIAPARCH(1,d,1)-t
'*******************************************************
read(t=xls) ftse100_FIAPARCHBBM_t.xls 11
for !i = 1 to !sample
```

```
  ftse100VaR95L(!i,place) = @sqrt(var1(!i))*@qtdist(0.05,
    var11(!i))+var2(!i)
  vector(!ES_repeat) ES = na
  !step1 = (1-0.95)/(!ES_repeat+1)
  !step2 = 0.05
  for !kk = 1 to !ES_repeat
    !step2 = !step2-!step1
    ES(!kk) = @sqrt(var1(!i))*@qtdist(!step2,var11
      (!i))+var2(!i)
  next
  ftse100ES951(!i,place) = @mean(ES)
Statusline Sample ftse100 !model !i
next
!model = !model +1
for !i = 1 to !sample
  if   ftse100(!i)<ftse100VaR95L(!i,place) then
                ftse100result95L(!i,place) = 1
  else
                ftse100result95L(!i,place) = 0
  endif
next

freeze(test) ftse100result951.stats
average95L(1,place) = test(5,place+1)
delete test
' _____
'Kupiec test
smpl @all
'Create the vector for the probabilities of default
vector(1) probab
probab(1)=0.05
'How many forecasts have you made?
vector(1) T
T(1)=!sample
'What is the average (expected)?
vector(1) average
average(1)= probab(1)*T(1)
'The number of brakes for long
vector(1) NLong = na
NLong(1)=@round(average951(1,place)*!sample)
'Vectors for LR tests
vector(1) LRunLong = na
series kupiecresult1=-2*log(((1-probab(1))^(t(1)-NLong
  (1)))*(probab(1)^NLong(1))*1e+150)
series kupiecresult2=2*log(((1-NLong(1)/t(1))^(t(1)-
```

```
  NLong(1)))*((NLong(1)/t(1))^NLong(1))*1e+150)
LRunLong(1)=kupiecresult1(1)+kupiecresult2(1)
Kupiec95L(1,place) = 1-@cchisq(LRunLong(1),1)
delete average
delete NLong
delete LRunLong
delete kupiecresult1
delete kupiecresult2
'_____
'Christoffersen test
smpl @all
'Create the vector for the conditional coverage test
vector(1) LRcc=na
'Create the coef for the independence
'The number of observations with value 0 followed by 0
coef(1) n00=0
'The number of observations with value 0 followed by 1
coef(1) n01=0
'The number of observations with value 1 followed by 1
coef(1) n11=0
'The number of observations with value 1 followed by 0
coef(1) n10=0
'Create the probability to go to 0 given that we are at 0
coef(1) pr00(1)=0
'Create the probability to go to 1 given that we are at 0
coef(1) pr01(1)=0
'Create the probability to go to 0 given that we are at 1
coef(1) pr10(1)=0
'Create the probability to go to 1 given that we are at 1
coef(1) pr11(1)=0
'Create the series for the independence coverage
series LRin
'Estimate the nij
for !i=1 to !sample-1
  if ftse100result95l(!i,place)=0 and ftse100result95L
    (!i+1,place)=0 then
    n00(1)=n00(1)+1
  else
    n00(1)=n00(1)
  endif
  if ftse100result95l(!i,place)=0 and ftse100result95L
    (!i+1,place) =1 then
    n01(1)=n01(1)+1
  else
    n01(1)=n01(1)
```

```
   endif
   if ftse100result95l(!i,place)=1 and ftse100result95L
     (!i+1,place)=1 then
     n11(1)=n11(1)+1
   else
     n11(1)=n11(1)
   endif
   if ftse100result95l(!i,place)=1 and ftse100result95L
     (!i+1,place)=0 then
     n10(1)=n10(1)+1
   else
     n10(1)=n10(1)
   endif
next
pr01(1)=n01(1)/(n00(1)+n01(1))
pr11(1)=n11(1)/(n10(1)+n11(1))
series LRin1=(((1-pr01(1))^n00(1))*(pr01(1)^n01(1))
  *((1-pr11(1))^n10(1))*pr11(1)^n11(1))*1e+150
coef(1) p2 = (n01(1)+n11(1))/(n01(1)+n00(1)+n10(1)
  +n11(1))
series LRin2 = (((1-p2(1))^(n00(1)+n10(1)))*(p2(1)^(n01
  (1)+n11(1))))*1e+150
LRcc(1) = -2*(log(lrin2(1)/lrin1(1)))
Christ95L(1,place) = 1-@cchisq(LRcc(1),1)
'_____
'MSE Loss function
for !i = 1 to !sample
  if ftse100result95l(!i,place) = 1 then
    ftse100loss95l(!i,place) = (ftse100(!i)-ftse100es95l
      (!i,place))^2
  else
    ftse100loss95l(!i,place) = 0
  endif
next
place = place + 1
for !i = 1 to 11
delete var!i
next
'***********************************************************
'AR(1)-GARCH(1,1)-ged
'***********************************************************
read(t=xls) ftse100_GARCH_ged.xls 8
for !i = 1 to !sample
  ftse100VaR95L(!i,place) = @sqrt(var1(!i))*@qged(0.05,
    var8(!i))+var2(!i)
```

```
 vector(!ES_repeat) ES = na
 !step1 = (1-0.95)/(!ES_repeat+1)
 !step2 = 0.05
 for !kk = 1 to !ES_repeat
   !step2 = !step2-!step1
   ES(!kk) = @sqrt(var1(!i))*@qged(!step2,var8(!i))
     +var2(!i)
 next
 ftse100ES951(!i,place) = @mean(ES)
Statusline Sample ftse100 !model !i
next
!model = !model +1
for !i = 1 to !sample
  if   ftse100(!i)<ftse100VaR95L(!i,place) then
               ftse100result95L(!i,place) = 1
  else
               ftse100result95L(!i,place) = 0
  endif
next

freeze(test) ftse100result95l.stats
average95L(1,place) = test(5,place+1)
delete test
'_____
'Kupiec test
smpl @all
'Create the vector for the probabilities of default
vector(1) probab
probab(1)=0.05
'How many forecasts have you made?
vector(1) T
T(1)=!sample
'What is the average (expected)?
vector(1) average
average(1)= probab(1)*T(1)
'The number of brakes for long
vector(1) NLong = na
NLong(1)=@round(average951(1,place)*!sample)
'Vectors for LR tests
vector(1) LRunLong = na
series kupiecresult1=-2*log((((1-probab(1))^(t(1)-NLong
  (1)))*(probab(1)^NLong(1))*1e+150)
series kupiecresult2=2*log((((1-NLong(1)/t(1))^(t(1)-
  NLong(1)))*((NLong(1)/t(1))^NLong(1))*1e+150)
LRunLong(1)=kupiecresult1(1)+kupiecresult2(1)
```

```
Kupiec95L(1,place) = 1-@cchisq(LRunLong(1),1)
delete average
delete NLong
delete LRunLong
delete kupiecresult1
delete kupiecresult2
'_____
'Christoffersen test
smpl @all
'Create the vector for the conditional coverage test
vector(1) LRcc=na
'Create the coef for the independence
'The number of observations with value 0 followed by 0
coef(1) n00-0
'The number of observations with value 0 followed by 1
coef(1) n01=0
'The number of observations with value 1 followed by 1
coef(1) n11=0
'The number of observations with value 1 followed by 0
coef(1) n10=0
'Create the probability to go to 0 given that we are at 0
coef(1) pr00(1)=0
'Create the probability to go to 1 given that we are at 0
coef(1) pr01(1)=0
'Create the probability to go to 0 given that we are at 1
coef(1) pr10(1)=0
'Create the probability to go to 1 given that we are at 1
coef(1) pr11(1)=0
'Create the series for the independence coverage
series LRin
'Estimate the nij
for !i=1 to !sample-1
  if ftse100result95l(!i,place)=0 and ftse100result95L
    (!i+1,place)=0 then
    n00(1)=n00(1)+1
  else
    n00(1)=n00(1)
  endif
  if ftse100result95l(!i,place)=0 and ftse100result95L
    (!i+1,place) =1 then
    n01(1)=n01(1)+1
  else
    n01(1)=n01(1)
  endif
  if ftse100result95l(!i,place)=1 and ftse100result95L
```

```
    (!i+1,place)=1 then
      n11(1)=n11(1)+1
    else
      n11(1)=n11(1)
    endif
    if ftse100result95l(!i,place)=1 and ftse100result95L
      (!i+1,place)=0 then
      n10(1)=n10(1)+1
    else
      n10(1)=n10(1)
    endif
  next
  pr01(1)=n01(1)/(n00(1)+n01(1))
  pr11(1)=n11(1)/(n10(1)+n11(1))
  series LRin1=(((1-pr01(1))^n00(1))*(pr01(1)^n01(1))*
    ((1-pr11(1))^n10(1))*pr11(1)^n11(1))*1e+150
  coef(1) p2 = (n01(1)+n11(1))/(n01(1)+n00(1)+n10(1)
      +n11(1))
  series LRin2 = (((1-p2(1))^(n00(1)+n10(1)))*(p2(1)^
    (n01(1)+n11(1))))*1e+150
  LRcc(1) = -2*(log(lrin2(1)/lrin1(1)))
  Christ95L(1,place) = 1-@cchisq(LRcc(1),1)
  '_____
  'MSE Loss function
  for !i = 1 to !sample
    if ftse100result95l(!i,place) = 1 then
      ftse100loss95l(!i,place) = (ftse100(!i)-ftse100es95l
        (!i,place))^2
    else
      ftse100loss95l(!i,place) = 0
    endif
  next
  place = place + 1
  for !i = 1 to 8
  delete var!i
  next
  '*********************************************************
  'AR(1)-GARCH(1,1)-n
  '*********************************************************
  read(t=xls) ftse100_GARCH_n.xls 7
  for !i = 1 to !sample
    ftse100VaR95L(!i,place) = @sqrt(var1(!i))*@qnorm(0.05)
      +var2(!i)
    vector(!ES_repeat) ES = na
    !step1 = (1-0.95)/(!ES_repeat+1)
```

```
!step2 = 0.05
for !kk = 1 to !ES_repeat
  !step2 = !step2-!step1
  ES(!kk) = @sqrt(var1(!i))*@qnorm(!step2)+var2(!i)
next
ftse100ES951(!i,place) = @mean(ES)
Statusline Sample ftse100 !model !i
next
!model = !model +1
for !i = 1 to !sample
  if   ftse100(!i)<ftse100VaR95L(!i,place) then
                  ftse100result95L(!i,place) = 1
  else
                  ftse100result95L(!i,place) = 0
  endif
next
freeze(test) ftse100result95l.stats
average95L(1,place) = test(5,place+1)
delete test
'_____
'Kupiec test
smpl @all
'Create the vector for the probabilities of default
vector(1) probab
probab(1)=0.05
'How many forecasts have you made?
vector(1) T
T(1)=!sample
'What is the average (expected)?
vector(1) average
average(1)= probab(1)*T(1)
'The number of brakes for long
vector(1) NLong = na
NLong(1)=@round(average95l(1,place)*!sample)
'Vectors for LR tests
vector(1) LRunLong = na
series kupiecresult1=-2*log(((1-probab(1))^(t(1)-
  NLong(1)))*(probab(1)^NLong(1))*1e+150)
series kupiecresult2=2*log(((1-NLong(1)/t(1))^(t(1)-
  NLong(1)))*((NLong(1)/t(1))^NLong(1))*1e+150)
LRunLong(1)=kupiecresult1(1)+kupiecresult2(1)
Kupiec95L(1,place) = 1-@cchisq(LRunLong(1),1)
delete average
delete NLong
delete LRunLong
```

```
delete kupiecresult1
delete kupiecresult2
'_____
'Christoffersen test
smpl @all
'Create the vector for the conditional coverage test
vector(1) LRcc=na
'Create the coef for the independence
'The number of observations with value 0 followed by 0
coef(1) n00=0
'The number of observations with value 0 followed by 1
coef(1) n01=0
'The number of observations with value 1 followed by 1
coef(1) n11=0
'The number of observations with value 1 followed by 0
coef(1) n10=0
'Create the probability to go to 0 given that we are at 0
coef(1) pr00(1)=0
'Create the probability to go to 1 given that we are at 0
coef(1) pr01(1)=0
'Create the probability to go to 0 given that we are at 1
coef(1) pr10(1)=0
'Create the probability to go to 1 given that we are at 1
coef(1) pr11(1)=0
'Create the series for the independence coverage
series LRin
'Estimate the nij
for !i=1 to !sample-1
  if ftse100result95l(!i,place)=0 and ftse100result95L
    (!i+1,place)=0 then
    n00(1)=n00(1)+1
  else
    n00(1)=n00(1)
  endif
  if ftse100result95l(!i,place)=0 and ftse100result95L
     (!i+1,place) =1 then
    n01(1)=n01(1)+1
  else
    n01(1)=n01(1)
  endif
  if ftse100result95l(!i,place)=1 and ftse100result95L
    (!i+1,place)=1 then
    n11(1)=n11(1)+1
  else
    n11(1)=n11(1)
```

```
   endif
   if ftse100result95l(!i,place)=1 and ftse100result95L
     (!i+1,place)=0 then
     n10(1)=n10(1)+1
   else
     n10(1)=n10(1)
   endif
next
pr01(1)=n01(1)/(n00(1)+n01(1))
pr11(1)=n11(1)/(n10(1)+n11(1))
series LRin1=(((1-pr01(1))^n00(1))*(pr01(1)^n01(1))*
  ((1-pr11(1))^n10(1))*pr11(1)^n11(1))*1e+150
coef(1) p2 = (n01(1)+n11(1))/(n01(1)+n00(1)+n10(1)
  +n11(1))
series LRin2 = (((1-p2(1))^(n00(1)+n10(1)))*(p2(1)^
  (n01(1)+n11(1))))*1e+150
LRcc(1) = -2*(log(lrin2(1)/lrin1(1)))
Christ95L(1,place) = 1-@cchisq(LRcc(1),1)
'_____
'MSE Loss function
for !i = 1 to !sample
  if ftse100result95l(!i,place) = 1 then
    ftse100loss95l(!i,place) = (ftse100(!i)-ftse100es95l
      (!i,place))^2
  else
    ftse100loss95l(!i,place) = 0
  endif
next
place = place + 1
for !i = 1 to 7
delete var!i
next
'************************************************
'AR(1)-GARCH(1,1)-skT
'************************************************
read(t=xls) ftse100_GARCH_skt.xls 9
!pi = @acos(-1)
for !i = 1 to !sample
  coef(1) asym = @exp(var8(!i))
  coef(1)  tdf1 = var9(!i)
  coef(1) m = ((asym(1)) - (1/(asym(1))))*@gamma((tdf1(1)
    -1)/2)*@sqrt(tdf1(1)-2)/(@sqrt(!pi)*@gamma
    (tdf1(1)/2))
  coef(1) s = @sqrt(((asym(1))^2 + (1/(asym(1))^2) - 1 -
    m(1)^2))
```

```
scalar p_value = 0.05
scalar freedom = tdf1(1)
  if p_value < (1/(1+asym(1)^2)) then
  scalar quant_95 = ((1/asym(1))*(@qtdist((p_value/2)*
    (1+asym(1)^2),freedom))-m(1))/s(1)

else
  scalar quant_95 = (-asym(1)*@qtdist(((1-p_value)/2)*
    (1+asym(1)^(-2)),freedom)-m(1))/s(1)
  endif
ftse100VaR95L(!i,place) = @sqrt(var1(!i))*quant_95
  +var2(!i)
vector(!ES_repeat) ES = na
!step1 = (1-0.95)/(!ES_repeat+1)
!step2 = 0.05
for !kk = 1 to !ES_repeat
  !step2 = !step2-!step1
  if !step2 < (1/(1+asym(1)^2)) then
    scalar quant = ((1/asym(1))*(@qtdist((!step2/2)*
      (1+asym(1)^2),freedom))-m(1))/s(1)
    ES(!kk) = quant*var1(1) + var2(!i)
  else
    scalar quant = (-asym(1)*@qtdist(((1-!step2)/2)*
      (1+asym(1)^(-2)),freedom)-m(1))/s(1)
    ES(!kk) = quant*var1(1) + var2(!i)
  endif
next
ftse100ES95l(!i,place) = @mean(ES)
Statusline Sample ftse100 !model !i
next
!model = !model +1
for !i = 1 to !sample
  if   ftse100(!i)<ftse100VaR95L(!i,place) then
                ftse100result95L(!i,place) = 1
  else
                ftse100result95L(!i,place) = 0
  endif
next
freeze(test) ftse100result95l.stats
average95L(1,place) = test(5,place+1)
delete test
'_____

'Kupiec test
smpl @all
'Create the vector for the probabilities of default
```

```
vector(1) probab
probab(1)=0.05
'How many forecasts have you made?
vector(1) T
T(1)=!sample
'What is the average (expected)?
vector(1) average
average(1)= probab(1)*T(1)
'The number of brakes for long
vector(1) NLong = na
NLong(1)=@round(average95l(1,place)*!sample)
'Vectors for LR tests
vector(1) LRunLong = na
series kupiecresult1=-2*log(((1-probab(1))^(t(1)-
  NLong(1)))*(probab(1)^NLong(1))*1e+150)
series kupiecresult2=2*log(((1-NLong(1)/t(1))^(t(1)-
  NLong(1)))*((NLong(1)/t(1))^NLong(1))*1e+150)
LRunLong(1)=kupiecresult1(1)+kupiecresult2(1)
Kupiec95L(1,place) = 1-@cchisq(LRunLong(1),1)
delete average
delete NLong
delete LRunLong
delete kupiecresult1
delete kupiecresult2
'_____
'Christoffersen test
smpl @all
'Create the vector for the conditional coverage test
vector(1) LRcc=na
'Create the coef for the independence
'The number of observations with value 0 followed by 0
coef(1) n00=0
'The number of observations with value 0 followed by 1
coef(1) n01=0
'The number of observations with value 1 followed by 1
coef(1) n11=0
'The number of observations with value 1 followed by 0
coef(1) n10=0
'Create the probability to go to 0 given that we are at 0
coef(1) pr00(1)=0
'Create the probability to go to 1 given that we are at 0
coef(1) pr01(1)=0
'Create the probability to go to 0 given that we are at 1
coef(1) pr10(1)=0
'Create the probability to go to 1 given that we are at 1
```

```
coef(1) pr11(1)=0
'Create the series for the independence coverage
series LRin
'Estimate the nij
for !i=1 to !sample-1
  if ftse100result95l(!i,place)=0 and ftse100result95L
    (!i+1,place)=0 then
    n00(1)=n00(1)+1
  else
    n00(1)=n00(1)
  endif
  if ftse100result95l(!i,place)=0 and ftse100result95L
    (!i+1,place) =1 then
    n01(1)=n01(1)+1
  else
    n01(1)=n01(1)
  endif
  if ftse100result95l(!i,place)=1 and ftse100result95L
    (!i+1,place)=1 then
    n11(1)=n11(1)+1
  else
    n11(1)=n11(1)
  endif
  if ftse100result95l(!i,place)=1 and ftse100result95L
    (!i+1,place)=0 then
    n10(1)=n10(1)+1
  else
    n10(1)=n10(1)
  endif
next
pr01(1)=n01(1)/(n00(1)+n01(1))
pr11(1)=n11(1)/(n10(1)+n11(1))
series LRin1=(((1-pr01(1))^n00(1))*(pr01(1)^n01(1))*
  ((1-pr11(1))^n10(1))*pr11(1)^n11(1))*1e+150
coef(1) p2 = (n01(1)+n11(1))/(n01(1)+n00(1)+n10(1)
  +n11(1))
series LRin2 = (((1-p2(1))^(n00(1)+n10(1)))*
  (p2(1)^(n01(1)+n11(1))))*1e+150
LRcc(1) = -2*(log(lrin2(1)/lrin1(1)))
Christ95L(1,place) = 1-@cchisq(LRcc(1),1)
'_____
'MSE Loss function
for !i = 1 to !sample
  if ftse100result95l(!i,place) = 1 then
    ftse100loss95l(!i,place) = (ftse100(!i)-ftse100es95l
```

```
        (!i,place))^2
    else
      ftse100loss95l(!i,place) = 0
    endif
next
place = place + 1
for !i = 1 to 9
delete var!i
next
'********************************************************
'AR(1)-GARCH(1,1)-t
'********************************************************
read(t=xls) ftse100_GARCH_t.xls 8
for !i = 1 to !sample
    ftse100VaR95L(!i,place) = @sqrt(var1(!i))
      *@qtdist(0.05,var8(!i))+var2(!i)
    vector(!ES_repeat) ES = na
    !step1 = (1-0.95)/(!ES_repeat+1)
    !step2 = 0.05
    for !kk = 1 to !ES_repeat
      !step2 = !step2-!step1
      ES(!kk) = @sqrt(var1(!i))*@qtdist(!step2,
        var8(!i))+var2(!i)
    next
    ftse100ES95l(!i,place) = @mean(ES)
Statusline Sample ftse100 !model !i
next
!model = !model +1
for !i = 1 to !sample
    if   ftse100(!i)<ftse100VaR95L(!i,place) then
                    ftse100result95L(!i,place) = 1
    else
                    ftse100result95L(!i,place) = 0
    endif
next
freeze(test) ftse100result95l.stats
average95L(1,place) = test(5,place+1)
delete test
'_____
'Kupiec test
smpl @all
'Create the vector for the probabilities of default
vector(1) probab
probab(1)=0.05
'How many forecasts have you made?
```

```
vector(1) T
T(1)=!sample
'What is the average (expected)?
vector(1) average
average(1)= probab(1)*T(1)
'The number of brakes for long
vector(1) NLong = na
NLong(1)=@round(average95l(1,place)*!sample)
'Vectors for LR tests
vector(1) LRunLong = na
series kupiecresult1=-2*log(((1-probab(1))^(t(1)
  -NLong(1)))*(probab(1)^NLong(1))*1e+150)
series kupiecresult2=2*log(((1-NLong(1)/t(1))
  ^(t(1)-NLong(1)))*((NLong(1)/t(1))^NLong(1))*1e+150)
LRunLong(1)=kupiecresult1(1)+kupiecresult2(1)
Kupiec95L(1,place) = 1-@cchisq(LRunLong(1),1)
delete average
delete NLong
delete LRunLong
delete kupiecresult1
delete kupiecresult2
'_____
'Christoffersen test
smpl @all
'Create the vector for the conditional coverage test
vector(1) LRcc=na
'Create the coef for the independence
'The number of observations with value 0 followed by 0
coef(1) n00=0
'The number of observations with value 0 followed by 1
coef(1) n01=0
'The number of observations with value 1 followed by 1
coef(1) n11=0
'The number of observations with value 1 followed by 0
coef(1) n10=0
'Create the probability to go to 0 given that we are at 0
coef(1) pr00(1)=0
'Create the probability to go to 1 given that we are at 0
coef(1) pr01(1)=0
'Create the probability to go to 0 given that we are at 1
coef(1) pr10(1)=0
'Create the probability to go to 1 given that we are at 1
coef(1) pr11(1)=0
'Create the series for the independence coverage
series LRin
```

```
'Estimate the nij
for !i=1 to !sample-1
  if ftse100result95l(!i,place)=0 and ftse
    100result95L(!i+1,place)=0 then
      n00(1)=n00(1)+1
  else
      n00(1)=n00(1)
  endif
  if ftse100result95l(!i,place)=0 and ftse
    100result95L(!i+1,place) =1 then
      n01(1)=n01(1)+1
  else
      n01(1)=n01(1)
  endif
  if ftse100result95l(!i,place)=1 and ftse
    100result95L(!i+1,place)=1 then
      n11(1)=n11(1)+1
  else
      n11(1)=n11(1)
  endif
  if ftse100result95l(!i,place)=1 and ftse
    100result95L(!i+1,place)=0 then
      n10(1)=n10(1)+1
  else
      n10(1)=n10(1)
  endif
next
pr01(1)=n01(1)/(n00(1)+n01(1))
pr11(1)=n11(1)/(n10(1)+n11(1))
series LRin1=(((1-pr01(1))^n00(1))*(pr01(1)^n01(1))
  *((1-pr11(1))^n10(1))*pr11(1)^n11(1))*1e+150
coef(1) p2 = (n01(1)+n11(1))/(n01(1)+n00(1)
  +n10(1)+n11(1))
series LRin2 = (((1-p2(1))^(n00(1)+n10(1)))*(p2(1)
  ^(n01(1)+n11(1))))*1e+150
LRcc(1) = -2*(log(lrin2(1)/lrin1(1)))
Christ95L(1,place) = 1-@cchisq(LRcc(1),1)
'_____

'MSE Loss function
for !i = 1 to !sample
  if ftse100result95l(!i,place) = 1 then
    ftse100loss951(!i,place) = (ftse100(!i)-ftse
      100es951(!i,place))^2
  else
    ftse100loss951(!i,place) = 0
```

```
    endif
next
place = place + 1
for !i = 1 to 8
delete var!i
next
'********************************************************
'AR(1)-TARCH(1,1)-ged
'********************************************************
read(t=xls) ftse100_GJR_ged.xls 9
for !i = 1 to !sample
  ftse100VaR95L(!i,place) = @sqrt(var1(!i))*
    @qged(0.05,var9(!i))+var2(!i)
  vector(!ES_repeat) ES = na
  !step1 = (1-0.95)/(!ES_repeat+1)
  !step2 = 0.05
  for !kk = 1 to !ES_repeat
    !step2 = !step2-!step1
    ES(!kk) = @sqrt(var1(!i))*@qged(!step2,
      var9(!i))+var2(!i)
  next
  ftse100ES95l(!i,place) = @mean(ES)
  Statusline Sample ftse100 !model !i
next
!model = !model +1
for !i = 1 to !sample
  if    ftse100(!i)<ftse100VaR95L(!i,place) then
                  ftse100result95L(!i,place) = 1
  else
                  ftse100result95L(!i,place) = 0
  endif
next
freeze(test) ftse100result95l.stats
average95L(1,place) = test(5,place+1)
delete test
'_____
'Kupiec test
smpl @all
'Create the vector for the probabilities of default
vector(1) probab
probab(1)=0.05
'How many forecasts have you made?
vector(1) T
T(1)=!sample
'What is the average (expected)?
```

```
vector(1) average
average(1)=probab(1)*T(1)
'The number of brakes for long
vector(1) NLong = na
NLong(1)=@round(average95l(1,place)*!sample)
'Vectors for LR tests
vector(1) LRunLong = na
series kupiecresult1=-2*log(((1-probab(1))
  ^(t(1)-NLong(1)))*(probab(1)^NLong(1))*1e+150)
series kupiecresult2=2*log(((1-NLong(1)/t(1))
   ^(t(1)-NLong(1)))*((NLong(1)/t(1))
   ^NLong(1))*1e+150)
LRunLong(1)=kupiecresult1(1)+kupiecresult2(1)
Kupiec95L(1,place) = 1-@cchisq(LRunLong(1),1)
delete average
delete NLong
delete LRunLong
delete kupiecresult1
delete kupiecresult2
'_____

'Christoffersen test
smpl @all
'Create the vector for the conditional coverage test
vector(1) LRcc=na
'Create the coef for the independence
'The number of observations with value 0 followed by 0
coef(1) n00=0
'The number of observations with value 0 followed by 1
coef(1) n01=0
'The number of observations with value 1 followed by 1
coef(1) n11=0
'The number of observations with value 1 followed by 0
coef(1) n10=0
'Create the probability to go to 0 given that we are at 0
coef(1) pr00(1)=0
'Create the probability to go to 1 given that we are at 0
coef(1) pr01(1)=0
'Create the probability to go to 0 given that we are at 1
coef(1) pr10(1)=0
'Create the probability to go to 1 given that we are at 1
coef(1) pr11(1)=0
'Create the series for the independence coverage
series LRin
'Estimate the nij
for !i=1 to !sample-1
```

```
if ftse100result95l(!i,place)=0 and ftse100result95L
  (!i+1,place)=0 then
    n00(1)=n00(1)+1
else
    n00(1)=n00(1)
endif
if ftse100result95l(!i,place)=0 and ftse
  100result95L(!i+1,place) =1 then
    n01(1)=n01(1)+1
else
    n01(1)=n01(1)
endif
if ftse100result95l(!i,place)=1 and ftse
  100result95L(!i+1,place)=1 then
    n11(1)=n11(1)+1
else
    n11(1)=n11(1)
endif
if ftse100result95l(!i,place)=1 and ftse
  100result95L(!i+1,place)=0 then
    n10(1)=n10(1)+1
else
    n10(1)=n10(1)
endif
next
pr01(1)=n01(1)/(n00(1)+n01(1))
pr11(1)=n11(1)/(n10(1)+n11(1))
series LRin1=(((1-pr01(1))^n00(1))*(pr01(1)
  ^n01(1))*((1-pr11(1))^n10(1))*pr11(1)^n11
  (1))*1e+150
coef(1) p2 = (n01(1)+n11(1))/(n01(1)+n00(1)+
  n10(1)+n11(1))
series LRin2 = (((1-p2(1))^(n00(1)+n10(1)))*(p2(1)
  ^(n01(1)+n11(1))))*1e+150
LRcc(1) = -2*(log(lrin2(1)/lrin1(1)))
Christ95L(1,place) = 1-@cchisq(LRcc(1),1)
'_____
'MSE Loss function
for !i = 1 to !sample
  if ftse100result95l(!i,place) = 1 then
    ftse100loss95l(!i,place) = (ftse100(!i)-ftse
      100es95l(!i,place))^2
  else
    ftse100loss95l(!i,place) = 0
  endif
```

```
next
place = place + 1
for !i = 1 to 9
delete var!i
next
'*********************************************************
'AR(1)-TARCH(1,1)-n
'*********************************************************
read(t=xls) ftse100_GJR_n.xls 8
for !i = 1 to !sample
  ftse100VaR95L(!i,place) = @sqrt(var1(!i))
    *@qnorm(0.05)+var2(!i)
  vector(!ES_repeat) ES = na
  !step1 = (1-0.95)/(!ES_repeat+1)
  !step2 = 0.05
  for !kk = 1 to !ES_repeat
    !step2 = !step2-!step1
    ES(!kk) = @sqrt(var1(!i))*@qnorm(!step2)+var2(!i)
  next
  ftse100ES951(!i,place) = @mean(ES)
  Statusline Sample ftse100 !model !i
next
!model = !model +1
for !i = 1 to !sample
  if   ftse100(!i)<ftse100VaR95L(!i,place) then
                  ftse100result95L(!i,place) = 1
  else
                  ftse100result95L(!i,place) = 0
  endif
next
freeze(test) ftse100result951.stats
average95L(1,place) = test(5,place+1)
delete test
' _____

'Kupiec test
smpl @all
'Create the vector for the probabilities of default
vector(1) probab
probab(1)=0.05
'How many forecasts have you made?
vector(1) T
T(1)=!sample
'What is the average (expected)?
vector(1) average
average(1) = probab(1)*T(1)
```

```
'The number of brakes for long
vector(1) NLong = na
NLong(1)=@round(average95l(1,place)*!sample)
'Vectors for LR tests
vector(1) LRunLong = na
series kupiecresult1=-2*log(((1-probab(1))
  ^(t(1)-NLong(1)))*(probab(1)^NLong(1))*1e+150)
series kupiecresult2=2*log(((1-NLong(1)/t(1))
  ^(t(1)-NLong(1)))*((NLong(1)/t(1))^NLong(1))*1e+150)
LRunLong(1)=kupiecresult1(1)+kupiecresult2(1)
Kupiec95L(1,place) = 1-@cchisq(LRunLong(1),1)
delete average
delete NLong
delete LRunLong
delete kupiecresult1
delete kupiecresult2
'_____

'Christoffersen test
smpl @all
'Create the vector for the conditional coverage test
vector(1) LRcc=na
'Create the coef for the independence
'The number of observations with value 0 followed by 0
coef(1) n00=0
'The number of observations with value 0 followed by 1
coef(1) n01=0
'The number of observations with value 1 followed by 1
coef(1) n11=0
'The number of observations with value 1 followed by 0
coef(1) n10=0
'Create the probability to go to 0 given that we are at 0
coef(1) pr00(1)=0
'Create the probability to go to 1 given that we are at 0
coef(1) pr01(1)=0
'Create the probability to go to 0 given that we are at 1
coef(1) pr10(1)=0
'Create the probability to go to 1 given that we are at 1
coef(1) pr11(1)=0
'Create the series for the independence coverage
series LRin
'Estimate the nij
for !i=1 to !sample-1
  if ftse100result95l(!i,place)=0 and ftse
    100result95L(!i+1,place)=0 then
      n00(1)=n00(1)+1
```

```
  else
     n00(1)=n00(1)
  endif
  if ftse100result95l(!i,place)=0 and ftse
    100result95L(!i+1,place) =1 then
       n01(1)=n01(1)+1
  else
     n01(1)=n01(1)
  endif
  if ftse100result95l(!i,place)=1 and ftse
    100result95L(!i+1,place)=1 then
       n11(1)=n11(1)+1
  else
     n11(1)=n11(1)
  endif
  if ftse100result95l(!i,place)=1 and ftse
    100result95L(!i+1,place)=0 then
       n10(1)=n10(1)+1
  else
     n10(1)=n10(1)
  endif
next
pr01(1)=n01(1)/(n00(1)+n01(1))
pr11(1)=n11(1)/(n10(1)+n11(1))
series LRin1=(((1-pr01(1))^n00(1))*(pr01(1)
  ^n01(1))*((1-pr11(1))^n10(1))*pr11(1)
  ^n11(1))*1e+150
coef(1) p2 = (n01(1)+n11(1))/(n01(1)+n00(1)+
  n10(1)+n11(1))
series LRin2 = (((1-p2(1))^(n00(1)+n10(1)))*(p2(1)
  ^(n01(1)+n11(1))))*1e+150
LRcc(1) = -2*(log(lrin2(1)/lrin1(1)))
Christ95L(1,place) = 1-@cchisq(LRcc(1),1)
'_____
'MSE Loss function
for !i = 1 to !sample
  if ftse100result95l(!i,place) = 1 then
    ftse100loss95l(!i,place) = (ftse100(!i)-ftse
      100es95l(!i,place))^2
  else
    ftse100loss95l(!i,place) = 0
  endif
next
place = place + 1
for !i = 1 to 8
```

```
delete var!i
next
'*********************************************************
'AR(1)-TARCH(1,1)-skT
'*********************************************************
read(t=xls) ftse100_GJR_skt.xls 10
!pi = @acos(-1)
for !i = 1 to !sample
  coef(1) asym = @exp(var9(!i))
  coef(1) tdf1 = var10(!i)
  coef(1) m = ((asym(1)) - (1/(asym(1))))
    *@gamma(((tdf1(1)-1)/2)*@sqrt(tdf1(1)-2)/
    (@sqrt(!pi)*@gamma(tdf1(1)/2))
  coef(1) s = @sqrt(((asym(1))^2 + (1/(asym(1))^2) -
    1 - m(1)^2))
  scalar p_value = 0.05
  scalar freedom = tdf1(1)
    if p_value < (1/(1+asym(1)^2)) then
    scalar quant_95 = ((1/asym(1))*(@qtdist((p_value/2)
      *(1+asym(1)^2),freedom))-m(1))/s(1)

  else
    scalar quant_95 = (-asym(1)*@qtdist(((1-p_value)/2)
      *(1+asym(1)^(-2)),freedom)-m(1))/s(1)
    endif
  ftse100VaR95L(!i,place) = @sqrt(var1(!i))
    *quant_95+var2(!i)
  vector(!ES_repeat) ES = na
  !step1 = (1-0.95)/(!ES_repeat+1)
  !step2 = 0.05
  for !kk = 1 to !ES_repeat
    !step2 = !step2-!step1
    if !step2 < (1/(1+asym(1)^2)) then
      scalar quant = ((1/asym(1))*(@qtdist((!step2/2)
        *(1+asym(1)^2),freedom))-m(1))/s(1)
      ES(!kk) = quant*var1(1) + var2(!i)
    else
      scalar quant = (-asym(1)*@qtdist(((1-!step2)/2)
        *(1+asym(1)^(-2)),freedom)-m(1))/s(1)
      ES(!kk) = quant*var1(1) + var2(!i)
    endif
  next
  ftse100ES95l(!i,place) = @mean(ES)
  Statusline Sample ftse100 !model !i
next
```

```
!model = !model +1
for !i = 1 to !sample
  if   ftse100(!i)<ftse100VaR95L(!i,place) then
                    ftse100result95L(!i,place) = 1
  else
                    ftse100result95L(!i,place) = 0
  endif
next
freeze(test) ftse100result95l.stats
average95L(1,place) = test(5,place+1)
delete test
'_____
'Kupiec test
smpl @all
'Create the vector for the probabilities of default
vector(1) probab
probab(1)=0.05
'How many forecasts have you made?
vector(1) T
T(1)=!sample
'What is the average (expected)?
vector(1) average
average(1)= probab(1)*T(1)
'The number of brakes for long
vector(1) NLong = na
NLong(1)=@round(average95l(1,place)*!sample)
'Vectors for LR tests
vector(1) LRunLong = na
series kupiecresult1=-2*log(((1-probab(1))
  ^(t(1)-NLong(1)))*(probab(1)^NLong(1))*1e+150)
series kupiecresult2=2*log(((1-NLong(1)/t(1))
  ^(t(1)-NLong(1)))*((NLong(1)/t(1))^NLong(1))*1e+150)
LRunLong(1)=kupiecresult1(1)+kupiecresult2(1)
Kupiec95L(1,place) = 1-@cchisq(LRunLong(1),1)
delete average
delete NLong
delete LRunLong
delete kupiecresult1
delete kupiecresult2
'_____
'Christoffersen test
smpl @all
'Create the vector for the conditional coverage test
vector(1) LRcc=na
'Create the coef for the independence
```

```
'The number of observations with value 0 followed by 0
coef(1) n00=0
'The number of observations with value 0 followed by 1
coef(1) n01=0
'The number of observations with value 1 followed by 1
coef(1) n11=0
'The number of observations with value 1 followed by 0
coef(1) n10=0
'Create the probability to go to 0 given that we are at 0
coef(1) pr00(1)=0
'Create the probability to go to 1 given that we are at 0
coef(1) pr01(1)=0
'Create the probability to go to 0 given that we are at 1
coef(1) pr10(1)=0
'Create the probability to go to 1 given that we are at 1
coef(1) pr11(1)=0
'Create the series for the independence coverage
series LRin
'Estimate the nij
for !i=1 to !sample-1
  if ftse100result95l(!i,place)=0 and ftse100result95L
    (!i+1,place)=0 then
      n00(1)=n00(1)+1
  else
      n00(1)=n00(1)
  endif
  if ftse100result95l(!i,place)=0 and ftse
    100result95L(!i+1,place) =1 then
      n01(1)=n01(1)+1
  else
      n01(1)=n01(1)
  endif
  if ftse100result95l(!i,place)=1 and ftse
    100result95L(!i+1,place)=1 then
      n11(1)=n11(1)+1
  else
      n11(1)=n11(1)
  endif
  if ftse100result95l(!i,place)=1 and ftse
    100result95L(!i+1,place)=0 then
      n10(1)=n10(1)+1
  else
      n10(1)=n10(1)
  endif
next
```

```
pr01(1)=n01(1)/(n00(1)+n01(1))
pr11(1)=n11(1)/(n10(1)+n11(1))
series LRin1=(((1-pr01(1))^n00(1))*(pr01(1)
  ^n01(1))*((1-pr11(1))^n10(1))*pr11(1)^n11(1))*1e+150
coef(1) p2 = (n01(1)+n11(1))/(n01(1)+n00(1)+
  n10(1)+n11(1))
series LRin2 = (((1-p2(1))^(n00(1)+n10(1)))*(p2(1)
  ^(n01(1)+n11(1)))))*1e+150
LRcc(1) = -2*(log(lrin2(1)/lrin1(1)))
Christ95L(1,place) = 1-@cchisq(LRcc(1),1)
'_____
'MSE Loss function
for !i = 1 to !sample
  if ftse100result95l(!i,place) = 1 then
    ftse100loss95l(!i,place) = (ftse100(!i)-ftse
      100es95l(!i,place))^2
  else
    ftse100loss95l(!i,place) = 0
  endif
next
place = place + 1
for !i = 1 to 10
delete var!i
next
'*********************************************************
'AR(1)-TARCH(1,1)-t
'*********************************************************
read(t=xls) ftse100_GJR_t.xls 9
for !i = 1 to !sample
  ftse100VaR95L(!i,place) = @sqrt(var1(!i))
    *@qtdist(0.05,var9(!i))+var2(!i)
  vector(!ES_repeat) ES = na
  !step1 = (1-0.95)/(!ES_repeat+1)
  !step2 = 0.05
  for !kk = 1 to !ES_repeat
    !step2 = !step2-!step1
    ES(!kk) = @sqrt(var1(!i))*@qtdist(!step2,
      var9(!i))+var2(!i)
  next
  ftse100ES95l(!i,place) = @mean(ES)
  Statusline Sample ftse100 !model !i
next
!model = !model +1
for !i = 1 to !sample
  if  ftse100(!i)<ftse100VaR95L(!i,place) then
```

```
                         ftse100result95L(!i,place) = 1
   else
                         ftse100result95L(!i,place) = 0
   endif
next
freeze(test) ftse100result95l.stats
average95L(1,place) = test(5,place+1)
delete test
'_____
'Kupiec test
smpl @all
'Create the vector for the probabilities of default
vector(1) probab
probab(1)=0.05
'How many forecasts have you made?
vector(1) T
T(1)=!sample
'What is the average (expected)?
vector(1) average
average(1)= probab(1)*T(1)
'The number of brakes for long
vector(1) NLong = na
NLong(1)=@round(average95l(1,place)*!sample)
'Vectors for LR tests
vector(1) LRunLong = na
series kupiecresult1=-2*log(((1-probab(1))
  ^(t(1)-NLong(1)))*(probab(1)^NLong(1))*1e+150)
series kupiecresult2=2*log(((1-NLong(1)/t(1))
  ^(t(1)-NLong(1)))*((NLong(1)/t(1))^NLong(1))*1e+150)
LRunLong(1)=kupiecresult1(1)+kupiecresult2(1)
Kupiec95L(1,place) = 1-@cchisq(LRunLong(1),1)
delete average
delete NLong
delete LRunLong
delete kupiecresult1
delete kupiecresult2
'_____
'Christoffersen test
smpl @all
'Create the vector for the conditional coverage test
vector(1) LRcc=na
'Create the coef for the independence
'The number of observations with value 0 followed by 0
coef(1) n00=0
'The number of observations with value 0 followed by 1
```

```
coef(1) n01=0
'The number of observations with value 1 followed by 1
coef(1) n11=0
'The number of observations with value 1 followed by 0
coef(1) n10=0
'Create the probability to go to 0 given that we are at 0
coef(1) pr00(1)=0
'Create the probability to go to 1 given that we are at 0
coef(1) pr01(1)=0
'Create the probability to go to 0 given that we are at 1
coef(1) pr10(1)=0
'Create the probability to go to 1 given that we are at 1
coef(1) pr11(1)=0
'Create the series for the independence coverage
series LRin
'Estimate the nij
for !i=1 to !sample-1
  if ftse100result95l(!i,place)=0 and ftse
    100result95L(!i+1,place)=0 then
      n00(1)=n00(1)+1
  else
      n00(1)=n00(1)
  endif
  if ftse100result95l(!i,place)=0 and ftse
    100result95L(!i+1,place) =1 then
      n01(1)=n01(1)+1
  else
      n01(1)=n01(1)
  endif
  if ftse100result95l(!i,place)=1 and ftse
    100result95L(!i+1,place)=1 then
      n11(1)=n11(1)+1
  else
    n11(1)=n11(1)
  endif
  if ftse100result95l(!i,place)=1 and ftse
    100result95L(!i+1,place)=0 then
      n10(1)=n10(1)+1
  else
      n10(1)=n10(1)
  endif
next
pr01(1)=n01(1)/(n00(1)+n01(1))
pr11(1)=n11(1)/(n10(1)+n11(1))
series LRin1=(((1-pr01(1))^n00(1))*(pr01(1)
```

```
 ^n01(1))*((1-pr11(1))^n10(1))*pr11(1)
 ^n11(1))*1e+150
coef(1) p2 = (n01(1)+n11(1))/(n01(1)+
 n00(1)+n10(1)+n11(1))
series LRin2 = (((1-p2(1))^(n00(1)+n10(1)))*(p2(1)
 ^(n01(1)+n11(1))))*1e+150
LRcc(1) = -2*(log(lrin2(1)/lrin1(1)))
Christ95L(1,place) = 1-@cchisq(LRcc(1),1)
'_____
'MSE Loss function
for !i = 1 to !sample
  if ftse100result95l(!i,place) = 1 then
    ftse100loss95l(!i,place) = (ftse100(!i)-ftse
      100es95l(!i,place))^2
  else
    ftse100loss95l(!i,place) = 0
  endif
next
place = place + 1
for !i = 1 to 9
delete var!i
next
save chapter8.var.es.ftse100.output.wf1
```

• *chapter8_sp500_options.partA.prg*

```
load chapter8_sp500_options.wf1
smpl @all
series _temp = 1
!length = @obs(_temp)
delete _temp
!ssize = 500
!g01 = 0
matrix(!length-!ssize,8) estg01
!g11 = 0
matrix(!length-!ssize,9) estg11
!g21 = 0
matrix(!length-!ssize,11) estg21
!e01 = 0
matrix(!length-!ssize,9) este01
!e11 = 0
matrix(!length-!ssize,10) este11
!e21 = 0
matrix(!length-!ssize,12) este21
```

```
!t01 = 0
matrix(!length-!ssize,10) estt01
!t11 = 0
matrix(!length-!ssize,11) estt11
!t21 = 0
matrix(!length-!ssize,13) estt21
for !i = 1 to !length-!ssize
  statusline !i
  smpl @first+!i-1 @first+!i+!ssize-2
equation tempg01.arch(1,0,m=10000,h, deriv=f)
  sp500 sp500(-1)
  tempg01.makegarch garchg01
  tempg01.makeresid resg01
!ht = garchg01(!i+!ssize-1)
!c1 = @coefs(1)
!c2 = @coefs(2)
!c3 = @coefs(3)
!h_hat = !c2 + (!c3*(resg01(!ssize + !i -1)^2) )
!y_hat = (!c1*sp500(!ssize + !i -1))
!res = sp500(!ssize + !i) - !y_hat
estg01(!i,1) = !y_hat
if !h_hat>=0 then
  estg01(!i,2) = !h_hat
else
  estg01(!i,2) = estg01(!i-1,2)
endif
estg01(!i,3) = !res
estg01(!i,4) =  (!res^2) / estg01(!i,2)
estg01(!i,5) = estg01(!i,4) + !g01
!g01 = estg01(!i,5)
estg01(!i,6) = !c1
estg01(!i,7) = !c2
estg01(!i,8) = !c3

equation tempg11.arch(1,1,m=10000,h, deriv=f)
    sp500 sp500(-1)
  tempg11.makegarch garchg11
  tempg11.makeresid resg11
!ht = garchg11(!i+!ssize-1)
!c1 = @coefs(1)
!c2 = @coefs(2)
!c3 = @coefs(3)
!c4 = @coefs(4)
!h_hat = !c2 + (!c3*(resg11(!ssize + !i -1)^2) ) + (!c4*!ht)
!y_hat = (!c1*sp500(!ssize + !i -1))
```

```
!res = sp500(!ssize + !i) - !y_hat
estg11(!i,1) = !y_hat
if !h_hat>=0 then
  estg11(!i,2) = !h_hat
else
estg11(!i,2) = estg11(!i-1,2)
endif
estg11(!i,3) = !res
estg11(!i,4) = (!res^2) / estg11(!i,2)
estg11(!i,5) = estg11(!i,4) + !g11
!g11 = estg11(!i,5)
estg11(!i,6) = !c1
estg11(!i,7) = !c2
estg11(!i,8) = !c3
estg11(!i,9) = !c4

equation tempg21.arch(1,2,m=10000,h,
  deriv=f) sp500 sp500(-1)
  tempg21.makegarch garchg21
  tempg21.makeresid resg21
!ht = garchg21(!i+!ssize-1)
!ht2 = garchg21(!i+!ssize-2)
!c1 = @coefs(1)
!c2 = @coefs(2)
!c3 = @coefs(3)
!c4 = @coefs(4)
!c5 = @coefs(5)
!h_hat = !c2 + (!c3*(resg21(!ssize + !i -1)^2) ) +
  (!c4*!ht) + (!c5*!ht2)
!y_hat = (!c1*sp500(!ssize + !i -1))
!res = sp500(!ssize + !i) - !y_hat
estg21(!i,1) = !y_hat
if !h_hat>=0 then
  estg21(!i,2) = !h_hat
else
  estg21(!i,2) = estg21(!i-1,2)
endif
estg21(!i,3) = !res
estg21(!i,4) = (!res^2) / estg21(!i,2)
estg21(!i,5) = estg21(!i,4) + !g21
!g21 = estg21(!i,5)
estg21(!i,6) = !c1
estg21(!i,7) = !c2
estg21(!i,8) = !c3
estg21(!i,9) = !c4
```

```
estg21(!i,10) = !c5
estg21(!i,11) = garchg21(!ssize + !i -1)

equation tempe01.arch(1,0,e,m=10000,h,
   deriv=f) sp500 sp500(-1)
   tempe01.makegarch garche01
   tempe01.makeresid rese01
!ht = garche01(!i+!ssize-1)
!ht2 = garche01(!i+!ssize-2)
!c1 = @coefs(1)
!c2 = @coefs(2)
!c3 = @coefs(3)
!c4 = @coefs(4)
!logh_hat = !c2 + (!c3*@abs(rese01(!ssize +
   !i -1)/(sqr(!ht)))) + (!c4*(rese01(!ssize +
   !i -1)/(sqr(!ht))))
!h_hat = @exp(!logh_hat)
!y_hat = (!c1*sp500(!ssize + !i -1))
!res = sp500(!ssize + !i) - !y_hat
este01(!i,1) = !y_hat
if !h_hat>=0 then
   este01(!i,2) = !h_hat
else
   este01(!i,2) = este01(!i-1,2)
endif
este01(!i,3) = !res
este01(!i,4) = (!res^2) / este01(!i,2)
este01(!i,5) = este01(!i,4) + !e01
!e01 = este01(!i,5)
este01(!i,6) = !c1
este01(!i,7) = !c2
este01(!i,8) = !c3
este01(!i,9) = !c4

equation tempe11.arch(1,1,e,m=10000,h, deriv=f)
   sp500 sp500(-1)
   tempe11.makegarch garche11
   tempe11.makeresid rese11
!ht = garche11(!i+!ssize-1)
!ht2 = garche11(!i+!ssize-2)
!c1 = @coefs(1)
!c2 = @coefs(2)
!c3 = @coefs(3)
!c4 = @coefs(4)
!c5 = @coefs(5)
```

```
!logh_hat = !c2 + (!c3*@abs(rese11(!ssize +
  !i -1)/(sqr(!ht)))) + (!c4*(rese11(!ssize +
  !i -1)/(sqr(!ht)))) + (!c5*@log(!ht))
!h_hat = @exp(!logh_hat)
!y_hat = (!c1*sp500(!ssize + !i -1))
!res = sp500(!ssize + !i) - !y_hat
este11(!i,1) = !y_hat
if !h_hat>=0 then
  este11(!i,2) = !h_hat
else
  este11(!i,2) = este11(!i-1,2)
endif
este11(!i,3) = !res
este11(!i,4) = (!res^2) / este11(!i,2)
este11(!i,5) = este11(!i,4) + !e11
!e11 = este11(!i,5)
este11(!i,6) = !c1
este11(!i,7) = !c2
este11(!i,8) = !c3
este11(!i,9) = !c4
este11(!i,10) = !c5

equation tempe21.arch(1,2,e,m=10000,h, deriv=f)
  sp500 sp500(-1)
  tempe21.makegarch garche21
  tempe21.makeresid rese21
!ht = garche21(!i+!ssize-1)
!ht2 = garche21(!i+!ssize-2)
!c1 = @coefs(1)
!c2 = @coefs(2)
!c3 = @coefs(3)
!c4 = @coefs(4)
!c5 = @coefs(5)
!c6 = @coefs(6)
!logh_hat = !c2 + (!c3*@abs(rese21(!ssize +
  !i -1)/(sqr(!ht)))) + (!c4*(rese21(!ssize +
  !i -1)/(sqr(!ht)))) + (!c5*@log(!ht)) +
  (!c6*@log(!ht2))
!h_hat = @exp(!logh_hat)
!y_hat = (!c1*sp500(!ssize + !i -1))
!res = sp500(!ssize + !i) - !y_hat
este21(!i,1) = !y_hat
if !h_hat>=0 then
  este21(!i,2) = !h_hat
else
```

```
  este21(!i,2) = este21(!i-1,2)
endif
este21(!i,3) = !res
este21(!i,4) = (!res^2) / este21(!i,2)
este21(!i,5) = este21(!i,4) + !e21
!e21 = este21(!i,5)
este21(!i,6) = !c1
este21(!i,7) = !c2
este21(!i,8) = !c3
este21(!i,9) = !c4
este21(!i,10) = !c5
este21(!i,11) = !c6
este21(!i,12) = garche21(!ssize + !i -1)

equation tempt01.arch(1,0,t,m=10000,h,
  deriv=f) sp500 sp500(-1)
  tempt01.makegarch garcht01
  tempt01.makeresid rest01
!ht = garcht01(!i+!ssize-1)
!ht2 = garcht01(!i+!ssize-2)
!c1 = @coefs(1)
!c2 = @coefs(2)
!c3 = @coefs(3)
!c4 = @coefs(4)
if rest01(!ssize + !i -1) < 0 then
!D = 1
else
!D = 0
endif
!h_hat = !c2 + (!c3*(rest01(!ssize + !i -1)^2) ) +
  (!c4*!D*(rest01(!ssize + !i -1)^2))
!y_hat = (!c1*sp500(!ssize + !i -1))
!res = sp500(!ssize + !i) - !y_hat
estt01(!i,1) = !y_hat
if !h_hat>=0 then
  estt01(!i,2) = !h_hat
else
  estt01(!i,2) = estt01(!i-1,2)
endif
estt01(!i,3) = !res
estt01(!i,4) = (!res^2) / estt01(!i,2)
estt01(!i,5) = estt01(!i,4) + !t01
!t01 = estt01(!i,5)
estt01(!i,6) = !c1
estt01(!i,7) = !c2
estt01(!i,8) = !c3
```

```
estt01(!i,9) = !c4
series avDt01 = 1*(rest01<0)
estt01(!i,10) = @mean(avDt01)
delete avDt01

equation tempt11.arch(1,1,t,m=10000,h,
  deriv=f) sp500 sp500(-1)
  tempt11.makegarch garcht11
  tempt11.makeresid rest11
!ht = garcht11(!i+!ssize-1)
!ht2 = garcht11(!i+!ssize-2)
!c1 = @coefs(1)
!c2 = @coefs(2)
!c3 = @coefs(3)
!c4 = @coefs(4)
!c5 = @coefs(5)
if rest11(!ssize + !i -1) < 0 then
!D = 1
else
!D = 0
endif
!h_hat = !c2 + (!c3*(rest11(!ssize + !i -1)^2) ) +
  (!c4*!D*(rest11(!ssize + !i -1)^2)) + (!c5*!ht)
!y_hat = (!c1*sp500(!ssize + !i -1))
!res = sp500(!ssize + !i) - !y_hat
estt11(!i,1) = !y_hat
if !h_hat>=0 then
  estt11(!i,2) = !h_hat
else
  estt11(!i,2) = estt11(!i-1,2)
endif
estt11(!i,3) = !res
estt11(!i,4) = (!res^2) / estt11(!i,2)
estt11(!i,5) = estt11(!i,4) + !t11
!t11 = estt11(!i,5)
estt11(!i,6) = !c1
estt11(!i,7) = !c2
estt11(!i,8) = !c3
estt11(!i,9) = !c4
estt11(!i,10) = !c5
series avDt11 = 1*(rest11<0)
estt11(!i,11) = @mean(avDt11)
delete avDt11

equation tempt21.arch(1,2,t,m=10000,h,
  deriv=f) sp500 sp500(-1)
```

```
  tempt21.makegarch garcht21
  tempt21.makeresid rest21
!ht = garcht21(!i+!ssize-1)
!ht2 = garcht21(!i+!ssize-2)
!c1 = @coefs(1)
!c2 = @coefs(2)
!c3 = @coefs(3)
!c4 = @coefs(4)
!c5 = @coefs(5)
!c6 = @coefs(6)
if rest21(!ssize + !i -1) < 0 then
!D = 1
else
!D = 0
endif
!h_hat = !c2 + (!c3*(rest21(!ssize + !i -1)^2) ) +
  (!c4*!D*(rest21(!ssize + !i -1)^2)) + (!c5*!ht) +
  (!c6*!ht2)
!y_hat = (!c1*sp500(!ssize + !i -1))
!res = sp500(!ssize + !i) - !y_hat
estt21(!i,1) = !y_hat
if !h_hat>=0 then
  estt21(!i,2) = !h_hat
else
  estt21(!i,2) = estt21(!i-1,2)
endif
estt21(!i,3) = !res
estt21(!i,4) = (!res^2) / estt21(!i,2)
estt21(!i,5) = estt21(!i,4) + !t21
!t21 = estt21(!i,5)
estt21(!i,6) = !c1
estt21(!i,7) = !c2
estt21(!i,8) = !c3
estt21(!i,9) = !c4
estt21(!i,10) = !c5
estt21(!i,11) = !c6
series avDt21 = 1*(rest21<0)
estt21(!i,12) = @mean(avDt21)
delete avDt21
estt21(!i,13) = garcht21(!ssize + !i -1)
next

delete garchg01
delete garchg11
delete garchg21
```

```
delete garcht01
delete garcht11
delete garcht21
delete garche01
delete garche11
delete garche21
delete resg01
delete resg11
delete resg21
delete rest01
delete rest11
delete rest21
delete rese01
delete resc11
delete rese21
delete tempg01
delete tempg11
delete tempg21
delete tempt01
delete tempt11
delete tempt21
delete tempe01
delete tempe11
delete tempe21

save chapter8_sp500_options.partA.wf1
```

- *chapter8_sp500_options.partB.prg*

```
load chapter8_sp500_options.partA.wf1
smpl @all
series _temp = 1
!day = @obs(_temp)
!models =9
!ssize =500
delete _temp
matrix(!day,!models) Calls=na
matrix(!day,!models) Puts=na
matrix(!day,!models) Straddles=na
matrix(!day,!models) Volatility=na
!d1 = 0
!d2 = 0
!pi = @acos(-1)

for !ahead=1 to !day
```

```
statusline !ahead
if ahead(!ahead)>0 then

  vector(ahead(!ahead)) h
  h(1)=estg01(!ahead-!ssize,7) + estg01(!ahead-!ssize,8)
    *estg01(!ahead-!ssize,2)
  for !r=2 to ahead(!ahead)
    h(!r) = estg01(!ahead-!ssize,7) +
      estg01(!ahead-!ssize,8)*h(!r-1)
  next
  volatility(!ahead,1)=sqr(@sum(h)*252/ahead(!ahead))
  delete h

  vector(ahead(!ahead)) h
  h(1)=estg11(!ahead-!ssize,7)+(estg11(!ahead-!ssize,8)
    +estg11(!ahead-!ssize,9)) * estg11(!ahead-!ssize,2)
  for !r=2 to ahead(!ahead)
    h(!r) = estg11(!ahead-!ssize,7) + (estg11(!ahead-
      !ssize,8)+estg11(!ahead-!ssize,9)) * h(!r-1)
  next
  volatility(!ahead,2)=sqr(@sum(h)*252/ahead(!ahead))
  delete h

  vector(ahead(!ahead)) h
  h(1)=estg21(!ahead-!ssize,7) + (estg21(!ahead-!ssize,8)
    +estg21(!ahead-!ssize,9))*estg21(!ahead-!ssize,2)
    + estg21(!ahead-!ssize,10)*estg21(!ahead-!ssize,11)
  h(2)=estg21(!ahead-!ssize,7) + (estg21(!ahead-!ssize,8)
    +estg21(!ahead-!ssize,9))*h(1)  + estg21
    (!ahead-!ssize,10)*estg21(!ahead-!ssize,2)
  for !r=3 to ahead(!ahead)
    h(!r) = estg21(!ahead-!ssize,7)
      + (estg21(!ahead-!ssize,8)+estg21(!ahead-!ssize,9))
      *h(!r-1) + estg21(!ahead-!ssize,10)*h(!r-2)
  next
  volatility(!ahead,3)=sqr(@sum(h)*252/ahead(!ahead))
  delete h

  vector(ahead(!ahead)) h
  h(1) = estt01(!ahead-!ssize,7) + (estt01
    (!ahead-!ssize,8) + estt01(!ahead-!ssize,9)*estt01
    (!ahead-!ssize,10)) * estt01(!ahead-!ssize,2)
    for !r=2 to ahead(!ahead)
  h(!r) = estt01(!ahead-!ssize,7) + (estt01
    (!ahead-!ssize,8) + estt01(!ahead-!ssize,9)
    *estt01(!ahead-!ssize,10)) * h(!r-1)
```

```
  next
volatility(!ahead,4)=sqr(@sum(h)*252/ahead(!ahead))
delete h

vector(ahead(!ahead)) h
h(1)=estt11(!ahead-!ssize,7) + (estt11(!ahead-!ssize,8)
  + estt11(!ahead-!ssize,9)*estt11(!ahead-!ssize,11)
  + estt11(!ahead-!ssize,10)) * estt11(!ahead-!ssize,2)
  for !r=2 to ahead(!ahead)
h(!r) = estt11(!ahead-!ssize,7) + (estt11
  (!ahead-!ssize,8) + estt11(!ahead-!ssize,9)*estt11
  (!ahead-!ssize,11) + estt11(!ahead-!ssize,10))
  * h(!r-1)
  next
volatility(!ahead,5)=sqr(@sum(h)*252/ahead(!ahead))
delete h

vector(ahead(!ahead)) h
h(1)=estt21(!ahead-!ssize,7) + (estt21(!ahead-!ssize,8)
  + estt21(!ahead-!ssize,9)*estt21(!ahead-!ssize,12)
  + estt21(!ahead-!ssize,10)) * estt21(!ahead-!ssize,2)
  + (estt21(!ahead-!ssize,11)
  * estt21(!ahead-!ssize,13))
h(2)=estt21(!ahead-!ssize,7) + (estt21(!ahead-!ssize,8)
  + estt21(!ahead-!ssize,9)*estt21(!ahead-!ssize,12)
  + estt21(!ahead-!ssize,10)) * h(1) + (estt21
  (!ahead-!ssize,11) * estt21(!ahead-!ssize,2))
  for !r=3 to ahead(!ahead)
h(!r) = estt21(!ahead-!ssize,7) + (estt21
  (!ahead-!ssize,8) + estt21(!ahead-!ssize,9)
  *estt21(!ahead-!ssize,12) + estt21(!ahead-!ssize,10))
  * h(!r-1) + (estt21(!ahead-!ssize,11) * h(!r-2))
  next
volatility(!ahead,6)=sqr(@sum(h)*252/ahead(!ahead))
delete h

vector(ahead(!ahead)) h
for !r=1 to ahead(!ahead)
  h(!r) = @exp(este01(!ahead-!ssize,7)
    + este01(!ahead-!ssize,8)*sqr(2/!pi))
next
volatility(!ahead,7)=sqr(@sum(h)*252/ahead(!ahead))
delete h

vector(ahead(!ahead)) h
h(1)= @exp(este11(!ahead-!ssize,7) + este11
```

```
          (!ahead-!ssize,8)*sqr(2/!pi) + este11
          (!ahead-!ssize,10)*log(este11(!ahead-!ssize,2)))
      for !r=2 to ahead(!ahead)
        h(!r) = @exp(este11(!ahead-!ssize,7)
          + este11(!ahead-!ssize,8)*sqr(2/!pi) + este11
          (!ahead-!ssize,10)*log(h(!r-1)))
      next
      volatility(!ahead,8)=sqr(@sum(h)*252/ahead(!ahead))
      delete h

      vector(ahead(!ahead)) h
      h(1)= @exp(este21(!ahead-!ssize,7) + este21
          (!ahead-!ssize,8)*sqr(2/!pi) + este21
          (!ahead-!ssize,10)*log(este21(!ahead-!ssize,2))
          + este21(!ahead-!ssize,11)*log(este21
          (!ahead-!ssize,12)))
      h(2)= @exp(este21(!ahead-!ssize,7) + este21
          (!ahead-!ssize,8)*sqr(2/!pi) + este21
          (!ahead-!ssize,10)*log(h(1)) + este21
          (!ahead-!ssize,11)*log(este21(!ahead-!ssize,2)))
      for !r=3 to ahead(!ahead)
        h(!r) = @exp(este21(!ahead-!ssize,7) + este21
          (!ahead-!ssize,8)*sqr(2/!pi) + este21
          (!ahead-!ssize,10)*log(h(!r-1)) + este21
          (!ahead-!ssize,11)*log(h(!r-2)))
      next
      volatility(!ahead,9)=sqr(@sum(h)*252/ahead(!ahead))
      delete h

    endif
  next
  for !j=1 to !models
    for !i=1 to !day
      statusline !j !i
      if K(!i)>0 then
        !d1 = ( log(S(!i)*exp(-q(!i)*t252(!i))/K(!i))
          + (rf(!i)*t365(!i)) + (Volatility(!i,!j)
          ^2*t252(!i)/2) ) / (Volatility(!i,!j)
          *sqr(t252(!i)))
        !d2 = !d1 - (Volatility(!i,!j)*sqr(t252(!i)))
        Calls(!i,!j) = S(!i)*exp(-q(!i)*t252(!i))
          *@cnorm(!d1) - K(!i)*exp(-rf(!i)*t365(!i))
          *@cnorm(!d2)
        Puts(!i,!j) = K(!i)*exp(-rf(!i)*t365(!i))
          *@cnorm(-!d2) - S(!i)*exp(-q(!i)*t252(!i))
```

```
      *@cnorm(-!d1)
      Straddles(!i,!j) = Calls(!i,!j) + Puts(!i,!j)
    else
      Calls(!i,!j) = na
      Puts(!i,!j) = na
      Straddles(!i,!j) = na
    endif
  next
next
save chapter8_sp500_options.partB.wf1
```

- *chapter8_sp500_options.partC.prg*

```
load chapter8_sp500_options.partB.wf1
!models=9
!ssize = 500
smpl @all
series _temp = 1
!day = @obs(_temp)
delete _temp

vector(3) filter
  filter(1)=0
  filter(2)=2
  filter(3)=3.5
vector(2) cost
  cost(1)=0
  cost(2)=2
for !x=1 to 3
  for !p=1 to 2
    matrix(!day,!models) Str_Ret_F{!x}C{!p}
  next
next
for !x=1 to 3
  for !p=1 to 2
    for !j=1 to !models
      for !i=1 to !day
      statusline filter !x cost !p model !j trading day !i
  if Straddles(!i,!j)<> na then
    if Straddles(!i,!j)>Straddle_obs(!i)+filter(!x) then
    Str_Ret_F{!x}C{!p}(!i,!j) = (next_Straddle(!i)-
    Straddle_obs(!i)-Cost(!p))/Straddle_obs(!i)
    else
      if Straddles(!i,!j)<Straddle_obs(!i)-filter(!x) then
        Str_Ret_F{!x}C{!p}(!i,!j) = (-next_Straddle(!i)
```

```
       +Straddle_obs(!i)-Cost(!p))/Straddle_obs(!i)
     else
        Str_Ret_F{!x}C{!p}(!i,!j) = na
     endif
  endif
 else
  Str_Ret_F{!x}C{!p}(!i,!j) = na
 endif
    next
   next
  next
next
save chapter8_sp500_options.partC.wf1
```

- *chapter8_duan's_option_pricing_formula.prg*

```
create chapter8_Duan_Option_Pricing.wf1 u 1 1
smpl @all
'Days to expiration
scalar day=30
'asset price
scalar s=60
'strike price
scalar k=65
'Number of Monte Carlo Simulations
scalar MC_total=20000
'parameter estimates of the ARCH model
scalar c_0=0.00011 'risk-free interest rate
scalar alpha_0=0.00001
scalar alpha=0.2
scalar beta=0.71
scalar c_star=0.01
scalar sigma=0.00027 'current variance estimate

matrix(MC_total,day) random_normal
matrix(MC_total,day) random_variance
matrix(MC_total,day) random_returns
vector(MC_total) price_at_maturity
vector(MC_total) payoff_LRNVR
for !ahead=1 to day
  for !MC=1 to MC_total
    random_normal(!MC,!ahead) = @nrnd
  next
next
```

```
!ahead=1
for !MC=1 to MC_total
    random_variance(!MC,!ahead) = alpha_0+alpha
    *(@sqrt(sigma)*random_normal(!MC,!ahead)
    -c_star*@sqrt(sigma))^2+beta*sigma
next
for !ahead=2 to day
 for !MC=1 to MC_total
    random_variance(!MC,!ahead) = alpha_0
      +alpha*(@sqrt(random_variance(!MC,!ahead-1))
      *random_normal(!MC,!ahead)-c_
      star*@sqrt(random_variance(!MC,!ahead-1)))^2
      +beta*random_variance(!MC,!ahead-1)
 next
next
for !ahead=1 to day
 for !MC=1 to MC_total
    random_returns(!MC,!ahead) = c_0-0.5
      *random_variance(!MC,!ahead)
      +@sqrt(random_variance(!MC,!ahead))
      *random_normal(!MC,!ahead)
 next
next
for !MC=1 to MC_total
    price_at_maturity(!MC) = s*exp(@sum(@rowextract
    (random_returns,!MC)))
    vector(2) exp_payoff
    exp_payoff(1)=price_at_maturity(!MC)-k
    exp_payoff(2)=0
    payoff_LRNVR(!MC)=@max(exp_payoff)
next
scalar ARCH_call_price=exp(-c_0*day)*@sum(payoff_LRNVR)/
    MC_total
scalar ARCH_put_price=ARCH_call_price-s+k*exp(-c_0*day)
```

9

Implied volatility indices and ARCH models

9.1 Implied volatility

Latane and Rendleman (1976) introduced the notion of *implied volatility*. Implied volatility is the standard deviation of the return on the asset, which would have to be fed into a theoretical option pricing formula to yield a theoretical value identical to the price of the option in the marketplace, assuming all other inputs are known.

Consider, for example, the BS option pricing formula in (8.49). The option price is a function of the market price of the underlying asset at time t, $S(t)$, the time to maturity, $T-t$, the exercise price at maturity day, K, the riskless interest rate, rf, the dividend yield, γ, and the average volatility over the life of the option, σ. The values $S(t)$, $T-t$, K, rf and γ are directly observed. In contrast, volatility is an unobserved input. If we equate the observed market price of the option, $C_t^{(T-t)}$, to the pricing formula, i.e. $C_t^{(T-t)} = BS(S(t), \gamma, K, rf, T, \sigma^2)$, we can solve for σ. Thus, the obtained volatility estimate is called *option implied volatility*.

A number of studies characterize implied volatility measures as less informative than volatility estimated from asset returns, because they induce important biases and contain misspecification problems (see, for example, Blair et al., 2001, and the references therein).

In 1993, the Chicago Board of Options Exchange (CBOE) published the first implied volatility index. Implied volatility indices have been considered by market participants as important tools for measuring investors' sentiment and market volatility. The computation of implied volatility indices takes into account the latest advances in financial theory, eliminating measurement errors that had characterized implied volatility measures. The CBOE implied volatility index based on S&P500

stock index options, named VIX, is the most widely known and traded volatility index. On 12 November 2007, the CBOE launched the VXV volatility index which measures the three-month S&P500 index option price implied volatility rather than the one-month implied volatility measured by VIX. Among other implied volatility indices are the CBOE DJIA volatility index (VXD) based on Dow Jones Industrial Average index options, the CBOE NASDAQ100 volatility index (VXN) based on NASDAQ100 index options, the Montréal Exchange volatility index (MVX) based on iShares of the CDN S&P/TSX 60 Fund, the German Futures and Options Exchange[1] volatility index (VDAX) based on DAX30 index options, the French Exchange Market MONEP volatility indices VX1 and VX6 based on CAC40 index options, the Swiss[2] Market volatility index (VSMI) based on all options of the SMI index, the Euronext[3] AEX (VAEX), BEL20 (VBEL) and CAC40 (VCAC) volatility indices.

9.2   The VIX index

In 1993 the CBOE launched the VIX index, considered by market participants to be the world's premier benchmark of stock market volatility. The VIX index reflects the standard deviation implied by a composite hypothetical option that has 30 calendar (about 22 trading) days to expiration. Therefore, it measures the expectation of market participants for the volatility of the next 30 calendar days as conveyed by stock index option prices. It is calculated as 100 times the square root of the expected 30-day variance of the S&P500 rate of return. The variance is annualized and VIX expresses volatility in percentage points. Specifically,

$$VIX_{index,t} = 100\sqrt{\frac{365}{30}} \times (\text{expected 30-day implied standard deviation}), \quad (9.1)$$

where the expected 30-day implied standard deviation is estimated by the forward price of a strip of options expiring in 30 days. Since 30-day options are usually not available, a 30-day expected variance is inter- or extrapolated as a weighted average of the forward prices of option strips in the two nearest-term expiration months. For details of the construction of CBOE volatility indices, see Carr and Madan (1998), Demeterfi et al. (1999), Fleming et al. (1995) and Whaley (1993) as well as the CBOE VIX white paper at http://www.cboe.com/micro/vix/vixwhite.pdf and the CFE tutorial at http://cfe.cboe.com/education/vixprimer/About.aspx. Although the VIX index was introduced in 2003, its daily prices date back and are available to 1990. Historical data are available on the CBOE website.[4]

On 22 September 2003 CBOE announced a new approach to the computation of its volatility index incorporating the latest advances in financial theory and practice,

---

[1] Available at http://deutsche-boerse.com/dbag/dispatch/en/isg/gdb_navigation/market_data_analytics.
[2] Available at http://www.swx.com/trading/products/indices/other_indices/vsmi_indices_en.html.
[3] Available at http://www.euronext.com/editorial/wide/editorial-3955-EN.html.
[4] http://www.cboe.com/vix.

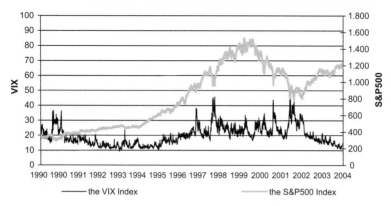

Figure 9.1   Plots of the VIX and the S&P500 indices on a daily basis from 2 January 1990 to 5 April 2005.

renaming the original VIX index as the VXO index. The changes provide a practical standard for trading and hedging volatility.[5]

VXO uses options on the S&P100 index, whereas VIX is based on S&P500 index options. The S&P500 index is the primary US stock market benchmark and underlies the most active stock index derivatives. VXO is calculated from the Black–Scholes option pricing formula, but VIX is independent of any model. VIX derives the market expectation of volatility directly from index option prices rather than on the basis of an algorithm that involves backing implied volatilities out of an option pricing formula. VXO uses eight near-the-money options on the S&P100 index, whereas VIX uses nearly all of the available S&P500 index options. Also, VIX pools information from option prices over a wide range of strike prices, thereby capturing the whole volatility skew.

Figure 9.1 plots the observed values of the VIX index along with the S&P500 closing prices over the period from 2 January 1990 to 22 April 2004. The VIX index was downloaded from the CBOE website[6] ($VIX_{index,t}$). Note that the VIX index has an inverse relationship to the market. As an increase in VIX usually signifies financial turmoil, it is referred to by market participants as a *fear index* or *investor fear gauge.* This inverse relationship between market and volatility is the leverage effect that has already been observed in other measures of volatility as well, e.g. the naive estimate of volatility discussed in Section 1.3, the realized volatility dealt with in Chapter 7, or the conditional volatility estimate from an asymmetric ARCH process.

---

[5] On 26 March 2004 VIX futures were launched on the CBOE Futures Exchange (CFE). For studies on VIX futures, see Brenner et al. (2007) and Zhang and Zhu (2006). VIX options were launched on 24 February 2006.

[6] http://www.cboe.com/micro/vix/historical.aspx.

9.3   The implied volatility index as an explanatory variable

Since, as noted above, implied volatility indices appear to eliminate the biases and misspecification problems that characterized the implied volatility measures, they are regarded as informative variables for forecasting the next day's volatility. Blair et al. (2001) estimate the GJR model treating the VXO index and the realized volatility on the basis of 5-minute S&P100 returns as explanatory variables in the variance specification. They note that the inclusion of the VXO index appears to lead to more accurate forecasts than does the inclusion of either the interday volatility extracted from daily S&P100 log-returns or the intraday realized volatility. Koopman et al. (2005) employ ARCH and stochastic volatility models for the daily S&P100 log-returns, and ARFIMA and unobserved components models for the 5-minute S&P100 prices and point out that the ARCH based volatility forecasts are apparently less accurate than forecasts obtained with the inclusion of the VXO index. Moreover, including realized or implied volatility as an explanatory variable in an ARCH model appears to lead to more accurate volatility forecasts. Note also that the findings of Christensen and Prabhala (1998), Day and Lewis (1992) and Fleming (1998) constitute evidence that implied volatility indices provide incremental information compared to ARCH models. However, Becker et al. (2007) note that the apparent advantage of implied volatility may be due to the shortcomings of the individual ARCH models used in the comparisons. Their findings do not indicate that the VIX index contains additional information in forecasting volatility beyond that provided by model based volatility forecasts. This should by no means be interpreted as implying that the VIX provides less accurate volatility forecasts. On the contrary, Becker et al. (2007) noted that the VIX index provides forecasts of equivalent accuracy to those obtained by a range of volatility forecasting models. However, if the VIX index is compared to any single model, it might be found to provide superior forecasts. If, on the other hand, a wide range of volatility models are considered, the VIX index does not appear to provide any incremental information of future volatility. Corrado and Truong (2007) compared the price range to the CBOE implied volatility indices for the S&P500 index (VIX), the S&P100 index (VXO), the NASDAQ100 index (VXN), and the Dow Jones Industrials index (VXD) and concluded that the implied volatility indices provide volatility forecasts with efficiency similar that of the price range.

To illustrate the above issues, we give an example using VIX index closing prices to obtain one-step-ahead volatility forecasts and we examine whether these offer more insight in comparison to the corresponding forecasts on the basis of some ARCH model taking no account of information from the VIX index.

We compare, in particular, S&P500 log-return one-step-ahead volatility forecasts obtained at time t with and without the incorporation of information on the closing price of the VIX index at time $t-1$. The volatility forecasts considered in the comparison are obtained with the use of an AR(0)-EGARCH(1,2) model (see Chapter 4), but with two different variance specifications – one taking no account of

VIX index based information, and one incorporating such information. In particular, the following two versions of the AR(0)-EGARCH(1,2) model are contested: the original version,

$$y_t = c_0 + \varepsilon_t,$$
$$\varepsilon_t = \sigma_t z_t,$$
$$\log(\sigma_t^2) = a_0 + \sum_{i=1}^{2}(a_i|z_{t-i}| + \gamma_i z_{t-i}) + b_1 \log(\sigma_{t-1}^2), \qquad (9.2)$$
$$z_t \stackrel{i.i.d.}{\sim} N(0,1);$$

and the VIX information enriched version,

$$y_t = c_0 + \varepsilon_t,$$
$$\varepsilon_t = \sigma_t z_t,$$
$$\log(\sigma_t^2) = a_0 + \sum_{i=1}^{2}(a_i|z_{t-i}| + \gamma_i z_{t-i}) + b_1 \log(\sigma_{t-1}^2) + \delta \log(VIX_{t-1}^2), \qquad (9.3)$$
$$z_t \stackrel{i.i.d.}{\sim} N(0,1).$$

The data set used is composed of the observed series of the S&P500 log-returns over the period from 2 January 1990 to 5 April 2005 (a total of $T = 3845$ trading days), and the observed series of the VIX closing prices over the same period.[7] The variable VIX_t^2 is computed as $VIX_t^2 = VIX_{index,t}^2/252$. Table 9.1 provides the estimated parameters of the two models. Note that the VIX coefficient δ differs statistically significantly from zero at any of the usual levels of significance. Both the Schwarz Bayesian criterion, SBC, and the average squared distance between in-sample estimated conditional variance and squared residuals, $\bar{\Psi}_{(SE)}$, exhibit smaller (though not noticeably) values for the model using $\log(VIX_{t-1}^2)$ as an exogenous variable in the conditional variance specification. It appears, therefore, from the in-sample findings that the VIX index cannot be deduced to provide incremental information that is not available in the past values of the conditional variance through which the variance of the original version of the model is specified.

In the remainder of this section, we provide some details on parts of the program named *chapter9.arch_with_vix.prg* (given in the Appendix to this chapter) that refer to various stages of the estimation of model parameters, forecasts and forecasting accuracy. The estimation of the models and of the loss function $\bar{\Psi}_{(SE)}$ is carried out by the first part of the program.

```
load chapter9.VIX.wf1
smpl @all
series vix=vix_index/@sqrt(252)
'_____In-sample_____
equation model1.arch(2,1,e,m=10000,h) sp500_ret c
```

---

[7] In Chapter 4, the S&P500 data set was utilized from 4 April 1988. In the present application the starting date is 2 January 1990, since the VIX index is available from that date.

```
equation model2.arch(2,1,e,m=10000,h) sp500_ret
  c @ log(vix(-1)^2)
vector(2) mse
for !j=1 to 2
  model{!j}.makegarch garch0{!j}
  series model{!j}_cond_stdev = @sqrt(garch0{!j})
  model{!j}.makeresid res0{!j}
  series mse0{!j}
    for !i=1 to 3845
      mse0{!j}(!i)=(garch0{!j}(!i)-res0{!j}(!i)^2)^2
    next
mse(!j)=@mean(mse0{!j})
next
group group1 vix model1_cond_stdev model2_cond_stdev
group1.scat(m)
```

Table 9.1   Parameter estimates of the AR(0)-EGARCH(1,2) model, with and without $\log\left(VIX_{t-1}^2\right)$ as an exogenous variable in the conditional variance equation, for the daily log-returns of the S&P500 index in the period from 2 January 1990 to 5 April 2005

Parameters	AR(0)-EGARCH(1,2)	AR(0)-EGARCH(1,2) with VIX as exogenous
c_0	0.031	0.026
	(2.45)	(2.1)
a_0	−0.089	−0.223
	(−6.87)	(−4.55)
a_1	−0.032	−0.087
	(−0.72)	(−1.89)
γ_1	−0.21	−0.18
	(−7.05)	(−6.21)
a_2	0.14	0.16
	(3.06)	(3.31)
γ_2	0.13	0.02
	(4.30)	(0.56)
b_1	0.98	0.74
	(343)	(12.7)
δ	—	0.29
		(4.30)
SBC	2.609	2.587
$\bar{\Psi}_{(SE)}$	5.364	5.248

Numbers in parentheses represent standard error ratios.

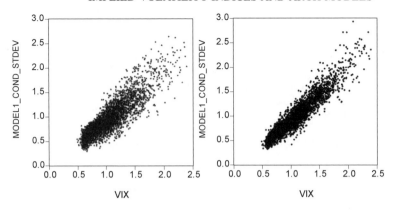

Figure 9.2   Scatterplots of VIX_t with the in-sample conditional standard deviation of AR(0)-EGARCH(1,2) without (model 1) and with the variable $\log\left(VIX_{t-1}^2\right)$ (model 2) as an exogenous variable in the variance equation, for the period from 2 January 1990 to 5 April 2005.

The workfile named *chapter9.VIX.wfl*, which contains the values of the S&P500 log-returns (series *sp500_ret*) and of the $VIX_{index,t}$ closing prices (series *vix_index*) for 3845 trading days, is loaded. The series *vix* contains the values of VIX_t. The equation object *model2* is the AR(0)-EGARCH(1,2) model with $\log\left(VIX_{t-1}^2\right)$ as an exogenous variable in the conditional variance. After the @ symbol, the exogenous variables for the conditional variance specification can be defined. In the series *mse01* and *mse02* the values of $\Psi_{(SE)t(i)} = \left(\hat{\sigma}_{t+1(i)}^2 - \hat{\varepsilon}_{t+1(i)}^2\right)^2$, for $i = 1, 2$, are computed for the equation object *model1* (without VIX_t as an exogenous variable) and *model2* (with VIX_t as an exogenous variable), respectively. The vector *mse* contains the values of $\bar{\Psi}_{(SE)(i)} = T^{-1}\sum_{t=1}^{T}\Psi_{(SE)t(i)}$, for $i = 1, 2$. The command *group* defines a group object that contains the series *vix* (VIX_t), *model1_cond_stdev* ($\hat{\sigma}_{t+1(1)}$) and *model2_cond_stdev* ($\hat{\sigma}_{t+1(2)}$). The command *group1.scat(m)* displays the scatterplots with the *vix* series on the horizontal axis and the series *model1_cond_stdev* and *model2_cond_stdev* on the vertical axis. Figure 9.2 shows the scatterplots obtained.

The second part of the program *chapter9.arch_with_vix.prg* carries out the estimation of the one-step-ahead volatility for the S&P500 equity index based on a rolling sample of $\bar{T} = 3000$. The command lines corresponding to the AR(0)-EGARCH(2,1) model with the VIX as an exogenous variable are given below.

```
'____Out-of-sample____
!ssize = 3000
!length = 3845
matrix(!length-!ssize,3) este12_VIX
for !i = 1 to !length-!ssize
  smpl !i !i+!ssize-1
```

```
equation tempe12_VIX.arch(2,1,e,m=10000,h)
  sp500_ret c @ log(vix(-1)^2)
smpl !i+!ssize !i+!ssize
tempe12_VIX.forecast y_hat se_hat h_hat
este12_VIX(!i,1) = h_hat(!i+!ssize) 'one-day-ahead
  conditional variance forecast
este12_VIX(!i,2) =sp500_ret(!i+!ssize) - y_hat(!i+
  !ssize) 'one-day-ahead residual
este12_VIX(!i,3) = (este12_VIX(!i,1)-
  este12_VIX(!i,2)^2)^2
next
smpl @all
vector(2) mse_out_of_sample
mse_out_of_sample(2) =@mean(@columnextract
  (este12_VIX,3))
```

For each model $\tilde{T} = 845$ (or *!length-!ssize*), forecasts are produced. For each iteration $!i = 1,...,845$, the AR(0)-EGARCH(1,2) model with VIX as an exogenous variable (equation object *tempe12_VIX*) is estimated. The command *tempe12_VIX. forecast y_hat se_hat h_hat* saves the one-step-ahead forecasts of the conditional mean, the standard errors and the conditional variance in the series named *y_hat*, *se_hat* and *h_hat*, respectively. In the third column of matrix *este12_VIX*, the estimated values of $\left(\sigma^2_{t+1|t(2)} - \varepsilon^2_{t+1|t(2)}\right)^2$ are saved. The out-of-sample loss function, $\bar{\Psi}_{(SE)(2)} = \tilde{T}^{-1} \sum_{t=1}^{\tilde{T}} \left(\sigma^2_{t+1|t(2)} - \varepsilon^2_{t+1|t(2)}\right)^2$, for the equation object *tempe12_VIX* is saved in the second cell of the vector *mse_out_of_sample*. The value of the loss function $\bar{\Psi}_{(SE)}$ is estimated as 5.52 for the model without the variable $\log\left(VIX^2_{t-1}\right)$ and 5.54 for the model with the variable $\log\left(VIX^2_{t-1}\right)$ as an exogenous variable in the variance equation. It appears, therefore, that the out-of-sample and the in-sample evaluation do not provide appreciably different results.

It should be noted at this point that testing whether the two models provide forecasts of statistically equivalent forecasting accuracy, and selecting a proxy for the unobserved true volatility are very important. Our two models differ in their values of of $\bar{\Psi}_{(SE)}$, but one needs to test whether the observed difference is statistically significant. Only then can one infer whether the forecasts produced by the two models are of equivalent predictive accuracy. Methods for investigating such issues are presented in Chapter 10.

As concerns the volatility proxy, we used the squared demeaned log-return for measuring the unobserved volatility. However, as noted in Chapter 2, y_t^2 and ε_t^2 are noisy estimators of the unobserved volatility. Figure 9.3 depicts the one-day-ahead conditional variance forecasts of the models and the squared log-returns of the S&P500 index. From this figure we note only minute differences between the forecasts by the two models, but, more interestingly, how noisy the daily volatility proxy y_t^2 is. Using realized volatility as a proxy for the true volatility, Degiannakis and Floros (2010) note, on the basis of the loss function $\bar{\Psi}_{(SE)}$, that

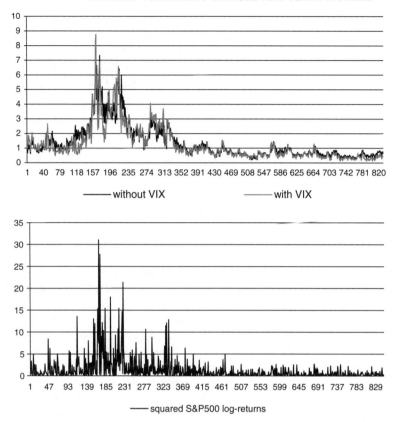

Figure 9.3   One-day-ahead conditional variance forecasts of the AR(0)-EGARCH (1,2) model without and with the variable $log\left(VIX_{t-1}^{2}\right)$ as an exogenous variable in the variance equation, and squared S&P500 log-returns over the period from 26 November 2001 to 5 April 2005.

the inclusion of information based on the VIX index appears to lead to forecasts of greater accuracy. Recalling from Chapter 7 that the realized volatility is a more accurate proxy for the actual variance, one might wonder whether its use in the place of the noisy proxy y_t^2 in our example would lead to different results. The implications of the choice of a volatility proxy in the accuracy of forecasts are illustrated in Chapter 10.

9.4   ARFIMAX model for implied volatility index

Degiannakis (2008b) estimates an ARFIMAX model for forecasting implied volatility with the VIX index as the dependent variable, and the ARCH conditional variance and the intraday realized volatility as exogenous variables. He notes that neither the conditional variance estimated from an ARCH model nor the realized volatility can provide any incremental information in forecasting VIX.

In what follows, we employ a similar approach in the case of the ARFIMAX model of Chapter 7. In particular, we consider the ARFIMAX specification

$$(1-c_1L)(1-L)\tilde{d}\left(\log(VIX_t)-\beta_0-\beta_1y_{t-1}-\delta\log(\hat{\sigma}_{t-1})\right) = (1+d_1L)\varepsilon_t, \varepsilon_t \sim N\left(0, \sigma_\varepsilon^2\right),$$

$$(9.4)$$

with the variable $\log(VIX_t)$ as the dependent variable, and the S&P500 log-returns, y_{t-1}, and the logarithmic transformation of the conditional standard deviation extracted from the AR(0)-EGARCH(2,1) model, $\log(\hat{\sigma}_{t-1})$, as exogenous variables. The model is estimated for both $\delta = 0$ (without any incremental information available from an ARCH standard deviation), and for $\delta \neq 0$ (including any incremental information available from the in-sample conditional standard deviation as estimated by the AR(0)-EGARCH(2,1) model), using the data set of Section 9.3. Thus, one-day-ahead forecasts are computed for $\tilde{T} = 845$ points in time, based on a rolling sample of $\breve{T} = 3000$ trading days. All the necessary data are available in *chapter9.ARFIMAX_for_VIX.xls*, which contains the values of the variables $\log(VIX_t)$, y_{t-1} and $\log(\hat{\sigma}_{t-1})$ for $T = 3845$ trading days. Here, $\hat{\sigma}_{t-1}$ is the in-sample conditional standard deviation of the AR(0)-EGARCH(1,2) model, without the variable VIX_t as an exogenous variable, which has been computed in the previous section. The Ox program *chapter9.ARFIMAX_ for_VIX.ox* (see the Appendix to this chapter) carries out the rolling estimation of the models and saves the results in the Excel files *chapter9.ARFIMAX_for_VIX_ output_exog_arch.xls*, for the case $\delta \neq 0$, and *chapter9.ARFIMAX_for_VIX_ output_without_arch.xls*, for the case $\delta = 0$. Note that the first columns of each of these files contain the rolling estimates of the vectors of coefficients $\left(c_1^{(t)}, \tilde{d}^{(t)}, \beta_0^{(t)}, \beta_1^{(t)}, \delta^{(t)}, d_1^{(t)}\right)$, for $t = \breve{T}+1, \breve{T}+2, \ldots, \breve{T}+\tilde{T}$, while the last column contains the one-day-ahead forecasts of the variable VIX_t, i.e. $VIX_{t+1|t}$. Their penultimate columns refer to the values of VIX_t. So the values of $VIX_{t+1|t}$ and VIX_t are displayed in the same row of each Excel file. For example, in the first row, VIX_t refers to the value of the index at $t = 3000$, while $VIX_{t+1|t}$ refers to the forecast value of VIX_t at $t = 3001$ computed on the basis of information available at $t = 3000$.

The value of the out-of-sample loss function, $\Psi_{(SE)} = \sum_{t=1}^{\tilde{T}}\left(VIX_{t+1|t}-VIX_{t+1}\right)^2$, is equal to 5.427 for the model without the term $\log(\hat{\sigma}_{t-1})$ (corresponding to $\delta = 0$) and 5.430 for the model with the term $\log(\hat{\sigma}_{t-1})$ (corresponding to $\delta \neq 0$), thus implying that the two models provide almost identical VIX_t forecasts. It appears, therefore, that the conditional standard deviation estimated from an ARCH model does not offer any incremental information in forecasting the next day's VIX price. The strong relationship between the index VIX_t and the in-sample conditional standard deviation obtained by the AR(0)-EGARCH(1,2) model is displayed by the scatterplot of Figure 9.4. Note that the scatterplot of $VIX_{index,t}$ against $VIX_{index,t-1}$ (Figure 9.5) displays a much stronger relationship.

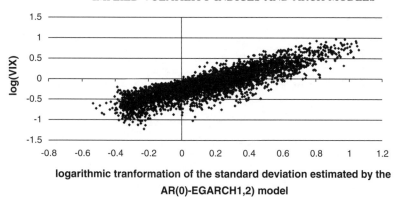

Figure 9.4   Scatterplot of VIX$_t$ against the in-sample conditional standard deviation of model AR(0)-EGARCH(1,2), over the period from 26 November 2001 to 5 April 2005.

In a paper on the influence of ARCH modelling in finance, Engle (2002a) noted that the answer is not simple, but depends upon the statistical approach to the evaluation of forecasts. The literature appears to agree that implied volatilities lead to more accurate forecasts of future volatility than ARCH models do. This might not be plausibly defendable in the case of the VIX index that measures the market's volatility forecast 30 calendar days forward and can thus only loosely be connected to the observed behaviour of the actual S&P500 volatility.

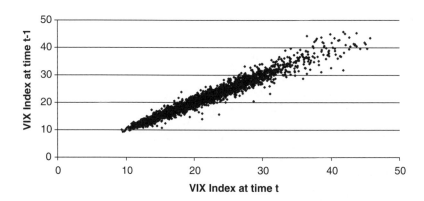

Figure 9.5   Scatterplot of VIX$_{index,t}$ against VIX$_{index,t-1}$, over the period from 26 November 2001 to 5 April 2005.

Appendix

EViews 6

- *chapter9.ARCH_with_VIX.prg*

```
load chapter9.VIX.wf1
smpl @all
series vix=vix_index/@sqrt(252)

'_____In-sample_____
equation model1.arch(2,1,e,m=10000,h) sp500_ret c
equation model2.arch(2,1,e,m=10000,h) sp500_ret c @ log
  (vix(-1)^2)
vector(2) mse
for !j=1 to 2
model{!j}.makegarch garch0{!j}
series model{!j}_cond_stdev = @sqrt(garch0{!j})
model{!j}.makeresid res0{!j}
series mse0{!j}
  for !i=1 to 3845
    mse0{!j}(!i)=(garch0{!j}(!i)-res0{!j}(!i)^2)^2
  next
mse(!j)=@mean(mse0{!j})
next
group group1 vix model1_cond_stdev model2_cond_stdev
group1.scat(m)


'_____Out-of-sample_____
!ssize = 3000
!length = 3845
matrix(!length-!ssize,3) este12_noVIX
matrix(!length-!ssize,3) este12_VIX

for !i = 1 to !length-!ssize
  statusline sample !i
  smpl !i !i+!ssize-1
  equation tempe12_noVIX.arch(2,1,e,m=10000,h) sp500_ret c
  smpl !i+!ssize !i+!ssize
  tempe12_noVIX.forecast y_hat se_hat h_hat
  este12_noVIX(!i,1) = h_hat(!i+!ssize) 'one-day-ahead
    conditional variance forecast
  este12_noVIX(!i,2) = sp500_ret(!i+!ssize) - y_hat(!i+
    !ssize) 'one-day-ahead residual
  este12_noVIX(!i,3) = (este12_noVIX(!i,1)-este12_noVIX
    (!i,2)^2)^2
```

```
  smpl !i !i+!ssize-1
  equation tempe12_VIX.arch(2,1,e,m=10000,h) sp500_ret c
    @ log(vix(-1)^2)
  smpl !i+!ssize !i+!ssize
  tempe12_VIX.forecast y_hat se_hat h_hat
  este12_VIX(!i,1) = h_hat(!i+!ssize) 'one-day-ahead
    conditional variance forecast
  este12_VIX(!i,2) =sp500_ret(!i+!ssize) - y_hat(!i+!ssize) '
    one-day-ahead residual
  este12_VIX(!i,3) = (este12_VIX(!i,1) -este12_VIX(!i,2)^2)^2
next
smpl @all
vector(2) mse_out_of_sample
mse_out_of_sample(1)=@mean(@columnextract(este12_noVIX,3))
mse_out_of_sample(2)=@mean(@columnextract(este12_VIX,3))

save chapter9.VIX.output.wf1
```

Ox Metrics

- *chapter9.ARFIMAX_for_VIX.ox*

```
#include <oxstd.h>
#include <oxfloat.h>
#import <packages/arfima104/arfima>
main()
{
  decl rolling_sample=2999; //It's the constant rolling
    sample minus 1
  decl ii;
  decl VIX_output;
////ARFIMAX model for VIX with ARCH standard deviation as
  exogenous variable////
VIX_output = zeros(845,8);
for (ii=1; ii<845+1;++ii)
  {
    decl arfima;
    arfima = new Arfima();
    arfima.Load("Chapter9.ARFIMAX_for_VIX.xls");
    arfima.Select(Y_VAR, {"logVIX",0,0 } );
    arfima.Deterministic(FALSE);
    arfima.Select(X_VAR, {"Constant", 0, 0 } );
    arfima.Select(X_VAR, {"SP500_logreturns_lag", 0, 0 } );
    arfima.Select(X_VAR, {"logsigma_lag_ARCH", 0, 0 } );
    arfima.ARMA(1,1);
```

```
arfima.SetMethod(M_MAXLIK);//use the exact maximum
    likelihood
arfima.SetSelSample(ii, 1, ii+rolling_sample, 1);
arfima.Estimate();
decl prmter=arfima.GetPar();
decl logVIX=arfima.GetGroupLag(Y_VAR, 0, 0);
decl fore=arfima.Forecast(1);
decl unbRVfore=exp(fore+0.5*arfima.GetSigma2());
VIX_output[ii-1][0]=prmter[0][0];
VIX_output[ii-1][1]=prmter[1][0];
VIX_output[ii-1][2]=prmter[2][0];
VIX_output[ii-1][3]=prmter[3][0];
VIX_output[ii-1][4]=prmter[4][0];
VIX_output[ii-1][5]=prmter[5][0];
VIX_output[ii-1][6]=exp(logVIX[rolling_sample][0]);
    // VIX at time t
VIX_output[ii-1][7]=unbRVfore[0][0];//unbiased out-
    of-sample of VIX[t+1|t]
delete arfima;
}
savemat("Chapter9.ARFIMAX_for_VIX_output_exog_arch.
    xls",VIX_output);
////ARFIMAX model for VIX without ARCH standard deviation as
    exogenous variable////
VIX_output = zeros(845,7);
for (ii=1; ii<845+1;++ii)
    {
    decl arfima;
    arfima = new Arfima();
    arfima.Load("Chapter9.ARFIMAX_for_VIX.xls");
    arfima.Select(Y_VAR, {"logVIX",0,0 } );
    arfima.Deterministic(FALSE);
    arfima.Select(X_VAR, {"Constant", 0, 0 } );
    arfima.Select(X_VAR, {"SP500_logreturns_lag", 0, 0 } );
    arfima.ARMA(1,1);
    arfima.SetMethod(M_MAXLIK);//use the exact maximum
        likelihood
    arfima.SetSelSample(ii, 1, ii+rolling_sample, 1);
    arfima.Estimate();
    decl prmter=arfima.GetPar();
    decl logVIX=arfima.GetGroupLag(Y_VAR, 0, 0);
    decl fore=arfima.Forecast(1);
    decl unbRVfore=exp(fore+0.5*arfima.GetSigma2());
    VIX_output[ii-1][0]=prmter[0][0];
    VIX_output[ii-1][1]=prmter[1][0];
```

```
      VIX_output[ii-1][2]=prmter[2][0];
      VIX_output[ii-1][3]=prmter[3][0];
      VIX_output[ii-1][4]=prmter[4][0];
      VIX_output[ii-1][5]=exp(logVIX[rolling_sample][0]);
        // VIX at time t
      VIX_output[ii-1][6]=unbRVfore[0][0];//unbiased out-
        of-sample of VIX[t+1|t]
      delete arfima;
  }
savemat("Chapter9.ARFIMAX_for_VIX_output_without_arch.
  xls",VIX_output);
}
```

10

ARCH model evaluation and selection

This chapter gives a concise presentation of aspects of ARCH model evaluation and selection among a set of competitors. At the same time, it provides a series of handy programs for exploring the various theoretical aspects of model evaluation and selection.

The evaluation of models is viewed in terms of information criteria in Section 10.1.1 (Akaike, Schwarz Bayesian, Hannan and Quinn, and Shibata criteria), and statistical loss functions in Section 10.1.2 (squared error, absolute error, heteroscedasticity-adjusted squared error, heteroscedasticity-adjusted absolute error, logarithmic error and Gaussian likelihood loss functions). Loss functions that are dependent upon the aims of a specific usage/application are discussed in Section 10.1.6. A series of EViews programs that simulate data from an ARCH process, estimate several ARCH models, including the one used to generate the data, and then assess the ability of some of the mentioned model evaluation methods in 'picking' the true data-generating process is given in Section 10.1.4.

Simulated evidence indicates that loss function distributions are in general asymmetric with extreme outliers, thus suggesting the use of median values instead of mean values for the loss functions (Section 10.1.2). In Section 10.3.3 the use of the mean or the median value of the loss functions is investigated.

Sections 10.1.2 and 10.1.3 deal with two important aspects in the evaluation of volatility models. Since the actual variance is not observed, one cannot compare it directly to a forecast. Thus, a proxy measure for the actual volatility is required. As noted in Section 10.1.2, the price range appears to be more informative as a proxy than the squared daily returns and less informative than the realized volatility. Moreover, the price range proxy is apparently superior to the realized variance when it is constructed on the basis of a small number of intraday observations. These

ARCH Models for Financial Applications   Evdokia Xekalaki and Stavros Degiannakis
© 2010 John Wiley & Sons, Ltd

findings are of practical value in model evaluation, where the choice of an appropriate proxy for actual variance is crucial. A noisy volatility proxy can lead to an evaluation appreciably different from what would be obtained if the true volatility were used. An additional question that arises in evaluating candidate models for volatility prediction is that of consistent ranking of competing models in their predictive ability. The combination of a non-consistent ranking method and a noisy volatility proxy could lead one to favour an inappropriate model. This question is discussed in Section 10.1.3. It is noted that the mean squared error loss function ensures the sufficient condition for consistent ranking.

Section 10.2 looks at model comparisons in terms of various statistical tests – in particular, the Diebold and Mariano, Harvey et al., Granger and Newbold, White, Hansen's superior predictive ability (SPA), and Degiannakis and Xekalaki's standardized prediction error criterion (SPEC) tests for testing whether a model provides statistically significantly more accurate volatility forecasts than its competitors. Hansen's (2005) SPA test appears to be more powerful and less sensitive to the inclusion of poor and irrelevant competitors. In Section 10.4 the SPA criterion is used to compare the forecasting performance of a model against its competitors, in terms of a loss function that is based on the measure of expected shortfalls. Degiannakis and Xekalaki's (2005) SPEC has the advantage that its loss function is of a known distributional form.

In Section 10.3.1 an application is provided in which the trader using the ARCH models picked by the SPEC model selection method achieves higher cumulative returns.

10.1   Evaluation of ARCH models

The evaluation of an economic forecast has always been of great interest. Numerous forecast evaluation criteria have been proposed in the literature. However, none is generally acceptable. Cox (1930), for example, assigned a grade for definiteness (from 0 to +1 in steps of $\frac{1}{4}$) and a grade for correctness (from −1 to +1, again in steps of $\frac{1}{4}$) for the monthly forecasts of the major indices of business activity. The grade given for adequacy of a monthly forecast was obtained by multiplying the score for correctness by the score for definiteness. The interested reader is referred to Hamilton (1991) for an annotated bibliography of 316 papers relevant to model validation, with an emphasis on work of interest to statisticians.

Earlier in this book, we too made use of some of the commonly used evaluation criteria. In particular, in Sections 2.3 and 2.6, we applied the SBC and AIC information criteria to select the model best fitting the data. Towards the same aim, the average squared distance between the in-sample estimated conditional variance and the squared residuals was considered as a criterion in Section 2.3. We also used criteria associated with the predictive ability of the models. So, in Section 4.1 we employed the average squared distance between the one-day-ahead conditional variance and the squared log-returns, while in Sections 8.3 and 8.5.4 we made inference based on one-step-ahead value-at-risk and expected shortfall forecasts as well as on τ-day-ahead straddle price forecasts. All of these criteria are key tools in model evaluation techniques. There are two general types of evaluation criteria or techniques in the literature according to whether these are associated with the descriptive performance of a model (in-sample evaluation) or with the predictive performance of the model (out-of-sample evaluation).

10.1.1 Model evaluation viewed in terms of information criteria

Most of the methods used in the literature for model selection are based on evaluating the ability of the models to describe the data. The most frequently utilized in-sample methods of model evaluation are the information criteria. Information criteria based on estimation of the Kullback and Leibler (1951) discrepancy have been applied by researchers to determine the appropriate number of lags in time series models. Standard model selection criteria, such as Akaike's (1973) information criterion (AIC), Schwarz's (1978) Bayesian criterion (SBC), the Hannan and Quinn (1979) criterion (HQ) and the Shibata (1980) criterion (SH), have also been used in the literature on ARCH models for selecting the appropriate model specification, despite the fact that their statistical properties in the ARCH context are unknown. These criteria are defined as follows:

$$SBC = -2T^{-1}L_T\left(\{y_t\}; \hat{\psi}^{(T)}\right) + \breve{\psi}T^{-1}\log(T), \tag{10.1}$$

$$AIC = -2T^{-1}L_T\left(\{y_t\}; \hat{\psi}^{(T)}\right) + 2T^{-1}\breve{\psi}, \tag{10.2}$$

$$HQ = -2T^{-1}L_T\left(\{y_t\}; \hat{\psi}^{(T)}\right) + 2\breve{\psi}T^{-1}\log(\log(T)), \tag{10.3}$$

$$SH = -2T^{-1}L_T\left(\{y_t\}; \hat{\psi}^{(T)}\right) + \log\left(\frac{T + 2\breve{\psi}}{T}\right). \tag{10.4}$$

where $L_T(.)$ is the maximized value of the model log-likelihood function, $\hat{\psi}^{(T)}$ is the maximum likelihood estimator of the parameter vector ψ based on a sample of size T, and $\breve{\psi}$ denotes the dimension of ψ. Hecq (1996), based on a set of Monte Carlo simulations, demonstrated how the information criteria behave in the presence of ARCH effects. In small-sample situations, the SBC is in general the best-performing criterion. Brooks and Burke (2003) investigated the applicability of information criteria in ARCH models with autoregressive conditional mean. They provided evidence that information criteria do not lead to an improvement in the conditional mean forecasting compared to models with fixed numbers of parameters, but they do lead to modestly improved volatility forecasts compared to fixed-order models.

10.1.2 Model evaluation viewed in terms of statistical loss functions

The construction of loss functions is mainly used in model evaluation. When we focus on estimation of means, the loss function of choice is typically the mean squared error. Because of high non-linearity in volatility models there are a variety of statistical functions that measure the distance between actual and estimated volatility. Denoting the forecasting variance over a τ-day period measured at day t by $\sigma_{t+1|t}^{2(\tau)}$, and the actual, but unobservable, variance over the same period τ by $\sigma_{t+1}^{2(\tau)}$, the following out-of-sample loss functions, $\Psi_t^{(\tau)}$, have been considered in literature: the squared error (SE),

$$\Psi_{(SE)t+1}^{(\tau)} = \left(\sigma_{t+1|t}^{2(\tau)} - \sigma_{t+1}^{2(\tau)} \right)^2; \tag{10.5}$$

the absolute error (AE),

$$\Psi_{(AE)t+1}^{(\tau)} = \left| \sigma_{t+1|t}^{2(\tau)} - \sigma_{t+1}^{2(\tau)} \right|; \tag{10.6}$$

the heteroscedasticity-adjusted squared error (HASE),

$$\Psi_{(HASE)t+1}^{(\tau)} = \left(1 - \frac{\sigma_{t+1}^{2(\tau)}}{\sigma_{t+1|t}^{2(\tau)}} \right)^2; \tag{10.7}$$

the heteroscedasticity-adjusted absolute error (HAAE),

$$\Psi_{(HAAE)t+1}^{(\tau)} = \left| 1 - \frac{\sigma_{t+1}^{2(\tau)}}{\sigma_{t+1|t}^{2(\tau)}} \right|; \tag{10.8}$$

the logarithmic error (LE),

$$\Psi_{(LE)t+1}^{(\tau)} = \left(\log \left(\frac{\sigma_{t+1}^{2(\tau)}}{\sigma_{t+1|t}^{2(\tau)}} \right) \right)^2; \tag{10.9}$$

and the Gaussian likelihood loss function (GLLS),

$$\Psi_{(GLLS)t+1}^{(\tau)} = \log \left(\sigma_{t+1|t}^{2(\tau)} \right) + \frac{\sigma_{t+1}^{2(\tau)}}{\sigma_{t+1|t}^{2(\tau)}}. \tag{10.10}$$

The SE and AE have been widely used in the literature (see Heynen and Kat, 1994; West and Cho, 1995; Yu, 2002; Brooks and Persand, 2003b). The HASE and HAAE functions were considered by Walsh and Tsou (1998) and Andersen et al. (1999a), the LE function was utilized by Pagan and Schwert (1990) and Saez (1997), while the

GLLS was mentioned in Bollerslev et al. (1994). Pagan and Schwert (1990) used statistical criteria to compare the in-sample and out-of-sample performance of parametric and non-parametric ARCH models. Heynen and Kat (1994) investigated the predictive performance of ARCH and stochastic volatility models and Hol and Koopman (2000) compared the predictive ability of stochastic volatility and implied volatility models. Andersen et al. (1999a) applied heteroscedasticity-adjusted statistics to examine the forecasting performance of intraday returns. See also Bollerslev and Ghysels (1996), Degiannakis and Xekalaki (2007a) and Angelidis and Degiannakis (2008). The loss functions can be also computed for the τ-day-ahead standard deviation, $\sigma_{t+1}^{(\tau)}$, instead of the variance, $\sigma_{t+1}^{2(\tau)}$, making the selection of the most accurate model even more complicated.

Obviously, the same loss functions can be used in evaluating the in-sample accuracy of volatility estimation. For the in-sample loss functions, we just replace $\sigma_{t+1|t}^{2(\tau)}$ (the variance forecast for a period of τ days, from $t+1$ to $t+\tau$, given the information available at time t) by $\hat{\sigma}_{t+1}^{2(\tau)}$ (the in-sample conditional variance for a period of τ days, from $t+1$ to $t+\tau$, based on the entire available data set T).

The general notation of a loss function, which measures the distance between model i's forecasting variance over a τ-day period given information available on day t, $\sigma_{t+1|t}^{2(\tau)}$, and the actual variance over the same period τ, $\sigma_{t+1}^{2(\tau)}$, is $\Psi_{(.)t+1(i)}^{(\tau)}$. For example, the SE loss function for the forecasting variance of model i over a 10-day period is denoted by $\Psi_{(SE)t+1(i)}^{(10)}$. For notational convenience, when it is obvious from the context, the superscript (τ) for $\tau = 1$ will be supressed. When no reference to a specific loss function (or model) needs to be made, the parentheses to the left or right of the time index in the subscript will also be supressed. The average of a loss function will be denoted by $\bar{\Psi}_{(.)(i)}^{(\tau)} = \tilde{T}^{-1} \sum_{t=1}^{\tilde{T}} \Psi_{(.)t(i)}^{(\tau)}$. For example, the SE loss function for the forecasting variance of model i over a one-day period is denoted by $\bar{\Psi}_{(SE)(i)} \equiv \bar{\Psi}_{(SE)(i)}^{(1)}$. We have already applied $\bar{\Psi}_{(SE)}$ for in-sample evaluation in Section 2.3 and for out-of-sample evaluation in Sections 4.1, 4.2 and 6.1. The actual variance $\sigma_{t+1}^{2(\tau)}$ was represented by the squared residuals or the squared log-returns.

10.1.2.1 Proxy measures for the actual variance

The fact that the actual variance for a period of τ days, from $t+1$ to $t+\tau$, $\sigma_{t+1}^{2(\tau)}$, is not observed does not allow us to compare forecasts directly to $\sigma_{t+1}^{2(\tau)}$. Thus, we require a proxy measure for the actual volatility, $\tilde{\sigma}_{t+1}^{2(\tau)}$. As mentioned in Chapter 2, the daily squared demeaned log-returns (i.e. the squared residuals of the estimated model) are the most commonly used proxy measures for the actual daily volatility, since, if the estimated model is correctly specified, then the squared residual constitutes an unbiased estimator of the daily variance. Recall that in Sections 4.1, 4.2 and 6.1 the squared log-returns were used as proxy measures for the actual volatility. In general, for a period of τ days, an unbiased proxy measure for the unobserved variance can be computed as

$$\tilde{\sigma}_{t+1}^{2(\tau)} = \sum_{i=1}^{\tau} \hat{\varepsilon}_{t+i}^2 = \sum_{i=1}^{\tau} \left(y_{t+i} - \hat{y}_{t+i} \right)^2. \tag{10.11}$$

It is also customary to proxy the actual variance by the sum of squared daily returns or the variance of the daily returns given by

$$\tilde{\sigma}_{t+1}^{2(\tau)} = \sum_{i=1}^{\tau} y_{t+i}^2, \tag{10.12}$$

and

$$\tilde{\sigma}_{t+1}^{2(\tau)} = \sum_{i=1}^{\tau} \left(y_{t+i} - \bar{y}_{t+1}^{(\tau)}\right)^2, \tag{10.13}$$

respectively, where $\bar{y}_{t+1}^{(\tau)} = \tau^{-1} \sum_{i=1}^{\tau} y_{t+i}$ is the average of log-returns.[1] However, as noted in Chapter 2, these proxies are noisy since the daily squared residual is the product of the actual variance multiplied by the square of a randomly distributed process, $\hat{\varepsilon}_{t+1}^2 = \sigma_{t+1}^{2(1)} z_{t+1}^2$. A more accurate proxy for measuring the actual variance is the realized volatility, discussed in Chapter 7, as a proxy for the integrated volatility:

$$\tilde{\sigma}_{t+1}^{2(\tau)} \equiv \sigma_{t+1}^{2(RV)(\tau)} = \sum_{i=1}^{\tau} \sum_{j=1}^{m-1} y_{((j+1)/m),t+i}^2, \tag{10.14}$$

where $y_{((j+1)/m),t} = \log(P_{((j+1)/m),t}) - \log(P_{(j/m),t})$. For $m = 1$, i.e. if one observation is available per day, naturally (10.14) reduces to (10.12). French et al. (1987), Schwert (1989a, 1990) and Schwert and Seguin (1990) considered using high-frequency data to compute volatility at a lower frequency to compute the monthly standard deviation by the sample standard deviation of the daily log-returns. Such computations can be made in terms of (10.14). So, for $\tau = 1$ and $m = 22$, the one-month variance with 22 intramonth (daily) observations is computed, while for $\tau = 22$ and $m = 1$ the 22 trading days variance with one intraday observation is obtained. Recall also the discussion, in Section 7.1, of techniques due to Martens (2002), Engle and Sun (2005), Zhang et al. (2005) and Martens and van Dijk (2007) for measuring realized volatility more accurately.

Even when detailed intraday data sets are not available, intraday high and low prices are recorded in business newspapers and technical analysis reports. Therefore, the price range in equation (2.117) can be treated as a proxy for measuring the actual daily variance. Note that the price range is more informative than the squared daily returns and less informative than the realized volatility. Assume that the instantaneous logarithmic price, $\log(P(t))$, evolves as a continuous random walk with a diffusion constant σ^2, as given by equation (7.1). Parkinson (1980) noted that

---

[1] If the true data-generating model does not have a conditional mean specification, e.g. $y_t = \sigma_t z_t$, $\sigma_t = g(\theta|I_{t-1})$, $z_t \stackrel{i.i.d.}{\sim} f(w; 0, 1)$, then (10.11) coincides with (10.12). If the true data-generating model has an AR(0) conditional mean without any exogenous variables, e.g. $y_t = c_0 + \sigma_t z_t$, $\sigma_t = g(\theta|I_{t-1})$, $z_t \stackrel{i.i.d.}{\sim} f(w; 0, 1)$, then (10.11) is equivalent to (10.13).

under this assumption,

$$E\left(\log\left(\frac{\max(P_t)}{\min(P_t)}\right)\right) = \sqrt{\frac{8}{\pi}}\sigma(t) \quad \text{and} \quad E\left(\left(\frac{\max(P_t)}{\min(P_t)}\right)^2\right) = 4\log(2)\sigma^2(t).$$

Therefore, the variance proxy for a period of τ days that is based on the price range is computed as in equation (2.117):

$$\tilde{\sigma}_{t+1}^{2(\tau)} \equiv Range_{t+1}^{(\tau)} = \frac{1}{4\log2}\left(\log\left(\frac{\max(P_\tau; t+1 \leq \tau \leq t+\tau)}{\min(P_\tau; t+1 \leq \tau \leq t+\tau)}\right)\right)^2. \qquad (10.15)$$

10.1.2.2 Median values of loss functions

In the majority of studies on model evaluation for mean forecasts, the average value of some loss function is used. In the volatility forecasting literature, the use of the median value of the loss function considered has been proposed. In what follows, we will generate data from an AR(1)-GARCH(1,1) model as defined by

$$\begin{aligned}
y_t &= 0.001 + 0.2y_{t-1} + \varepsilon_t, \\
\varepsilon_t &= \sigma_t z_t, \\
\sigma_t^2 &= 0.0001 + 0.12\varepsilon_{t-1}^2 + 0.8\sigma_{t-1}^2 \\
z_t &\sim N(0,1).
\end{aligned} \qquad (10.16)$$

The program named *chapter10.sim_ar1g11_loss_functions.prg* carries out the necessary computations.[2] The simulated series contains 4000 values. Using a rolling sample of 3000 observations, the one-step-ahead residuals and conditional variances are estimated for the remaining 1000 points in time. The loss functions given by (10.5)–(10.10) are computed on the basis of (10.11) and (10.12) as proxy measures for the actual variance. The results are qualitatively similar for $\tilde{\sigma}_{t+1}^{2(1)} = (y_{t+1}-\hat{y}_{t+1})^2$ and $\tilde{\sigma}_{t+1}^{2(1)} = y_{t+1}^2$. Table 10.1 provides the mean and median values of the loss functions. Figure 10.1 provides graphs of the histograms of the AE, HASE, LE and GLLS loss functions for $\tilde{\sigma}_{t+1}^{2(1)} = y_{t+1}^2$. As can be seen from the graphs, the distributions of the loss functions are asymmetric with extreme outliers, suggesting the computation of the median instead of the mean of the evaluation criteria. The loss functions AE and HAAE that are based on absolute differences have less asymmetric distributions, while the HASE and LE loss functions are highly asymmetrically distributed.

The program named *chapter10.sim_ar1_loss_functions.prg* carries out the computation of the loss functions associated with the conditional mean prediction of a first-order autoregressive model with constant variance:

$$\begin{aligned}
y_t &= 0.1 + 0.6y_{t-1} + \varepsilon_t, \\
\varepsilon_t &\sim N(0,1).
\end{aligned} \qquad (10.17)$$

---

[2] The parameters *!c0*, *!c1*, *!a0*, *!a1* and *!b1* in the program can be set at different values by the reader to investigate whether the results hold for any set of appropriate parameter values. Recall that each time a program simulating series is run, the series produced are not the same.

Table 10.1 Mean and median values of the loss functions associated with an AR(1)-GARCH(1,1) simulated model

Loss function	Proxy of actual variance			
	$\tilde{\sigma}^{2(1)}_{t+1} = y^2_{t+1}$		$\tilde{\sigma}^{2(1)}_{t+1} = \left(y_{t+1} - \hat{y}_{t+1}\right)^2$	
	Mean	Median	Mean	Median
$\Psi_{(SE)t+1}$	4.1E-06	7.3E-07	3.5E-06	7.1E-07
$\Psi_{(AE)t+1}$	0.0012	0.0009	0.0012	0.0008
$\Psi_{(HASE)t+1}$	2.202	0.648	2.041	0.645
$\Psi_{(HAAE)t+1}$	0.993	0.805	0.976	0.803
$\Psi_{(LE)t+1}$	7.582	1.291	6.568	1.307
$\Psi_{(GLLS)t+1}$	−5.717	−6.200	−5.729	−6.181

Table 10.2 presents the mean and the median values of loss functions modified for the evaluation of the conditional mean forecasts. Figure 10.2 provides histograms of these loss functions. The loss functions are asymmetric even for conditional mean forecasts. However, the magnitude of the asymmetry in the distribution of the loss functions is much less than in the distribution of the loss functions of variance forecasts.

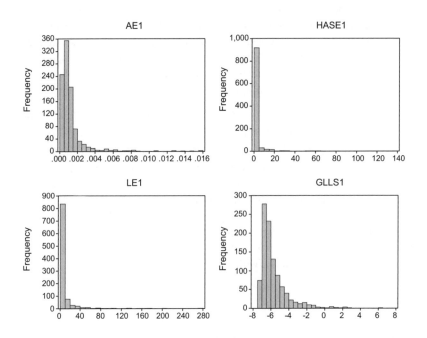

Figure 10.1   Frequency histograms of the loss functions associated with an AR(1)-GARCH(1,1) simulated process, with $\tilde{\sigma}^{2(1)}_{t+1} = y^2_{t+1}$ as proxy of actual variance.

Table 10.2 Mean and median values of the loss functions associated with an AR(1) simulated model with constant variance

Loss function	Mean	Median
SE: $\left(y_{t+1\mid t} - y_{t+1}\right)^2$	1.001	0.465
AE: $\left\lvert y_{t+1\mid t} - y_{t+1}\right\rvert$	0.799	0.682
LE: $\left(\log\left(y_{t+1}^2/y_{t+1\mid t}^2\right)\right)^2$	10.074	2.893
Percentage AE: $\left\lvert y_{t+1}^{-1}\right\rvert\left\lvert y_{t+1\mid t} - y_{t+1}\right\rvert$	4.329	0.783

The program named *chapter10.sim_ar1g11_loss_functions_b.prg* carries out the computation of the theoretical values of the loss functions,

$$\tilde{\sigma}_{t+1}^{2(1)} \equiv \varepsilon_{t+1}^2 = \left(y_{t+1} - 0.001 - 0.2y_t\right)^2,$$
$$\sigma_{t+1\mid t}^{2(1)} \equiv \sigma_{t+1}^2 = 0.0001 + 0.12\varepsilon_t^2 + 0.8\sigma_t^2,$$
$$\sigma_1^2 = 0.0001/(1 - 0.12 - 0.8),$$
$$\varepsilon_1^2 = \sigma_1^2 z_1^2,$$

(10.18)

for $z_t \sim N(0,1)$. From Table 10.3, one may note that the theoretical values of the loss functions are very close to the estimated values in Table 10.1. Therefore, the asymmetry is observed in theoretically computed loss functions as well as in the

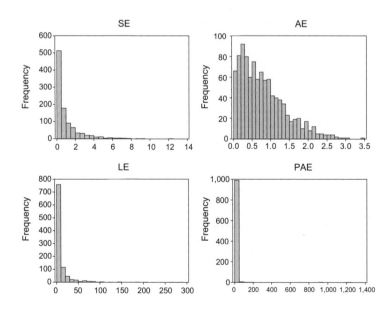

Figure 10.2 Frequency histograms of the loss functions associated with an AR(1) simulated process with constant variance.

Table 10.3 Mean and median values of the theoretical values of the loss functions associated with an AR(1)-GARCH(1,1) model

Loss function	Proxy of actual variance $\tilde{\sigma}_{t+1}^{2(1)} = (y_{t+1} - 0.001 - 0.2y_t)^2$	
	Mean	Median
$\Psi_{(SE)t+1}$	4.5E-06	7.9E-07
$\Psi_{(AE)t+1}$	0.0013	0.0009
$\Psi_{(HASE)t+1}$	2.063	0.695
$\Psi_{(HAAE)t+1}$	0.986	0.834
$\Psi_{(LE)t+1}$	6.683	1.315
$\Psi_{(GLLS)t+1}$	-5.670	-6.112

loss functions that are based on volatility forecasts. Moreover, the median values of Table 10.1 are slightly closer to those of Table 10.3 than the mean values are. Each time a program is run, we get different results as the processes are simulated. Try to run the programs several times and check that the results are robust.

The Mincer–Zarnowitz regression approach
Mincer and Zarnowitz (1969) introduced a regression approach as an alternative method to using loss functions for evaluating volatility forecasts. For the one-step-ahead forecasts $y_{t+1|t}$ and a loss function Ψ_{t+1}, we would like the expectation of the forecast error conditional on the most recent available information to be zero (i.e. the forecast error to be perfectly related to the forecast). Thus

$$E\left(\frac{\partial \Psi_{t+1}}{\partial y_{t+1|t}} | I_t\right) = 0, \tag{10.19}$$

and the parameters β_0, β_1 of the regression model

$$\frac{\partial \Psi_{t+1}}{\partial y_{t+1|t}} = \beta_0 + \beta_1 y_{t+1|t} + \varepsilon_{t+1}^*, \quad \varepsilon_{t+1}^* \sim N(0, \sigma^2), \tag{10.20}$$

must be equal to zero. The Mincer–Zarnowitz (MZ) evaluation method is based on testing the null hypothesis $H_0 : \beta_0 = \beta_1 = 0$. When volatility is the forecasting object, the MZ regression model evaluates volatility forecasts by regressing the proxy for the true, but unobservable, variance on the model forecasts increased by a constant. In the case of one-step-ahead volatility forecasts and the squared error loss function in (10.5), we would like

$$E\left(\frac{\partial \Psi_{(SE)t+1}}{\partial \sigma_{t+1|t}^2} | I_t\right) = 0 \Rightarrow E\left(\sigma_{t+1|t}^2\right) = \sigma_{t+1}^2.$$

The MZ forecast evaluation regression model would then be $-(\sigma^2_{t+1|t} - \sigma^2_{t+1}) = \beta_0 + \beta_1 \sigma^2_{t+1|t} + \varepsilon^*_{t+1}$ or, equivalently,

$$\tilde{\sigma}^2_{t+1} = \beta_0 + (\beta_1 + 1)\sigma^2_{t+1|t} + \varepsilon^*_{t+1}. \tag{10.21}$$

The null hypothesis to be tested would again be $H_0 : \beta_0 = \beta_1 = 0$ against the alternative H_1 : at least one of β_0, β_1 has a non-zero value. Note that the regression coefficients should be estimated taking into consideration the White heterosce-dastic consistent (HC) standard errors (White, 1980). In EViews, the HC variance–covariance matrix is straightforwardly computed. In the *Equation Estimation* dialogue box, select the *Options* tab, check the box labelled *Heteroskedasticity Consistent Covariance* and press the *White* radio button. In the case of the evaluation of multi-step-ahead volatility forecasts, the MZ regression model becomes

$$\tilde{\sigma}^{2(\tau)}_{t+1} = \beta_0 + (\beta_1 + 1)\sigma^{2(\tau)}_{t+1|t} + \varepsilon^*_{t+1}. \tag{10.22}$$

For τ-step-ahead forecasts, the parameters should be estimated taking into conside-ration the Newey–West heteroscedastic and autocorrelated consistent (HAC) standard errors (Newey and West, 1987), since the overlapping nature of forecasts induces autocorrelation. In EViews, the HAC variance–covariance matrix is computed if we check the box labelled *Heteroskedasticity Consistent Covariance* and press the *Newey-West* radio button.

Modifications of the regression based evaluation technique that have been proposed in the literature amount to transformations of the variances. Pagan and Schwert (1990) proposed the logarithmic transformation of $\tilde{\sigma}^{2(\tau)}_{t+1}$ and $\sigma^{2(\tau)}_{t+1|t}$ in order to decrease the sensitivity to extreme values of the proxy variable:

$$\log\left(\tilde{\sigma}^{2(\tau)}_{t+1}\right) = \beta_0 + (\beta_1 + 1)\log\left(\sigma^{2(\tau)}_{t+1|t}\right) + \varepsilon^*_{t+1}. \tag{10.23}$$

An alternative modification, considered by Jorion (1995) among others, amounts to substituting the variances by the corresponding standard deviations:

$$\tilde{\sigma}^{(\tau)}_{t+1} = \beta_0 + (\beta_1 + 1)\sigma^{(\tau)}_{t+1|t} + \varepsilon^*_{t+1}. \tag{10.24}$$

10.1.3 Consistent ranking

The simplest proxies for the one-day unobserved variance, σ^2_{t+1}, is the squared log-return, y^2_{t+1}, and the squared demeaned log-return or model residual, ε^2_{t+1}. It has already been noted that $E(\varepsilon^2_{t+1}|I_t) = \sigma^2_{t+1}$, i.e. the daily-based volatility proxy is conditionally unbiased, with $V(\varepsilon^2_{t+1}|I_t) = \sigma^4_{t+1}(Ku-1)$, but it is a noisy estimator. Andersen and Bollerslev (1998a), Andersen et al. (2005b), Hansen and Lunde (2006) and Patton (2006) provided analytical and simulated examples in which the use of an unbiased volatility proxy appears to lead to an evaluation appreciably different from what would be obtained if the true volatility were used.

Consider the AR(1)-GARCH(1,1) model in (10.16) as the true data-generating process. Then $\tilde{\sigma}^2_{t+1} = \varepsilon^2_{t+1}$ is a conditionally unbiased volatility proxy, i.e.

$E(\varepsilon^2_{t+1}|I_t) = E((y_{t+1} - c^{(t)}_0 - c^{(t)}_1 y_t)^2 | I_t) = \sigma^2_{t+1}$. Assume now that a forecaster is able to provide perfect one-step-ahead variance forecasts, i.e., predictions $\sigma^2_{t+1|t}$, which coincide with the true variance σ^2_{t+1} at each point in time. Naturally, for such ideal forecasts, the estimated values of the MZ regression models in (10.21), (10.23) and (10.24) would be expected to be in support of the null hypothesis $H_0 : \beta_0 = \beta_1 = 0$. The program named *chapter10.sim_MZ_regression.prg* generates the data process in (10.16) 1000 times. Each time, the above regression models are respectively estimated in the forms

$$\varepsilon^2_{t+1} = \beta_0 + (\beta_1 + 1)\sigma^2_{t+1} + \varepsilon^*_{t+1}, \tag{10.25}$$

$$\log(\varepsilon^2_{t+1}) = \beta_0 + (\beta_1 + 1)\log(\sigma^2_{t+1}) + \varepsilon^*_{t+1}, \tag{10.26}$$

$$|\varepsilon_{t+1}| = \beta_0 + (\beta_1 + 1)\sigma_{t+1} + \varepsilon^*_{t+1}. \tag{10.27}$$

The values of the estimated parameters are saved in the matrix *MZ*. In particular, $\{\beta^{(t)}_0, \beta^{(t)}_1 + 1\}^{1000}_{t=1}$ are stored in the first two columns of *MZ* in the case of model (10.25), in the next two columns in the case of model (10.27) and in the last two columns in the case of model (10.26). The program is also run for $b_1 = 0.20$ and 0.50.[3] Table 10.4 presents the average values of the 1000 parameter estimates. As mentioned above, for such ideal forecasts, one would expect evidence in support of the null hypothesis $H_0 : \beta_0 = \beta_1 = 0$. However, the evidence obtained on the basis of the modified MZ regression models (10.23) and (10.24) calls for its rejection. This is quite paradoxical. Nevertheless, simulation studies have interestingly led to findings pointing to this being rather an intrinsic feature of such methods. In particular, Patton (2005) obtained ordinary least squares estimates for the coefficients β_0 and

Table 10.4   Average values of the estimates of coefficients $\{\beta^{(t)}_0\}^{1000}_{t=1}$ and $\{\beta^{(t)}_1 + 1\}^{1000}_{t=1}$ in the Mincer–Zarnowitz regression models. The data-generating process is $\varepsilon_{t+1} = \sigma_{t+1} z_{t+1}$, $\sigma^2_{t+1} = 0.0001 + 0.12\varepsilon^2_t + b_1\sigma^2_t$, $z_{t+1} \sim N(0, 1)$

Mincer–Zarnowitz regression function		$b_1 = 0.80$	$b_1 = 0.50$	$b_1 = 0.20$	Theoretical values
$\varepsilon^2_{t+1} = \beta_0 +$	β_0	6E-05	8E-06	2E-06	0.00
$(\beta_1 + 1)\sigma^2_{t+1}$	$(\beta_1 + 1)$	0.95	0.97	0.98	1.00
$|\varepsilon_{t+1}| = \beta_0 +$	β_0	6E-04	2E-04	5E-05	0.00
$(\beta_1 + 1)\sigma_{t+1}$	$(\beta_1 + 1)$	0.779	0.784	0.794	0.798
$\log(\varepsilon^2_{t+1}) = \beta_0 +$	β_0	−1.38	−1.39	−1.22	−1.27
$(\beta_1 + 1)\log(\sigma^2_{t+1})$	$(\beta_1 + 1)$	0.98	0.99	1.01	1.00

---

[3] Recall that the GARCH(1,1) model has the form $\sigma^2_t = a_0 + a_1\varepsilon^2_{t-1} + b_1\sigma^2_{t-1}$.

$(\beta_1 + 1)$ of various MZ models on different distributional assumptions for $\varepsilon_{t+1}|I_t$. For $\varepsilon_{t+1}|I_t \sim N(0, \sigma_{t+1}^2)$, the parameter vector $(\beta_0, \beta_1 + 1)$ of the MZ regression models in (10.25), (10.26) and (10.27) was estimated to be equal to $(0, 1)$, $(-1.27, 1)$, and $(0, \sqrt{2/\pi}) \approx (0, 0.798)$, respectively, while in the case of the latter model, if $\varepsilon_{t+1}|I_t \sim t(0, \sigma_{t+1}^2; v)$, the estimate obtained for $(\beta_0, \beta_1 + 1)$ was

$$\left(0, \Gamma\left(\frac{v+1}{2}\right) \frac{2\sqrt{v-2}}{\sqrt{\pi}(v-1)\Gamma(v/2)}\right).$$

Similar evidence is provided by findings of studies on comparative evaluations of volatility forecasting models based on the coefficient of multiple determination, R^2, from the MZ regression which is most commonly used as a measure of variability in the variance proxy, $\tilde{\sigma}_{t+1}^2$, which is explained by the variance forecasts, $\sigma_{t+1|t}^2$ (see Boudoukh et al., 1997; Brailsford and Faff, 1996; Frennberg and Hansson, 1996). In all of the studies, very low values of R^2 are reported either for in-sample or for out-of-sample volatility forecasts. Andersen and Bollerslev (1998a) showed that if the AR(0)-GARCH(1,1) model is the true data-generating process, i.e. if $\varepsilon_t = \sigma_t z_t$ and $\sigma_t^2 = a_0 + a_1 \varepsilon_{t-1}^2 + b_1 \sigma_{t-1}^2$, with $z_t \sim N(0, 1)$, then the true value of R^2 is

$$R^2 = a_1^2(1 - b_1^2 - 2a_1b_1)^{-1}. \tag{10.28}$$

Employing this result in the case of the model in (10.16), the true value of R^2 is 0.086. The reader wishing to obtain a numerical approximation may run the program *chapter10.sim_R2_regression.prg*. This program generates 1000 runs of an AR(0)-GARCH(1,1) process, for $a_0 = 0.0001$, $a_1 = 0.12$ and $b_1 = 0.80$. The estimated R^2 values for the MZ regression model in (10.21) are saved in matrix R2. The average of the obtained estimates of the true value of R^2 is 0.085, a finding seemingly in support of the assertion. It appears, therefore, that a low value of R^2 should rather be treated as the result of an intrinsic trait of MZ regression models than be taken as necessarily indicating model inadequacy.

In the literature, the above-discussed deficiency of the MZ approach has been addressed by replacing a noisy volatility proxy, such as the squared daily log-returns, in our case, by a more informative one, such as the realized volatility. It appears that this proxy generally allows for higher values of the MZ regression coefficient of multiple determination. So, for example, Nelson (1990b) noted that the stochastic difference equation representation of an ARCH process convergences to a stochastic differential equation as the length of the discrete time intervals goes to zero.[4] The resulting continuous time AR(0)-GARCH(1,1) diffusion is given by

$$d\log(P(t)) = \sigma(t)dW_1(t),$$
$$d\sigma^2(t) = a_0^*(a_1^* - \sigma^2(t))dt + \sqrt{2a_0^* b_1^* \sigma(t)}dW_2(t), \tag{10.29}$$

---

[4] For example, in the continuous time limit, the GARCH(1,1) process is inverted gamma distributed and the EGARCH(1,1) process is log-normally distributed.

with $W_1(t)$ and $W_2(t)$ denoting independent standard Wiener processes and with parameters a_0^*, a_1^*, b_1^* related to those of the initial model as follows:

$$a_0^* = -\log(a_1 + b_1), \tag{10.30}$$

$$a_1^* = \frac{a_0}{1 - a_1 - b_1}, \tag{10.31}$$

and

$$b_1^* = \frac{2(\log(a_1 + b_1))^2}{\frac{(1 - (a_1 + b_1)^2)(1 - b_1)^2}{a_1(1 - b_1(a_1 + b_1))} + 6\log(a_1 + b_1) + 2(\log(a_1 + b_1))^2 + 4(1 - a_1 - b_1)} \tag{10.32}$$

(see Andersen and Bollerslev, 1998a; Drost and Werker, 1996). Andersen and Bollerslev (1998a) showed that the R^2 estimator of the MZ regression model for the evaluation of volatility forecasts increases monotonically towards

$$R^2_{(m \to \infty)} = 1 - E\left(\sigma_{t+1}^2 - \int_0^1 \sigma^2(t+m)dm\right)^2 V\left(\int_0^1 \sigma^2(t+m)dm\right)^{-1}, \tag{10.33}$$

with a sampling frequency m, which is close to 0.50. Recall from Chapter 7 that the realized variance converges in probability to the integrated variance as $m \to \infty$, that is,

$$\underset{m \to \infty}{\text{plim}}\left(\sum_{j=1}^{m-1}(\log(P_{((j+1)/m),t+1}) - \log(P_{(j/m),t+1}))^2\right) = \int_0^1 \sigma^2(t+m)dm. \tag{10.34}$$

Andersen and Bollerslev (1998a) and Andersen et al. (2006) simulated the continuous time AR(0)-GARCH(1,1) diffusion for $a_0^* = 0.035$, $a_1^* = 0.636$ and $b_1^* = 0.296$.

In what follows, a simulation comparison of findings obtained through the use of the GARCH(1,1) model to those obtained through the use of the continuous time ARCH diffusion analogue of the discrete process considered before is presented. In particular, starting with the discrete time AR(0)-GARCH(1,1) process with parameters $a_0 = 0.001$, $a_1 = 0.12$ and $b_1 = 0.80$, we obtain the continuous time AR(0)-GARCH(1,1) diffusion given below with stochastic volatility parameters calculated as in (10.30)–(10.32):

$$\begin{aligned}\log(P(t+dt)) &= \log(P(t)) + \sigma(t)\sqrt{dt}W_1(t),\\ \sigma^2(t+dt) &= 0.00108dt + \sigma^2(t)(1 - 0.083dt + \sqrt{0.084dt}W_2(t)),\end{aligned} \tag{10.35}$$

where, as before, $W_1(t)$ and $W_2(t)$ denote independent standard normal variables and $dt = 1/1000$. Then, using the program named *chapter10.sim.continuousgarch.prg* we simulate 100 sequences of 3 200 000 observations from this continuous time

AR(0)-GARCH(1,1) diffusion. We define a data set of 3200 trading days for each of which there will thus be 1000 intraday log-returns available. The part of the program that carries out the simulation of the diffusion is as follows:

```
!alpha0 = 0.083
!alpha1 = 0.013
!beta1 = 0.505
!dt = 1/1000
series z1 = nrnd
series z2 = nrnd
h(1) = !alpha0*!alpha1*!dt
y(1) = sqr(h(1)*!dt)*z1(1)
p(1) = 1000
for !i = 2 to !length
  h(!i) = !alpha0*!alpha1*!dt + h(!i-1)*(1-!alpha0*!dt+
    (sqr(2*!beta1*!alpha0*!dt))*z2(!i))
  y(!i) = sqr(h(!i)*!dt)*z1(!i)
  p(!i) = p(!i-1)*exp(y(!i))
next
```

Next, the daily log-return, $y_t = \sum_{j=1}^{1000} y_{(j/1000),t}$, is computed by the command *series sum_y_1000*. Each cell, $(1 + ((!j-1)/T(6)))$, corresponds to a trading day, t.

```
series sum_y_1000
for !j=1 to !length-T(6)+1 step T(6)
  !sum=0
  for !t=!j to T(6)-1+!j
    sum_y_1000(1+((!j-1)/T(6))) = y(!t) + !sum
    !sum = sum_y_1000(1+((!j-1)/T(6)))
  next
next
```

The price-range,

$$Range_t = \frac{1}{4\log 2} \left(\log \left(\frac{\max(P_{(j/1000),t}; \, 1 \leq j \leq 1000)}{\min(P_{(j/1000),t}; \, 1 \leq j \leq 1000)} \right) \right)^2,$$

is computed by the command *series range*. Also, the cell $(1 + ((!j-1)/T(6)))$ corresponds to the tth trading day.

```
series range
  for !j=1 to !length-T(6)+1 step T(6)
    !min=p(!j)
    !max=p(!j)
    for !t=!j to T(6)-1+!j
```

```
if p(!t)<!min then
   !min=p(!t)
endif
if p(!t)>!max then
   !max=p(!t)
endif
next
range(1+((!j-1)/T(6)))=(log(!max/!min))^2/(4*log(2))
next
```

The AR(0)-GARCH(1,1) model is estimated, for $T = 3000$ simulated daily data $y_t = \sum_{j=1}^{1000} y_{(j/1000),t}$, as $y_t = \sigma_t z_t$, $\sigma_t^2 = a_0 + a_1 y_{t-1}^2 + b_1 \sigma_{t-1}^2$ and $z_t \sim N(0,1)$, discarding the first 200 simulated daily log-returns. The estimates of the conditional variance, $\hat{\sigma}_t^2 = a_0^{(T)} + a_1^{(T)} y_{t-1}^2 + b_1^{(T)} \hat{\sigma}_{t-1}^2$, are stored in the series *garchg11*.

```
smpl 201 3200
equation g11.arch(1,1,m=10000,h, deriv=f) sum_y_1000
g11.makegarch garchg11
```

In the next lines, the series *y{!Ti}* contains the log-returns for aggregation frequencies $m = 500, 100, 20, 10, 2, 1$, where $T(!i)=1000/m$. Cell $1 + ((!t-1)/T(!i))$ corresponds to the *t*th trading day. Figure 10.3 presents the steps we follow to compute the log-returns, $y_t = \sum_{j=1}^{m} y_{(j/m),t}$, for day *t* with a sampling frequency of $m = 100$.

```
smpl @all
for !i=1 to 6
  !Ti = T(!i)
  series sum_sq_y{!Ti}
  series y{!Ti}
  for !j=1 to !length-T(6)+1 step T(6)
    statusline !rep !i !j
      for !t=!j to T(6)-1+!j step T(!i)
        y{!Ti}(1+((!t-1)/T(!i))) = log(p(!t+(T(!i)-1))) -
          log(p(!t))
  next
next
```

The next lines of the program compute the realized variance, $\sigma_t^{2(RV)} = \sum_{j=1}^{m} \left(y_{(j/m),t}\right)^2$, for $m = 500, 100, 20, 10, 2, 1$. The *!i*th cell of vector $T(!i)$ contains the sampling frequency $T(!i) = 1000/m$.

```
for !j=1 to !length/T(6)
  !sum2=0
  for !t=((!j-1)*T(6)/T(!i))+1 to !j*T(6)/T(!i)
```

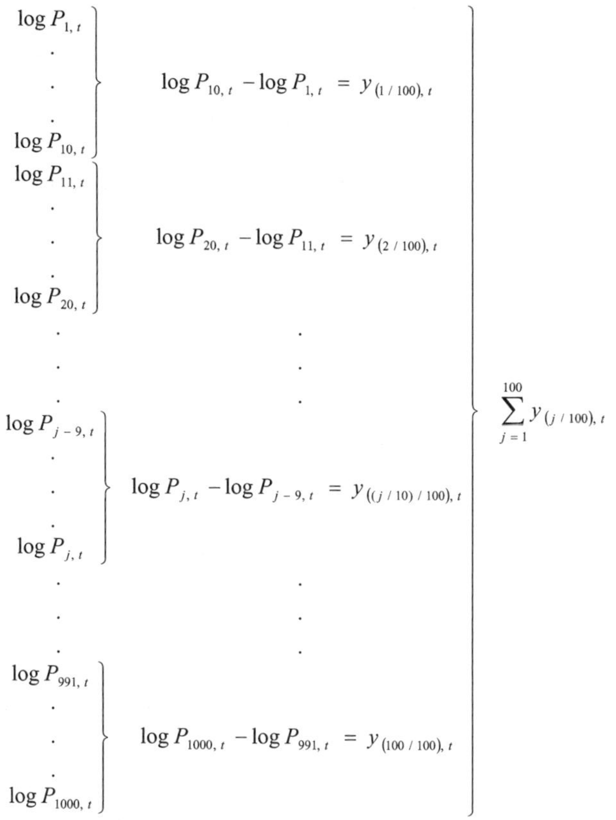

$$\left. \begin{array}{c} \log P_{1,\,t} \\ . \\ . \\ . \\ \log P_{10,\,t} \end{array} \right\} \quad \log P_{10,\,t} - \log P_{1,\,t} \; = \; y_{(1\,/\,100),\,t}$$

$$\left. \begin{array}{c} \log P_{11,\,t} \\ . \\ . \\ . \\ \log P_{20,\,t} \end{array} \right\} \quad \log P_{20,\,t} - \log P_{11,\,t} \; = \; y_{(2\,/\,100),\,t}$$

$$\left. \begin{array}{c} \log P_{j-9,\,t} \\ . \\ . \\ . \\ \log P_{j,\,t} \end{array} \right\} \quad \log P_{j,\,t} - \log P_{j-9,\,t} \; = \; y_{((j\,/\,10)\,/\,100),\,t}$$

$$\left. \begin{array}{c} \log P_{991,\,t} \\ . \\ . \\ . \\ \log P_{1000,\,t} \end{array} \right\} \quad \log P_{1000,\,t} - \log P_{991,\,t} \; = \; y_{(100\,/\,100),\,t}$$

$$\left. \rule{0pt}{8em}\right\} \; \sum_{j=1}^{100} y_{(j\,/\,100),\,t}$$

Figure 10.3   Determination of log-return for day t, $y_t = \sum_{j=1}^{m} y_{(j/m),t}$, when 1000 observations are available at day t and a sampling frequency of $m = 100$ is required.

```
      sum_sq_y{!Ti}(!j) = y{!Ti}(!t)^2 + !sum2
      !sum2 = sum_sq_y{!Ti}(!j)
   next
 next
```

Finally, the MZ methods in (10.22)–(10.24) are applied for the evaluation of the in-sample variance, $\hat{\sigma}_t^2$ (stored in the *garchg11* series), instead of the one-step-ahead variance forecast, $\sigma_{t+1|t}^2$. The proxy measures for the actual volatility, $\tilde{\sigma}_t^2$, are obtained by $\sigma_t^{2(RV)} = \sum_{j=1}^{m} \left(y_{(j/m),t}\right)^2$, for $m = 500, 100, 20, 10, 2, 1$ (contained in the *sum_sq_y{!Ti}* series) and the price range $Range_t$ (stored in the *range* series). The coefficients of multiple determination for the regressions (10.22), (10.23) and (10.24) form the columns of the matrices *MZ1(!rep,!i)*, *MZ2(!rep,!i)* and *MZ3(!rep,!i)*, respectively, with their *i*th columns, in particular, containing the estimates of the value of R^2 obtained on the basis of $\tilde{\sigma}_t^2 = \sum_{j=1}^{m_i} \left(y_{(j/m_i),t}\right)^2$ as a proxy measure for the actual volatility for $i = 1, \ldots, 6$. Here, the m_i denotes the ith term in the sequence of values 500, 100, 20, 10, 2, 1. The seventh (last) columns of the matrices contain the estimates of the value of R^2 obtained on the basis of the price range as a proxy for the actual variance. The simulation of the 3 200 000 observations is repeated 100 times. The !rep rows of the matrices contain the estimates of the value of R^2 corresponding to the 100 independently simulated continuous time AR(0)-GARCH(1,1) diffusions.

```
smpl 201 3200
equation MZ1_{!i}reg.ls(h) sum_sq_y{!Ti} c garchg11
MZ1(!rep,!i) = MZ1_{!i}reg.@r2
equation MZ2_{!i}reg.ls(h) sqr(sum_sq_y{!Ti})
c sqr(garchg11)
MZ2(!rep,!i) = MZ2_{!i}reg.@r2
equation MZ3_{!i}reg.ls(h) log(sum_sq_y{!Ti})
c log(garchg11)
MZ3(!rep,!i) = MZ3_{!i}reg.@r2
next
equation MZ1_7reg.ls(h) range c garchg11
MZ1(!rep,7) = MZ1_7reg.@r2
equation MZ2_7reg.ls(h) sqr(range) c sqr(garchg11)
MZ2(!rep,7) = MZ2_7reg.@r2
equation MZ3_7reg.ls(h) log(range) c log(garchg11)
MZ3(!rep,7) = MZ3_7reg.@r2
next
```

The findings are summarized in Table 10.5(a), which presents the average estimates of the coefficient of multiple determination, R^2, for the 100 MZ regression models. R^2 appears to increase monotonically with the sampling frequency (m). The price range proxy is apparently superior to the realized variance measure when it is constructed on the basis of a small number of intraday observations. The opposite appears to hold when the realized volatility is constructed on the basis of a large number of intraday observations. Table 10.5(b) presents averages of the estimates of R^2 associated with the simulated continuous AR(0)-GARCH(1,1) time diffusion with parameters as reported in Andersen and Bollerslev (1998a), i.e. with $a_0^* = 0.035$, $a_1^* = 0.636$ and $b_1^* = 0.296$. A comparison of the tabulated figures of both parts of the Table 10.5 indicates a qualitative agreement of the findings.

Therefore, the construction of volatility proxies based on intraday returns allows for more accurate volatility measurement and for more meaningful qualitative assessments of the daily volatility forecasts. However, even when high-frequency data are available (for $m \rightarrow \infty$), the sampling frequency should be as high as needed for the market microstructure features not to induce bias to the volatility estimators.

Hansen and Lunde (2006) noted that the substitution of a noisy proxy such as the squared daily returns for the true but unobservable conditional variance can result in the choice of an inferior model as the best one. In contrast, the realized volatility as a proxy variable does not appear to lead to an inferior model. Moreover, Hansen and

Table 10.5 Average estimates of the coefficients of multiple determination, R^2, of the 100 Mincer–Zarnowitz regressions

(a) Volatility proxy	$\tilde{\sigma}_t^2 = \beta_0 + (\beta_1 + 1)\hat{\sigma}_t^2$	$\tilde{\sigma}_t = \beta_0 + (\beta_1 + 1)\hat{\sigma}_t$	$\log(\tilde{\sigma}_t^2) = \beta_0 + (\beta_1 + 1)\log(\hat{\sigma}_t^2)$
$\tilde{\sigma}_t^2 \equiv \sum_{j=1}^{500} \left(y_{(j/500),t}\right)^2$	0.388	0.392	0.354
$\tilde{\sigma}_t^2 \equiv \sum_{j=1}^{100} \left(y_{(j/100),t}\right)^2$	0.372	0.378	0.340
$\tilde{\sigma}_t^2 \equiv \sum_{j=1}^{20} \left(y_{(j/20),t}\right)^2$	0.317	0.322	0.280
$\tilde{\sigma}_t^2 \equiv \sum_{j=1}^{10} \left(y_{(j/10),t}\right)^2$	0.270	0.273	0.227
$\tilde{\sigma}_t^2 \equiv \sum_{j=1}^{2} \left(y_{(j/2),t}\right)^2$	0.123	0.119	0.070
$\tilde{\sigma}_t^2 \equiv y_t^2$	0.073	0.068	0.027
$\tilde{\sigma}_t^2 \equiv Range_t$	0.202	0.219	0.189

(continued)

Table 10.5 *(Continued)*

(b)

Volatility proxy	$\tilde{\sigma}_t^2 = \beta_0 + (\beta_1+1)\hat{\sigma}_t^2$	$\tilde{\sigma}_t = \beta_0 + (\beta_1+1)\hat{\sigma}_t$	$\log(\tilde{\sigma}_t^2) = \beta_0 + (\beta_1+1)\log(\hat{\sigma}_t^2)$
$\tilde{\sigma}_t^2 \equiv \sum_{j=1}^{500}(y_{(j/500),t})^2$	0.451	0.451	0.428
$\tilde{\sigma}_t^2 \equiv \sum_{j=1}^{100}(y_{(j/100),t})^2$	0.427	0.427	0.403
$\tilde{\sigma}_t^2 \equiv \sum_{j=1}^{20}(y_{(j/20),t})^2$	0.338	0.337	0.309
$\tilde{\sigma}_t^2 \equiv \sum_{j=1}^{10}(y_{(j/10),t})^2$	0.267	0.265	0.232
$\tilde{\sigma}_t^2 \equiv \sum_{j=1}^{2}(y_{(j/2),t})^2$	0.102	0.097	0.060
$\tilde{\sigma}_t^2 \equiv y_t^2$	0.056	0.052	0.023
$\tilde{\sigma}_t^2 \equiv Range_t$	0.182	0.198	0.186

The data-generating process is the continuous time diffusion $\log(P(t+dt)) = \log(P(t)) + \sigma(t)\sqrt{dt}W_1(t)$ with $\sigma^2(t+dt) = 0.00108dt + \sigma^2(t)(1-0.083dt + \sqrt{0.084dt}W_2(t))$, for (a) and $\sigma^2(t+dt) = 0.0223dt + \sigma^2(t)(1-0.035dt + \sqrt{0.021dt}W_2(t))$ for (b). The daily log-returns are computed as $y_t = \sum_{j=1}^{1000} y_{(j/1000),t} = \log P_{1000,t} - \log P_{1,t} = \log P(t+1) - \log P(t)$. The in-sample conditional variance, $\hat{\sigma}_t^2 = a_0^{(T)} + a_1^{(T)} y_{t-1}^2 + b_1^{(T)} \hat{\sigma}_{t-1}^2$, is estimated from the AR(0)-GARCH(1,1) model, represented by $y_t = \sigma_t z_t$, $\sigma_t^2 = a_0 + a_1 y_{t-1}^2 + b_1 \sigma_{t-1}^2$ and $z_t \sim N(0,1)$.

Lunde (2006) derived conditions which ensure that the ranking of any two variance forecasts by a loss function is the same whether the ranking is done via the true and unobserved variance, $\sigma_{t+1}^{2(\tau)}$, or via a conditionally unbiased volatility proxy, $\tilde{\sigma}_{t+1}^{2(\tau)}$. In particular,

$$E\left(\Psi\left(\sigma_{t+1}^{2(\tau)}, \sigma_{t+1|t(A)}^{2(\tau)}\right)\right) < E\left(\Psi\left(\sigma_{t+1}^{2(\tau)}, \sigma_{t+1|t(B)}^{2(\tau)}\right)\right) \Leftrightarrow$$

$$E\left(\Psi\left(\tilde{\sigma}_{t+1}^{2(\tau)}, \sigma_{t+1|t(A)}^{2(\tau)}\right)\right) < E\left(\Psi\left(\tilde{\sigma}_{t+1}^{2(\tau)}, \sigma_{t+1|t(B)}^{2(\tau)}\right)\right). \tag{10.36}$$

They showed that a sufficient condition for a consistent ranking is that $\partial^2\Psi\left(\sigma_{t+1}^{2(\tau)}, \sigma_{t+1|t}^{2(\tau)}\right)/\partial\left(\sigma_{t+1}^{2(\tau)}\right)^2$ does not depend on $\sigma_{t+1|t}^{2(\tau)}$, a condition that ensures that the true variance is the optimal forecast under the chosen loss function. Note that the average of the SE loss function and the MZ regression do ensure the equivalence of the ranking of volatility models that is induced by the true volatility and its proxy. They concluded their study by noting that a non-consistent ranking method in combination with a noisy volatility proxy would mistakenly lead one to favour an inappropriate model. Patton (2006) extended the work by Hansen and Lunde (2006). He derived necessary and sufficient conditions on a function for it to yield rankings of competing volatility forecasts that are robust to the presence of noise in the volatility proxy. He proposed a family of robust loss functions that includes the SE (10.5) and GLLS (10.10) loss functions. Although the rankings obtained from a robust loss function remain invariant in the presence of noise in the proxy, the actual level of expected loss from using a proxy will be higher than that from using the true conditional variance. Andersen et al. (2005b) provided a method for estimating the level of expected loss that would have been obtained using the true latent variable of interest.

10.1.4 Simulation, estimation and evaluation

In this section, we provide a series of EViews programs that generate data from an ARCH process, estimate a set of ARCH models, including the ARCH process from which the series is generated, and then assess the ability of some of the aforementioned methods of model evaluation in distinguishing the true data-generating process. The programs can be modified by the reader interested in looking into the aspects discussed earlier or in exploring various side themes.

The first program, named *chapter10.sim.compare.models.prg*, estimates four ARCH models with dependent variable the series of simulated daily log-returns,

$$y_t = \sum_{j=1}^{1000} y_{(j/1000),t} = \log P_{1000,t} - \log P_{1,t}$$

which has been saved in *sum_y_1000* in workfile *chapter10.sim.continuousgarch. wf1*. Apart from for the AR(0)-GARCH(1,1) model, which is to be treated as the data-generating process, we choose to estimate the models AR(0)-IGARCH(1,1), AR (0)-GARCH(2,2), AR(0)-GARCH(1,2) and AR(0)-GARCH(2,1). This selection is arbitrary. The reader may make other preference based choices in part A of the program. In part C, five evaluation criteria are defined: (1) the R^2 coefficient of the MZ regression, (2) the R^2 coefficient of the MZ log-regression, (3) the mean squared error loss function, (4) the mean absolute error loss function, and (5) the Schwarz Bayesian criterion. Each evaluation criterion measures the ability of the five models in estimating the in-sample daily volatility. Apart from the volatility estimate, the evaluation criteria require a measure for the actual variance. Thus, in part B, five volatility proxies are utilized: the realized volatility, $\sigma_t^{2(RV)} = \sum_{j=1}^{m} \left(y_{(j/m),t}\right)^2$, for intraday observations and the price range, $Range_t$. Naturally, one would expect any evaluation technique to pick the AR(0)-GARCH(1,1) model as it is the data-generating process. Table 10.6 presents the rating of the models for each evaluation criterion. As m increases, AR(0)-GARCH(1,1) model, is judged as most adequate by the MZ regression R^2 coefficient. The SBC criterion also points at the true model as the most adequate. However, the true model appears not to be picked by all criteria and for any volatility proxy. It should be noted, though, that running the program only once could lead to misleading conclusions. On another run of the program, the ranking of the models could be different.

Therefore, in order to get a clearer picture of the situation, we construct the following program, which runs the procedures mentioned above for a sequence of 100 simulated continuous time AR(0)-GARCH(1,1) diffusions. The program named *chapter10.sim.compare.models.repeated.prg* gives us a clearer view as concerns the ability of the evaluation criteria in distinguishing the true data-generating model. The matrix named *correct_model_selected* contains the numbers of times the AR(0)-GARCH(1,1) model was picked by the criteria. Each column, *!criteria* in number, refers to the evaluation criterion and each row, *!proxies* in number, refers to the volatility proxy used. Table 10.7 provides the entries of the matrix *correct_model_-selected*. As the number of intraday observations, m, increases, the number of times an evaluation criterion selects the AR(0)-GARCH(1,1) model increases as well. The AE loss function is apparently unable to pick the true model as frequently as the other evaluation criteria. Note also that the SBC appears to exhibit a superior performance. It has identified the true data-generating process 94% of the time.

The program should create 100 sets of series, each comprising 3 200 000 observations. In order to avoid memory problems and to keep the size of the workfile to a reasonable size, we save the series in matrix form. For example, the five volatility proxies of each iteration, *!rep*, are stored in matrices denoted by *proxy_{!rep}*:

```
for !rep = 1 to !repeat
matrix(3200,!proxies) proxy_{!rep}
   for !i = 1 to 3200
      proxy_{!rep}(!i,1) = sum_sq_y1000(!i)
      proxy_{!rep}(!i,2) = sum_sq_y100(!i)
      proxy_{!rep}(!i,3) = sum_sq_y10(!i)
      proxy_{!rep}(!i,4) = sum_sq_y2(!i)
      proxy_{!rep}(!i,5) = range(!i)
   next
next
```

Table 10.6 Values of five criteria: (1) the R^2 coefficient of the Mincer–Zarnowitz regression, (2) the R^2 coefficient of the Mincer–Zarnowitz log-regression, (3) the mean squared error loss function, (4) the mean absolute error loss function and (5) the Schwarz Bayesian criterion, used to evaluate the ability of five models in estimating the in-sample variance

Volatility Proxy	R^2 of Mincer–Zarnowitz regression	R^2 of Mincer–Zarnowitz log-regression	Average of the $\Psi_{(SE)t+1}$ loss function (times 1000)	Average of the $\Psi_{(AE)t+1}$ loss function (times 1000)	SBC criterion
$\tilde{\sigma}_t^2 \equiv y_t^2$	M1: 0.0449 M2: 0.0288 M3: 0.0454 **M4: 0.0459** M5: 0.0454	M1: 0.01581 M2: 0.01165 **M3: 0.01584** M4: 0.01564 M5: 0.01572	M1: 0.392 M2: 0.404 M3: 0.391 **M4: 0.391** M5: 0.391	M1: 11.487 M2: 11.647 M3: 11.487 **M4: 11.484** M5: 11.486	
$\tilde{\sigma}_t^2 \equiv \sum_{j=1}^{10} (y_{(j/10),t})^2$	M1: 0.1745 M2: 0.1276 M3: 0.1736 M4: 0.1743 **M5: 0.1746**	**M1: 0.1658** M2: 0.1121 M3: 0.1632 M4: 0.1644 M5: 0.1654	M1: 0.086 M2: 0.096 M3: 0.086 M4: 0.086 **M5: 0.086**	M1: 6.097 M2: 6.517 M3: 6.097 **M4: 6.093** M5: 6.094	
$\tilde{\sigma}_t^2 \equiv \sum_{j=1}^{100} (y_{(j/100),t})^2$	**M1: 0.2348** M2: 0.1708 M3: 0.2312 M4: 0.2334 M5: 0.2345	**M1: 0.2521** M2: 0.1713 M3: 0.2473 M4: 0.2502 M5: 0.2515	**M1: 0.042** M2: 0.052 M3: 0.043 M4: 0.042 M5: 0.042	**M1: 4.602** M2: 5.105 M3: 4.613 M4: 4.608 M5: 4.604	**M1: −1.6880** M2: −1.6519 M3: −1.6832 M4: −1.6854 M5: −1.6854

(continued)

Table 10.6 (Continued)

Volatility Proxy	R^2 of Mincer–Zarnowitz regression	R^2 of Mincer–Zarnowitz log-regression	Average of the $\Psi_{(SE)I+1}$ loss function (times 1000)	Average of the $\Psi_{(AE)I+1}$ loss function (times 1000)	SBC criterion
$\tilde{\sigma}_t^2 \equiv \sum_{j=1}^{500}(y_{(j/500),t})^2$	M1: **0.2448** M2: 0.1749 M3: 0.2420 M4: 0.2438 M5: 0.2447	M1: **0.2606** M2: 0.1761 M3: 0.2565 M4: 0.2593 M5: 0.2603	M1: 0.049 M2: 0.060 M3: 0.049 M4: **0.049** M5: 0.049	M1: 5.886 M2: 6.017 M3: 5.888 M4: **5.879** M5: 5.883	
$\tilde{\sigma}_t^2 \equiv Range_t$	M1: 0.1206 M2: 0.0820 M3: 0.1210 M4: **0.1218** M5: 0.1213	M1: **0.1310** M2: 0.0846 M3: 0.1305 M4: 0.1304 M5: 0.1308	M1: 0.111 M2: 0.122 M3: 0.111 M4: **0.111** M5: 0.111	M1: 6.810 M2: 7.154 M3: 6.803 M4: **6.803** M5: 6.807	

M1, M2, M3, M4 and M5 denote the AR(0)-GARCH(1,1)-N, AR(0)-IGARCH(1,1)-N, AR(0)-GARCH(2,2)-N, AR(0)- GARCH (1,2)-N and AR(0)-GARCH(2,1)-N models, respectively. The evaluation is carried out for five different volatility proxies, $\tilde{\sigma}_t^2$. The data generating process is the continuous time AR(0)-GARCH(1,1) diffusion:$\log(P(t+dt)) = \log(P(t)) + \sigma(t)\sqrt{dt}W_1(t)$ and $\sigma^2(t+dt) = 0.00108dt + \sigma^2(t)(1-0.083dt + \sqrt{0.084dt}W_2(t))$, with 1000 intraday observations, or $dt = 1/1000$. The model regarded as most suitable by the evaluation criterion is shown in bold. The SBC criterion is the same for any proxy as its computation does not depend on it.

Table 10.7   Numbers of times an evaluation criterion picks the AR(0)-GARCH(1,1) model (out of a total of 100 iterations). Each evaluation criterion is computed for five volatility proxies, $\tilde{\sigma}_t^2$. The true data-generating process is the continuous time AR(0)-GARCH(1,1) diffusion. The value of the SBC criterion is the same for any proxy as its computation does not depend on it

Volatility proxy	R^2 of Mincer–Zarnowitz regression	R^2 of Mincer–Zarnowitz log-regression	$\Psi_{(SE)t+1}$ loss function	$\Psi_{(AE)t+1}$ loss function	SBC criterion
$\tilde{\sigma}_t^2 \equiv y_t^2$	19	29	21	34	94
$\tilde{\sigma}_t^2 \equiv \sum_{j=1}^{10} \left(y_{(j/10),t}\right)^2$	52	58	54	52	94
$\tilde{\sigma}_t^2 \equiv \sum_{j=1}^{100} \left(y_{(j/100),t}\right)^2$	61	67	61	59	94
$\tilde{\sigma}_t^2 \equiv \sum_{j=1}^{500} \left(y_{(j/500),t}\right)^2$	64	68	63	37	94
$\tilde{\sigma}_t^2 \equiv Range_t$	34	51	38	45	94

Thus, only the 3200 observations of each series are kept, rather than the 3 200 000 cells per series. Then, after changing the range of the file from 3 200 000 observations to 3200 observations, the matrices named *proxy_{!rep}* are transformed back into the series named *proxy{!prox}_{!rep}*:

```
range 1 3200
for !rep = 1 to !repeat
for !prox = 1 to !proxies
   for !i=1 to 3200
      series proxy{!prox}_{!rep}(!i) = proxy_{!rep}
         (!i,!prox)
   next
 next
next
```

The third program, *chapter10.sim.compare.models.repeated.b.prg*, produces a sequence of 100 simulated discrete time AR(0)-GARCH(1,1) processes.[5] It works in the same way as the second program, but it does not allow the use of intraday based volatility proxies. As far as concerns the ability of the evaluation criteria in

---

[5] This program helps us to verify whether there is any bias in the inference from the second program due to any simulation inaccuracy. In order to reduce the sampling variation, antithetic control variables are strongly suggested in similar simulation studies. For details, see Geweke (1996).

Table 10.8   Numbers of times, out of a total of 100 iterations, an evaluation criterion picks the AR(0)-GARCH(1,1) model, in the in-sample evaluation. The volatility proxy is $\tilde{\sigma}_t^2 \equiv y_t^2$. The true data generating process is the discrete time AR(0)-GARCH-(1,1) process

Volatility proxy	R^2 of Mincer–Zarnowitz regression	R^2 of Mincer–Zarnowitz log-regression	$\Psi_{(SE)t+1}$ loss function	$\Psi_{(AE)t+1}$ loss function	SBC criterion
$\tilde{\sigma}_t^2 \equiv y_t^2$	26	22	28	22	100

distinguishing the true data-generating process, the results appear to be qualitatively similar to those of the second program, thus indicating that whether the data-producing process is discrete or continuous has no appreciable effect on the results. The matrix *correct_model_selected* contains the numbers of times the AR (0)-GARCH(1,1) model was picked by the criteria. It is tabulated in Table 10.8. The SBC appears, on the basis of this table, to exhibit a superior performance, as it identifies the true data-generating process in all of the iterations.

The fourth program, called *chapter10.sim.compare.models.repeated.outofsample.prg*, provides the out-of-sample evaluation for the same models. The same evaluation criteria are computed except for the SBC criterion which is replaced by the GLLS loss function. The program helps identify whether the evaluation criteria behave differently in in-sample and out-of-sample evaluations. Table 10.9 provides the frequencies with which the AR(0)-GARCH(1,1) model has been judged as more accurate by the evaluation criteria, on the basis of the out-of-sample evaluation. The true data-generating model has been picked with a higher frequency in terms of the out-of-sample evaluation than in terms of the in-sample evaluation.

Of course, the results obtained only pertain to a specific data-generating process, the AR(0)-GARCH(1,1), for a small number (five) of candidate models, and a small number (five) of evaluation criteria. The failure of the evaluation criteria to pick the AR(0)-GARCH(1,1) model as the most appropriate one with a higher frequency may be thought of as associated with the particular characteristics of the five models used for the estimation of volatility. Simulating another ARCH process and/or estimating models with other forms of volatility estimating components might lead to different results.

Table 10.9   Numbers of times, out of a total of 100 iterations, an evaluation criterion picks the AR(0)-GARCH(1,1) model, in the out-of-sample evaluation. The volatility proxy is $\tilde{\sigma}_t^2 \equiv y_t^2$. The true data generating process is the discrete time AR(0)-GARCH-(1,1) process

Volatility proxy	R^2 of Mincer–Zarnowitz regression	R^2 of Mincer–Zarnowitz log-regression	$\Psi_{(SE)t+1}$ loss function	$\Psi_{(AE)t+1}$ loss function	$\Psi_{(GLLS)t+1}$ loss function
$\tilde{\sigma}_t^2 \equiv y_t^2$	44	48	53	27	57

Table 10.10   Number of times, out of a total of 100 iterations, an evaluation criterion picks the AR(0)-GJR(1,1) model, in the in-sample evaluation. The volatility proxy is $\tilde{\sigma}_t^2 \equiv y_t^2$. The true data generating process is the discrete time AR(0)-GJR(1,1) process

Volatility proxy	R^2 of Mincer–Zarnowitz regression	R^2 of Mincer–Zarnowitz log-regression	$\Psi_{(SE)t+1}$ loss function	$\Psi_{(AE)t+1}$ loss function	SBC criterion
Data generating model: $\sigma_t^2 = 0.0001 + 0.12\varepsilon_{t-1}^2 + 0.5d(\varepsilon_{t-1} < 0)\varepsilon_{t-1}^2 + 0.2\sigma_{t-1}^2$, with $d(\varepsilon_{t-i} < 0) = 1$ if $\varepsilon_{t-i} < 0$, and $d(\varepsilon_{t-i} < 0) = 0$ otherwise					
$\tilde{\sigma}_t^2 \equiv y_t^2$	100	100	98	100	100
Data generating model: $\sigma_t^2 = 0.0001 + 0.12\varepsilon_{t-1}^2 + 0.3d(\varepsilon_{t-1} < 0)\varepsilon_{t-1}^2 + 0.5\sigma_{t-1}^2$					
$\tilde{\sigma}_t^2 \equiv y_t^2$	96	100	90	100	100

Let us, for example, run the program *chapter10.sim.compare.models.repeated. c.prg*, which is a modification of the program *chapter10.sim.compare.models. repeated.b.prg*. Instead of the AR(0)-GARCH(1,1) process, the program simulates the AR(0)-GJR(1,1) process, $\sigma_t^2 = a_0 + a_1\varepsilon_{t-1}^2 + \gamma_1 d(\varepsilon_{t-1} < 0)\varepsilon_{t-1}^2 + b_1\sigma_{t-1}^2$, for $a_0 = 0.0001$, $a_1 = 0.12$, $\gamma_1 = 0.5$, $b_1 = 0.2$. It then estimates the AR(0)-GJR(1,1), AR(0)-IGARCH(1,1), AR(0)-GARCH(2,2), AR(0)-GARCH(1,2) and AR(0)-GARCH(2,1) models and computes the same evaluation criteria as before. Table 10.10 summarizes the new situation in terms of the numbers of times the AR(0)-GJR(1,1) model has been selected as the most adequate model. The evaluation criteria have performed far better; in almost all the cases the data-generating model has been picked as the most adequate. The reason is that when we generate data from the GJR(1,1) process, the simulated volatility process is characterized by the leverage effect, which is not captured by the rest of the competing models. Thus, the evaluation criteria can easily identify the true data generating process. However, even in the case of a model picked as being the one minimizing some loss function, the question of whether the model provides statistically significantly more accurate volatility estimates than its competitors should be checked. Formal methods for comparative evaluation of the predictive abilities of competing forecasting models will be discussed in Section 10.2.

Note that a rerun of the program starting with generating data from an AR(0)-GJR (1,1) process with different parameters, and in particular, $a_0 = 0.0001$, $a_1 = 0.12$, $\gamma_1 = 0.3$, $b_1 = 0.5$, yielded similar results (second part of Table 10.10).

10.1.5 Point, interval and density forecasts

The loss functions discussed in the preceding subsection lead to point estimates of future volatility values. As the actual variance is not observed, the evaluation of such point estimates is not simple. Thus, interval estimates become of central importance in

ARCH modelling. VaR forecasting can be regarded as a one-sided interval estimate of future volatility. In Section 8.1.2, for example, we constructed one-sided interval forecasts to obtain VaR forecasts. The VaR forecast, which is simply a specified quantile of the forecasted return distribution, is directly compared with the actual log-return. Therefore, instead of evaluating the volatility prediction based on a proxy of the actual variance, the prediction interval is evaluated based on the actual value of the dependent random variable. The accuracy of the VaR forecasts was evaluated in terms of back-testing measures (Section 8.1.5) and VaR loss functions (Section 8.1.6).

Berkowitz (2001) and Diebold et al. (1998) proposed the evaluation of the entire forecast density, rather than a scalar or interval, as a forecast density provides a complete description of the probabilities that are attached to all possible values of the outcome variables. For a review of the forecast evaluation literature the interested reader is referred to Diebold and Lopez (1996) and West (2006). An excellent illustration of point, interval and density forecasts evaluation can be found in Clements (2005).

10.1.6 Model evaluation viewed in terms of loss functions based on the use of volatility forecasts

The question of model selection was touched upon in Section 2.3 on the basis of various goodness-of-fit criteria, and in Section 4.1 in terms of loss function based criteria. The latter were measures of the distance between actual and estimated values. However, the choice of a loss function is dependent upon the aims of a particular application. For example, in Chapter 8, model evaluation was performed according to whether volatility was used in forecasting one-day-ahead VaR (Section 8.1.6) or in pricing next day's straddles (Section 8.5.4). All of the studies that aim at volatility forecasting construct their own framework for comparing a set of candidate models. We mention only a few examples. West et al. (1993) developed a criterion based on the decisions of a risk averse investor. Engle et al. (1993) and Xekalaki and Degiannakis (2005), concerned with pricing options, developed a loss function that measures the profitability of a trading strategy. As a forecast is made for helping the decision makers to improve their decisions, Granger and Pesaran (2000a, 2000b) considered linking forecast evaluation to decision making. Other examples are the loss functions proposed by Lopez (1999), Sarma et al. (2003) and Angelidis and Degiannakis (2007) for model evaluation based on VaR and ES forecasting (see Sections 8.1.6 and 8.2.2).

González-Rivera et al. (2004) considered comparing the performance of various volatility models on the basis of economic and statistical loss functions. Their findings revealed that there does not exist a unique model that can be regarded as performing best across various loss functions. Brooks and Persand (2003b) also noted that the forecasting accuracy of various methods considered in the literature appears to be highly sensitive to the measure used to evaluate them. Hence, different loss functions may point towards different models as the most appropriate in volatility forecasting. Angelidis and Degiannakis (2008) considered evaluating the ability of intra- and inter-day volatility models to generate the most accurate forecasts within the framework of VaR forecasting, option pricing prediction, and measuring the distance between actual variance and its one-day-ahead forecast. They concluded that the choice of a model is a function of the selection criterion implemented and noted that selecting one model that would do well according to all criteria appears not to be possible.

Having arrived at this point, we should be clear about the various evaluation criteria of ARCH models. We have classified them into two categories: in-sample criteria (referring to a model's ability to describe the data and based on the estimated volatility), and out-of-sample criteria (referring to a model's predictive ability and based on the volatility forecasts generated by them). What is most important, however, for the users of an ARCH model is to have a well-defined idea of what the research aims are so as to be able to decide what to use the model for (estimation or forecasting), and a clear understanding of what type of model selection criterion is appropriate.

Let us take the more intriguing case of a user interested in using a model for forecasting purposes. If the interest is in obtaining one-step-ahead VaR forecasts, the available models should be evaluated in terms of their ability to produce one-step-ahead VaR forecasts. If, on the other hand, the interest is in using volatility forecasts for the purpose of trading options, the candidate models should be evaluated with respect to the ability to generate higher returns from trading options. If, in addition, one is interested in obtaining a forecast of the variance itself, the competing models should be compared with respect to how close to the true value of the variance the predicted value is. This is where the question of consistent ranking of the competing models in terms of their predictive ability comes in. This question can be dealt with only after we define (i) the variable that will be used as a proxy for the true value of the variance, which is never observed, and (ii) a distance measure. It should be noted that the use of a proxy variable often creates difficulties in the evaluation of models as it contains noise. For instance, simulation findings (Andersen and Bollerslev, 1998a; Andersen et al., 2005b; Hansen and Lunde, 2006; Patton, 2006) indicate that there may be a mismatch between the ranking of models obtained through using a proxy variable for the variance and the one obtained through using its true value. So, a model may rank first in terms of a loss function comparing its forecasts to the values of a proxy variable, but it may rank much lower in terms of the same loss function when the true value of the variance is considered. This mismatch is less frequently noted in the case of evaluation criteria that satisfy Hansen and Lunde's (2006) criteria such as the average squared error (SE) loss function and the MZ regression. Further, it has been noted that the number of mismatches reduces with the noise of the proxy variable. This explains why the use of realized volatility instead of the square of log-returns appears to reduce mismatches in the ranking with respect to the proxy variable for the true variance and the true variance itself. As Patton (2006) says:

> the stated goal of forecasting the conditional variance is not consistent with the use of some loss functions when an imperfect volatility proxy is employed. However, these loss functions are not themselves inherently invalid or inappropriate: if the forecast user's preferences are indeed

described by a 'non-robust' loss function, then this simply implies that the object of interest to that forecast user is not the conditional variance but rather some other quantity.

A forecaster is at liberty to choose any loss function to evaluate the accuracy of the forecasts obtained by the various candidate models. Therefore, the accuracy of the conditional variance forecast is only indirectly measured in terms of that of another variable (depending on the evaluation criterion chosen).

10.2   Selection of ARCH models

In previous chapters we considered using statistical loss functions and information criteria for selecting a model. For example, in Sections 2.3 and 4.1 the distance between the estimated value of the conditional variance and that of its proxy measure was employed. In Section 10.1 several measures for in-sample and out-of-sample model evaluation were presented, though none was generally acceptable.

In other application contexts, we discussed methods of model selection according to the utility of the forecasts. For example, in Section 8.1.5 we referred to back-testing measures for testing the accuracy of the VaR forecasts. In Sections 8.1.6 and 8.2.2 we presented statistical loss functions which evaluate the ability of models to forecast the VaR and ES, in particular. In Section 8.5.2 we considered assessing the forecasting performance of a set of ARCH models by devising rules to trade straddle options. In what follows we will look at model comparisons in terms of statistical tests for testing whether a model provides statistically significantly more accurate volatility forecasts than its competitors.

10.2.1   The Diebold–Mariano test

Diebold and Mariano (1995) proposed a unified method for testing the null hypothesis of no difference in the forecasting accuracy of two competing models. Consider the loss functions $\Psi_{t(A)}^{(\tau)}$ and $\Psi_{t(B)}^{(\tau)}$, at time t, of models A and B respectively, and the loss differential $\Psi_{t(A,B)}^{(\tau)} = \Psi_{t(A)}^{(\tau)} - \Psi_{t(B)}^{(\tau)}$. The loss functions can be of any form, not necessarily symmetric or quadratic. In the case of loss functions such as those given in (10.5)–(10.10), a negative value of $\Psi_{t(A,B)}^{(\tau)}$ indicates that the predictive performance of model A is superior to that of model B. The null hypothesis, that models A and B are of equivalent predictive ability, is tested against the alternative hypothesis that model A is of superior predictive ability. In terms of the loss functions considered, these are formulated as

$$H_0 : E\left(\Psi_{t(A)}^{(\tau)} - \Psi_{t(B)}^{(\tau)}\right) = 0, \quad H_1 : E\left(\Psi_{t(A)}^{(\tau)} - \Psi_{t(B)}^{(\tau)}\right) < 0. \tag{10.37}$$

If $\Psi_{t(A,B)}^{(\tau)}$ is a covariance-stationary short-memory process, the sample mean loss differential $\overline{\Psi}_{(A,B)}^{(\tau)} = \tilde{T}^{-1} \sum_{t=1}^{\tilde{T}} \Psi_{t(A,B)}^{(\tau)}$ is asymptotically normally distributed,

$$\sqrt{\tilde{T}}\left(\overline{\Psi}_{(A,B)}^{(\tau)}-\mu\right)\sim N(0,2\pi f_d(0)),\tag{10.38}$$

where $f_d(0)=(2\pi)^{-1}\sum_{i=-\infty}^{\infty}\gamma_d(i)$ is the spectral density of the loss differential at frequency zero. A consistent estimate of the variance of $\overline{\Psi}_{(A,B)}^{(\tau)}$ is based on the sum of the sample autocovariances:

$$\hat{V}\left(\overline{\Psi}_{(A,B)}^{(\tau)}\right)=\tilde{T}^{-1}2\pi\hat{f}_d(0)$$

$$=\tilde{T}^{-2}\left(\sum_{t=1}^{\tilde{T}}\left(\Psi_{t(A,B)}^{(\tau)}-\overline{\Psi}_{(A,B)}^{(\tau)}\right)^2+2\sum_{i=1}^{\tau-1}\sum_{t=i+1}^{\tilde{T}}\left(\Psi_{t(A,B)}^{(\tau)}-\overline{\Psi}_{(A,B)}^{(\tau)}\right)\right.$$

$$\left.\times\left(\Psi_{t-i(A,B)}^{(\tau)}-\overline{\Psi}_{(A,B)}^{(\tau)}\right)\right).\tag{10.39}$$

Therefore, the Diebold–Mariano (DM) statistic for testing the null hypothesis is given by

$$DM_{(A,B)}=\frac{\overline{\Psi}_{(A,B)}^{(\tau)}}{\sqrt{\hat{V}(\overline{\Psi}_{(A,B)}^{(\tau)})}}\sim N(0,1),\tag{10.40}$$

and is approximately normally distributed for large samples $(\tilde{T})$. The DM statistic may also arise in the context of regression models based on the regression of $\Psi_{t(A,B)}^{(\tau)}$ on a constant with HAC standard errors in the sense of Newey and West (1987).[6] Also, Sarma et al. (2003) and Angelidis et al. (2004) considered evaluating the VaR forecasting accuracy by implementing Diebold–Mariano tests.

In Chapter 6, a set of ARCH models were estimated by the G@RCH package, and one-step-ahead daily volatility forecasts were obtained. Among the models considered, the AR(1)-FIAPARCH$(1,d,1)$-t model led to a minimum value of the mean squared error loss function, $\Psi_{(SE)}$. However, we could not tell whether this model was of equivalent forecasting ability to any of the remaining competing models. On the other hand, an in-sample evaluation of the models on the basis of the SBC and AIC information criteria pointed towards the AR(1)-GJR(1,1)-t and the AR(1)-APARCH(1,1)-t models, respectively. We may now apply the DM test for testing

---

[6] The DM statistic is directly derived in EViews. Regress the series $\Psi_{t(A,B)}^{(\tau)}$ on a constant and activate the Newey–West method. In the *Equation Estimation* dialogue box, select the *Options* tab, check the box labelled *Heteroskedasticity Consistent Covariance* and press the *Newey-West* radio button. In equation (10.39) for a τ-step-ahead loss differential, $\tau-1$ sample autocovariances are considered. In EViews the Newey–West covariance matrix is obtained by taking a weighted sum of the autocovariances in order to ensure that the variance–covariance matrix is positive definite:

$$\sum_{i=1}^{floor(4(\tilde{T}/100)^{2/9})}\sum_{t=i+1}^{\tilde{T}}w_i(\Psi_{t(A,B)}^{(\tau)}-\overline{\Psi}_{(A,B)}^{(\tau)}))\Psi_{t-i(A,B)}^{(\tau)}-\overline{\Psi}_{(A,B)}^{(\tau)}),\quad\text{for }w_i=\left(1-\frac{i}{(4(\tilde{T}/100)^{2/9})+1}\right),$$

instead of

$$2\sum_{i=1}^{\tau-1}\sum_{t=i+1}^{\tilde{T}}(\Psi_{t(A,B)}^{(\tau)}-\overline{\Psi}_{(A,B)}^{(\tau)})(\Psi_{t-i(A,B)}^{(\tau)}-\overline{\Psi}_{(A,B)}^{(\tau)}).$$

Table 10.11   Values of the MSE loss functions and of the DM and mDM test statistics for testing the null hypothesis that the AR(1)-FIAPARCH$(1, d, 1)$-t model is of equal predictive ability as model B

Model B	MSE	DM statistic	p-value	mDM statistic
AR(1)-FIAPARCH(1,1)-skT	8.357	−0.6790	0.497	−0.6788
AR(1)-GJR(1,1)-t	8.470	−1.5434	0.123	−1.5428
AR(1)-APARCH(1,1)-t	8.481	−1.5244	0.128	−1.5239
AR(1)-IGARCH(1,1)-skt	8.887	−3.5906	0.000	−3.5894

the null hypothesis that the AR(1)-FIAPARCH$(1, d, 1)$-t model is of the same predictive ability as the AR(1)-GJR(1,1)-t model, against the alternative hypothesis that the AR(1)-FIAPARCH$(1, d, 1)$-t model is of superior predictive ability. Designating AR(1)-FIAPARCH$(1, d, 1)$-t as model A and AR(1)-GJR(1,1)-t as model B, we regress

$$\Psi_{(SE)t+1(A,B)} \equiv \Psi_{(SE)t+1(A)} - \Psi_{(SE)t+1(B)} \equiv \left(\sigma^2_{t+1|t(A)} - y^2_{t+1}\right)^2 - \left(\sigma^2_{t+1|t(B)} - y^2_{t+1}\right)^2,$$
(10.41)

on a constant with HAC standard errors.[7] The results are summarized in Table 10.11. The observed value of the DM statistic (-1.54) with a p-value of 0.123 indicates that the null hypothesis is a plausible hypothesis. Table 10.11 also provides the results of testing model AR(1)-FIAPARCH$(1, d, 1)$-t for equivalence of predictive ability with each of the models AR(1)-APARCH(1,1)-t, AR(1)-FIAPARCH(1,1)-skT and AR(1)-IGARCH(1,1)-skT. The evidence provided by the tabulated observed values of the DM statistic and their associated p-values reveals that the models suggested by the information criteria can plausibly be assumed to have equivalent predictive ability to the AR(1)-FIAPARCH$(1, d, 1)$-t model. Only in the case of model AR(1)-IGARCH(1,1)-skt does the null hypothesis appear not to be plausible as the evidence is in support of the alternative hypothesis that the AR(1)-IGARCH(1,1)-skt model is of inferior predictive ability in comparison to the AR(1)-FIAPARCH$(1, d, 1)$-t model.

Ferreira and Lopez (2005) considered applying the DM test in a VaR framework for an international interest rate portfolio and noted that volatility forecasts from multivariate models appear to perform as well as those from computationally simpler ARCH models. Sadorsky (2006) was led to a similar conclusion. Using the DM test to forecast petroleum futures volatility, he pointed out that single-equation ARCH models seem to perform better than state space models, vector autoregressive models and bivariate ARCH models. The DM test has been widely applied in various other application contexts, such as forecasting inflation (e.g. Kapetanios et al., 2006; Moser

---

[7] Eviews workfile *chapter10.Dmtest.wf1* contains the series with the MSE loss functions, $\Psi_{(SE)t+1(i)}$, for each model $i = 1, \ldots, 4$, and the regression of $\Psi_{(SE)t+1(A,B)}$ on a constant with HAC standard errors.

et al., 2007), modelling VaR of intraday cocoa futures data (Taylor, 2004) and forecasting exchange rates with neural network ensembles (Zhang and Berardi, 2001). The interested reader is also referred to Wilhelmsson (2006), Espasa and Albacete (2007), Muñoz et al. (2007), Angelidis and Degiannakis (2008) and Bhattacharya and Thomakos (2008).

In conclusion, we observe that any comparative model evaluation should be handled with care. The most suitable loss function for volatility forecasting comparison is the mean squared error loss function for the conditional variance, $\overline{\Psi}_{(SE)}^{(\tau)}$. For example, Patton (2005) demonstrated that, in the framework of the DM test, the choice of an inappropriate loss function, such as the mean squared error loss function based on the standard deviation $\sigma_{t+1}^{(\tau)}$, instead of the variance $\sigma_{t+1}^{2(\tau)}$, would lead to the rejection of the value of the true conditional variance in favour of a volatility forecast equal to the true value of the conditional variance multiplied by $2/\pi$. McCracken (2000) provided evidence that the normality of the asymptotic distribution of the DM statistic is not valid if the competing models are nested.

10.2.2 The Harvey–Leybourne–Newbold test

Harvey et al. (1997) proposed a modification of the DM test to circumvent the shortcoming revealed by Diebold and Mariano's findings concerning the test's behaviour that point to a high percentage of rejections of true null hypotheses for small values of $\tilde{T}$ and volatility forecasts of more than one-step-ahead. They proposed using an approximately unbiased estimator of $\hat{V}(\overline{\Psi}_{(A,B)}^{(\tau)})$. The modified DM (mDM) statistic is computed as:

$$mDM_{(A,B)} = \sqrt{(\tilde{T} + 1 - 2\tau + \tilde{T}^{-1}\tau(\tau-1)\tilde{T}^{-1})} DM_{(A,B)}. \qquad (10.42)$$

Harvey et al. (1997) suggested comparing the statistic with critical values from Student's t distribution with $\tilde{T}-1$ degrees of freedom. Clark (1999) provided simulation evidence in favour of the mDM statistic as it suffers distortions of relatively limited size in the case of small samples.[8] In Table 10.11, the values of the mDM statistic are also presented. For one-step-ahead forecasts and a large number of out-of-sample observations, there is no differences between the performance of the DM and mDM statistics.

10.2.3 The Morgan–Granger–Newbold test

Granger and Newbold (1977) extended Morgan's (1940) approach to give the Morgan–Granger–Newbold (MGN) test. The MGN test statistic is based on an orthogonalizing transformation of $(\sigma_{t+1|t(A)}^2 - \sigma_{t+1}^2)$ and $(\sigma_{t+1|t(B)}^2 - \sigma_{t+1}^2)$, and is used for testing whether $(\sigma_{t+1|t(A)}^2 - \sigma_{t+1}^2) - (\sigma_{t+1|t(B)}^2 - \sigma_{t+1}^2)$ and

---

[8] Van Dijk and Franses (2003) proposed a weighted average loss differential modification of the DM statistic that puts more weight on some crucial observations relative to less important ones.

$(\sigma_{t+1|t(A)}^2 - \sigma_{t+1}^2) + (\sigma_{t+1|t(B)}^2 - \sigma_{t+1}^2)$ are autocorrelated. The test statistic is given by

$$MGN_{(A,B)} = \frac{r}{\sqrt{(\tilde{T}-1)^{-1}(1-r^2)}},$$ (10.43)

where

$$r = \frac{\sum_{t=1}^{\tilde{T}} \left(\sigma_{t+1|t(A)}^2 - \sigma_{t+1|t(B)}^2\right)\left(\sigma_{t+1|t(A)}^2 + \sigma_{t+1|t(B)}^2 - 2\sigma_{t+1}^2\right)}{\sqrt{\sum_{t=1}^{\tilde{T}} \left(\sigma_{t+1|t(A)}^2 - \sigma_{t+1|t(B)}^2\right)^2 \sum_{t=1}^{\tilde{T}} \left(\sigma_{t+1|t(A)}^2 + \sigma_{t+1|t(B)}^2 - 2\sigma_{t+1}^2\right)^2}}.$$

The $MGN_{(A,B)}$ statistic is Student t distributed with $\tilde{T}-1$ degrees of freedom. The use of the MGN test is limited to the case of one-step-ahead forecasts and presupposes that the quantities $(\sigma_{t+1|t}^2 - \sigma_{t+1}^2)$ are normally distributed with zero mean.

Finally, an interesting non-parametric test for comparing pairwise forecasting ability across models was proposed by Pesaran and Timmermann (1992).

10.2.4 White's reality check for data snooping

The DM test framework involves a simple comparison between two models. It does not allow simultaneous juxtaposition of more than two models. White (2000) provided a framework for a composite hypothesis, the *reality check* (RC) for data snooping. In White's framework, the predictive ability of a seemingly best-performing forecasting model, from a large set of potential models, is tested against that of a given benchmark model. The expected loss or utility associated with the forecasts produced by the models is measured and the model that renders the loss function minimum (or the utility function maximum) is compared to a benchmark model. González-Rivera et al. (2004) considered using the RC test with Black and Scholes' (1973) option pricing function, West et al's. (1993) utility function, Koenker and Bassett's (1978) loss function for measuring the goodness of fit of VaR, and a predictive likelihood function. Sullivan et al. (1999) applied the RC methodology to evaluate technical trading rules. Based on the RC test, Bao et al. (2006) compared the one-step-ahead predictive ability of a set of VaR models in terms of the empirical coverage probability and the predictive quantile loss for Asian stock markets that suffered during the 1997–1998 financial crisis. An interesting finding was that, although the adequate VaR models generally behave in a similar manner before and after a crisis, during the crisis period they behave quite differently.

10.2.5 Hansen's superior predictive ability test

Hansen (2005) proposed the superior predictive ability test that can be used for comparing the performance of two or more forecasting models. As in the case of the DM test, the forecasts are evaluated using a pre-specified loss function, and the forecast model that generates the smallest expected loss is regarded to be the best-performing one. The SPA test compares the forecasting performance of a benchmark model against its M competitors. Compared to the RC test, the SPA test is more powerful and less sensitive to the inclusion of poor and irrelevant competitors.

Let $\Psi_{t(i)}^{(\tau)}$ be the value of a loss function of model i at time t. The null hypothesis that the benchmark model i^* is not outperformed by competing model i, for $i = 1, \ldots, M$, is tested against the alternative hypothesis that the benchmark model is inferior to one or more of the competing models. The hypotheses of the test can be formulated as:

$$H_0 : E\left(\Psi_{t(i^*,1)}^{(\tau)} \cdots \Psi_{t(i^*,M)}^{(\tau)}\right)' \leq 0, \quad H_1 : E\left(\Psi_{t(i^*,1)}^{(\tau)} \cdots \Psi_{t(i^*,M)}^{(\tau)}\right)' > 0, \quad (10.44)$$

where $\Psi_{t(i^*,i)}^{(\tau)} = \Psi_{t(i^*)}^{(\tau)} - \Psi_{t(i)}^{(\tau)}$, i^* refers to the benchmark model and i refers to competing model i, $i = 1, \ldots, M$. Rejection of the null hypothesis implies that the benchmark model cannot be plausibly assumed to be superior to its M competitors. The test is based on the statistic

$$SPA_{(i^*)} = \max_{i=1,\ldots,M} \frac{\sqrt{M}\,\overline{\Psi}_{(i^*,i)}^{(\tau)}}{\sqrt{V\left(\sqrt{M}\,\overline{\Psi}_{(i^*,i)}^{(\tau)}\right)}}, \quad (10.45)$$

where $\overline{\Psi}_{(i^*,i)}^{(\tau)} = \tilde{T}^{-1}\sum_{t=1}^{\tilde{T}} \Psi_{t(i^*,i)}^{(\tau)}$. The value of $V\left(\sqrt{M}\,\overline{\Psi}_{(i^*,i)}^{(\tau)}\right)$ and the p-values of the $SPA_{(i^*)}$ statistic are estimated by Politis and Romano's (1994) stationary bootstrap method. The RC test can be interpreted as the unstandardized version of the SPA test:

$$RC_{(i^*)} = \max_{i=1,\ldots,M} \sqrt{M}\,\overline{\Psi}_{(i^*,i)}^{(\tau)}. \quad (10.46)$$

On his website, Hansen provides a program which estimates the SPA test using Ox Metrics. The reader can download the program and an instruction manual from http://www.stanford.edu./~prhansen/software/SPA.html.[9] In Section 10.2.1 we applied the DM test for testing the null hypothesis that AR(1)-FIAPARCH$(1, d, 1)$-t is of the same predictive ability as a competing model B. Let us regard the AR(1)-FIAPARCH$(1, d, 1)$-t model as our benchmark model and employ the SPA test to compare its forecasting performance with that of its M competitors. Using the program *chapter10.SPA.ftse100.ox* given in the Appendix to the chapter, the SPA test is carried out to examine the null hypothesis that the benchmark model AR(1)-FIAPARCH$(1, d, 1)$-t is not outperformed by the four

---

[9] Hansen and Lunde developed a package named Mulcom (package for multiple comparisons), which also contains the SPA and RC tests as well as the model confidence set methodology of Hansen et al. (2005). It is available at http://www.hha.dk/~alunde/MULCOM/MULCOM.HTM. Note that the Mulcom package works fully with Ox 4 or later versions, whereas for the SPA for Ox package Ox 3.1 or an earlier version is suggested.

Table 10.12   Output produced by the program Chapter10.SPA.ftse100.ox

Ox version 3.10 (Windows) (C) J.A. Doornik, 1994–2002				

Ox version 3.10 (Windows) (C) J.A. Doornik, 1994–2002
The benchmark model is AR(1)-FIAPARCH(1,d,1)-t
------TEST FOR SUPERIOR PREDICTIVE ABILITY------
SPA version 1.14, September 2003. (C) 2001, 2002, 2003 Peter Reinhard Hansen

Number of models: l = 4
Sample size: n = 1435
Loss function: mse_spa
Test Statistic: TestStatScaledMax
Bootstrap parameters: B = 10000 (resamples), q = 0.5 (dependence)

	Model Number	Sample Loss	t-stat	"p-value"
Benchmark	0.00000	8.35534	-	-
Most Significant	1.00000	8.35677	−0.62194	0.74073
Best	1.00000	8.35677	−0.62194	0.74073
Model_25%	2.00000	8.46983	−1.37710	0.91431
Median	3.00000	8.48103	−1.51353	0.93031
Model_75%	3.00000	8.48103	−1.51353	0.93031
Worst	4.00000	8.88651	−3.47608	0.99950

		Lower	Consistent	Upper
SPA p-values:		0.65250	0.95520	0.96110
Critical values:	10%	0.73825	1.66685	1.76086
	5%	1.05288	1.97305	2.05976
	1%	1.66155	2.55983	2.62792

rival models. Table 10.12 summarizes the output of the program. The p-value termed consistent is indicative of whether there is evidence against the null hypothesis. A high (low) p-value indicates evidence in support of the hypothesis that the benchmark model is superior (inferior) to one or more of the rival models. As the p-value of the test is 0.9552, it appears that there is evidence supporting the hypothesis that the forecasting ability of the AR(1)-FIAPARCH($1, d, 1$)-t model is superior to its four competitors.

The SPA package allows the user to choose between the SPA and RC tests. The *TestStat[0]* option estimates the SPA test statistic, whereas the *TestStat[1]* option estimates the RC test statistic. More details about the SPA program are given in the manual by P.R. Hansen et al. (2003). The reader interested in using the program may find the latter authors' comments about the use of the critical values produced in the output quite useful. When the null hypothesis is not rejected, large critical values indicate that the data set being analysed is not very informative about the hypothesis of interest, and the SPA test may lack power.

Hansen and Lunde (2005a) compared 330 ARCH models in terms of the SPA test and the RC test. They noted that the RC test is unable to distinguish between models that perform well and those that perform poorly. The SPA, which is more powerful, provided evidence that the GARCH(1,1) model exhibited inferior performance compared to other models in forecasting volatility of stock returns (IBM). In contrast, in the case of the exchange rate data (Deutschmark–dollar), they observed that 'nothing beats a GARCH(1,1) model'. Koopman et al. (2005) applied the SPA test to compare the forecasting ability of four classes of volatility models – stochastic volatility and ARCH models for the daily log-returns, and unobserved components and ARFIMA models with the realized volatility as dependent variable. Lanne (2006) utilized the SPA and DM tests for testing the forecasting performance of a multiplicative mixture model for the realized volatility of Deutschmark–dollar and yen–dollar returns. Based on the SPA test, Bruneau et al. (2007) evaluated forecasts of inflation for France and Aiolfi and Favero (2005) investigated the predictability of excess returns in the US stock market.

10.2.6 The standardized prediction error criterion

Degiannakis and Xekalaki (2005) developed a model selection method based on the comparative evaluation of candidate ARCH models in terms of a criterion judging their ability to predict future values of the dependent variable and its volatility forecasts. As reflected by its name, their standardized prediction error criterion (SPEC), being an extension to ARCH set-ups of Xekalaki et al.'s (2003) criterion for linear regression models, utilizes the standardized prediction errors of the competing ARCH models to rank their predictive performance. According to the SPEC, in particular, models with smaller sums of squared standardized one-step-ahead prediction errors are considered to have a better ability to predict future values of the dependent variable and its volatility forecasts.

Assume that we are interested in comparing the predictive ability of two ARCH models, as represented by (1.6):

Model A	Model B
$y_{t(A)} = \mu_{t(A)} + \varepsilon_{t(A)}$	$y_{t(B)} = \mu_{t(B)} + \varepsilon_{t(B)}$
$\mu_{t(A)} = \mu(\theta_{(A)}\|I_{t-1})$	$\mu_{t(B)} = \mu(\theta_{(B)}\|I_{t-1})$
$\varepsilon_{t(A)} = \sigma_{t(A)} z_{t(A)}$	$\varepsilon_{t(B)} = \sigma_{t(B)} z_{t(B)}$
$\sigma_{t(A)} = g(\theta_{(A)}\|I_{t-1})$	$\sigma_{t(B)} = g(\theta_{(B)}\|I_{t-1})$
$z_{t(A)} \overset{i.i.d.}{\sim} N(0,1)$	$z_{t(B)} \overset{i.i.d.}{\sim} N(0,1)$

The models can be compared through testing for equivalence in their forecasting abilities. The models can be compared through testing for equivalence in their forecasting abilities. Hypotheses referring to notions such as 'forecasting equivalence' can only be formulated implicitly in terms of parameters, measures or properties closely linked with them conceptually.

The closest description of our hypothesis here – adopting the rationale of Xekalaki et al. (2003) and Degiannakis and Xekalaki (2005) – would naturally be in terms of mean squared prediction errors. In particular, our hypothesis that models A and B are of equivalent predictive ability would most closely be formulated as

H_0: models A and B have equal mean squared prediction errors, which can be tested against the alternative hypothesis

H_1: model A has lower mean squared prediction error than model B.

The test proposed by Degiannakis and Xekalaki (2005) for testing the above hypotheses is based on a statistic that compares the standardized one-step-ahead prediction errors, $z_{t+1|t} \equiv (y_t - y_{t+1|t})\sigma_{t+1|t}^{-1}$, given by $\sum_{t=1}^{\widehat{T}} z_{t+1|t(B)}^2 / \sum_{t=1}^{\widehat{T}} z_{t+1|t(A)}^2$. The null hypothesis is rejected if

$$\frac{\sum_{t=1}^{\widehat{T}} z_{t+1|t(B)}^2}{\sum_{t=1}^{\widehat{T}} z_{t+1|t(A)}^2} > CGR\left(\frac{\widehat{T}}{2}, \rho, a\right), \qquad (10.47)$$

where $CGR(\widehat{T}/2, \rho, a)$ is the $100(1-a)$th percentile of the correlated gamma ratio (CGR) distribution with parameters $\widehat{T}/2$ and ρ, defined by Xekalaki et al. (2003). The parameter ρ is the correlation coefficient between $z_{t+1|t(A)}$ and $z_{t+1|t(B)}$, i.e. $\rho \equiv Cor\left(z_{t+1|t(A)}, z_{t+1|t(B)}\right)$. An excerpt from the percentiles of the CGR distribution is given in Table 10.13. Detailed tables of its percentage points and graphs depicting its probability density function are given in Degiannakis and Xekalaki (2005).

10.2.6.1 Theoretical motivation for the SPEC method

The CGR statistic is the ratio of the standardized one-step-ahead prediction errors for a pair of competing models. Let $\theta_{(i)}^{(t)}$ denote the vector of unknown parameters to be estimated for model i at time t. Under the assumption of constancy of the parameters over time, $(\theta_{(i)}^{(1)}) = (\theta_{(i)}^{(2)}) = \ldots = (\theta_{(i)}^{(T)}) = (\theta_{(i)})$, the estimated standardized one-step-ahead prediction errors $z_{t+1|t(i)}, z_{t+2|t+1(i)}, \ldots, z_{\widehat{T}+1|\widehat{T}(i)}$ are asymptotically independently standard normally distributed:

$$z_{t+1|t(i)} \equiv (y_t - y_{t+1|t(i)})\sigma_{t+1|t(i)}^{-1} \sim N(0, 1), \qquad (10.48)$$

for $t = 1, 2, \ldots, \widehat{T}$. Equation (10.48) is valid for all conditional variance functions with consistent estimators of the parameters.[10]

According to Slutsky's theorem, if $\text{plim} z_{t+1|t(i)} = z_{t+1} \sim N(0, 1)$ and $g(z_{t+1|t(i)}) = \sum_{t=1}^{\widehat{T}} z_{t+1|t(i)}^2$, which is a continuous function, then $\text{plim} \sum_{t=1}^{\widehat{T}} z_{t+1|t(i)}^2 = \sum_{t=1}^{\widehat{T}} z_{t+1}^2$. As convergence in probability implies convergence in distribution, $\sum_{t=1}^{\widehat{T}} z_{t+1|t(i)}^2 \xrightarrow{d} \sum_{t=1}^{\widehat{T}} z_{t+1}^2 \sim \chi_{\widehat{T}}^2$. Hence, as the $z_{t+1|t(i)}$

---

[10] For example, Bollerslev (1986), Nelson (1991) and Glosten et al. (1993) noted that, under sufficient regularity conditions, the maximum likelihood estimators of their model parameters are consistent and asymptotically normal.

Table 10.13 Percentage points of the correlated gamma ratio distribution for $a = 0.05$

$$\int_0^z f_{(CGR)}(x; 0.5\,\widehat{T}, \rho)dx = \int_0^z \frac{(1-\rho^2)^{\widehat{T}/2}}{\Gamma(0.5)\widehat{T}^2/\Gamma(\widehat{T})} x^{\widehat{T}/2-1}(1+x)^{-\widehat{T}} \left[1 - \left(\frac{2\rho}{x+1}\right)^2 x\right]^{-(\widehat{T}+1)/2} dx = 1-a$$

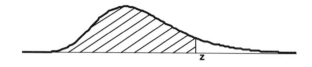

$\widehat{T}/2$	ρ									
	0.000	0.050	0.100	0.150	0.200	0.250	0.300	0.350	0.400	0.450
1	19.202	19.158	19.020	18.808	18.500	18.109	17.620	17.060	16.400	15.663
2	6.388	6.377	6.342	6.283	6.202	6.097	5.968	5.816	5.640	5.441
3	4.284	4.277	4.257	4.224	4.177	4.117	4.043	3.956	3.855	3.740
4	3.438	3.433	3.419	3.396	3.362	3.320	3.267	3.205	3.133	3.051
5	2.978	2.975	2.963	2.945	2.919	2.885	2.844	2.795	2.739	2.674
6	2.687	2.684	2.674	2.659	2.637	2.609	2.575	2.535	2.487	2.434
7	2.484	2.481	2.473	2.460	2.441	2.417	2.388	2.353	2.312	2.265
8	2.333	2.331	2.324	2.312	2.296	2.275	2.249	2.218	2.182	2.141
9	2.217	2.215	2.209	2.198	2.184	2.164	2.141	2.113	2.081	2.044
10	2.124	2.122	2.117	2.107	2.093	2.076	2.055	2.029	2.000	1.966
11	2.048	2.046	2.041	2.032	2.019	2.003	1.984	1.960	1.933	1.902
12	1.984	1.982	1.977	1.969	1.957	1.943	1.924	1.902	1.877	1.848
13	1.929	1.928	1.923	1.915	1.905	1.891	1.874	1.853	1.829	1.802
14	1.882	1.881	1.876	1.869	1.859	1.846	1.830	1.810	1.788	1.762
15	1.841	1.840	1.835	1.829	1.819	1.807	1.791	1.773	1.752	1.727
16	1.804	1.803	1.799	1.793	1.784	1.772	1.757	1.740	1.720	1.697
17	1.772	1.771	1.767	1.761	1.752	1.741	1.727	1.711	1.691	1.669
18	1.743	1.742	1.738	1.732	1.724	1.713	1.700	1.684	1.666	1.644
19	1.717	1.716	1.712	1.706	1.698	1.688	1.675	1.660	1.643	1.622
20	1.693	1.692	1.688	1.683	1.675	1.665	1.653	1.638	1.621	1.602
21	1.671	1.670	1.667	1.661	1.654	1.644	1.633	1.619	1.602	1.583
22	1.651	1.650	1.647	1.642	1.635	1.625	1.614	1.600	1.584	1.566
23	1.632	1.631	1.629	1.624	1.617	1.608	1.597	1.584	1.568	1.550
24	1.615	1.614	1.612	1.607	1.600	1.591	1.581	1.568	1.553	1.536
25	1.599	1.599	1.596	1.591	1.585	1.576	1.566	1.553	1.539	1.522
26	1.585	1.584	1.581	1.577	1.570	1.562	1.552	1.540	1.526	1.501
27	1.571	1.570	1.567	1.563	1.557	1.549	1.539	1.527	1.514	1.498
28	1.558	1.557	1.555	1.550	1.544	1.536	1.527	1.515	1.502	1.487
29	1.546	1.545	1.542	1.538	1.532	1.525	1.516	1.504	1.491	1.476
30	1.534	1.534	1.531	1.527	1.521	1.514	1.505	1.494	1.481	1.466

(continued)

Table 10.13 (*Continued*)

| | | | | | ρ | | | | | |
k	0.500	0.550	0.600	0.650	0.700	0.750	0.800	0.850	0.900	0.950
1	14.835	13.920	12.910	11.830	10.650	9.392	8.041	6.596	5.049	3.368
2	5.217	4.969	4.696	4.397	4.072	3.719	3.336	2.919	2.456	1.923
3	3.611	3.467	3.309	3.135	2.944	2.736	2.507	2.255	1.971	1.633
4	2.959	2.856	2.742	2.616	2.478	2.327	2.159	1.973	1.760	1.503
5	2.601	2.520	2.429	2.330	2.220	2.098	1.964	1.813	1.640	1.428
6	2.373	2.305	2.229	2.145	2.053	1.951	1.837	1.709	1.560	1.377
7	2.213	2.154	2.088	2.016	1.935	1.846	1.747	1.634	1.503	1.340
8	2.094	2.042	1.984	1.919	1.847	1.768	1.679	1.578	1.460	1.312
9	2.002	1.954	1.902	1.843	1.779	1.706	1.625	1.533	1.425	1.290
10	1.927	1.884	1.836	1.783	1.723	1.657	1.582	1.497	1.397	1.272
11	1.866	1.826	1.782	1.732	1.677	1.616	1.546	1.467	1.374	1.256
12	1.815	1.778	1.736	1.690	1.638	1.581	1.516	1.442	1.354	1.243
13	1.771	1.736	1.697	1.654	1.605	1.551	1.490	1.420	1.337	1.232
14	1.733	1.700	1.663	1.622	1.577	1.525	1.467	1.401	1.322	1.222
15	1.700	1.669	1.634	1.595	1.551	1.502	1.447	1.384	1.309	1.213
16	1.670	1.641	1.607	1.570	1.529	1.482	1.430	1.369	1.297	1.205
17	1.644	1.616	1.584	1.548	1.509	1.464	1.414	1.356	1.287	1.198
18	1.620	1.593	1.563	1.529	1.491	1.448	1.399	1.344	1.277	1.192
19	1.599	1.573	1.544	1.511	1.474	1.433	1.386	1.333	1.269	1.186
20	1.580	1.554	1.526	1.495	1.459	1.420	1.375	1.323	1.261	1.181
21	1.562	1.538	1.510	1.480	1.446	1.407	1.364	1.313	1.253	1.176
22	1.545	1.522	1.496	1.466	1.433	1.396	1.354	1.305	1.247	1.171
23	1.530	1.508	1.482	1.454	1.421	1.385	1.344	1.297	1.240	1.167
24	1.516	1.494	1.470	1.442	1.411	1.376	1.336	1.290	1.234	1.163
25	1.503	1.482	1.458	1.431	1.401	1.367	1.328	1.283	1.229	1.159
26	1.491	1.470	1.447	1.421	1.391	1.358	1.320	1.276	1.224	1.156
27	1.480	1.460	1.437	1.411	1.382	1.350	1.313	1.270	1.219	1.153
28	1.469	1.449	1.427	1.402	1.374	1.343	1.307	1.265	1.215	1.150
29	1.459	1.440	1.418	1.394	1.366	1.336	1.300	1.260	1.211	1.147
30	1.450	1.431	1.410	1.386	1.359	1.329	1.294	1.255	1.207	1.144

are asymptotically standard normal variables, the statistic $\sum_{t=1}^{\widehat{T}} z_{t+1|t(i)}^2$ is asymptotically χ^2 distributed with $\widehat{T}$ degrees of freedom:

$$\sum_{t=1}^{\widehat{T}} z_{t+1|t(i)}^2 \xrightarrow{d} \chi_{\widehat{T}}^2. \tag{10.49}$$

According to Kibble (1941), if, for $t = 1, 2, \ldots, \widehat{T}$, $z_{t+1|t(A)}$ and $z_{t+1|t(B)}$ are standard normally distributed variables, following jointly the bivariate standard normal distribution, then the joint distribution of $\left(\frac{1}{2}\sum_{t=1}^{\widehat{T}} z_{t+1|t(A)}^2, \frac{1}{2}\sum_{t=1}^{\widehat{T}} z_{t+1|t(B)}^2\right)$ is the bivariate gamma distribution with probability density function given by:

$$f_{(BG)}(x, y; \widehat{T}, \rho) = \frac{\exp(-(x+y)/(1-\rho^2))}{\Gamma(\widehat{T}/2)(1-\rho^2)^{\widehat{T}/2}}$$

$$\times \sum_{i=0}^{\infty} \left(\frac{(\rho/(1-\rho^2))^{2i}}{\Gamma(i+1)\Gamma(i+(\widehat{T}/2))} (xy)^{(\widehat{T}/2)-1-i} \right), \qquad (10.50)$$

where $x, y > 0$, $\Gamma(.)$ is the gamma function and $\rho \equiv Cor(z_{t+1|t(A)}, z_{t+1|t(B)})$.[11] Xekalaki et al. (2003) showed that, when the joint distribution of $\left(\frac{1}{2}\sum_{t=1}^{\widehat{T}} z_{t+1|t(A)}^2, \frac{1}{2}\sum_{t=1}^{\widehat{T}} z_{t+1|t(B)}^2 \right)$ is Kibble's bivariate gamma, the distribution of the ratio $\frac{1}{2}\sum_{t=1}^{\widehat{T}} z_{t+1|t(B)}^2 / \frac{1}{2}\sum_{t=1}^{\widehat{T}} z_{t+1|t(A)}^2$ is defined by the p.d.f.

$$f_{(CGR)}(x; 0.5\widehat{T}, \rho) = \frac{(1-\rho^2)^{\widehat{T}/2}}{(\Gamma(0.5\widehat{T}))^2/\Gamma(\widehat{T})} x^{\widehat{T}/2-1}(1+x)^{-\widehat{T}} \left[1 - \left(\frac{2\rho}{x+1}\right)^2 x \right]^{-(\widehat{T}+1)/2},$$

$$(10.51)$$

where $x > 0$. Therefore, in the case of comparing two ARCH models,

$$x \equiv \frac{\sum_{t=1}^{\widehat{T}} z_{t+1|t(B)}^2}{\sum_{t=1}^{\widehat{T}} z_{t+1|t(A)}^2} \sim CGR\left(\frac{\widehat{T}}{2}, \rho\right). \qquad (10.52)$$

Xekalaki et al. (2003) referred to the distribution in (10.51) as the correlated gamma ratio distribution.

The standardized prediction errors of linear regression models can also be used to compare their predictive accuracy. Consider the regression model in the form $y_t = x_t'\beta + \varepsilon_t$, where x_t is a $k \times 1$ vector of explanatory variables included in the information set I_{t-1}, β is a $k \times 1$ vector of unknown parameters and $\varepsilon_t \sim N(0, \sigma^2)$. In such a case, the variable $z_{t+1|t}$ is defined as the standardized distance between the predicted and the observed value of the dependent random variable, or $z_{t+1|t} \equiv (y_{t+1} - x_{t+1}'\beta^{(t)}) V(x_{t+1}'\beta^{(t)})^{-1/2}$, where

$$V(x_{t+1}'\beta^{(t)}) = (Y_t - X_t\beta^{(t)})'(Y_t - X_t\beta^{(t)})(1 + x_{t+1}(X_t'X_t)^{-1}x_{t+1}')(t-k)^{-1},$$

$\beta^{(t)}$ is the least squares estimator of β at time t, Y_t is the $(t \times 1)$ vector of t observations on the dependent variable y_t, and X_t is the $(t \times k)$ matrix whose rows comprise the k-dimensional vectors x_t of the explanatory variables, i.e.

$$X_t = \begin{bmatrix} X_{t-1} \\ x_t' \end{bmatrix}, \quad Y_t = \begin{bmatrix} Y_{t-1} \\ y_t \end{bmatrix}.$$

---

[11] In the case of the standardized one-step-ahead prediction errors from ARCH models, $x \equiv \frac{1}{2}\sum_{t=1}^{\widehat{T}} z_{t+1|t(A)}^2$ and $y \equiv \frac{1}{2}\sum_{t=1}^{\widehat{T}} z_{t+1|t(B)}^2$.

10.2.6.2 The poly-model SPEC

According to the SPEC model selection method, the models that are considered as having a better ability to predict future values are those with lower sum of squared standardized one-step-ahead prediction errors. Assume, in particular, that a set of candidate ARCH models $i = 1, \ldots, M$ is available and that the most suitable model is sought for predicting conditional volatility:

$$y_{t(i)} = \mu_{t(i)} + \varepsilon_{t(i)},$$
$$\mu_{t(i)} = \mu(\theta_{(i)}|I_{t-1}),$$
$$\varepsilon_{t(i)} = \sigma_{t(i)}z_{t(i)}, \qquad\qquad (10.53)$$
$$\sigma_{t(i)} = g(\theta_{(i)}|I_{t-1}),$$
$$z_{t(i)} \overset{i.i.d.}{\sim} N(0,1).$$

The most suitable model for forecasting one-step-ahead volatility at time $\widehat{T} + 1$ is the model, i, with the minimum value of $\sum_{t=1}^{\widehat{T}} z_{t+1|t(i)}^2$. Therefore, the use of the SPEC algorithm can be repeated for each of a sequence of points in time at which we are looking for the most suitable of the M competing models for obtaining a volatility forecast for the next point in time. At time $\widehat{T}$, selecting a strategy for the most appropriate model to forecast volatility at time $\widehat{T} + 1$ could naturally amount to selecting the model which, at time $\widehat{T}$, has the lowest standardized one-step-ahead prediction errors. Table 10.14 provides a stepwise representation of the SPEC algorithm for selecting the most suitable of the M candidate models at each of a series of points in time, using a rolling sample of $\widehat{T}$ observations.

Model i will be chosen at time $j = \widehat{T}$ for forecasting volatility at time $j = \widehat{T} + 1$ if it is the one that corresponds to the cell of column $\widehat{T}$ that has the minimum value of $\sum_{t=1}^{\widehat{T}} z_{t+1|t(i)}^2$.

By this method, even in the case of a non-normal conditional distribution for the residuals, the ARCH model with the lowest sum of squared standardized one-step-ahead forecasting errors appears to be the most appropriate in forecasting one-step-ahead volatility. Hence, the applicability of the SPEC algorithm in the case of non-normally distributed conditional innovations and of more flexible volatility specifications deserves investigation. Looking into such aspects in the light of Politis's (2004, 2006, 2007b) approach (in terms of transformation of innovations to normally distributed empirical ratios by dividing the ARCH process by a time-localized measure of standard deviation), might be a possibility.

In the context of a simulated options market, Xekalaki and Degiannakis (2005) noted that the SPEC algorithm, for $\widehat{T} = 5$, exhibits a better performance than any other comparative method of model selection in forecasting one-day-ahead conditional variance. Its advantage in predicting realized volatility, for forecast horizons ranging from 1 day ahead to 100 days ahead, has also been examined by Degiannakis and Xekalaki (2007a). The question of whether a trader using models for volatility

Table 10.14 Estimation steps required at time j for each model i by the SPEC model selection algorithm. At time j, select the model i with the minimum value for the sum of the squares of the $\widehat{T}$ most recent standardized one-step-ahead prediction errors, $\sum_{t=1}^{T} z_{t+1|t(i)}^2$

Model	Time				
	$j = \widehat{T}$	$j = \widehat{T}+1$	$\ldots$	$j = \widehat{T}+j'$	$\ldots$	$j = \widetilde{T}$
$i = 1$	$\displaystyle\sum_{t=1}^{\widehat{T}} z_{t+1|t(1)}^2$	$\displaystyle\sum_{t=2}^{\widehat{T}+1} z_{t+1|t(1)}^2$	$\ldots$	$\displaystyle\sum_{t=j'+1}^{\widehat{T}+j'} z_{t+1|t(1)}^2$	$\ldots$	$\displaystyle\sum_{t=\widetilde{T}-\widehat{T}+1}^{\widetilde{T}} z_{t+1|t(1)}^2$
$i = 2$	$\displaystyle\sum_{t=1}^{\widehat{T}} z_{t+1|t(2)}^2$	$\displaystyle\sum_{t=2}^{\widehat{T}+1} z_{t+1|t(2)}^2$	$\ldots$	$\displaystyle\sum_{t=j'+1}^{\widehat{T}+j'} z_{t+1|t(2)}^2$	$\ldots$	$\displaystyle\sum_{t=\widetilde{T}-\widehat{T}+1}^{\widetilde{T}} z_{t+1|t(2)}^2$
$\ldots$	$\ldots$	$\ldots$	$\ddots$	$\ldots$	$\ddots$	$\ldots$
$i = M$	$\displaystyle\sum_{t=1}^{\widehat{T}} z_{t+1|t(M)}^2$	$\displaystyle\sum_{t=2}^{\widehat{T}+1} z_{t+1|t(M)}^2$	$\ldots$	$\displaystyle\sum_{t=j'+1}^{\widehat{T}+j'} z_{t+1|t(M)}^2$	$\ldots$	$\displaystyle\sum_{t=\widetilde{T}-\widehat{T}+1}^{\widetilde{T}} z_{t+1|t(M)}^2$

forecasts picked by the SPEC algorithm makes profits from option pricing was investigated by Degiannakis and Xekalaki (2008). Adopting a similar approach in Section 10.3, we apply the SPEC algorithm as well as other model selection methods that are based on the loss functions of Section 10.1.2.

10.2.7 Forecast encompassing tests

Let $y_{t+\tau|t(i)}$ denote the forecast produced by model $i = 1, 2$ for the variable y_{t+1}. A composite forecast can be created, given by $y_{t+\tau|t(3)} = (1-w)y_{t+\tau|t(1)} + wy_{t+\tau|t(2)}$, for $0 \leq w \leq 1$. The one-step-ahead error of the combined[12] forecast can be written as $\varepsilon_{t+\tau|t(3)} = \varepsilon_{t+\tau|t(1)} - w(\varepsilon_{t+\tau|t(1)} - \varepsilon_{t+\tau|t(2)})$, for $\varepsilon_{t+\tau|t(i)} = y_{t+\tau|t(i)} - y_{t+\tau}$. Granger and Newbold (1973) refer to a forecast $y_{t+\tau|t(1)}$ as being *conditionally efficient* with respect to the competing forecast $y_{t+\tau|t(2)}$, if the hypothesis $H_0 : w = 0$ is not rejected in favour of the alternative $H_1 : w > 0$. In such a case, $y_{t+\tau|t(2)}$ is encompassed by $y_{t+\tau|t(1)}$. Based on the technique of Diebold and Mariano (1995), Harvey et al. (1998) tested the null hypothesis that $y_{t+\tau|t(1)}$ encompasses $y_{t+\tau|t(2)}$, using the statistic

$$FoEn_{(1,2)} = \frac{\tilde{T}^{-1}\sum_{t=1}^{\tilde{T}}\left(\varepsilon_{t+\tau|t(1)}\left(\varepsilon_{t+\tau|t(1)}-\varepsilon_{t+\tau|t(2)}\right)\right)}{\sqrt{V\left(\tilde{T}^{-1}\sum_{t=1}^{\tilde{T}}\left(\varepsilon_{t+\tau|t(1)}\left(\varepsilon_{t+\tau|t(1)}-\varepsilon_{t+\tau|t(2)}\right)\right)\right)}}. \tag{10.54}$$

The choice of this statistic derives from the fact that if the null hypothesis is true, then the covariance between $\varepsilon_{t+\tau|t(1)}$ and $(\varepsilon_{t+\tau|t(1)} - \varepsilon_{t+\tau|t(2)})$ will be zero or negative. In contrast, if $y_{t+\tau|t(1)}$ does not encompass $y_{t+\tau|t(2)}$, then the covariance between $\varepsilon_{t+\tau|t(1)}$ and $(\varepsilon_{t+\tau|t(1)} - \varepsilon_{t+\tau|t(2)})$ will be positive. The $FoEn_{(1,2)}$ statistic is asymptotically normally distributed. A consistent estimate of the variance in its denominator can be obtained as in the case of the DM statistic. Harvey et al.'s (1997) modification to the DM statistic as in (10.42) applied to $FoEn_{(1,2)}$ yields

$$mFoEn_{(1,2)} = \sqrt{\frac{\tilde{T} + 1 - 2\tau + \tilde{T}^{-1}\tau(\tau-1)}{\tilde{T}}} FoEn_{(1,2)}. \tag{10.55}$$

This modified version of $mFoEn_{(1,2)}$ is compared to critical values of Student's t distribution with $\tilde{T}-1$ degrees of freedom rather than the standard normal distribution. The forecast encompassing statistics can be applied to test whether $\sigma^2_{t+\tau|t(1)}$ encompasses $\sigma^2_{t+\tau|t(2)}$ by replacing, for example, $\varepsilon_{t+\tau|t(i)}$ by $(\sigma^2_{t+\tau|t(i)} - \varepsilon^2_{t+\tau|t(i)})$. More information about forecast encompassing tests can be found in Chong and Hendry (1986), Clark and McCracken (2001), Clements and Hendry (1993), Ericsson (1992) and the references therein. Sollis (2005) considered the use of forecast encompassing tests to evaluate one-step-ahead forecasts of the S&P500 composite index returns and variance from ARCH models with macroeconomic variables in both the conditional mean and the conditional variance. Awartani and Corradi (2005), employing forecast encompassing tests as well as White's RC and Diebold and Mariano's DM tests, noted

---

[12] For more about the fundamental ideas on the combination of forecasts, see Bates and Granger (1969), Clemen (1989), Granger (1989) and Newbold and Granger (1974).

that the GARCH(1,1) model appears to exhibit inferior performance in comparison to asymmetric ARCH models.

10.3 Application of loss functions as methods of model selection

In Section 8.5.4 the volatility forecasting performance of nine ARCH models was assessed in terms of measuring the rate of return from trading straddles. In what follows, we compare model selection methods that allow traders flexibility as to the choice of the model to use for prediction at each of a sequence of points in time. The comparisons will be carried out on the basis of cash flows from trading options based on volatility forecasts as a means of measuring the performance of the model selection methods under study.

10.3.1 Applying the SPEC model selection method

The loss function of the SPEC algorithm will be computed at the first stage. Recall that the SPEC model selection method results in picking the model with the lowest sum of squared standardized one-step-ahead prediction errors.

The value of $\sum_{t=1}^{\widehat{T}} z_{t+1|t}^2$ is computed for various values of $\widehat{T}$ and, in particular, for $\widehat{T} = 5(10)55$. Thus, it is implicitly assumed that there are six traders, each using, on each trading day, the ARCH model picked by the SPEC algorithm to forecast volatility and straddle prices for the next trading day. The SPEC algorithm will be run for selecting the most suitable of nine candidate models at each of a series of points in time. These are the AR(1)-GARCH(p, 1)-N, AR(1)-EGARCH(p,1)-N and AR(1)-GJR(p,1)-N models for $p = 0, 1, 2$. The program that carries out the necessary computations is *chapter10.SPEC.prg* and is given in the Appendix to this chapter. For models $i = 1, 2, \ldots, 9$ and for each point in time $j = \widehat{T}, \widehat{T} + 1, \ldots, \tilde{T}$,[13] the values of the statistic $\sum_{t=1}^{\widehat{T}} (y_{t+1} - y_{t+1|t(i)})^2 / \sigma_{t+1|t(i)}^2$ are saved in the matrices denoted by *sum_z{!w}* for $!w = 5,15,25,\ldots,55$. Columns $!i$, $!i = 1,\ldots,9$, of the matrices *sum_z{!w}* contain the values of the statistic for models i, $i = 1, 2, \ldots, 9$. The minimum value of $\sum_{t=1}^{\widehat{T}} (y_{t+1} - y_{t+1|t(i)})^2 / \sigma_{t+1|t(i)}^2$, at each point in time, is stored in the matrix *min_z*. Columns $!k$, $!k = 1,\ldots,6$, correspond to the different values of $\widehat{T}$, for $\widehat{T} = 5, 15, 25, \ldots, 55$. Rows $!j$, $!j = 1,\ldots,1161$, correspond to the 1161 trading days. *Model_min_z* refers to the number indexing the model with the minimum value of $\sum_{t=1}^{\widehat{T}} (y_{t+1} - y_{t+1|t(i)})^2 / \sigma_{t+1|t(i)}^2$ at each point in time. The columns of the matrix denoted by *straddle_spec* contain the straddle prices based on the volatility forecasts of the model picked by the SPEC algorithm, i.e. the model with the minimum value of $\sum_{t=1}^{\widehat{T}} (y_{t+1} - y_{t+1|t(i)})^2 / \sigma_{t+1|t(i)}^2$ at each point in time. For example,

---

[13] Recall that $\widehat{T}$ denotes the number of observations for the computation of the SPEC statistic, whereas $\tilde{T}$ denotes the number of observations for out-of-sample forecasting.

the cell of the matrix named *straddle_spec*, corresponding to its 1082nd row and first column, stores the straddle price based on the volatility forecast of the AR(1)-EGARCH(0,1)-N model for the 1082nd trading day, or 11 March 1998. Also, in the matrix denoted by *model_min_z*, the model picked, for the 1082nd trading day, corresponds to its 582nd row and first column, which is the seventh model, i.e. AR(1)-EGARCH(0,1)-N.

Table 10.15 presents, for each trader following the SPEC model selection strategy, the rate of return from trading straddles on a daily basis. The rate of return is computed according to (8.62).[14] With transaction costs of $2 and a filter of $3.50, the trader utilizing the SPEC algorithm with $\widehat{T} = 35$ (symbolically denoted by SPEC(35)) achieves the highest rate of return. In particular, the agent basing the choice of models on the SPEC(35) algorithm makes 1.99% per day by trading for 329 days, with a t-ratio of 1.93. Also the trader using the ARCH models picked by the SPEC($\widehat{T}$) model selection method for $\widehat{T} = 5, 35, 45, 55$ has achieved higher cumulative returns than those of any other trader always using the same ARCH model.[15]

However, the SPEC algorithm appears not to have picked, for any value of $\widehat{T}$, models that generate cumulative returns higher than those generated by the AR(1)-GJR (0,1)-N model. This may be attributed to the fact that the SPEC algorithm is based on the assumption that $z_{t+1|t(i)} \equiv (y_{t+1} - y_{t+1|t(i)})\sigma_{t+1|t(i)}^{-1}$ is asymptotically standard normally distributed, an assumption that is valid for any model with consistent parameter estimators. Therefore, the SPEC method has been designed to apply in one-step-ahead forecasting. As the distribution of $z_{t+\tau|t(i)}$, for $\tau > 1$, is not common to all the ARCH models, a unified evaluation framework for a set of ARCH models over a longer time period, e.g. over the lifetime of the option, is not available. In the present application, however, the SPEC method of model selection, which compares the one-step-ahead forecasting ability of a set of models, was evaluated in longer-than-one-day-ahead forecasting. Even so, the SPEC method has exhibited a satisfactory performance.

The models picked by the SPEC algorithm are presented in Table 10.16. For example, the AR(1)-GARCH(0,1)-N model was picked by SPEC(35) on 19% of trading days. The SPEC(35) selection algorithm picked the GARCH, the GJR and the EGARCH models on 41.5%, 18.4%, and 40.1% of cases, respectively. It is noticeable that the classical AR(1)-GARCH(1,1)-N model is less frequently selected.

10.3.2 Applying loss functions as methods of model selection

The loss functions presented in (10.5)–(10.10), like any relevant measure of the closeness of the forecasts to the corresponding realized values, can be used as a means of model selection. In this section we will look at how these functions perform when

---

[14] Matrices *spec_Str_Ret_F{!x}C{!p}*, for $\{!x\} = 1,2,3$ and $\{!p\} = 1,2$ are constructed in a similar manner to matrices *str_Ret_F{!x}C{!p}*, in Section 8.5.4. For example, *spec_Str_Ret_F3C2* stores in row *!i* and column *!j* the rate of return from trading straddles according to the volatility forecast produced by the model picked by the SPEC($\widehat{T}$) algorithm, for $\widehat{T} = 10(!j - 1) + 5$, on the *!i*th trading day, taking into consideration a filter of $F_{il} = \$3.50$, or $\{!x\}=3$, and a transaction cost of $X = \$2.00$, or $\{!p\} = 2$.

[15] Recall from Section 8.5.4, that, for $F_{il} = \$3.50$ and $X = \$2.00$, the trader using the AR(1)-GJR(0,1)-N model makes a daily profit of 1.8% with a t-ratio of 1.64.

Table 10.15 Daily rates of return from trading straddles on the S&P500 index based on the ARCH models picked by the SPEC model selection algorithm. Applying the SPEC model selection algorithm, $\sum_{t=1}^{T} \hat{z}_{t+1}^2 I|t(t)$ was estimated for various values for $\hat{T}$, and, in particular, $\hat{T} = 5(10)55$, e.g., SPEC(5), corresponds to the SPEC model selection algorithm for $\hat{T} = 5$

Model	$0 transaction cost $0.00 filter				$2 transaction cost $2.00 filter				$2 transaction cost $3.50 filter			
	Mean	Stand. Dev.	t ratio	Days	Mean	Stand. Dev.	t ratio	Days	Mean	Stand. Dev.	t ratio	Days
SPEC(5)	3.97%	17.9%	4.75	456	1.00%	18.5%	1.04	374	1.97%	18.7%	1.91	327
SPEC(15)	3.87%	17.9%	4.63	456	0.71%	18.8%	0.74	376	1.49%	18.8%	1.43	323
SPEC(25)	3.87%	17.9%	4.63	456	0.74%	18.5%	0.79	389	1.33%	18.4%	1.33	340
SPEC(35)	**4.05%**	**17.8%**	**4.85**	**456**	**1.20%**	**18.8%**	**1.23**	**371**	**1.99%**	**18.7%**	**1.93**	**329**
SPEC(45)	4.00%	17.8%	4.78	456	1.12%	18.8%	1.15	372	1.96%	18.8%	1.88	326
SPEC(55)	4.02%	17.8%	4.81	456	1.16%	18.9%	1.18	370	1.91%	18.8%	1.85	328

Table 10.16 Percentages of ARCH models picked by the SPEC($\widehat{T}$) algorithm for trading straddles on the S&P500 index with transaction costs of $2.00 and a $3.5 filter, classified by the conditional variance specification (14 March 1996 to 2 June 2000)

| | $\widehat{T} = 5(10)55$ | | | | | |
	$\widehat{T} = 5$	$\widehat{T} = 15$	$\widehat{T} = 25$	$\widehat{T} = 35$	$\widehat{T} = 45$	$\widehat{T} = 55$
1. AR(1)-GARCH(0,1)	14.2	15.8	15.4	19.3	24.5	30.0
2. AR(1)-GARCH(1,1)	5.1	5.1	6.0	9.2	10.6	12.9
3. AR(1)-GARCH(2,1)	14.9	8.7	11.7	12.9	10.8	9.4
4. AR(1)-GJR (0,1)	2.7	1.4	0.7	0.0	0.0	0.0
5. AR(1)-GJR (1,1)	18.3	18.3	17.4	15.4	17.4	16.7
6. AR(1)-GJR (2,1)	6.9	7.8	6.2	3.0	0.7	0.4
7. AR(1)-EGARCH(0,1)	19.3	28.2	31.4	29.1	27.1	25.9
8. AR(1)-EGARCH(1,1)	12.4	13.1	9.4	9.0	6.6	4.3
9. AR(1)-EGARCH(2,1)	6.2	1.6	1.8	2.0	2.3	0.5
Total trading days	564	564	564	564	564	564

Conditional Variance

they are used as criteria for model selection in comparison to the SPEC method. As in the previous section, the comparisons will be made in an applied financial context. In particular, the performance of the loss functions defined by (10.5)–(10.10) will be judged in the context of a hypothetical options market against the performance of the SPEC criterion by comparing daily rates of returns that hypothetical traders achieve from trading straddles when they use volatility forecasts produced by models picked by the criterion of their choice.

We have the following set-up: seven traders in all, each using one of the seven available criteria to pick one of the ARCH models considered in the previous section that will provide them, at each of a sequence of points in time, with the volatility forecast they need in order to trade straddles.

In applying the SPEC model selection algorithm before, the sum of squared standardized one-step-ahead prediction errors, $\sum_{t=1}^{T} \varepsilon_{t+1|t(i)}^2 / \sigma_{t+1|t(i)}^2$, was estimated for various values of $\widehat{T}$. Therefore, each of the remaining six model selection criteria as defined by (10.5)–(10.10) can be computed for the same values of $\widehat{T}$, and, in particular, $\widehat{T} = 5(10)55$. Selecting a strategy based on any of several competing methods of model selection naturally amounts to selecting the ARCH model that, at each of a sequence of points in time, has the lowest value of the average of the $\widehat{T}$ most recently computed loss functions. Consider, for example, a trader who utilizes as a method of model selection the SE loss function given by (10.5). This trader would select at each point in time the model with the lowest average value of the $\widehat{T}$ most recently squared differences between one-day-ahead variance forecasts and squared one-day ahead prediction errors, or $\min_i \widehat{T}^{-1} \sum_{t=1}^{T} (\sigma_{t+1|t(i)}^2 - \varepsilon_{t+1|t(i)}^2)^2$.

The program named *chapter10.loss_functions_as_model_selection_methods.prg* computes, at each point in time, the values of the loss functions in (10.5)–(10.10), their $\widehat{T}$ most recent average values, for $\widehat{T} = 5(10)55$, and the cumulative rate of returns from trading straddles based on the volatility forecast of the ARCH model with the minimum average value of the $\widehat{T}$ most recently estimated loss functions in (10.5)–(10.10). The values of the loss functions in (10.5)–(10.10) are saved in the columns of the matrices named *lossF{!ii}*, *!ii* = 1,...,6. In the illustration that follows, consider for example the SE loss function in (10.5). Column *i* of *loss_F1* contains the values of the squared error loss function for model *i*, *i* = 1, ..., 9. Each of the *t* = 1, ..., 1161 rows of *loss_F1* contains the values of the squared error loss function for each point in time. Thus, the *t*th row and *i*th column cell contains the value of the loss function $(\sigma_{t+1|t(i)}^2 - \varepsilon_{t+1|t(i)}^2)^2$. The matrix *min_lossF1* contains the minimum values of $\sum_{t=1}^{T} (\sigma_{t+1|t(i)}^2 - \varepsilon_{t+1|t(i)}^2)^2$ at each point in time. Columns *!k*, *!k* = 1,...,6, correspond to the different values of $\widehat{T}$, for $\widehat{T} = 5, 15, 25, \ldots, 55$, and rows *!j*, *!j* = 1,...,1161, correspond to the 1161 trading days. *Model_min_lossF1* refers to the number indexing the model with the minimum value of $\sum_{t=1}^{T} (\sigma_{t+1|t(i)}^2 - \varepsilon_{t+1|t(i)}^2)^2$ at each point in time. The matrix *straddle_lossF1* contains the straddle prices based on the volatility forecast of the model picked by the F1 algorithm (the sum of the squared error loss function), i.e. the model with the minimum value of $\sum_{t=1}^{T} (\sigma_{t+1|t(i)}^2 - \varepsilon_{t+1|t(i)}^2)^2$ at each point in time. For example, the cell in the

Table 10.17   Daily rates of return from trading straddles on the S&P500 index based on the ARCH models picked by the averages of the loss functions (10.5)–(10.10) with a transaction cost of $2.00 and a $3.50 filter. The column headed 'sample size' refers to the sample size, $\widehat{T}$, for which the corresponding model selection method leads to the highest rate of return

Model selection method	Sample size ($\widehat{T}$)	Mean (%)	Stand. dev. (%)	t ratio	Days
$\bar{\Psi}_{(SE)}$	5	1.20	18.16	1.25	355
$\bar{\Psi}_{(AE)}$	45	1.04	18.24	1.06	347
$\bar{\Psi}_{(HASE)}$	45	1.31	18.08	1.35	345
$\bar{\Psi}_{(HAAE)}$	45	1.58	18.28	1.61	347
$\bar{\Psi}_{(LE)}$	35	0.87	18.23	0.89	342
$\bar{\Psi}_{(GLLS)}$	25	0.81	18.02	0.85	359

1082nd row and the first column of $straddle\_lossF1$ contains the straddle price based on the volatility forecast produced by the AR(1)-EGARCH(2,1)-N model for the 1082nd trading day, 11 March 1998. In the matrix $model\_min\_lossF1$, the model picked, for the 1082nd trading day, corresponds to its 582nd row and first column, which is the ninth model, i.e. AR(1)-EGARCH(2,1)-N.

Taking into acccount a transaction cost of $2 and a filter of $3.50, Table 10.17 presents the daily rate of returns from trading straddles on the S&P500 index based on six model selection methods that pick, at each point in time, the ARCH model with the minimum of the average value of the (10.5)–(10.10) loss functions. For example, a trader who selects the volatility forecast models according to the SE algorithm with $\widehat{T} = 5$, makes 1.20% per day trading for 355 days, with a t-ratio of 1.25.[16] None of the six methods appears to have rendered higher returns than the SPEC(35) algorithm. Moreover, the highest returns in Table 10.17 appear to be lower than the lowest return in Table 10.15, excluding the case of the HAAE loss function. Only the agent who selects models according to the average value of the HAAE loss function appears to achieve higher returns than an agent who employs the SPEC algorithm with the worst performance, SPEC(25). Even the AR(1)-GJR(0,1)-N model, which yields a daily profit of 1.8%, leads to a higher rate of return than that of the models picked by the methods of model selection that are based on the average value of the loss functions given by (10.5)–(10.10). In none of the cases does an agent picking at each point in time the model with the minimum value of the average of any one of the loss functions (10.5)–(10.10) appear to have achieved returns higher than those of a trader who is based only on a single ARCH model, e.g. the AR(1)-GJR(0,1)-N model.

The comparison of the SPEC algorithm with a set of other model evaluation criteria reveals that traders using the SPEC algorithm for deciding which model's forecasts to use at any given point in time appear to achieve higher cumulative profits

---

[16] The values in the first row of Table 10.17 have been computed as the descriptive statistics of the first column of the matrix $lossF1\_str\_ret\_f3c2$. The values in the last row of Table 10.17 have been computed as the descriptive statistics of the third column of $lossF6\_str\_ret\_f3c2$.

than those using other methods of model selection or only a single model all the time. This is an indication of the superiority of the loss function of the SPEC algorithm over other loss functions in terms of the ability to select the models that would produce accurate volatility estimations for option pricing predictions. Of course, this is only an indication. We could probably find data sets or models for which the SPEC method may not pick models with the best forecasting performance.

In Xekalaki and Degiannakis (2005) the comparative evaluation in an option pricing framework was considered by simulating an options market. The simulated options market has been considered to avoid the bias induced by the use of actual option prices. The results showed that the SPEC algorithm for $\widehat{T} = 5$ gave the highest rate of return. One may therefore infer that the evidence is rather in support of the assumption that the increase in profits is due to improved volatility prediction and that the SPEC model selection algorithm offers a potential tool for picking the model that would yield the best volatility prediction. If the increase in profits were random, the SPEC algorithm would not yield the highest profits in both the simulated market and the present study based on real-world options data.

10.3.3 Median values of loss functions as methods of model selection

In Section 10.1.2.2 we simulated the distribution of all six loss functions and noted that they appear to be asymmetric with extreme outliers. Such asymmetries may suggest the use of median values instead of average values for the loss functions. With the help of the program named *chapter10.median_loss_functions_as_model_selection_methods.prg*, given in the Appendix to this chapter, we compute the median values of the loss functions and proceed with similar comparisons to those made above. Taking into acccount a transaction cost of $2 and a filter of $3.5, Table 10.18 presents the returns based on the six model selection methods that pick, at each point in time, the ARCH model with the minimum value for the median of the loss functions (10.5)–(10.10). The model with the minimum value of the median of a squared loss function will also render a minimum value for the median of the absolute

Table 10.18   Daily rates of return from trading straddles on the S&P500 index based on the ARCH models picked by the medians of the loss functions (10.5)–(10.10) with a transaction cost of $2.00 and a $3.50 filter

Model selection method	Sample size ($\widehat{T}$)	Mean (%)	Stand. dev. (&)	t ratio	Days
Median of $\Psi_{(SE)t+1}$	55	0.87	18.44	0.88	354
Median of $\Psi_{(AE)t+1}$	55	0.87	18.44	0.88	354
Median of $\Psi_{(HASE)t+1}$	55	1.90	18.57	1.87	333
Median of $\Psi_{(HAAE)t+1}$	55	1.90	18.57	1.87	333
Median of $\Psi_{(LE)t+1}$	5	0.97	18.51	0.98	349
Median of $\Psi_{(GLLS)t+1}$	55	1.15	18.66	1.10	319

transformation of the same loss function. Thus, an agent whose criterion is based on the minimum of the median value of the SE loss function, picks the same model as does an agent whose selection is made according to the minimum of the median value of the AE loss function. Of course, the same holds for the HASE and HAAE functions. An agent who selects the models with the minimum value for the median of the loss function appears to achieve higher returns than an agent who selects the models with the minimum value for the mean of the same loss function in four out of the six cases.

10.4   The SPA test for VaR and expected shortfall

In Chapter 8, for the FTSE100 index, the one-day-ahead expected shortfall was computed by 12 models for $\tilde{T} = 1435$ trading days based on a rolling sample of $\tilde{T} = 3000$ days. A two-stage back-testing method was applied. In the first stage, the

Table 10.19   Output produced by the program Chapter10.SPA.and.ES.ftse100.ox

Ox version 3.10 (Windows) (C) J.A. Doornik, 1994–2002
The benchmark model is AR(1)-FIAPARCH(1,d,1)-t
- - - - TEST FOR SUPERIOR PREDICTIVE ABILITY - - - -
SPA version 1.14, September 2003. (C) 2001, 2002, 2003 Peter Reinhard Hansen

Number of models: 1 = 9
Sample size: n = 1435
Loss function: mse_spa
Test Statistic: TestStatScaledMax
Bootstrap parameters: B = 10000 (resamples), q = 0.5 (dependence)

	Model Number	Sample Loss	t-stat	"p-value"
Benchmark	0.00000	0.01492	-	-
Most Significant	1.00000	0.01951	−1.93016	0.96310
Best	9.00000	0.01926	−1.96452	0.96580
Model_25%	2.00000	0.02101	−1.99318	0.96650
Median	5.00000	0.02296	−2.14297	0.97200
Model_75%	3.00000	0.06166	−4.51128	0.99990
Worst	4.00000	0.06342	−4.79282	0.99980
		Lower	Consistent	Upper
SPA p-values:		0.72320	0.72320	0.99940
Critical values:	10%	−0.38738	−0.38738	1.82899
	5%	−0.13550	−0.13550	2.04493
	1%	0.26460	0.26460	2.47037

statistical accuracy of the models was examined. Two models failed to pass the first stage. In the second stage, the loss function $\bar{\Psi}_{(ES)}$ was computed to investigate which model, of those that pass the first stage, minimized the loss function. The lowest value of the function $\bar{\Psi}_{(ES)}$ was achieved by the AR(1)-FIAPARCH(1,d,1)-t model, but on the available evidence we could not make any inference about its predictive performance. Angelidis and Degiannakis (2007) considered utilizing the SPA test presented in Section 10.2.5 for evaluating the VaR or the expected shortfall forecasts. Adopting their approach, we proceeded with applying the SPA criterion for comparing the forecasting performance of the AR(1)-FIAPARCH(1,d,1)-t model against that of its nine competitors, in terms of the loss function $\bar{\Psi}_{(ES)}$. With the use of the program *chapter10.SPA.and.ES.ftse100.ox*, given in the Appendix to this chapter, the SPA test is performed for testing the null hypothesis that the benchmark model AR(1)-FIAPARCH(1,d,1)-t is not outperformed by its nine rival models. Table 10.19 summarizes the output of the program. As reflected by the resulting p-value (0.72), it appears that the hypothesis that the AR(1)-FIAPARCH(1,d,1)-t model is superior to its nine competitors is quite plausible.

The SPA criterion can also be utilized to test the forecasting performance of a benchmark model against its competitors, in terms of a VaR loss function, such as those presented in Section 8.1.6, e.g. $\bar{\Psi}_{(VaR)}$.

Appendix

EViews 6

- *chapter10.sim_ar1g11_loss_functions.prg*

```
create chapter10.sim_ar1g11_loss_functions u 1 4000
series _temp = 1
!length = @obs(_temp)
delete _temp
!ssize = 3000
  !c0 = 0.001
  !c1 = 0.2
  !a0 = 0.0001
  !a1 = 0.12
  !b1 = 0.8
series z = nrnd
series e
series h
series y
h(1) = !a0/(1-!a1-!b1)
e(1) = sqr(h(1))*z(1)
y(1) = !c0/(1-!c1)
for !i = 2 to !length
  h(!i) = !a0 + (!a1*(e(!i-1)^2)) + (!b1*h(!i-1))
  e(!i) = sqr(h(!i))*z(!i)
  y(!i) = !c0 + (!c1*y(!i-1)) + e(!i)
next

matrix(!length-!ssize,8) est_ar1g11
for !i = 1 to !length-!ssize
  statusline !i
  smpl !i !i+!ssize-1
  equation ar1_g11.arch(1,1,m=10000,h) y c y(-1)
  ar1_g11.makegarch garch_ar1g11
  ar1_g11.makeresid res_ar1g11
  !ht = garch_ar1g11(!i+!ssize-1)
  !c0 = @coefs(1)
  !c1 = @coefs(2)
  !a0 = @coefs(3)
  !a1 = @coefs(4)
  !b1 = @coefs(5)
  !res = y(!ssize + !i) - !c0 - !c1*y(!ssize + !i - 1)
  !h_hat = !a0 + (!a1*(res_ar1g11(!ssize + !i -1)^2) ) +
    (!b1*!ht)
  est_ar1g11(!i,1) = y(!ssize + !i)
```

```
  est_ar1g11(!i,2) = !res
  est_ar1g11(!i,3) = !h_hat
  est_ar1g11(!i,4) = !c0
  est_ar1g11(!i,5) = !c1
  est_ar1g11(!i,6) = !a0
  est_ar1g11(!i,7) = !a1
  est_ar1g11(!i,8) = !b1
next
vector(!length-!ssize) se1
vector(!length-!ssize) ae1
vector(!length-!ssize) hase1
vector(!length-!ssize) haae1
vector(!length-!ssize) le1
vector(!length-!ssize) glls1
for !i=1 to !length-!ssize
  se1(!i) = (est_ar1g11(!i,3) - est_ar1g11(!i,1)^2)^2
  ae1(!i) = @abs(est_ar1g11(!i,3) - est_ar1g11(!i,1)^2)
  hase1(!i) = (1-(est_ar1g11(!i,1)^2/est_ar1g11(!i,3)))^2
  haae1(!i) = @abs(1-(est_ar1g11(!i,1)^2/
    est_ar1g11(!i,3)))
  le1(!i) = (log(est_ar1g11(!i,1)^2/est_ar1g11(!i,3)))^2
  glls1(!i) = log(est_ar1g11(!i,3)) + (est_ar1g11(!i,1)
    ^2/est_ar1g11(!i,3))
next
vector(!length-!ssize) se2
vector(!length-!ssize) ae2
vector(!length-!ssize) hase2
vector(!length-!ssize) haae2
vector(!length-!ssize) le2
vector(!length-!ssize) glls2
for !i=1 to !length-!ssize
  se2(!i) = (est_ar1g11(!i,3) - est_ar1g11(!i,2)^2)^2
  ae2(!i) = @abs(est_ar1g11(!i,3) - est_ar1g11(!i,2)^2)
  hase2(!i) = (1-(est_ar1g11(!i,2)^2/est_ar1g11(!i,3)))^2
  haae2(!i) = @abs(1-(est_ar1g11(!i,2)
    ^2/est_ar1g11(!i,3)))
  le2(!i) = (log(est_ar1g11(!i,2)^2/est_ar1g11(!i,3)))^2
  glls2(!i) = log(est_ar1g11(!i,3)) + (est_ar1g11(!i,2)
    ^2/est_ar1g11(!i,3))
next
smpl @all
matrix(6,4) stats
stats(1,1) = @mean(se1)
stats(2,1) = @mean(ae1)
stats(3,1) = @mean(hase1)
```

```
stats(4,1) = @mean(haae1)
stats(5,1) = @mean(le1)
stats(6,1) = @mean(glls1)
stats(1,2) = @median(se1)
stats(2,2) = @median(ae1)
stats(3,2) = @median(hase1)
stats(4,2) = @median(haae1)
stats(5,2) = @median(le1)
stats(6,2) = @median(glls1)
stats(1,3) = @mean(se2)
stats(2,3) = @mean(ae2)
stats(3,3) = @mean(hase2)
stats(4,3) = @mean(haae2)
stats(5,3) = @mean(le2)
stats(6,3) = @mean(glls2)
stats(1,4) = @median(se2)
stats(2,4) = @median(ae2)
stats(3,4) = @median(hase2)
stats(4,4) = @median(haae2)
stats(5,4) = @median(le2)
stats(6,4) = @median(glls2)
```

- *chapter10.sim_ar1_loss_functions.prg*

```
create chapter10.sim_ar1_loss_functions u 1 4000
series _temp = 1
!length = @obs(_temp)
delete _temp
!ssize = 3000
  !c0 = 0.1
  !c1 = 0.6
series z = nrnd
series y
y(1) = !c0/(1-!c1)
for !i = 2 to !length
y(!i) = !c0 + (!c1*y(!i-1)) + z(!i)
next

matrix(!length-!ssize,4) est_ar1
for !i = 1 to !length-!ssize
  statusline !i
  smpl !i !i+!ssize-1
  equation ar1.ls y c y(-1)
  ar1.makeresid res_ar1
  !c0 = @coefs(1)
```

```
  !c1 = @coefs(2)
  est_ar1(!i,1) = y(!ssize + !i)
  est_ar1(!i,2) = !c0 + !c1*y(!ssize + !i - 1)
  est_ar1(!i,3) = !c0
  est_ar1(!i,4) = !c1
next
vector(!length-!ssize) se
vector(!length-!ssize) ae
vector(!length-!ssize) le
vector(!length-!ssize) pae
for !i=1 to !length-!ssize
  se(!i) = (est_ar1(!i,2) - est_ar1(!i,1))^2
  ae(!i) = @abs(est_ar1(!i,2) - est_ar1(!i,1))
  le(!i) = (log(est_ar1(!i,1)^2/est_ar1(!i,2)^2))^2
  pae(!i) = @abs(est_ar1(!i,2) - est_ar1(!i,1)) /
    @abs(est_ar1(!i,1))
next
smpl @all
matrix(4,2) stats
stats(1,1) = @mean(se)
stats(2,1) = @mean(ae)
stats(3,1) = @mean(le)
stats(4,1) = @mean(pae)
stats(1,2) = @median(se)
stats(2,2) = @median(ae)
stats(3,2) = @median(le)
stats(4,2) = @median(pae)
```

- *chapter10.sim_ar1g11_loss_functions_b.prg*

```
create chapter10.sim_ar1g11_loss_functions_b u 1 4000
smpl @all
series _temp = 1
!length = @obs(_temp)
delete _temp
  !c0 = 0.001
  !c1 = 0.2
  !a0 = 0.0001
  !a1 = 0.12
  !b1 = 0.8
series z = nrnd
series e
series h
series y
h(1) = !a0/(1-!a1-!b1)
```

```
e(1) = sqr(h(1))*z(1)
y(1) = !c0/(1-!c1)
for !i = 2 to !length
  h(!i) = !a0 + (!a1*(e(!i-1)^2)) + (!b1*h(!i-1))
  e(!i) = sqr(h(!i))*z(!i)
  y(!i) = !c0 + (!c1*y(!i-1)) + e(!i)
next
vector(!length) se
vector(!length) ae
vector(!length) hase
vector(!length) haae
vector(!length) le
vector(!length) glls
for !i=1 to !length
  se(!i) = (h(!i) - e(!i)^2)^2
  ae(!i) = @abs(h(!i) - e(!i)^2)
  hase(!i) = (1-(e(!i)^2/h(!i)))^2
  haae(!i) = @abs(1-(e(!i)^2/h(!i)))
  le(!i) = (log(e(!i)^2/h(!i)))^2
  glls(!i) = log(h(!i)) + (e(!i)^2/h(!i))
next
matrix(6,2) stats
stats(1,1) = @mean(se)
stats(2,1) = @mean(ae)
stats(3,1) = @mean(hase)
stats(4,1) = @mean(haae)
stats(5,1) = @mean(le)
stats(6,1) = @mean(glls)
stats(1,2) = @median(se)
stats(2,2) = @median(ae)
stats(3,2) = @median(hase)
stats(4,2) = @median(haae)
stats(5,2) = @median(le)
stats(6,2) = @median(glls)
```

- *chapter10.sim_mz_regression.prg*

```
create chapter10.sim_MZ_regression u 1 4000
series _temp = 1
!length = @obs(_temp)
delete _temp
  !c0 = 0.001
  !c1 = 0.2
  !a0 = 0.0001
  !a1 = 0.12
```

```
  !b1 = 0.80 ' 0.20 ' 0.50
!repeat = 1000
matrix(!repeat,6) MZ=na

for !j = 1 to !repeat
  smpl @all
  statusline !j
  series z = nrnd
  series e = na
  series h = na
  series y = na
  h(1) = !a0/(1-!a1-!b1)
  e(1) = sqr(h(1))*z(1)
  y(1) = !c0/(1-!c1)
  for !i = 2 to !length
    h(!i) = !a0 + (!a1*(e(!i-1)^2)) + (!b1*h(!i-1))
    e(!i) = sqr(h(!i))*z(!i)
    y(!i) = !c0 + (!c1*y(!i-1)) + e(!i)
  next
  smpl 1000 4000
  equation MZ1_reg.ls(h) e^2 c h
  MZ(!j,1) = @coefs(1)
  MZ(!j,2) = @coefs(2)
  equation MZ2_reg.ls(h) @abs(e) c sqr(h)
  MZ(!j,3) = @coefs(1)
  MZ(!j,4) = @coefs(2)
  equation MZ3_reg.ls(h) log(e^2) c log(h)
  MZ(!j,5) = @coefs(1)
  MZ(!j,6) = @coefs(2)
next
```

- *chapter10.sim_R2_regression.prg*

```
create chapter10.sim_R2_regression u 1 40000
series _temp = 1
!length = @obs(_temp)
delete _temp
  !a0 = 0.0001
  !a1 = 0.12
  !b1 = 0.80
!repeat = 1000
matrix(!repeat,1) R2=na
for !j = 1 to !repeat
  smpl @all
  statusline !j
```

```
  series z = nrnd
  series e = na
  series h = na
  h(1) = !a0/(1-!a1-!b1)
  e(1) = sqr(h(1))*z(1)
  for !i = 2 to !length
    h(!i) = !a0 + (!a1*(e(!i-1)^2)) + (!b1*h(!i-1))
    e(!i) = sqr(h(!i))*z(!i)
  next
  smpl 1000 40000
  equation MZ1_reg.ls(h) e^2 c h
  R2(!j,1) = MZ1_reg.@r2
next
```

- *chatper10.sim.continuousgarch.prg*

```
create chapter10.sim.continuousgarch u 1 3200000
series _temp = 1
!length = @obs(_temp)
delete _temp
  !alpha0 = 0.083
  !alpha1 = 0.013
  !beta1 = 0.505
  !dt = 1/1000
  !repeat = 100
  vector(6) T
  T(1) = 2
  T(2) = 10
  T(3) = 50
  T(4) = 100
  T(5) = 500
  T(6) = 1000
matrix(!repeat,7) MZ1=na
matrix(!repeat,7) MZ2=na
matrix(!repeat,7) MZ3=na

for !rep = 1 to !repeat

  smpl @all
  series z1 = nrnd
  series z2 = nrnd
  series y = na
  series h = na
  series p = na
  h(1) = !alpha0*!alpha1*!dt
```

```
 y(1) = sqr(h(1)*!dt)*z1(1)
 p(1) = 1000
 for !i = 2 to !length
   h(!i) = !alpha0*!alpha1*!dt + h(!i-1)*(1-!alpha0*!dt
     +(sqr(2*!beta1*!alpha0*!dt))*z2(!i))
   y(!i) = sqr(h(!i)*!dt)*z1(!i)
   p(!i) = p(!i-1)*exp(y(!i))
next

series sum_y_1000
series range
for !j=1 to !length-T(6)+1 step T(6)
  !sum=0
  !min=p(!j)
  !max=p(!j)
  for !t=!j to T(6)-1+!j
  sum_y_1000(1+((!j-1)/T(6))) = y(!t) + !sum
  !sum = sum_y_1000(1+((!j-1)/T(6)))
  if p(!t) <!min then
     !min=p(!t)
  endif
  if p(!t)>!max then
     !max=p(!t)
  endif
next
range(1+((!j-1)/T(6)))=(log(!max/!min))^2/(4*log(2))
next

smpl 201 3200
equation g11.arch(1,1,m=10000,h, deriv=f) sum_y_1000
g11.makegarch garchg11

smpl @all
for !i=1 to 6
  !Ti = T(!i)
  series sum_sq_y{!Ti}
  series y{!Ti}
  for !j=1 to !length-T(6)+1 step T(6)
   statusline !rep !i !j
  for !t=!j to T(6)-1+!j step T(!i)
   y{!Ti}(1+((!t-1)/T(!i))) = log(p(!t+(T(!i)-1)))
     - log(p(!t))
  next
 next
```

```
for !j=1 to !length/T(6)
 !sum2=0
 for !t=((!j-1)*T(6)/T(!i))+1 to !j*T(6)/T(!i)
   sum_sq_y{!Ti}(!j) = y{!Ti}(!t)^2 + !sum2
   !sum2 = sum_sq_y{!Ti}(!j)
 next
next

smpl 201 3200
equation MZ1_{!i}reg.ls(h) sum_sq_y{!Ti} c garchg11
MZ1(!rep,!i) = MZ1_{!i}reg.@r2
equation MZ2_{!i}reg.ls(h) sqr(sum_sq_y{!Ti})
 c sqr(garchg11)
MZ2(!rep,!i) = MZ2_{!i}reg.@r2
equation MZ3_{!i}reg.ls(h) log(sum_sq_y{!Ti})
 c log(garchg11)
MZ3(!rep,!i) = MZ3_{!i}reg.@r2
next
equation MZ1_7reg.ls(h) range c garchg11
MZ1(!rep,7) = MZ1_7reg.@r2
equation MZ2_7reg.ls(h) sqr(range) c sqr(garchg11)
MZ2(!rep,7) = MZ2_7reg.@r2
equation MZ3_7reg.ls(h) log(range) c log(garchg11)
MZ3(!rep,7) = MZ3_7reg.@r2
next
smpl 201 3200
delete h
delete p
delete y
delete z1
delete z2
range 1 3200
save chapter10.sim.continuousgarch
```

- *chapter10.sim.compare.models.prg*

```
load chapter10.sim.continuousgarch.wf1
smpl 201 3200
for !i=1 to 3
  delete mz{!i}
  for !j=1 to 7
    delete mz{!i}_{!j}reg
  next
next
!models = 5
```

```
!proxies = 5
!criteria = 5
series _temp = 1
!length = @obs(_temp)
delete _temp
'_____Part A_____
equation model1.arch(1,1,m=10000,h, deriv=f) sum_y_1000
equation model2.arch(1,1,integrated,m=10000,h,
  deriv=f) sum_y_1000
equation model3.arch(2,2,m=10000,h, deriv=f) sum_y_1000
equation model4.arch(1,2,m=10000,h, deriv=f) sum_y_1000
equation model5.arch(2,1,m=10000,h, deriv=f) sum_y_1000
'_____Part B_____
rename sum_sq_y1000 proxy1
rename sum_sq_y100 proxy2
rename sum_sq_y10 proxy3
rename sum_sq_y2 proxy4
rename range proxy5
'_____Part C_____
for !prox = 1 to !proxies
matrix(!models,!criteria) criteria{!prox}
matrix(3200,!models) loss{!prox}SE = na
matrix(3200,!models) loss{!prox}AE = na
for !i = 1 to !models
  statusline !prox !i
  model{!i}.makegarch garch_model{!i}
  for !j=201 to 3200
    loss{!prox}SE(!j,!i) = (garch_model{!i}(!j)
      - proxy{!prox}(!j))^2 'Create the SE Loss function
    loss{!prox}AE(!j,!i) = @abs(garch_model{!i}(!j)
      - proxy{!prox}(!j)) 'Create the AE Loss function
  next
  equation reg{!prox}_{!i}.ls(h) proxy{!prox}
    c garch_model{!i} 'Mincer Zarnowitz regression
  equation logreg{!prox}_{!i}.ls(h) log(proxy{!prox})
    c log(garch_model{!i}) 'Mincer Zarnowitz log-regression
  criteria{!prox}(!i,1) = reg{!prox}_{!i}.@r2
  criteria{!prox}(!i,2) = logreg{!prox}_{!i}.@r2
  criteria{!prox}(!i,3) = (@mean(@columnextract
    (loss{!prox}SE,!i)))
  criteria{!prox}(!i,4) = (@mean(@columnextract
    (loss{!prox}AE,!i)))
  criteria{!prox}(!i,5) = model{!i}.@schwarz
    'Schwarz Information Criterion
 next
```

```
next
save chapter10.sim.compare.models
```

- *chapter10.sim.compare.models.repeated.prg*

```
create chapter10.sim.compare.models.repeated u 1 3200000
series _temp = 1
!length = @obs(_temp)
delete _temp
  !alpha0 = 0.083
  !alpha1 = 0.013
  !beta1 = 0.505
  !dt = 1/1000
  !repeat = 100
  vector(4) T
  T(1) = 2
  T(2) = 10
  T(3) = 100
  T(4) = 1000
  !models = 5
  !proxies = 5
  !criteria = 5
'_____Simulate continuous time AR(0)-GARCH(1,1)
  model_____
for !rep = 1 to !repeat
  smpl @all

  series z1 = nrnd
  series z2 = nrnd
  series y = na
  series h = na
  series p = na
  h(1) = !alpha0*!alpha1*!dt
  y(1) = sqr(h(1)*!dt)*z1(1)
  p(1) = 1000
  for !i = 2 to !length
    h(!i) = !alpha0*!alpha1*!dt + h(!i-1)*(1-!alpha0*!dt
      +(sqr(2*!beta1*!alpha0*!dt))*z2(!i))
    y(!i) = sqr(h(!i)*!dt)*z1(!i)
    p(!i) = p(!i-1)*exp(y(!i))
  next

series sum_y_1000
series range
for !j=1 to !length-T(4)+1 step T(4)
```

```
!sum=0
!min=p(!j)
!max=p(!j)
for !t=!j to T(4)-1+!j
sum_y_1000(1+((!j-1)/T(4))) = y(!t) + !sum
!sum = sum_y_1000(1+((!j-1)/T(4)))
if p(!t)<!min then
!min=p(!t)
endif
if p(!t)>!max then
  !max=p(!t)
endif
next
range(1+((!j-1)/T(4)))=(log(!max/!min))^2/(4*log(2))
  'Compute the range price
next

for !i=1 to 4
  statusline !rep !i PART A
    !Ti = T(!i)
    series sum_sq_y{!Ti}
    series y{!Ti}
    for !j=1 to !length-T(4)+1 step T(4)
      for !t=!j to T(4)-1+!j step T(!i)
        y{!Ti}(1+((!t-1)/T(!i))) = log(p(!t+(T(!i)-1))) -
          log(p(!t)) 'Compute the log-returns
      next
    next

for !j=1 to !length/T(4)
  !sum2=0
  for !t=((!j-1)*T(4)/T(!i))+1 to !j*T(4)/T(!i)
    sum_sq_y{!Ti}(!j) = y{!Ti}(!t)^2 + !sum2
      'Compute the realized volatility
    !sum2 = sum_sq_y{!Ti}(!j)
  next
next
next

smpl 201 3200

  equation model1_{!rep}.arch(1,1,m=10000,h,
    deriv=f) sum_y_1000
  equation model2_{!rep}.arch(1,1,integrated,m=10000,h,
    deriv=f) sum_y_1000
```

```
equation model3_{!rep}.arch(2,2,m=10000,h,
  deriv=f) sum_y_1000
equation model4_{!rep}.arch(1,2,m=10000,h,
  deriv=f) sum_y_1000
equation model5_{!rep}.arch(2,1,m=10000,h,
  deriv=f) sum_y_1000
for !i = 1 to !models
  model{!i}_{!rep}.makegarch garch_model{!i}
  matrix(3200,!models) garch_model_{!rep} 'save
    the series in a matrix form due to memory problems.
  for !j = 1 to 3200
    garch_model_{!rep}(!j,!i) = garch_model{!i}(!j)
  next
next
  matrix(3200,!proxies) proxy_{!rep} 'save the series in a
    matrix form due to memory problems. In the next loop the
    matrices are transformed into series
  for !i = 1 to 3200
  proxy_{!rep}(!i,1) = sum_sq_y1000(!i)
  proxy_{!rep}(!i,2) = sum_sq_y100(!i)
  proxy_{!rep}(!i,3) = sum_sq_y10(!i)
  proxy_{!rep}(!i,4) = sum_sq_y2(!i)
  proxy_{!rep}(!i,5) = range(!i)
next
next
range 1 3200
for !rep = 1 to !repeat
  statusline !rep PART B
  for !i = 1 to !models
    for !j=1 to 3200
      series garch_model{!i}_{!rep}(!j) = garch_model_
        {!rep}(!j,!i) 'transform the matrices into series
    next
next
  for !prox = 1 to !proxies
    for !i=1 to 3200
      series proxy{!prox}_{!rep}(!i) = proxy_{!rep}(!i,
        !prox) 'transform the matrices into series
    next
next

for !prox = 1 to !proxies
  matrix(!models,!criteria) criteria{!prox}_{!rep}
  matrix(3200,!models) loss{!prox}SE_{!rep} = na
  matrix(3200,!models) loss{!prox}AE_{!rep} = na
```

```
for !i = 1 to !models
  for !j=201 to 3200
    loss{!prox}SE_{!rep}(!j,!i) = (garch_model{!i}_
      {!rep}(!j) - proxy{!prox}_{!rep}(!j))^2 'Create the
      SE Loss function
    loss{!prox}AE_{!rep}(!j,!i) = @abs(garch_model{!i}_
      {!rep}(!j) - proxy{!prox}_{!rep}(!j)) 'Create the
      AE Loss function
  next
  equation reg{!prox}_{!i}_{!rep}.ls(h) proxy{!prox}_
    {!rep} c garch_model{!i}_{!rep} 'Mincer Zarnowitz
    regression
  equation logreg{!prox}_{!i}_{!rep}.ls(h) log(proxy
    {!prox}_{!rep}) c log(garch_model{!i}_{!rep})
    'Mincer Zarnowitz log-regression
  criteria{!prox}_{!rep}(!i,1) = reg{!prox}_{!i}_
    {!rep}.@r2
  criteria{!prox}_{!rep}(!i,2) = logreg{!prox}_
    {!i}_{!rep}.@r2
  criteria{!prox}_{!rep}(!i,3) = (@mean(@columnextract
    (loss{!prox}SE_{!rep},!i)))
  criteria{!prox}_{!rep}(!i,4) = (@mean(@columnextract
    (loss{!prox}AE_{!rep},!i)))
  criteria{!prox}_{!rep}(!i,5) = model{!i}_{!rep}.
    @schwarz ' Schwarz Information Criterion
  next
next

next
'_____delete to save space_____
for !rep = 1 to !repeat
  for !prox = 1 to !proxies
    for !i = 1 to !models
      delete reg{!prox}_{!i}_{!rep}
      delete logreg{!prox}_{!i}_{!rep}
    next
  next
next
for !rep = 1 to !repeat
  for !prox = 1 to !proxies
    delete loss{!prox}SE_{!rep}
    delete loss{!prox}AE_{!rep}
  next
next
for !rep = 1 to !repeat
```

```
  for !i = 1 to !models
    delete model{!i}_{!rep}
  next
next
```

_____

```
matrix(!proxies,!criteria) correct_model_selected
  'Compute the number of times the correct model was
  selected as best
for !prox = 1 to !proxies
scalar criterion1 = 0
scalar criterion2 = 0
scalar criterion3 = 0
scalar criterion4 = 0
scalar criterion5 = 0
  for !rep = 1 to !repeat
    for !crit = 1 to 2
      if criteria{!prox}_{!rep}(1,!crit)=@max
        (@columnextract(criteria{!prox}_{!rep},!crit))
        then
      criterion{!crit} = criterion{!crit} +1
      else
      criterion{!crit} = criterion{!crit}
      statusline criterion1
      endif
    correct_model_selected(!prox,!crit)=criterion{!crit}
    next
    for !crit = 3 to 5
      if criteria{!prox}_{!rep}(1,!crit)=@min
        (@columnextract(criteria{!prox}_{!rep},!crit))
        then
      criterion{!crit} = criterion{!crit} +1
      else
      criterion{!crit} = criterion{!crit}
      endif
    correct_model_selected(!prox,!crit)=criterion{!crit}
    next
  next
next
save chapter10.sim.compare.models.repeated
```

- *chapter10.sim.compare.models.repeated.b.prg*

```
create chapter10.sim.compare.models.repeated.b u 1 11000
series _temp = 1
!length = @obs(_temp)
```

```
delete _temp
  !a0 = 0.0001
  !a1 = 0.12
  !b1 = 0.8
!repeat = 100
!models = 5
!criteria = 5
`_____Simulate AR(0)-GARCH(1,1) model_____
for !rep = 1 to !repeat
  statusline iteration !rep
  smpl @all
  series z{!rep} = nrnd
  series h{!rep}
  series y{!rep}
  h{!rep}(1) = !a0/(1-!a1-!b1)
  y{!rep}(1) = sqr(h{!rep}(1))*z{!rep}(1)
  for !i = 2 to !length
    h{!rep}(!i) = !a0 + (!a1*(y{!rep}(!i-1)^2)) +
      (!b1*h{!rep}(!i-1))
    y{!rep}(!i) = sqr(h{!rep}(!i))*z{!rep}(!i)
  next
`_____

smpl 1001 11000
equation model1_{!rep}.arch(1,1,m=10000,h,
  deriv=f) y{!rep}
equation model2_{!rep}.arch(1,1,integrated,
  m=10000,h, deriv=f) y{!rep}
equation model3_{!rep}.arch(2,2,m=10000,h,
  deriv=f) y{!rep}
equation model4_{!rep}.arch(1,2,m=10000,h,
  deriv=f) y{!rep}
equation model5_{!rep}.arch(2,1,m=10000,h,
  deriv=f) y{!rep}
for !i = 1 to !models
  model{!i}_{!rep}.makegarch garch_model{!i}_{!rep}
next
  matrix(!models,!criteria) criteria_{!rep}
  matrix(11000,!models) lossSE_{!rep} = na
  matrix(11000,!models) lossAE_{!rep} = na
  for !i = 1 to !models
    for !j=1001 to 11000
    lossSE_{!rep}(!j,!i) = (garch_model{!i}_{!rep}(!j) -
      y{!rep}(!j)^2)^2 `Create the SE Loss function
    lossAE_{!rep}(!j,!i) = @abs(garch_model{!i}_
      {!rep}(!j) - y{!rep}(!j)^2) `Create the AE Loss
```

```
      function
  next
    equation reg_{!i}_{!rep}.ls(h) y{!rep}^2
      c garch_model{!i}_{!rep} 'Mincer Zarnowitz regression
    equation logreg_{!i}_{!rep}.ls(h) log(y{!rep}^2)
      c log(garch_model{!i}_{!rep}) 'Mincer Zarnowitz
      log-regression
    criteria_{!rep}(!i,1) = reg_{!i}_{!rep}.@r2
    criteria_{!rep}(!i,2) = logreg_{!i}_{!rep}.@r2
    criteria_{!rep}(!i,3) = (@mean(@columnextract
      (lossSE_{!rep},!i)))
    criteria_{!rep}(!i,4) = (@mean(@columnextract
      (lossAE_{!rep},!i)))
    criteria_{!rep}(!i,5) = model{!i}_{!rep}.@schwarz
      'Schwarz Information Criterion
  next

next

matrix(1,!criteria) correct_model_selected
  'Compute the number of times the correct
  model was selected as best
scalar criterion1 = 0
scalar criterion2 = 0
scalar criterion3 = 0
scalar criterion4 = 0
scalar criterion5 = 0
  for !rep = 1 to !repeat
    for !crit = 1 to 2
      if criteria_{!rep}(1,!crit)=@max(@columnextract
        (criteria_{!rep},!crit)) then
      criterion{!crit} = criterion{!crit} +1
      else
      criterion{!crit} = criterion{!crit}
      statusline criterion1
      endif
    correct_model_selected(1,!crit)=criterion{!crit}
    next
    for !crit = 3 to 5
      if criteria_{!rep}(1,!crit)=@min(@columnextract
        (criteria_{!rep},!crit)) then
      criterion{!crit} = criterion{!crit} +1
      else
      criterion{!crit} = criterion{!crit}
      endif
```

```
      correct_model_selected(1,!crit)=criterion{!crit}
   next
next
```

- *chapter10.sim.compare.models.repeated.outofsample.prg*

```
create chapter10.sim.compare.models.repeated.outofsample
   u 1 5000
series _temp = 1
!length = @obs(_temp)
delete _temp
!repeat = 100
!models = 5
!criteria = 5
!ssize = 4000

for !rep = 1 to !repeat
   statusline iteration !rep
   smpl @all
'_____Simulate AR(0)-GARCH(1,1) model_____
!a0 = 0.0001
!a1 = 0.12
!b1 = 0.8
series z{!rep} = nrnd
series h{!rep} = na
series y{!rep} = na
   h{!rep}(1) = !a0/(1-!a1-!b1)
   y{!rep}(1) = sqr(h{!rep}(1))*z{!rep}(1)
for !i = 2 to !length
   h{!rep}(!i) = !a0 + (!a1*(y{!rep}(!i-1)^2))
      + (!b1*h{!rep}(!i-1))
   y{!rep}(!i) = sqr(h{!rep}(!i))*z{!rep}(!i)
next
'_____Estimate AR(0)-GARCH(1,1) model_____
series garch_model1_{!rep}
for !i = 1 to !length-!ssize
   statusline !rep model1 !i
   smpl !i !i+!ssize-1
   equation m1_{!rep}.arch(1,1,m=10000,h) y{!rep}
   m1_{!rep}.makegarch garch_m1_{!rep}
   m1_{!rep}.makeresid res_m1_{!rep}
   !ht = garch_m1_{!rep}(!i+!ssize-1)
   !a0 = @coefs(1)
   !a1 = @coefs(2)
   !b1 = @coefs(3)
   !h_hat = !a0 + (!a1*(res_m1_{!rep}(!ssize
```

```
    + !i -1)^2) ) + (!b1*!ht)
  garch_model1_{!rep}(!i + !ssize) = !h_hat
next
'_____
'_____Estimate AR(0)-IGARCH(1,1) model_____
series garch_model2_{!rep}
for !i = 1 to !length-!ssize
  statusline !rep model2 !i
  smpl !i !i+!ssize-1
  equation m2_{!rep}.arch(1,1,integrated,
    m=10000,h) y{!rep}
  m2_{!rep}.makegarch garch_m2_{!rep}
  m2_{!rep}.makeresid res_m2_{!rep}
  !ht = garch_m2_{!rep}(!i+!ssize-1)
  !a1 = @coefs(1)
  !h_hat = (!a1*(res_m2_{!rep}(!ssize
    + !i -1)^2) ) + ((1-!a1)*!ht)
  garch_model2_{!rep}(!i + !ssize) = !h_hat
next
'_____
'_____Estimate AR(0)-GARCH(2,2) model_____
series garch_model3_{!rep}
for !i = 1 to !length-!ssize
  statusline !rep model3 !i
  smpl !i !i+!ssize-1
  equation m3_{!rep}.arch(2,2,m=10000,h) y{!rep}
  m3_{!rep}.makegarch garch_m3_{!rep}
  m3_{!rep}.makeresid res_m3_{!rep}
  !ht = garch_m3_{!rep}(!i+!ssize-1)
  !h2t = garch_m3_{!rep}(!i+!ssize-2)
  !a0 = @coefs(1)
  !a1 = @coefs(2)
  !a2 = @coefs(3)
  !b1 = @coefs(4)
  !b2 = @coefs(5)
  !h_hat = !a0 + (!a1*(res_m3_{!rep}(!ssize
    + !i -1)^2) ) + (!a2*(res_m3_{!rep}(!ssize
    + !i -2)^2) ) + (!b1*!ht) + (!b2*!h2t)
  garch_model3_{!rep}(!i + !ssize) = !h_hat
next
'_____
'_____Estimate AR(0)-GARCH(1,2) model_____
series garch_model4_{!rep}
for !i = 1 to !length-!ssize
  statusline !rep model4 !i
  smpl !i !i+!ssize-1
```

```
equation m4_{!rep}.arch(2,1,m=10000,h) y{!rep}
m4_{!rep}.makegarch garch_m4_{!rep}
m4_{!rep}.makeresid res_m4_{!rep}
!ht = garch_m4_{!rep}(!i+!ssize-1)
!a0 = @coefs(1)
!a1 = @coefs(2)
!a2 = @coefs(3)
!b1 = @coefs(4)
!h_hat = !a0 + (!a1*(res_m4_{!rep}(!ssize + !i -1)^2) )
   + (!a2*(res_m4_{!rep}(!ssize + !i -2)^2) ) + (!b1*!ht)
garch_model4_{!rep}(!i + !ssize) = !h_hat
next
'_____
'_____Estimate AR(0)-GARCH(2,1) model_____
series garch_model5_{!rep}
for !i = 1 to !length-!ssize
  statusline !rep model5 !i
  smpl !i !i+!ssize-1
  equation m5_{!rep}.arch(1,2,m=10000,h) y{!rep}
  m5_{!rep}.makegarch garch_m5_{!rep}
  m5_{!rep}.makeresid res_m5_{!rep}
  !ht = garch_m5_{!rep}(!i+!ssize-1)
  !h2t = garch_m5_{!rep}(!i+!ssize-2)
  !a0 = @coefs(1)
  !a1 = @coefs(2)
  !b1 = @coefs(3)
  !b2 = @coefs(4)
  !h_hat = !a0 + (!a1*(res_m5_{!rep}(!ssize + !i -1)^2) )
  + (!b1*!ht) + (!b2*!h2t)
  garch_model5_{!rep}(!i + !ssize) = !h_hat
next
'_____


matrix(!models,!criteria) criteria_{!rep}
matrix(!length,!models) lossSE_{!rep} = na
matrix(!length,!models) lossAE_{!rep} = na
matrix(!length,!models) lossGLLS_{!rep} = na
!ss = !ssize+1
smpl !ss !length
for !i = 1 to !models
 for !j=!ssize+1 to !length
  lossSE_{!rep}(!j,!i) = (garch_model{!i}_{!rep}(!j)
   - y{!rep}(!j)^2)^2 'Create the SE Loss function
  lossAE_{!rep}(!j,!i) = @abs(garch_model{!i}_{!rep}(!j)
   - y{!rep}(!j)^2) 'Create the AE Loss function
  lossGLLS_{!rep}(!j,!i) =log(garch_model{!i}_{!rep}(!j))
```

```
   + (y{!rep}(!j)^2 / garch_model{!i}_{!rep}(!j))
   'Create the GLLS Loss function
next
equation reg_{!i}_{!rep}.ls(h) y{!rep}^2
 c garch_model{!i}_{!rep} 'Mincer Zarnowitz regression
equation logreg_{!i}_{!rep}.ls(h) log(y{!rep}^2)
 c log(garch_model{!i}_{!rep})
 'Mincer Zarnowitz log-regression
criteria_{!rep}(!i,1) = reg_{!i}_{!rep}.@r2
criteria_{!rep}(!i,2) = logreg_{!i}_{!rep}.@r2
criteria_{!rep}(!i,3) = (@mean(@columnextract
 (lossSE_{!rep},!i)))
criteria_{!rep}(!i,4) = (@mean(@columnextract
 (lossAE_{!rep},!i)))
criteria_{!rep}(!i,5) = (@mean(@columnextract
 (lossGLLS_{!rep},!i)))
next

next

matrix(1,!criteria) correct_model_selected
  'Compute the number of times the correct
  model was selected as best
scalar criterion1 = 0
scalar criterion2 = 0
scalar criterion3 = 0
scalar criterion4 = 0
scalar criterion5 = 0
  for !rep = 1 to !repeat
    for !crit = 1 to 2
      if criteria_{!rep}(1,!crit)=@max(@columnextract
        (criteria_{!rep},!crit)) then
      criterion{!crit} = criterion{!crit} +1
      else
      criterion{!crit} = criterion{!crit}
      statusline criterion1
       endif
    correct_model_selected(1,!crit)=criterion{!crit}
    next
    for !crit = 3 to 5
      if criteria_{!rep}(1,!crit)=@min(@columnextract
        (criteria_{!rep},!crit)) then
      criterion{!crit} = criterion{!crit} +1
      else
      criterion{!crit} = criterion{!crit}
```

```
        endif
      correct_model_selected(1,!crit)=criterion{!crit}
   next
next
```

- *chapter10.SPEC.prg*

```
load chapter8_sp500_options.partc.wf1
smpl @all
series _temp = 1
!length = @obs(_temp)
delete _temp
!models =9
!ssize = 500
!sum=0
scalar firstT = 5
scalar lastT = 55
scalar stepT = 10
scalar numT = ((lastT - firstT)/stepT)+1
for !i=firstT to lastT step stepT
  scalar T{!i}=!i
next
matrix(!length-!ssize,!models) z
for !i=firstT to lastT step stepT
  matrix(!length-!ssize,!models) sum_z{!i}=na
next
matrix(!length-!ssize,numT) min_z=na
matrix(!length-!ssize,numT) model_min_z=na
for !i=1 to !length-!ssize
  z(!i,1) = estg01(!i,4)
  z(!i,2) = estg11(!i,4)
  z(!i,3) = estg21(!i,4)
  z(!i,4) = estt01(!i,4)
  z(!i,5) = estt11(!i,4)
  z(!i,6) = estt21(!i,4)
  z(!i,7) = este01(!i,4)
  z(!i,8) = este11(!i,4)
  z(!i,9) = este21(!i,4)
next

for !k=1 to numT
  !w=firstT + (stepT*(!k-1))
for !j=1 to !length-!ssize-T{!w}+1
  for !i=1 to !models
    statusline !k !j
```

```
  !sum=0
  for !t=!j to T{!w}-1+!j
    sum_z{!w}(T{!w}-1+!j,!i) = z(!t,!i) + !sum
    !sum = sum_z{!w}(T{!w}-1+!j,!i)
  next
 next
next
for !j=1 to !length-!ssize
  min_z(!j,!k)=sum_z{!w}(!j,1)
  for !i=1 to !models-1
   if sum_z{!w}(!j,!i+1)<min_z(!j,!k) then
    min_z(!j,!k)=sum_z{!w}(!j,!i+1)
   endif
  next
 next
for !j=1 to !length-!ssize
  for !i=1 to !models
   if sum_z{!w}(!j,!i)=min_z(!j,!k) then
    model_min_z(!j,!k)=!i
   endif
  next
 next
next
matrix(!length,numT) Straddle_spec=na
for !k=1 to numT
  for !j=1 to !length-!ssize
  if min_z(!j,!k)>0 then
  !m=model_min_z(!j,!k)
  Straddle_spec(!j+!ssize,!k)=Straddles(!j+!ssize,!m)
  else
  !m= na
  Straddle_spec(!j+!ssize,!k)= na
  endif
  next
next
vector(3) filter
  filter(1)=0
  filter(2)=2
  filter(3)=3.5
vector(2) cost
  cost(1)=0
  cost(2)=2
for !x=1 to 3
for !p=1 to 2
matrix(!length,numT) spec_Str_Ret_F{!x}C{!p}
```

```
next
next
for !x=1 to 3
  for !p=1 to 2
     for !j=1 to numT
       for !i=1 to !length
         statusline filter !x cost !p T !j trading day !i
         if Straddle_spec(!i,!j)<> na then
         if Straddle_spec(!i,!j)>Straddle_obs(!i)
           +filter(!x) then
       spec_Str_Ret_F{!x}C{!p}(!i,!j) = (next_Straddle(!i)
         -Straddle_obs(!i)-Cost(!p))/Straddle_obs(!i)
         else
         if Straddle_spec(!i,!j)<Straddle_obs(!i)
           -filter(!x) then
       spec_Str_Ret_F{!x}C{!p}(!i,!j) = (-next_Straddle
         (!i)+Straddle_obs(!i)-Cost(!p))/Straddle_obs(!i)
         else
         spec_Str_Ret_F{!x}C{!p}(!i,!j) = na
         endif
         endif
         else
         spec_Str_Ret_F{!x}C{!p}(!i,!j) = na
         endif
       next
     next
  next
next
for !i=firstT to lastT step stepT
  delete T{!i}
next
save chapter10.spec.wf1
```

- *chapter10.Loss_Functions_as_Model_Selection_Methods.prg*

```
load chapter10.spec.wf1
smpl @all
series _temp = 1
!length = @obs(_temp)
delete _temp
!models = 9
!num_lossF = 6
!ssize = 500
!sum=0
scalar firstT = 5
```

```
scalar lastT = 55
scalar stepT = 10
scalar numT = ((lastT - firstT)/stepT)+1
for !i=firstT to lastT step stepT
  scalar T{!i}=!i
next
for !ii=1 to !num_lossF
  matrix(!length-!ssize,!models) lossF{!ii}
  for !i=firstT to lastT step stepT
    matrix(!length-!ssize,!models) sum_lossF{!ii}_
      {!i}=na
  next
  matrix(!length-!ssize,numT) min_lossF{!ii}=na
  matrix(!length-!ssize,numT) model_min_lossF{!ii}=na
next
'Create the SE Loss function
for !i=1 to !length-!ssize
  lossF1(!i,1) = (estg01(!i,2) - estg01(!i,3)^2)^2
  lossF1(!i,2) = (estg11(!i,2) - estg11(!i,3)^2)^2
  lossF1(!i,3) = (estg21(!i,2) - estg21(!i,3)^2)^2
  lossF1(!i,4) = (estt01(!i,2) - estt01(!i,3)^2)^2
  lossF1(!i,5) = (estt11(!i,2) - estt11(!i,3)^2)^2
  lossF1(!i,6) = (estt21(!i,2) - estt21(!i,3)^2)^2
  lossF1(!i,7) = (este01(!i,2) - este01(!i,3)^2)^2
  lossF1(!i,8) = (este11(!i,2) - este11(!i,3)^2)^2
  lossF1(!i,9) = (este21(!i,2) - este21(!i,3)^2)^2
next
'Create the AE Loss function
for !i=1 to !length-!ssize
  lossF2(!i,1) = @abs(estg01(!i,2) - estg01(!i,3)^2)
  lossF2(!i,2) = @abs(estg11(!i,2) - estg11(!i,3)^2)
  lossF2(!i,3) = @abs(estg21(!i,2) - estg21(!i,3)^2)
  lossF2(!i,4) = @abs(estt01(!i,2) - estt01(!i,3)^2)
  lossF2(!i,5) = @abs(estt11(!i,2) - estt11(!i,3)^2)
  lossF2(!i,6) = @abs(estt21(!i,2) - estt21(!i,3)^2)
  lossF2(!i,7) = @abs(este01(!i,2) - este01(!i,3)^2)
  lossF2(!i,8) = @abs(este11(!i,2) - este11(!i,3)^2)
  lossF2(!i,9) = @abs(este21(!i,2) - este21(!i,3)^2)
next
'Create the HASE Loss function
for !i=1 to !length-!ssize
  lossF3(!i,1) = (1-(estg01(!i,3)^2/estg01(!i,2)))^2
  lossF3(!i,2) = (1-(estg11(!i,3)^2/estg11(!i,2)))^2
  lossF3(!i,3) = (1-(estg21(!i,3)^2/estg21(!i,2)))^2
  lossF3(!i,4) = (1-(estt01(!i,3)^2/estt01(!i,2)))^2
```

```
  lossF3(!i,5) = (1-(estt11(!i,3)^2/estt11(!i,2)))^2
  lossF3(!i,6) = (1-(estt21(!i,3)^2/estt21(!i,2)))^2
  lossF3(!i,7) = (1-(este01(!i,3)^2/este01(!i,2)))^2
  lossF3(!i,8) = (1-(este11(!i,3)^2/este11(!i,2)))^2
  lossF3(!i,9) = (1-(este21(!i,3)^2/este21(!i,2)))^2
next
'Create the HAAE Loss function
for !i=1 to !length-!ssize
  lossF4(!i,1) = @abs(1-(estg01(!i,3)^2/estg01(!i,2)))
  lossF4(!i,2) = @abs(1-(estg11(!i,3)^2/estg11(!i,2)))
  lossF4(!i,3) = @abs(1-(estg21(!i,3)^2/estg21(!i,2)))
  lossF4(!i,4) = @abs(1-(estt01(!i,3)^2/estt01(!i,2)))
  lossF4(!i,5) = @abs(1-(estt11(!i,3)^2/estt11(!i,2)))
  lossF4(!i,6) = @abs(1-(estt21(!i,3)^2/estt21(!i,2)))
  lossF4(!i,7) = @abs(1-(este01(!i,3)^2/este01(!i,2)))
  lossF4(!i,8) = @abs(1-(este11(!i,3)^2/este11(!i,2)))
  lossF4(!i,9) = @abs(1-(este21(!i,3)^2/este21(!i,2)))
next
'Create the LE Loss function
for !i=1 to !length-!ssize
  lossF5(!i,1) = (log(estg01(!i,3)^2/estg01(!i,2)))^2
  lossF5(!i,2) = (log(estg11(!i,3)^2/estg11(!i,2)))^2
  lossF5(!i,3) = (log(estg21(!i,3)^2/estg21(!i,2)))^2
  lossF5(!i,4) = (log(estt01(!i,3)^2/estt01(!i,2)))^2
  lossF5(!i,5) = (log(estt11(!i,3)^2/estt11(!i,2)))^2
  lossF5(!i,6) = (log(estt21(!i,3)^2/estt21(!i,2)))^2
  lossF5(!i,7) = (log(este01(!i,3)^2/este01(!i,2)))^2
  lossF5(!i,8) = (log(este11(!i,3)^2/este11(!i,2)))^2
  lossF5(!i,9) = (log(este21(!i,3)^2/este21(!i,2)))^2
next
'Create the GLLS Loss function
for !i=1 to !length-!ssize
  lossF6(!i,1) = log(estg01(!i,2)) + (estg01(!i,3)^2/
    estg01(!i,2))
  lossF6(!i,2) =log(estg11(!i,2)) + (estg11(!i,3)^2/
    estg11(!i,2))
  lossF6(!i,3) = log(estg21(!i,2)) + (estg21(!i,3)^2/
    estg21(!i,2))
  lossF6(!i,4) = log(estt01(!i,2)) + (estt01(!i,3)^2/
    estt01(!i,2))
  lossF6(!i,5) = log(estt11(!i,2)) + (estt11(!i,3)^2/
    estt11(!i,2))
  lossF6(!i,6) = log(estt21(!i,2)) + (estt21(!i,3)^2/
    estt21(!i,2))
  lossF6(!i,7) = log(este01(!i,2)) + (este01(!i,3)^2/
```

```
      este01(!i,2))
    lossF6(!i,8) = log(este11(!i,2)) + (este11(!i,3)^2/
      este11(!i,2))
    lossF6(!i,9) = log(este21(!i,2)) + (este21(!i,3)^2/
      este21(!i,2))
next
for !ii=1 to !num_lossF
 for !k=1 to numT
  !w=firstT + (stepT*(!k-1))
 for !j=1 to !length-!ssize-T{!w}+1
  for !i=1 to !models
  statusline Loss Function !ii T !k Day !j
  !sum=0
   for !t=!j to T{!w}-1+!j
    sum_lossF{!ii}_{!w}(T{!w}-1+!j,!i) = lossF{!ii}(!t,!i)
    + !sum
    !sum = sum_lossF{!ii}_{!w}(T{!w}-1+!j,!i)
   next
  next
 next
 for !j=1 to !length-!ssize
   min_lossF{!ii}(!j,!k)=sum_lossF{!ii}_{!w}(!j,1)
   for !i=1 to !models-1
   if sum_lossF{!ii}_{!w}(!j,!i+1)<min_lossF{!ii}(!j,!k)
     then
    min_lossF{!ii}(!j,!k)=sum_lossF{!ii}_{!w}(!j,!i+1)
    endif
   next
 next
 for !j=1 to !length-!ssize
   for !i=1 to !models
   if sum_lossF{!ii}_{!w}(!j,!i)=min_lossF{!ii}(!j,!k)
     then
    model_min_lossF{!ii}(!j,!k)=!i
    endif
   next
 next
 next
matrix(!length,numT) Straddle_lossF{!ii}=na
for !k=1 to numT
  for !j=1 to !length-!ssize
  if min_lossF{!ii}(!j,!k)<>na then
  !m=model_min_lossF{!ii}(!j,!k)
  Straddle_lossF{!ii}(!j+!ssize,!k)=Straddles(!j
    +!ssize,!m)
```

```
else
!m= na
Straddle_lossF{!ii}(!j+!ssize,!k)= na
endif
next
next
vector(3) filter
  filter(1)=0
  filter(2)=2
  filter(3)=3.5
vector(2) cost
  cost(1)=0
  cost(2)=2
for !x=1 to 3
for !p=1 to 2
 matrix(!length,numT) lossF{!ii}_Str_Ret_F{!x}C{!p}
next
next
for !x=1 to 3
  for !p=1 to 2
    for !j=1 to numT
  for !i=1 to !length
    statusline filter !x cost !p T !j trading day !i
    if Straddle_lossF{!ii}(!i,!j)<> na then
     if Straddle_lossF{!ii}(!i,!j)>Straddle_obs(!i)+filter
     (!x) then
    lossF{!ii}_Str_Ret_F{!x}C{!p}(!i,!j)
      = (next_Straddle(!i)-Straddle_obs(!i)-Cost(!p))/
      Straddle_obs(!i)
    else
     if Straddle_lossF{!ii}(!i,!j)<Straddle_obs(!i)-filter
     (!x) then
    lossF{!ii}_Str_Ret_F{!x}C{!p}(!i,!j) =
      (-next_Straddle(!i)+Straddle_obs(!i)-Cost(!p))/
      Straddle_obs(!i)
    else
    lossF{!ii}_Str_Ret_F{!x}C{!p}(!i,!j) = na
    endif
    endif
    else
    lossF{!ii}_Str_Ret_F{!x}C{!p}(!i,!j) = na
    endif
  next
    next
  next
```

```
 next
next
for !i=firstT to lastT step stepT
delete T{!i}
next
for !ii=1 to !num_lossF
 for !k=1 to numT
  !w=firstT + (stepT*(!k-1))
  delete sum_lossF{!ii}_{!w}
 next
next
save chapter10.Loss_Functions_As_Model_
  Selection_Methods.wf1
```

- *chapter10.Median_Loss_Functions_as_Model_Selection_Methods.prg*

```
load chapter10.spec.wf1
smpl @all
series _temp = 1
!length = @obs(_temp)
delete _temp
!models = 9
!num_lossF = 6
!ssize = 500
!sum=0
scalar firstT = 5
scalar lastT = 55
scalar stepT = 10
scalar numT = ((lastT - firstT)/stepT)+1
for !i=firstT to lastT step stepT
  scalar T{!i}=!i

next
for !ii=1 to !num_lossF
  matrix(!length-!ssize,!models) lossF{!ii}
for !i=firstT to lastT step stepT
  matrix(!length-!ssize,!models)
    median_lossF{!ii}_{!i}=na
  next
matrix(!length-!ssize,numT) min_lossF{!ii}=na
matrix(!length-!ssize,numT) model_min_lossF{!ii}=na
next
'Create the SE Loss function
for !i=1 to !length-!ssize
  lossF1(!i,1) = (estg01(!i,2) - estg01(!i,3)^2)^2
  lossF1(!i,2) = (estg11(!i,2) - estg11(!i,3)^2)^2
```

```
  lossF1(!i,3) = (estg21(!i,2) - estg21(!i,3)^2)^2
  lossF1(!i,4) = (estt01(!i,2) - estt01(!i,3)^2)^2
  lossF1(!i,5) = (estt11(!i,2) - estt11(!i,3)^2)^2
  lossF1(!i,6) = (estt21(!i,2) - estt21(!i,3)^2)^2
  lossF1(!i,7) = (este01(!i,2) - este01(!i,3)^2)^2
  lossF1(!i,8) = (este11(!i,2) - este11(!i,3)^2)^2
  lossF1(!i,9) = (este21(!i,2) - este21(!i,3)^2)^2
next

'Create the AE Loss function
for !i=1 to !length-!ssize
  lossF2(!i,1) = @abs(estg01(!i,2) - estg01(!i,3)^2)
  lossF2(!i,2) = @abs(estg11(!i,2) - estg11(!i,3)^2)
  lossF2(!i,3) = @abs(estg21(!i,2) - estg21(!i,3)^2)
  lossF2(!i,4) = @abs(estt01(!i,2) - estt01(!i,3)^2)
  lossF2(!i,5) = @abs(estt11(!i,2) - estt11(!i,3)^2)
  lossF2(!i,6) = @abs(estt21(!i,2) - estt21(!i,3)^2)
  lossF2(!i,7) = @abs(este01(!i,2) - este01(!i,3)^2)
  lossF2(!i,8) = @abs(este11(!i,2) - este11(!i,3)^2)
  lossF2(!i,9) = @abs(este21(!i,2) - este21(!i,3)^2)
next
'Create the HASE Loss function
for !i=1 to !length-!ssize
  lossF3(!i,1) = (1-(estg01(!i,3)^2/estg01(!i,2)))^2
  lossF3(!i,2) = (1-(estg11(!i,3)^2/estg11(!i,2)))^2
  lossF3(!i,3) = (1-(estg21(!i,3)^2/estg21(!i,2)))^2
  lossF3(!i,4) = (1-(estt01(!i,3)^2/estt01(!i,2)))^2
  lossF3(!i,5) = (1-(estt11(!i,3)^2/estt11(!i,2)))^2
  lossF3(!i,6) = (1-(estt21(!i,3)^2/estt21(!i,2)))^2
  lossF3(!i,7) = (1-(este01(!i,3)^2/este01(!i,2)))^2
  lossF3(!i,8) = (1-(este11(!i,3)^2/este11(!i,2)))^2
  lossF3(!i,9) = (1-(este21(!i,3)^2/este21(!i,2)))^2
next
'Create the HAAE Loss function
for !i=1 to !length-!ssize
  lossF4(!i,1) = @abs(1-(estg01(!i,3)^2/estg01(!i,2)))
  lossF4(!i,2) = @abs(1-(estg11(!i,3)^2/estg11(!i,2)))
  lossF4(!i,3) = @abs(1-(estg21(!i,3)^2/estg21(!i,2)))
  lossF4(!i,4) = @abs(1-(estt01(!i,3)^2/estt01(!i,2)))
  lossF4(!i,5) = @abs(1-(estt11(!i,3)^2/estt11(!i,2)))
  lossF4(!i,6) = @abs(1-(estt21(!i,3)^2/estt21(!i,2)))
  lossF4(!i,7) = @abs(1-(este01(!i,3)^2/este01(!i,2)))
  lossF4(!i,8) = @abs(1-(este11(!i,3)^2/este11(!i,2)))
  lossF4(!i,9) = @abs(1-(este21(!i,3)^2/este21(!i,2)))
next
'Create the LE Loss function
```

```
for !i=1 to !length-!ssize
  lossF5(!i,1) = (log(estg01(!i,3)^2/estg01(!i,2)))^2
  lossF5(!i,2) = (log(estg11(!i,3)^2/estg11(!i,2)))^2
  lossF5(!i,3) = (log(estg21(!i,3)^2/estg21(!i,2)))^2
  lossF5(!i,4) = (log(estt01(!i,3)^2/estt01(!i,2)))^2
  lossF5(!i,5) = (log(estt11(!i,3)^2/estt11(!i,2)))^2
  lossF5(!i,6) = (log(estt21(!i,3)^2/estt21(!i,2)))^2
  lossF5(!i,7) = (log(este01(!i,3)^2/este01(!i,2)))^2
  lossF5(!i,8) = (log(este11(!i,3)^2/este11(!i,2)))^2
  lossF5(!i,9) = (log(este21(!i,3)^2/este21(!i,2)))^2
next
'Create the GLLS Loss function
for !i=1 to !length-!ssize
  lossF6(!i,1) = log(estg01(!i,2)) + (estg01(!i,3)
    ^2/estg01(!i,2))
  lossF6(!i,2) =log(estg11(!i,2)) + (estg11(!i,3)
    ^2/estg11(!i,2))
  lossF6(!i,3) = log(estg21(!i,2)) + (estg21(!i,3)
    ^2/estg21(!i,2))
  lossF6(!i,4) = log(estt01(!i,2)) + (estt01(!i,3)
    ^2/estt01(!i,2))
  lossF6(!i,5) = log(estt11(!i,2)) + (estt11(!i,3)
    ^2/estt11(!i,2))
  lossF6(!i,6) = log(estt21(!i,2)) + (estt21(!i,3)
    ^2/estt21(!i,2))
  lossF6(!i,7) = log(este01(!i,2)) + (este01(!i,3)
    ^2/este01(!i,2))
  lossF6(!i,8) = log(este11(!i,2)) + (este11(!i,3)
    ^2/este11(!i,2))
  lossF6(!i,9) = log(este21(!i,2)) + (este21(!i,3)
    ^2/este21(!i,2))
next

for !ii=1 to !num_lossF
  for !k=1 to numT
!w=firstT + (stepT*(!k-1))
for !j=1 to !length-!ssize-T{!w}+1
  for !i=1 to !models
  statusline Loss Function !ii T !k Day !j
    series memory=na
    for !t=!j to T{!w}-1+!j
      memory(!t)=lossF{!ii}(!t,!i)
      median_lossF{!ii}_{!w}(T{!w}-1+!j,!i) =
        @median(memory)
    next
```

```
     delete memory
  next
next
for !j=1 to !length-!ssize
  min_lossF{!ii}(!j,!k)=median_lossF
    {!ii}_{!w}(!j,1)
  for !i=1 to !models-1
    if median_lossF{!ii}_{!w}(!j,!i+1)
      <min_lossF{!ii}(!j,!k) then
    min_lossF{!ii}(!j,!k)=median_lossF
      {!ii}_{!w}(!j,!i+1)
    endif
  next
next
for !j=1 to !length-!ssize
  for !i=1 to !models
    if median_lossF{!ii}_{!w}(!j,!i)
      =min_lossF{!ii}(!j,!k) then
    model_min_lossF{!ii}(!j,!k)=!i
    endif
  next
next
next
matrix(!length,numT) Straddle_lossF{!ii}=na
for !k=1 to numT
  for !j=1 to !length-!ssize
  if min_lossF{!ii}(!j,!k)<>na then
  !m=model_min_lossF{!ii}(!j,!k)
  Straddle_lossF{!ii}(!j+!ssize,!k)
  =Straddles(!j+!ssize,!m)
  else
  !m= na
  Straddle_lossF{!ii}(!j+!ssize,!k)= na
  endif
  next
next
vector(3) filter
  filter(1)=0
  filter(2)=2
  filter(3)=3.5
vector(2) cost
  cost(1)=0
  cost(2)=2
for !x=1 to 3
for !p=1 to 2
```

```
      matrix(!length,numT) lossF{!ii}_Str_Ret_F{!x}C{!p}
next
next
for !x=1 to 3
  for !p=1 to 2
    for !j=1 to numT
      for !i=1 to !length
        statusline filter !x cost !p T !j trading day !i
        if Straddle_lossF{!ii}(!i,!j)<> na then
        if Straddle_lossF{!ii}(!i,!j)>Straddle_obs(!i)
          +filter(!x) then
        lossF{!ii}_Str_Ret_F{!x}C{!p}(!i,!j)
          = (next_Straddle(!i)-Straddle_obs(!i)-
          Cost(!p))/Straddle_obs(!i)
        else
        if Straddle_lossF{!ii}(!i,!j)<Straddle_obs(!i)-
          filter(!x) then
        lossF{!ii}_Str_Ret_F{!x}C{!p}(!i,!j) = (-next_
          Straddle(!i)+Straddle_obs(!i)-Cost(!p))/
          Straddle_obs(!i)
        else
        lossF{!ii}_Str_Ret_F{!x}C{!p}(!i,!j) = na
        endif
        endif
        else
      lossF{!ii}_Str_Ret_F{!x}C{!p}(!i,!j) = na
      endif
    next
      next
  next
 next
next
for !i=firstT to lastT step stepT
  delete T{!i}
next
for !ii=1 to !num_lossF
  for !k=1 to numT
    !w=firstT + (stepT*(!k-1))
    delete median_lossF{!ii}_{!w}
  next
next
save chapter10.Median_Loss_Functions_As
  _Model_Selection_Methods.wf1
```

Ox Metrics

- ***chapter10.SPA.ftse100.ox***

```
//TestStat[0] -> the SPA hypothesis testing,
  "TestStatScaledMax"
//TestStat[1] -> the RC hypothesis testing,
  "TestStatMax"
#include <oxstd.h>
#include <oxfloat.h>
#import "spa_src"
mse(const y, const yhat) { return meanc((y-yhat).^2);}
decl dB=10000,    // number of resamples (default is 1000)
   dq=0.5,    // for bootstrapping    (default is .5)
   myseed = 12136,    // Seed for random number generated
   lossfct = {"mse"},
   Ofile  = "ftse100"; // Where to save the results.
main()
{
decl i,j,t,TestStat={"TestStatScaledMax",
  "TestStatMax"},res1,res2=<>,
mY   = loadmat("ftse100_squared_returns.xls"),
  // Dataset with Realized volatility
mYhat = loadmat("variance_forecasts.xls");
  // Dataset with volatility forecasts
for (i=0; i<sizerc(lossfct); ++i){
 res1=<>;
 ranseed(myseed);
 println("The benchmark model is AR(1)-
  FIAPARCH(1,d,1)-t");
 res1 = SPA(mY[][0], mYhat, lossfct[i] , TestStat[0], dB,
    dq);
 res2 |= res1;
 }
}
```

- ***chapter10.SPA.and.ES.ftse100.ox***

```
#include <oxstd.h>
#include <oxfloat.h>
#import "spa_src"
mse(const y, const yhat) { return meanc((y-yhat).^2);}
decl dB=10000,    // number of resamples (default is 1000)
   dq=0.5,    // for bootstrapping    (default is .5)
   myseed = 12136,    // Seed for random number generater
```

```
  lossfct = {"mse"},
  Ofile  = "ftse100951";    // Where to save the results.
main()
{
  decl i,j,t,TestStat={"TestStatScaledMax",
    "TestStatMax"},res1,res2=<>,
  mY  = loadmat("ftse100.xls"),   // Dataset with log-returns
  mYhat = loadmat("ftse100951.xls");/ Dataset with Expected
    Shortfalls
  for (i=0; i<sizerc(lossfct); ++i){
    res1=<>;
    ranseed(myseed);
    println("Data is ftse100951, Benchmark is AR(1)-
      FIAPARCH(1,d,1)-t.");
    res1'= SPA(mY[][0], mYhat, lossfct[i], TestStat[0], dB,
      dq);
    res2 |= res1;
  }
}
```

11

Multivariate ARCH models

The models that have been discussed so far are univariate. However, assets and markets affect each other not only in terms of expected returns, but also in terms of volatility. Thus, accurate estimation of time-varying covariances between asset returns has been crucial for asset pricing and risk management. The generalization of univariate models to a multivariate context leads to a straightforward application of ARCH models to portfolio selection and asset pricing theory.

Let the $(n \times 1)$ vector $\{\mathbf{y}_t\}$ refer to the multivariate discrete time real-valued stochastic process to be predicted, with $E_{t-1}(\mathbf{y}_t) \equiv \boldsymbol{\mu}_t$ denoting the conditional mean. The innovation process for the conditional mean $\boldsymbol{\varepsilon}_t \equiv \mathbf{y}_t - \boldsymbol{\mu}_t$ has an $(n \times n)$ conditional covariance matrix $V_{t-1}(\mathbf{y}_t) \equiv \mathbf{H}_t$. For a system of n regression equations, the natural extension of (1.5) to a multivariate framework can be written as

$$
\begin{aligned}
\mathbf{y}_t &= \mathbf{B}'\mathbf{x}_t + \boldsymbol{\varepsilon}_t, \\
\boldsymbol{\varepsilon}_t | I_{t-1} &\sim f[0, \mathbf{H}_t], \\
\mathbf{H}_t &= g(\mathbf{H}_{t-1}, \mathbf{H}_{t-2}, \ldots, \boldsymbol{\varepsilon}_{t-1}, \boldsymbol{\varepsilon}_{t-2}, \ldots),
\end{aligned}
\tag{11.1}
$$

where $\mathbf{B}$ is a $k \times n$ matrix of unknown parameters, $\mathbf{x}_t$ a $k \times 1$ vector of endogenous and exogenous explanatory variables included in the available information set, I_{t-1}, $f(.)$ the conditional multivariate density function of the innovation process and $g(.)$ a function of the lagged conditional covariance matrices and the innovation process. The innovation process can be written as

$$
\boldsymbol{\varepsilon}_t = \mathbf{H}_t^{1/2} \mathbf{z}_t,
\tag{11.2}
$$

where, analogously to the univariate case assumptions, $\mathbf{z}_t$ is an $(n \times 1)$ independently and identically distributed vector process such that $E(\mathbf{z}_t) = \mathbf{0}$ and $E(\mathbf{z}_t \mathbf{z}_t') = \mathbf{I}$. As in the case of the univariate ARCH process, the conditional mean, $\boldsymbol{\mu}_t$, is not limited to a linear regression specification, $\boldsymbol{\mu}_t \equiv \mathbf{B}'\mathbf{x}_t$. ARMA, ARFIMAX or other conditional

ARCH Models for Financial Applications   Evdokia Xekalaki and Stavros Degiannakis
© 2010 John Wiley & Sons, Ltd

mean specifications presented in Chapter 1 can be applied in their vector form. For textbooks referring to multivariate ARCH models, see, among others, Andersen et al. (2006, p. 841), Bollerslev et al. (1994, p. 3002), Brooks (2002, p. 506), Harris and Sollis (2003, p. 221), Lütkepohl (2005, p. 557), Lütkepohl and Krätzig (2004, p. 212) and Tsay (2002, p. 357). Multivariate ARCH models have also been surveyed by McAleer (2005), Bauwens et al. (2006) and Silvennoinen and Teräsvirta (2008). Bauwens et al. (2006) divided the multivariate ARCH models into three categories: (i) direct generalizations of the univariate models (e.g. VECH, BEKK, Factor-ARCH etc.); (ii) linear combinations of univariate ARCH models (e.g. Latent-Factor-ARCH, O-ARCH, GO-ARCH etc.); and (iii) non-linear combinations of univariate ARCH models (e.g. CCC-ARCH, DCC-ARCH, Copula-ARCH etc.). Silvennoinen and Teräsvirta (2008) considered four categories: the first includes specifications where $\mathbf{H}_t$ is modelled directly (e.g. VECH, BEKK etc.), the second contains parsimonious models (e.g. Factor ARCH, Latent-Factor-ARCH, O-ARCH, GO-ARCH etc.), the third consists of models constructed on the idea of modelling the conditional variances and correlations instead of the straightforward modelling of $\mathbf{H}_t$ (e.g. CCC-ARCH, DCC-ARCH etc.), and the fourth consists of models implemented via semi- and non-parametric approaches, which can offset the loss of efficiency of parametric estimators in the case of a misspecified conditional covariance matrix structure.

11.1   Model specifications

11.1.1   Symmetric model specifications

11.1.1.1   BEKK(p,q)

The natural multivariate extension of the GARCH(p, q) model in equation (2.5) is

$$\mathbf{H}_t = \mathbf{A}_0\mathbf{A}_0' + \sum_{i=1}^{q} \left(\mathbf{A}_i\boldsymbol{\varepsilon}_{t-i}\boldsymbol{\varepsilon}_{t-i}'\mathbf{A}_i'\right) + \sum_{j=1}^{p} \left(\mathbf{B}_j\mathbf{H}_{t-j}\mathbf{B}_j'\right), \qquad (11.3)$$

where $\mathbf{A}_0$ is a lower triangular matrix with $(n(n+1)/2)$ parameters, and $\mathbf{A}_i$ and $\mathbf{B}_j$ denote $(n \times n)$ matrices with n^2 parameters each. Based on earlier work by Baba et al. (1990), Engle and Kroner (1995) proposed model (11.3), which they referred to as the BEKK(p, q) model. This parameterization guarantees that $\mathbf{H}_t$ is positive definite, if $\mathbf{A}_0\mathbf{A}_0'$ is a positive definite matrix, and requires the estimation of $(n(n+1)/2) + n^2(q+p)$ parameters. For example, for $n = 3$, the BEKK(1,1) model contains 24 parameters for estimation. J. Lee (1999) investigated the output–inflation variability tradeoff using the bivariate BEKK model.

In order to estimate time-varying optimal hedge ratios in commodity markets, Moschini and Myers (2002) modified the BEKK model of (11.3) as follows:

$$\mathbf{H}_t = \boldsymbol{\Gamma}_t'\left(\mathbf{A}_0\mathbf{A}_0' + \sum_{i=1}^{q} \left(\mathbf{A}_i\boldsymbol{\varepsilon}_{t-i}\boldsymbol{\varepsilon}_{t-i}'\mathbf{A}_i'\right) + \sum_{j=1}^{p} \left(\mathbf{B}_j\mathbf{H}_{t-j}\mathbf{B}_j'\right)\right)\boldsymbol{\Gamma}_t. \qquad (11.4)$$

As Moschini and Myers noted, the covariance matrix is positive definite as long as $\mathbf{\Gamma}_t$ is a positive definite matrix.

The originally proposed BEKK$(p, q, \tilde{q})$ model is somewhat more general since it involves a summation over $\tilde{q}$ terms:

$$\mathbf{H}_t = \mathbf{A}_0\mathbf{A}_0' + \sum_{k=1}^{\tilde{q}}\sum_{i=1}^{q}\left(\mathbf{A}_{k,i}\mathbf{\varepsilon}_{t-i}\mathbf{\varepsilon}_{t-i}'\mathbf{A}_{k,i}'\right) + \sum_{k=1}^{\tilde{q}}\sum_{j=1}^{p}\left(\mathbf{B}_{k,j}\mathbf{H}_{t-j}\mathbf{B}_{k,j}'\right), \quad (11.5)$$

where $\mathbf{A}_0$ is a lower triangular matrix with $(n(n+1)/2)$ parameters and $\mathbf{A}_{k,i}$ and $\mathbf{B}_{k,j}$ are $(n \times n)$ matrices.

11.1.1.2 VECH(p,q)

A simpler expression for $\mathbf{H}_t$ can be obtained through the use of the $vech(.)$ operator that stacks the columns of an $(n \times n)$ square matrix from the diagonal downwards into an $(n(n+1)/2) \times 1$ vector. Equation (11.3) is thus rewritten as

$$vech(\mathbf{H}_t) = vech(\mathbf{A}_0\mathbf{A}_0') + \sum_{i=1}^{q}\left(\tilde{\mathbf{A}}_i vech(\mathbf{\varepsilon}_{t-i}\mathbf{\varepsilon}_{t-i}')\right) + \sum_{j=1}^{p}\left(\tilde{\mathbf{B}}_j vech(\mathbf{H}_{t-j})\right), \quad (11.6)$$

where $vech(\mathbf{A}_0\mathbf{A}_0')$ is an $(n(n+1)/2) \times 1$ parameter vector and $\tilde{\mathbf{A}}_i$ and $\tilde{\mathbf{B}}_j$ are parameter matrices of dimension $(n(n+1)/2 \times n(n+1)/2)$. Engle et al. (1986) published the first paper on multivariate ARCH models applying the VECH(0,2) model. However, in the VECH expression for the multivariate ARCH model, serious problems might arise: (i) the model might not yield a positive definite covariance matrix unless non-linear inequality restrictions are imposed; (ii) the number of parameters to be estimated is $(n(n+1)/2)(1 + (n(n+1)/2)(q+p))$, a very large number even for low values of n. For example, for $n = 3$, the VECH(1,1) model contains 78 parameters for estimation.

A number of models in the financial literature have dealt with imposing constraints in multivariate ARCH models in order to reduce the number of parameters to be estimated. These constraints have to be compatible with a positive definite conditional covariance matrix and must lead to tractable estimation procedures.

11.1.1.3 Diag-VECH(p,q)

Bollerslev et al. (1988) proposed the diagonal VECH(p, q), or Diag-VECH(p, q), model:

$$vech(\mathbf{H}_t) = vech(\mathbf{A}_0\mathbf{A}_0') + \sum_{i=1}^{q}\left(\tilde{\mathbf{A}}_i vech(\mathbf{\varepsilon}_{t-i}\mathbf{\varepsilon}_{t-i}')\right) + \sum_{j=1}^{p}\left(\tilde{\mathbf{B}}_j vech(\mathbf{H}_{t-j})\right), \quad (11.7)$$

where the matrices $\tilde{\mathbf{A}}_i$ and $\tilde{\mathbf{B}}_j$ are assumed diagonal. Thus, the number of parameters is reduced to $(n(n+1)/2)(1 + q + p)$, as no interaction is allowed between the different conditional variances and covariances. So for $n = 3$, for example, the diagonal VECH (1,1) model requires the estimation of 18 parameters. Bollerslev et al. (1988) used this model to analyse returns on bills, bonds and stocks, while Baillie and Myers (1991), Bera et al. (1991) and Myers (1991) employed it to estimate hedge ratios in commodity markets. Ding and Engle (2001) gave sufficient conditions for the diagonal multivariate GARCH(1,1) model to be positive definite and proposed four models which are nested in the multivariate diagonal multivariate GARCH(1,1) model.

11.1.1.4 Diag-BEKK(p, q)

The BEKK(p,q) model in (11.3) assumes that the $\mathbf{A}_i$ and $\mathbf{B}_j$ matrices contain n^2 parameters each. A less general version known as the Diag-BEKK(p,q) model is commonly applied:

$$\mathbf{H}_t = \mathbf{A}_0\mathbf{A}_0' + \sum_{i=1}^{q} \left(\mathbf{A}_i\boldsymbol{\varepsilon}_{t-i}\boldsymbol{\varepsilon}_{t-i}'\mathbf{A}_i'\right) + \sum_{j=1}^{p} \left(\mathbf{B}_j\mathbf{H}_{t-j}\mathbf{B}_j'\right), \tag{11.8}$$

where the matrices $\mathbf{A}_i$ and $\mathbf{B}_j$ are again restricted to being diagonal. The Diag-BEKK(p,q) model requires the estimation of $(n(n+1)/2) + n(q+p)$ parameters. The advantage is that the Diag-BEKK model is guaranteed to be positive definite and is regarded as a parsimonious version of the Diag-VECH model, as it requires the estimation of fewer parameters than the Diag-VECH model.

11.1.1.5 Scalar-BEKK(p, q)

The number of estimated parameters can be greatly reduced by assuming the same value for all the elements of the matrices $\mathbf{A}_i$ and $\mathbf{B}_j$. The scalar-BEKK(p, q) model is defined as:

$$\mathbf{H}_t = \mathbf{A}_0\mathbf{A}_0' + \sum_{i=1}^{q} \left(a_i\mathbf{i}\mathbf{i}'\boldsymbol{\varepsilon}_{t-i}\boldsymbol{\varepsilon}_{t-i}'\right) + \sum_{j=1}^{p} \left(b_j\mathbf{i}\mathbf{i}'\mathbf{H}_{t-j}\right), \tag{11.9}$$

where a_i and b_j are positive scalars and $\mathbf{i}$ is an $(n \times 1)$ vector of ones.

11.1.1.6 Factor ARCH

A special case of the BEKK model, for $p = q = 1$, is the factor ARCH model first proposed by Engle (1987). The Factor-ARCH(1,1) model was constructed to overcome the problem of having to estimate a vast number of parameters, while retaining the benefits of positive definiteness. The model has the form

$$\mathbf{H}_t = \mathbf{A} + \lambda\lambda'\left(\alpha(\mathbf{w}'\boldsymbol{\varepsilon}_{t-1})^2 + b\mathbf{w}'\mathbf{H}_{t-1}\mathbf{w}\right), \tag{11.10}$$

where $\mathbf{A}$ is an $(n \times n)$ positive definite matrix, α and b are scalars, and λ and $\mathbf{w}$ are $(n \times 1)$ vectors. The vector $\mathbf{w}$ can be considered as a vector of portfolio weights and is often conveniently restricted to the case $\mathbf{i}'\mathbf{w} = 1$, where $\mathbf{i}$ is a vector of ones. This model is a special case of the BEKK model, where the matrices $\mathbf{A}_1 = \sqrt{\alpha}\mathbf{w}\lambda'$ and

$\mathbf{B}_1 = \sqrt{b}\mathbf{w}\boldsymbol{\lambda}'$ have rank 1. The number of parameters to be estimated is $(n^2 + 5n)/2$. So for $n = 3$, for example, we have to estimate 12 parameters.

The model can be extended to allow for $\ddot{K}$ factors and a higher-order ARCH structure. The number of factors is intended to be much smaller than the number of assets, $\ddot{K} < n$, which makes the model feasible even for a large number of assets. So, the $\ddot{K}$ Factor-ARCH(p, q) model can be written as

$$\mathbf{H}_t = \mathbf{A} + \sum_{k=1}^{\ddot{K}} \boldsymbol{\lambda}_k \boldsymbol{\lambda}_k' \left(\sum_{i=1}^{q} \alpha_{k,i} \mathbf{w}_k' \boldsymbol{\varepsilon}_{t-i} \boldsymbol{\varepsilon}_{t-i}' \mathbf{w}_k + \sum_{j=1}^{p} b_{k,j} \mathbf{w}_k' \mathbf{H}_{t-j} \mathbf{w}_k \right), \qquad (11.11)$$

where $\boldsymbol{\lambda}_k$ are linearly independent $(n \times 1)$ vectors of factor weights and there are $\ddot{K}(2n + p + q) + n(n+1)/2$ free parameters. Engle et al. (1990b) and Ng et al. (1992) applied factor ARCH models to treasury bills and stock returns (see also Bollerslev and Engle, 1993; Lin, 1992).

The factor ARCH type specification assumes that the volatility dynamics is generated by observable factors, which are conditionally heteroscedastic and are defined as univariate GARCH(p, q) models. Therefore, the $\ddot{K}$ Factor-ARCH(p, q) model can alternatively be written as

$$\mathbf{H}_t = \mathbf{A} + \sum_{k=1}^{\ddot{K}} \boldsymbol{\lambda}_k \boldsymbol{\lambda}_k' f_{k,t},$$

$$f_{k,t} = \alpha_{k,0} + \sum_{i=1}^{q} \alpha_{k,i} \left(\mathbf{w}_k' \boldsymbol{\varepsilon}_{t-i} \right)^2 + \sum_{j=1}^{p} b_{k,j} f_{k,t-j}. \qquad (11.12)$$

Diebold and Nerlove (1989), Harvey et al. (1992) and King et al. (1994) proposed latent-factor ARCH models based on the assumption that only a few factors influence the conditional variances and covariances of asset returns, which are not functions of the information set.

11.1.1.7 O-ARCH

A factor ARCH model does not necessarily assume uncorrelated factors, whereas the class of orthogonal ARCH models restricts the factors to being uncorrelated in order to represent genuinely different common components driving the returns. In an O-ARCH model the observed data are generated by an orthogonal transformation of $m < n$ univariate ARCH-type processes. Alexander and Chibumba (1997) proposed the orthogonal ARCH, or O-ARCH(p, q, m), model:

$$\mathbf{y}_t = \boldsymbol{\mu}_t + \boldsymbol{\varepsilon}_t,$$

$$\boldsymbol{\varepsilon}_t = \mathbf{V}^{1/2} \mathbf{u}_t, \qquad (11.13)$$

$$\mathbf{u}_t = \mathbf{W} \mathbf{f}_t.$$

The innovation process vector is considered as the product of $\mathbf{V}^{1/2} = diag(v_1, v_2, \ldots, v_n)$ and $\mathbf{u}_t$, with $\{v_i\}_{i=1}^{n}$ denoting the population standard deviations of $\{\varepsilon_{i,t}\}_{i=1}^{n}$, the elements of the vector $\boldsymbol{\varepsilon}_t$. For the random vector process

$\mathbf{f}_t = \left(f_{1,t}, f_{2,t}, \ldots f_{m,t}\right)'$, we assume zero conditional mean and univariate ARCH-type conditional variance, or

$$E_{t-1}(\mathbf{f}_t) = 0,$$
$$V_{t-1}(\mathbf{f}_t) \equiv \mathbf{H}_{\mathbf{f},t} = diag\left(\sigma_{f_{1,t}}^2, \sigma_{f_{2,t}}^2, \ldots, \sigma_{f_{m,t}}^2\right), \qquad (11.14)$$
$$\sigma_{f_{i,t}}^2 = \left(1 - a_{i,1} - b_{i,1}\right) + a_{i,1} f_{i,t-1}^2 + b_{i,1} \sigma_{f_{i,t-1}}^2,$$

for $i = 1, 2, \ldots, m$. The matrix of the linear transformation, $\mathbf{W}$, is an orthogonal matrix defined as

$$\mathbf{W} = \mathbf{P}\mathbf{\Lambda}^{1/2}, \qquad (11.15)$$

where $\mathbf{\Lambda}^{1/2} = diag(l_1^{1/2}, l_2^{1/2}, \ldots, l_m^{1/2})$, with $l_1 \geq l_2 \geq \ldots \geq l_m$ being the eigenvalues of the unconditional correlation matrix of $\boldsymbol{\varepsilon}_t$ (or of the unconditional covariance matrix of $\mathbf{u}_t$), and $\mathbf{P}$ is the $(n \times m)$ matrix whose columns are the corresponding eigenvectors. Therefore, the conditional covariance matrix of the innovation process is computed as

$$V_{t-1}(\boldsymbol{\varepsilon}_t) \equiv \mathbf{H}_t = \mathbf{V}^{1/2} \mathbf{V}_t \mathbf{V}'^{1/2}, \qquad (11.16)$$

where

$$V_{t-1}(\mathbf{u}_t) \equiv \mathbf{V}_t = \mathbf{W}\mathbf{H}_{\mathbf{f},t}\mathbf{W}'. \qquad (11.17)$$

Alexander (1998, 2000, 2001, 2002) provided a series of empirical applications and supporting programs for O-ARCH estimation. However, the O-ARCH model has two important limitations: (i) the matrix $\mathbf{W}$ is assumed orthogonal, which only covers a very small subset of all possible invertible matrices; and (ii) even if $\mathbf{W}$ is orthogonal, its estimation is not always identifiable.

11.1.1.8 GO-ARCH

Van der Weide (2002) relaxed the orthogonality condition assumed in the O-ARCH model and proposed the generalized O-ARCH, or GO-ARCH, model, which has the same framework as (11.13) for $m \equiv n$ univariate ARCH-type processes:

$$\begin{aligned} \mathbf{y}_t &= \boldsymbol{\mu}_t + \boldsymbol{\varepsilon}_t, \\ \boldsymbol{\varepsilon}_t &= \mathbf{V}^{1/2}\mathbf{u}_t, \\ \mathbf{u}_t &= \mathbf{W}\mathbf{f}_t, \end{aligned} \qquad (11.18)$$

except that the matrix $\mathbf{W}$ is square and invertible and is defined as

$$\mathbf{W} = \mathbf{P}\mathbf{\Lambda}^{1/2}\mathbf{U}, \qquad (11.19)$$

where $\mathbf{U}$ is orthogonal. The matrix $\mathbf{U}$ is estimated conditionally on the information set as the product of $n(n-1)/2$ rotation matrices

$$\mathbf{U} = \prod_{j=i+1}^{n} \prod_{i=1}^{n-1} \mathbf{G}_{i,j}(\delta_{i,j}), \qquad (11.20)$$

for $-\pi \leq \delta_{i,j} \leq \pi$. $\mathbf{G}_{i,j}(\delta_{i,j})$ performs a rotation in the plane spanned by the ith and jth vectors of the canonical basis of $\mathfrak{R}^n$ over an angle $\delta_{i,j}$. The $\mathbf{G}_{i,j}$ matrices are $(n \times n)$ identity matrices, where the (i, i) th and (j, j) th elements are replaced by $\cos(\delta_{i,j})$, and the (i, j) th and the (j, i) th elements are replaced by $\sin(\delta_{i,j})$ and $-\sin(\delta_{i,j})$, respectively.

Vrontos et al. (2003) proposed the full factor ARCH model, where the matrix $\mathbf{W}$ is restricted to being triangular, an assumption that simplifies parameter estimation but rules out certain relationships between the ε_t and $\mathbf{f}_t$. Recently, Lanne and Saikkonen (2007) introduced the generalized orthogonal factor ARCH, or GOF-ARCH, model which allows for a reduced number of conditionally heteroscedastic factors and idiosyncratic shocks. The GOF-ARCH model is more parsimonious and easier to estimate than the factor ARCH and does not impose the restrictions (e.g. no idiosyncratic shocks) of the GO-ARCH specifications.

11.1.1.9 CCC-ARCH

The constant conditional correlation model, introduced by Bollerslev (1990), has instituted a popular new approach to modelling multivariate ARCH processes: univariate ARCH models are estimated for each asset and the correlation matrix is then estimated. The time-varying conditional covariances are parameterized to be proportional to the product of the corresponding conditional standard deviations. This assumption greatly simplifies the estimation of the model and reduces the computation cost.

Let us assume that the covariance matrix can be decomposed into $\mathbf{H}_t = \mathbf{\Sigma}_t^{1/2}\mathbf{C}_t\mathbf{\Sigma}_t^{1/2}$, where $\mathbf{\Sigma}_t^{1/2}$ is the diagonal matrix with conditional standard deviations along the diagonal, i.e. $\mathbf{\Sigma}_t^{1/2} = diag(\sigma_{1,t}, \sigma_{2,t}, \ldots, \sigma_{n,t})$, and $\mathbf{C}_t$ is the matrix of conditional correlations. The constant conditional correlation model assumes that the matrix of conditional correlations is a time-invariant matrix, C, so that the temporal variation of $\mathbf{H}_t$ can be determined solely by the conditional variances:

$$\mathbf{H}_t = \mathbf{\Sigma}_t^{1/2}\mathbf{C}\mathbf{\Sigma}_t^{1/2}. \tag{11.21}$$

$\mathbf{H}_t$ is positive definite if $\mathbf{C}$ is positive definite and the conditional variances are positive. The number of parameters to be estimated reduces to $(n(n-1)/2) + n(1+q+p)$. So, for $n = 3$ the CCC-GARCH(1,1) model requires the estimation of 12 parameters.

Several authors have considered this representation, among them Baillie and Bollerslev (1990), Brown and Ligeralde (1990), Cecchetti et al. (1988), Fornari et al. (2002), Kim (2000), Kroner and Claessens (1991), Kroner and Lastrapes (1991), Kroner and Sultan (1991, 1993), Lien and Tse (1998) and Park and Switzer (1995).

However, the constancy of conditional correlation has been challenged. Bera and Kim (1996), introducing the information matrix test, were led to the rejection of a constant correlation hypothesis for the US, European and Japanese stock markets. Tse (2000), proposing his Lagrange multiplier test for testing the same hypothesis, found evidence against it for Asian stock markets. Tsui and Yu (1999), adopting the information matrix test, came to the same conclusion for the Chinese stock market.

Longin and Solnik (1995) also found that the hypothesis of constant conditional correlation did not appear to be plausible in international equity returns against three alternative sources of variability of the correlation such as a time trend, the presence of threshold and asymmetry and the influence of information variables.

11.1.1.10 DCC-ARCH

As the hypothesis of constancy of correlation was found not to be supported in various applied contexts, Engle (2002b), Engle and Sheppard (2001) and Tse and Tsui (2002) considered a new form of multivariate ARCH model, the dynamic conditional correlation ARCH, or DCC-ARCH(p, q), model. The model is estimated in two steps. In the first, a series of univariate GARCH models are estimated. In the second, using the residuals resulting from the first step, the conditional correlation is estimated. The success of the DCC-ARCH model depends on the estimability of extremely large time-varying covariance matrices. Engle proposed the use of the decomposed covariance matrix $\mathbf{H}_t = \mathbf{\Sigma}_t^{1/2}\mathbf{C}_t\mathbf{\Sigma}_t^{1/2}$ with a time-varying correlation matrix of the form

$$\mathbf{C}_t = \mathbf{Q}_t^{*-1/2}\mathbf{Q}_t\mathbf{Q}_t^{*-1/2}. \tag{11.22}$$

The conditional variances, $\sigma_{i,t}^2$, are estimated as univariate GARCH(p_i, q_i) models, allowing for different lag lengths for each series $i = 1, 2, \ldots, n$,

$$\sigma_{i,t}^2 = a_{i,0} + \sum_{j=1}^{q_i}\left(a_{i,j}\varepsilon_{i,t-j}^2\right) + \sum_{j=1}^{p_i}\left(b_{i,j}\sigma_{i,t-j}^2\right). \tag{11.23}$$

The correlation matrix, $\mathbf{Q}_t = \left(q_{i,j,t}\right)$, is computed using

$$\mathbf{Q}_t = \left(1 - \sum_{m=1}^{q}a_m - \sum_{c=1}^{p}b_c\right)\bar{\mathbf{Q}} + \sum_{m=1}^{q}a_m(\mathbf{z}_{t-m}\mathbf{z}'_{t-m}) + \sum_{c=1}^{p}b_c\mathbf{Q}_{t-c}, \tag{11.24}$$

where $\mathbf{z}_t$ are the residuals standardized by their conditional standard deviation, i.e. $\mathbf{z}_t = (z_{1,t}, z_{2,t}, \ldots, z_{n,t})' = (\varepsilon_{1,t}\sigma_{1,t}^{-1}, \varepsilon_{2,t}\sigma_{2,t}^{-1}, \ldots, \varepsilon_{n,t}\sigma_{n,t}^{-1})'$, $\bar{\mathbf{Q}}$ is the unconditional covariance of the standardized residuals and $\mathbf{Q}_t^{*-1/2}$ is a diagonal matrix composed of the square roots of the inverse of the diagonal elements of $\mathbf{Q}_t$, i.e. $\mathbf{Q}_t^{*-1/2} = diag(q_{1,1,t}^{-1/2}, q_{2,2,t}^{-1/2}, \ldots, q_{n,n,t}^{-1/2})$. Engle and Sheppard (2001) proved the consistency and asymptotic normality of the two-step estimators as well as the positive definiteness of the covariance matrix. However, the two-step estimators are not fully efficient since they are limited information estimators. Engle and Sheppard also proposed a test of the null hypothesis of constant correlation against an alternative of dynamic conditional correlation.

Tse and Tsui (2002) suggested a different approximation in estimating the matrix $\mathbf{C}_t$:

$$\mathbf{C}_t = \left(1 - a_q - b_p\right)\bar{\mathbf{C}} + a_q\mathbf{C}_{t-1}^* + b_p\mathbf{C}_{t-1}, \tag{11.25}$$

where $\bar{\mathbf{C}}$ is an $(n \times n)$ symmetric positive definite matrix with 1s on the diagonal, $\mathbf{C}^*_{t-1}$ is an $(n \times n)$ correlation matrix of the t^* most recent standardized innovations, i.e. z_{t-i^*}, for $i^* = 1, 2, \ldots, t^*$, $t^* \geq n$, or the $\{c^*_{i,j,t-1}\}$ elements of $\mathbf{C}^*_{t-1}$ are computed as

$$c^*_{i,j,t-1} = \frac{\sum_{i^*=1}^{t^*} \left(\varepsilon_{i,t-i^*} \sigma^{-1}_{i,t-i^*} \right) \left(\varepsilon_{j,t-i} \sigma^{-1}_{j,t-i^*} \right)}{\sqrt{\sum_{i^*=1}^{t^*} \left(\varepsilon^2_{i,t-i^*} \sigma^{-2}_{i,t-i^*} \right) \sum_{i^*=1}^{t^*} \left(\varepsilon^2_{j,t-i} \sigma^{-2}_{j,t-i^*} \right)}}. \tag{11.26}$$

Tse and Tsui's (2002) DCC-ARCH model is also termed varying correlation ARCH, or VC-ARCH.

11.1.1.11 Corr-ARCH

Christodoulakis and Satchell (2002) considered an alternative extension of the constant conditional correlation model of Bollerslev (1990) and developed a bivariate ARCH model with time-varying conditional variances and correlations, named the correlated ARCH, or Corr-ARCH, model.

11.1.2 Asymmetric and long-memory model specifications

The multivariate ARCH models that have been presented so far, although simplifying the estimation and inference procedures, do not account for empirical regularities such as the asymmetric effects. In order to capture the leverage effect in a multivariate framework, Braun et al. (1995) introduced a bivariate version of the univariate EGARCH model. Sentana (1995) proposed a quadratic GARCH model and applied a multivariate version of it to UK stock returns. Kroner and Ng (1998), following Hentschel's (1995) approach, introduced a general multivariate ARCH model which has the BEKK, factor ARCH and CCC-ARCH models and their natural asymmetric extensions nested within it. Their model can be regarded as a multivariate extension of the univariate GJR model. Bekaert and Wu (2000), Ding and Engle (2001) and Tai (2001) have also modified multivariate ARCH models to accommodate asymmetric effects on conditional variances and covariances. Brunetti and Gilbert (1998) and Teyssiere (1997), based on Bollerslev's (1990) parameterization, proposed the bivariate constant correlation FIGARCH model and Brunetti and Gilbert (2000) applied the model to the crude oil market. Finally, Bayesian analysis of symmetric and asymmetric multivariate ARCH processes was considered in a number of articles such as Aguilar and West (2000).

Ñíguez and Rubia (2006) combined the O-ARCH model and the univariate HYGARCH model and proposed a multivariate extension in which a portfolio is formed by assets exhibiting long memory in volatility. Morana (2007) proposed a method of modelling the common long-memory components requiring neither the estimation of the fractional cointegration space nor the maximization of a frequency domain likelihood function.

454 ARCH MODELS FOR FINANCIAL APPLICATIONS

11.1.2.1 M-EGARCH(p,q)

Kawakatsu (2006) provided the multivariate expansion of Nelson's (1991) EGARCH model, named the matrix EGARCH(p,q), or M-EGARCH(p,q), model:

$$vech(\log \mathbf{H}_t - \mathbf{A}) = \sum_{i=1}^{q} (\mathbf{A}_i \mathbf{z}_{t-i}) + \sum_{i=1}^{q} (\mathbf{G}_i(|\mathbf{z}_{t-i}| - E|\mathbf{z}_{t-i}|))$$

$$+ \sum_{j=1}^{p} (\tilde{\mathbf{B}}_j vech(\log \mathbf{H}_{t-j} - \mathbf{A})), \qquad (11.27)$$

where $\mathbf{A}$ is an $(n \times n)$ symmetric parameter matrix, $\mathbf{A}_i$ and $\mathbf{G}_i$ are $(n(n+1)/2) \times n$ matrices and $\tilde{\mathbf{B}}_j$ is a parameter matrix of dimension $(n(n+1)/2) \times (n(n+1)/2)$. Note that in the case of the M-EGARCH model there is no need to ensure positive definiteness because the matrix $\log \mathbf{H}_t$ need not be positive definite. Since the model contains $(n(n+1)/2) + p(n(n+1)/2)^2 + qn^2(n+1)$ parameters, Kawakatsu (2006) considered parsimonious specifications that reduce the number of parameters that need to be estimated.

11.1.2.2 Asymmetric DCC-ARCH

Pelletier (2006) introduced a regime-switching model of constant correlations within each regime, while the dynamics enters through switching regimes. Silvennoinen and Teräsvirta (2005) proposed the smooth transition conditional correlation ARCH, or STCC-ARCH, model in which the conditional correlations change smoothly between two extreme states of constant correlations according to a transition variable. Silvennoinen and Teräsvirta (2007) proposed the double smooth transition conditional correlation ARCH model, or DSTCC-ARCH, which extends the STCC-ARCH model of Silvennoinen and Teräsvirta (2005) by including another variable according to which the correlations change smoothly between states of constant correlations. Franses and Hafner (2003) proposed an extended DCC model that allows for asset-specific heterogeneity in the correlation structure. Kwan et al. (2010) proposed the multivariate threshold ARCH with time-varying correlation, or VC-MTARCH model, which is a threshold extension to the VC-ARCH model of Tse and Tsui (2002). Billio et al. (2006) and Cappiello et al. (2006) provided extensions of the DCC model by introducing asymmetry in the correlation dynamics. McAleer et al. (2005) proposed the generalized autoregressive conditional correlation, or GARCC, model in which the standardized residuals follow a multivariate random coefficient autoregressive process. Billio and Caporin (2006) introduced a new DCC-type model, quadratic flexible DCC-ARCH, which generalizes Billio et al.'s (2006) flexible DCC-ARCH model by allowing for constant dynamics only among blocks of assets belonging to the same class of assets.

11.2 Maximum likelihood estimation

As for the univariate ARCH models, the method of maximum likelihood can be applied to jointly estimate the parameters of the mean and the variance equations.

Let us assume that the conditional distribution of vector $\mathbf{y}_t$ is multivariate normal with mean $\boldsymbol{\mu}_t$ and covariance matrix $\mathbf{H}_t$. For a multivariate ARCH model in the general form of (11.1), the log-likelihood function for a sample of T observations is:

$$L_T(\mathbf{y}_t; \theta) = -\frac{nT}{2}\log(2\pi) - \frac{1}{2}\sum_{t=1}^{T}\log|\mathbf{H}_t| - \frac{1}{2}\sum_{t=1}^{T}(\mathbf{y}_t - \boldsymbol{\mu}_t)'\mathbf{H}_t^{-1}(\mathbf{y}_t - \boldsymbol{\mu}_t), \quad (11.28)$$

where θ denotes the vector of the $\breve{\theta}$ parameters that need to be estimated.

Engle et al. (1990b) proposed a two-step maximum likelihood estimation method for the factor ARCH model, which is consistent but not efficient. Boswijk and van der Weide (2006) proposed a three-step non-linear least squares method for estimating the GO-ARCH model, noting that the GO-ARCH model is more attractive from a practical point of view. The new estimation procedure, although less efficient than the maximum likelihood procedure, is characterized by an increase in robustness. Thus, any misspecifications of the ARCH models of the independent components will have no effect on the estimation of the link matrix.

As in the case of univariate ARCH models, where the normality of conditional innovations is frequently rejected, the same holds for the multivariate expansion of the models. Bollerslev and Wooldridge's (1992) quasi-maximum likelihood estimator is a consistent estimator of the parameter vector θ even if the distribution of innovations is conditionally non-normal, provided that the conditional mean and conditional variance are correctly specified. For details about quasi-maximum likelihood estimators in multivariate ARCH models, see Gouriéroux (1997) and Jeantheau (1998). Jeantheau (1998) has proved the strong consistency of the quasi-maximum likelihood estimator of multivariate ARCH models.

However, as Bauwens and Laurent (2005) noted, in order to gain statistical efficiency, it is of primary importance to base modelling and inference on a more suitable distribution than the multivariate normal. Harvey et al. (1992) and Fiorentini et al. (2003) proposed the multivariate standardized Student t distribution in the estimation of multivariate ARCH models. To allow for both skewness and lepto-kurtosis in innovations, within a multivariate framework, Bauwens and Laurent (2005) introduced a multivariate skewed Student t distribution. Also, Mencia and Sentana (2005) proposed the generalized hyperbolic distribution. Bauwens and Laurent (2005) found that the multivariate skew Student t density provides better, or at least not worse, out-of-sample VaR forecasts than a symmetric density.

Let us assume that the $(n \times 1)$ vector of standardized innovations, $\mathbf{z}_t$, is an i.i.d. vector process with density function $f(\mathbf{z}_t; w)$, where w is the vector of the parameters of f to be estimated. For $\psi' = (\theta', w')$ denoting the whole set of the parameters that have to be estimated, the full-sample log-likelihood function for a sample of T observations can be written as

$$L_T(\{\mathbf{y}_t\}; \psi) = \sum_{t=1}^{T}\log\left(f\left(\mathbf{H}_t^{-1/2}(\theta)(\mathbf{y}_t - \boldsymbol{\mu}_t(\theta)); w\right)\right) + \sum_{t=1}^{T}\log|\mathbf{H}_t(\theta)|^{-1/2}.$$

$$(11.29)$$

Thus, in the case of choosing the multivariate standardized Student-t distribution, $f(\mathbf{z}_t; w)$ is as follows:

$$f_{(t)}(\mathbf{z}_t; v) = \frac{\Gamma((v+n)/2)}{\Gamma(v/2)(\pi(v-2))^{n/2}} \left(1 + \frac{\mathbf{z}_t' \mathbf{z}_t}{v-2}\right)^{-(v+n)/2}, \qquad (11.30)$$

where $\Gamma(.)$ is the gamma function and $w = (v)$ is the parameter to be estimated, for $v > 2$.

When the multivariate skewed Student t distribution is assumed for the standardized innovations, the following density function is applied:

$$f_{(skT)}(\mathbf{z}_t; v, g) = \frac{2^n \Gamma((v+n)/2)}{\Gamma(v/2)(\pi(v-2))^{n/2}} \left(1 + \frac{\tilde{\mathbf{z}}_t' \tilde{\mathbf{z}}_t}{v-2}\right)^{-(v+n)/2} \prod_{i=1}^{n} \left(\frac{g_i s_i}{1+g_i^2}\right) \quad (11.31)$$

where $\tilde{\mathbf{z}}_t = (\tilde{z}_{1,t}, \ldots, \tilde{z}_{n,t})'$, $\tilde{z}_{i,t} = (s_i z_{i,t} + m_i) g_i^{II_t}$, $m_i = \Gamma((v-1)/2)\sqrt{(v-2)}$ $(\Gamma(v/2)\sqrt{\pi})^{-1}(g_i - g_i^{-1})$, $s_i = \sqrt{g_i^2 + g_i^{-2} - m_i^2 - 1}$, $II_t = 1$ if $z_i < -m_i s_i^{-1}$, and $II_t = -1$ otherwise, g_i is the asymmetry parameter, v denotes the number of degrees of freedom of the distribution and $\Gamma(.)$ is the gamma function.

11.3 Estimating multivariate ARCH models using EViews 6

EViews incorporates the programs *bv_garch.prg* and *tv_garch.prg* for estimating bivariate and trivariate BEKK models, respectively, based on the logl tool.[1] In EViews 6 the estimation of multivariate ARCH models is available as a built-in system object. The multivariate models can be estimated via the rolling menus. EViews 6 allows the estimation of the Diag-BEKK, Diag-VECH and CCC-GARCH model specifications, assuming multivariate normal or multivariate Student t distributed innovations. The Diag-BEKK model is identical to the Diag-VECH model, where the coefficient matrices are matrices of rank 1.[2] EViews 6 does not restrict to the symmetric GARCH specification. The asymmetric modelling of the conditional variance as a univariate GJR specification is also available. Technical details about the estimation of the models are provided in the EViews 6.0 help document.

Let us now examine how to estimate four-dimensional multivariate ARCH models for the FTSE100 and S&P500 equity indices, the $/£ exchange rate and the gold price. The data are those considered in Chapter 2 for the period from 4 April 1988 to 5 April 2005. Figures 1.7–1.8 and 2.3–2.4 depict the log-return series of $\{y_{SP500,t}\}_{t=1}^{T}$, $\{y_{FTSE100,t}\}_{t=1}^{T}$, $\{y_{\$/£,t}\}_{t=1}^{T}$ and $\{y_{Gold,t}\}_{t=1}^{T}$, respectively.

---

[1] The programs are also available in earlier versions of EViews.

[2] EViews provides an option to estimate the Diag-VECH model, but displays the result in Diag-BEKK form. This Diag-BEKK model is identical to the Diag-VECH model, where the coefficient matrices are matrices of rank 1. For technical details, the reader is referred to the EViews technical discussion for multivariate ARCH models (*EViews 6 User's Guide II*, pp. 341–343).

We will estimate three model specifications, namely the Diag-BEKK(1,1) model with Student t distributed innovations, the Diag-VECH(1,1) model with GJR-type asymmetric effects and Student t distributed innovations, and the CCC-ARCH model with conditional variances regarded as univariate GJR specifications and standardized innovations Student t distributed. The program that carries out the necessary estimations, named *chapter11.multivariate.prg*, is given in the Appendix to this chapter.

The Diag-BEKK(1,1) model is estimated in the form

$$
\mathbf{y}_t = \begin{pmatrix} y_{FTSE100,t} \\ y_{Gold,t} \\ y_{SP500,t} \\ y_{\$/\pounds,t} \end{pmatrix} = \begin{pmatrix} \beta_{0,1} \\ \beta_{0,2} \\ \beta_{0,3} \\ \beta_{0,4} \end{pmatrix} + \varepsilon_t,
$$

(11.32)

$$
\varepsilon_t | I_{t-1} \sim t[0, \mathbf{H}_t],
$$
$$
\mathbf{H}_t = \mathbf{A}_0 \mathbf{A}_0' + \mathbf{A}_1 \varepsilon_{t-1} \varepsilon_{t-1}' \mathbf{A}_1' + \mathbf{B}_1 \mathbf{H}_{t-1} \mathbf{B}_1'.
$$

EViews provides five different ways to estimate the parameters in the matrices. Consider, for example, the matrix $\tilde{\mathbf{A}}_0 \equiv \mathbf{A}_0 \mathbf{A}_0'$. It can be estimated (a) without any restrictions as $\tilde{\mathbf{A}}_0$ (indefinite matrix), (b) as a matrix $\mathbf{A}_0$ with rank up to n based on the Cholesky factorization (full-rank matrix), (c) as a matrix $\mathbf{A}_0$ with all but the first column of coefficients equal to zero (rank-one matrix), (d) as a diagonal matrix $\tilde{\mathbf{A}}_0$, and (e) as a matrix $\tilde{\mathbf{A}}_0 \equiv a\mathbf{i}\mathbf{i}'$, where $\mathbf{i}$ is a vector of 1s and a is the only estimated parameter (scalar). Methods (a), (d) and (e) do not ensure a positive semi-definite conditional covariance matrix. The full-rank and rank-one methods ensure that the conditional covariance is positive semi-definite. In methods (a) and (b) the matrix contains $n(n+1)/2$ parameters for estimation, whereas the rank-one technique reduces the number of parameters to be estimated to n. For details about the reduction of the number of parameters in order to ensure a positive semi-definite conditional covariance matrix, see also Ding and Engle (2001).

There is also a sixth technique for the computation of the constant matrix $\tilde{\mathbf{A}}_0$, known as variance targeting. Variance targeting restricts the matrix of constants $\tilde{\mathbf{A}}_0$ to a function of the conditional variance parameters and the unconditional variance, $\tilde{\mathbf{A}}_0 \equiv \mathbf{H} \circ (\mathbf{i}\mathbf{i}' - \mathbf{A}_1 \mathbf{A}_1' - \mathbf{B}_1 \mathbf{B}_1')$, where $\mathbf{H}$ is the unconditional sample variance of the innovations. When the variance target is applied, the constant matrix is not estimated, reducing the number of estimated parameters.

As an illustration, we consider the estimation of the matrix $\tilde{\mathbf{A}}_0 \equiv \mathbf{A}_0 \mathbf{A}_0'$ without any restrictions, using the program *chapter11.multivariate.prg*. The argument c(indef) indicates the estimation of $\tilde{\mathbf{A}}_0$ as an indefinite matrix. The arguments c(fullrank), c(rank1), c(diag), c(scalar), and c(vt) render estimates of the matrix $\tilde{\mathbf{A}}_0$ as a full-ank matrix, a rank-one matrix, a diagonal matrix, a scalar or according to the variance targeting technique, respectively. Table 11.1 provides the conditional variance of (11.32) in an expanded form.

The conditional variance of the log-return, $y_{i,t}$, i.e. the ith diagonal element of $\mathbf{H}_t$, is computed as

$$
\sigma_{i,t}^2 = a_{0,i,i} + a_{1,i,i} \varepsilon_{i,t-1}^2 a_{1,i,i} + b_{1,i,i} \sigma_{i,t-1}^2 b_{1,i,i},
$$

(11.33)

Table 11.1 The conditional variance matrix in the Diag-BEKK(1,1) model in EViews 6.0, with the matrix of constants estimated without restrictions

$$
\begin{pmatrix}
\sigma_{1,t}^2 & \sigma_{1,2,t} & \sigma_{1,3,t} & \sigma_{1,4,t} \\
\sigma_{1,2,t} & \sigma_{2,t}^2 & \sigma_{3,2,t} & \sigma_{4,2,t} \\
\sigma_{1,3,t} & \sigma_{3,2,t} & \sigma_{3,t}^2 & \sigma_{4,3,t} \\
\sigma_{1,4,t} & \sigma_{4,2,t} & \sigma_{4,3,t} & \sigma_{4,t}^2
\end{pmatrix}
=
\begin{pmatrix}
a_{0,1,1} & a_{0,1,2} & a_{0,1,3} & a_{0,1,4} \\
a_{0,1,2} & a_{0,2,2} & a_{0,2,3} & a_{0,2,4} \\
a_{0,1,3} & a_{0,2,3} & a_{0,3,3} & a_{0,3,4} \\
a_{0,1,4} & a_{0,2,4} & a_{0,3,4} & a_{0,4,4}
\end{pmatrix}
$$

$$
+
\begin{pmatrix}
a_{1,1,1} & 0 & 0 & 0 \\
0 & a_{1,2,2} & 0 & 0 \\
0 & 0 & a_{1,3,3} & 0 \\
0 & 0 & 0 & a_{1,4,4}
\end{pmatrix}
\begin{pmatrix}
\varepsilon_{1,t-1}^2 & \varepsilon_{1,t-1}\varepsilon_{2,t-1} & \varepsilon_{1,t-1}\varepsilon_{3,t-1} & \varepsilon_{1,t-1}\varepsilon_{4,t-1} \\
\varepsilon_{1,t-1}\varepsilon_{2,t-1} & \varepsilon_{2,t-1}^2 & \varepsilon_{2,t-1}\varepsilon_{3,t-1} & \varepsilon_{2,t-1}\varepsilon_{4,t-1} \\
\varepsilon_{1,t-1}\varepsilon_{3,t-1} & \varepsilon_{2,t-1}\varepsilon_{3,t-1} & \varepsilon_{3,t-1}^2 & \varepsilon_{3,t-1}\varepsilon_{4,t-1} \\
\varepsilon_{1,t-1}\varepsilon_{4,t-1} & \varepsilon_{2,t-1}\varepsilon_{4,t-1} & \varepsilon_{3,t-1}\varepsilon_{4,t-1} & \varepsilon_{4,t-1}^2
\end{pmatrix}
\begin{pmatrix}
a_{1,1,1} & 0 & 0 & 0 \\
0 & a_{1,2,2} & 0 & 0 \\
0 & 0 & a_{1,3,3} & 0 \\
0 & 0 & 0 & a_{1,4,4}
\end{pmatrix}'
$$

$$
+
\begin{pmatrix}
b_{1,1,1} & 0 & 0 & 0 \\
0 & b_{1,2,2} & 0 & 0 \\
0 & 0 & b_{1,3,3} & 0 \\
0 & 0 & 0 & b_{1,4,4}
\end{pmatrix}
\begin{pmatrix}
\sigma_{1,t-1}^2 & \sigma_{1,2,t-1} & \sigma_{1,3,t-1} & \sigma_{1,4,t-1} \\
\sigma_{1,2,t-1} & \sigma_{2,t-1}^2 & \sigma_{2,3,t-1} & \sigma_{2,4,t-1} \\
\sigma_{1,3,t-1} & \sigma_{2,3,t-1} & \sigma_{3,t-1}^2 & \sigma_{3,4,t-1} \\
\sigma_{1,4,t-1} & \sigma_{2,4,t-1} & \sigma_{3,4,t-1} & \sigma_{4,t-1}^2
\end{pmatrix}
\begin{pmatrix}
b_{1,1,1} & 0 & 0 & 0 \\
0 & b_{1,2,2} & 0 & 0 \\
0 & 0 & b_{1,3,3} & 0 \\
0 & 0 & 0 & b_{1,4,4}
\end{pmatrix}'
$$

whereas the conditional covariance between the log-returns, $y_{i,t}$ and $y_{j,t}$, i.e. the (i,j) th element of $\mathbf{H}_t$ is computed as

$$\sigma_{i,j,t} = a_{0,i,j} + a_{1,i,i}\varepsilon_{i,t-1}\varepsilon_{j,t-1}a_{1,j,j} + b_{1,i,i}\sigma_{i,j,t-1}b_{1,j,j}. \tag{11.34}$$

Table 11.2 provides the estimated parameters of the Diag-BEKK(1,1) model and Figure 11.1 depicts plots of the conditional variances and conditional covariances. We now turn to the Diag-VECH(1,1) model, which is estimated in the form

$$\mathbf{y}_t = \begin{pmatrix} y_{FTSE100,t} \\ y_{Gold,t} \\ y_{SP500,t} \\ y_{\$/£,t} \end{pmatrix} + \varepsilon_t,$$

$$\varepsilon_t | I_{t-1} \sim t[0, \mathbf{H}_t],$$

$$vech(\mathbf{H}_t) = vech(\tilde{\mathbf{A}}_0) + \tilde{\mathbf{A}}_1 vech(\varepsilon_{t-1}\varepsilon'_{t-1}) + \tilde{\Gamma}_1 vech(\tilde{\varepsilon}_{t-1}\tilde{\varepsilon}'_{t-1}) + \tilde{\mathbf{B}}_1 vech(\mathbf{H}_{t-1});$$
$$\tag{11.35}$$

however, the parameter matrices are defined in a *vech* form,

$$\begin{aligned} vech(\mathbf{H}_t) &= vech(\tilde{\mathbf{A}}_0) + vech(\tilde{\mathbf{A}}_1) \circ vech(\varepsilon_{t-1}\varepsilon'_{t-1}) \\ &\quad + vech(\tilde{\Gamma}_1) \circ vech(\tilde{\varepsilon}_{t-1}\tilde{\varepsilon}'_{t-1}) + vech(\tilde{\mathbf{B}}_1) \circ vech(\mathbf{H}_{t-1}). \end{aligned} \tag{11.36}$$

The symbol $\circ$ denotes the Hadamard (elementwise) product. Table 11.3 presents the conditional variance of (11.36) in an expanded form. In total, 22 parameters are estimated:

$$vech(\mathbf{H}_t) = \begin{pmatrix} \sigma^2_{1,t} \\ \sigma_{1,2,t} \\ \sigma_{1,3,t} \\ \sigma_{1,4,t} \\ \sigma^2_{2,t} \\ \sigma_{3,2,t} \\ \sigma_{4,2,t} \\ \sigma^2_{3,t} \\ \sigma_{4,3,t} \\ \sigma^2_{4,t} \end{pmatrix} = \begin{pmatrix} a_{1,1} + \tilde{a}_{1,1}\varepsilon^2_{1,t-1} + \gamma_{1,1}\varepsilon^2_{1,t-1}d_{1,t-1} + \tilde{b}_{1,1}\sigma^2_{1,t-1} \\ a_{1,2} \\ a_{1,3} \\ a_{1,4} \\ a_{2,2} + \tilde{a}_{2,2}\varepsilon^2_{2,t-1} + \gamma_{2,2}\varepsilon^2_{2,t-1}d_{2,t-1} + \tilde{b}_{2,2}\sigma^2_{2,t-1} \\ a_{2,3} \\ a_{2,4} \\ a_{3,3} + \tilde{a}_{3,3}\varepsilon^2_{3,t-1} + \gamma_{3,3}\varepsilon^2_{3,t-1}d_{3,t-1} + \tilde{b}_{3,3}\sigma^2_{3,t-1} \\ a_{3,4} \\ a_{4,4} + \tilde{a}_{4,4}\varepsilon^2_{4,t-1} + \gamma_{4,4}\varepsilon^2_{4,t-1}d_{4,t-1} + \tilde{b}_{4,4}\sigma^2_{4,t-1} \end{pmatrix}$$
$$\tag{11.37}$$

If the model were estimated in the original form of (11.35), then $(n(n+1)/2)$ $(1+2q+p) = 40$ parameters would have been estimated:

Table 11.2 Estimated parameters of the Diag-BEKK(1,1) model, in EViews 6.0. The coefficient to standard error ratios are reported in brackets

$$\mathbf{y}_t = \begin{pmatrix} y_{FTSE100,t} \\ y_{Gold,t} \\ y_{SP500,t} \\ y_{\$/£,t} \end{pmatrix} = \begin{pmatrix} \beta_{0,1} \\ \beta_{0,2} \\ \beta_{0,3} \\ \beta_{0,4} \end{pmatrix} + \boldsymbol{\varepsilon}_t = \begin{pmatrix} 0.048[4.26] \\ -0.008[-1.01] \\ 0.064[5.82] \\ 0.009[1.39] \end{pmatrix} + \boldsymbol{\varepsilon}_t$$

$$vech(\mathbf{H}_t) = \begin{pmatrix} \sigma_{1,t}^2 \\ \sigma_{1,2,t} \\ \sigma_{1,3,t} \\ \sigma_{1,4,t} \\ \sigma_{2,t}^2 \\ \sigma_{3,2,t} \\ \sigma_{4,2,t} \\ \sigma_{3,t}^2 \\ \sigma_{4,3,t} \\ \sigma_{4,t}^2 \end{pmatrix} = \begin{pmatrix} a_{0,1,1} + a_{1,1,1}\varepsilon_{1,t-1}^2 a_{1,1,1} + b_{1,1,1}\sigma_{1,t-1}^2 b_{1,1,1} \\ a_{0,1,2} + a_{1,1,1}\varepsilon_{1,t-1}\varepsilon_{2,t-1}a_{1,2,2} + b_{1,1,1}\sigma_{1,2,t-1}b_{1,2,2} \\ a_{0,1,3} + a_{1,1,1}\varepsilon_{1,t-1}\varepsilon_{3,t-1}a_{1,3,3} + b_{1,1,1}\sigma_{1,3,t-1}b_{1,3,3} \\ a_{0,1,4} + a_{1,1,1}\varepsilon_{1,t-1}\varepsilon_{4,t-1}a_{1,4,4} + b_{1,1,1}\sigma_{1,4,t-1}b_{1,4,4} \\ a_{0,2,2} + a_{1,2,2}\varepsilon_{2,t-1}^2 a_{1,2,2} + b_{1,2,2}\sigma_{2,t-1}^2 b_{1,2,2} \\ a_{0,2,3} + a_{1,2,2}\varepsilon_{2,t-1}\varepsilon_{3,t-1}a_{1,3,3} + b_{1,2,2}\sigma_{2,3,t-1}b_{1,3,3} \\ a_{0,2,4} + a_{1,2,2}\varepsilon_{2,t-1}\varepsilon_{4,t-1}a_{1,4,4} + b_{1,2,2}\sigma_{2,4,t-1}b_{1,4,4} \\ a_{0,3,3} + a_{1,3,3}\varepsilon_{3,t-1}^2 a_{1,3,3} + b_{1,3,3}\sigma_{3,t-1}^2 b_{1,3,3} \\ a_{0,3,4} + a_{1,3,3}\varepsilon_{3,t-1}\varepsilon_{4,t-1}a_{1,4,4} + b_{1,3,3}\sigma_{3,4,t-1}b_{1,4,4} \\ a_{0,4,4} + a_{1,4,4}\varepsilon_{4,t-1}^2 a_{1,4,4} + b_{1,4,4}\sigma_{4,t-1}^2 b_{1,4,4} \end{pmatrix} =$$

$$= \begin{pmatrix} 0.011 + 0.195 \quad \varepsilon_{1,t-1}^2 0.195 + 0.976 \quad \sigma_{1,t-1}^2 0.976 \\ [5.25] \quad [18.6] \qquad\qquad [381] \\ -0.001 + 0.195\varepsilon_{1,t-1}\varepsilon_{2,t-1}0.185 + 0.976\,\sigma_{1,2,t-1}0.977 \\ [-2.45] \\ 0.002 + 0.195\varepsilon_{1,t-1}\varepsilon_{3,t-1}0.148 + 0.976\,\sigma_{1,3,t-1}0.987 \\ [3.90] \\ -0.0007 + 0.195\varepsilon_{1,t-1}\varepsilon_{4,t-1}0.164 + 0.976\,\sigma_{1,4,t-1}0.984 \\ [-2.44] \\ 0.004 + 0.185 \quad \varepsilon_{2,t-1}^2 0.185 + 0.977 \quad \sigma_{2,t-1}^2 0.977 \\ [6.07] \quad [22.2] \qquad\qquad [521] \\ -0.0005 + 0.185\varepsilon_{2,t-1}\varepsilon_{3,t-1}0.148 + 0.977\,\sigma_{2,3,t-1}0.987 \\ [-1.72] \\ 0.0005 + 0.185\varepsilon_{2,t-1}\varepsilon_{4,t-1}0.164 + 0.977\,\sigma_{2,4,t-1}0.984 \\ [2.39] \\ 0.003 + 0.148 \quad \varepsilon_{3,t-1}^2 0.148 + 0.987 \quad \sigma_{3,t-1}^2 0.987 \\ [4.51] \quad [18.6] \qquad\qquad [770] \\ -0.00006 + 0.148\varepsilon_{3,t-1}\varepsilon_{4,t-1}0.164 + 0.987\,\sigma_{3,4,t-1}0.984 \\ [-0.26] \\ 0.002 + 0.164 \quad \varepsilon_{4,t-1}^2 0.164 + 0.984 \quad \sigma_{4,t-1}^2 0.984 \\ [4.15] \quad [18.5] \qquad\qquad [558] \end{pmatrix}$$

Student t distribution degrees of freedom, $v = 6.215[23.1]$.

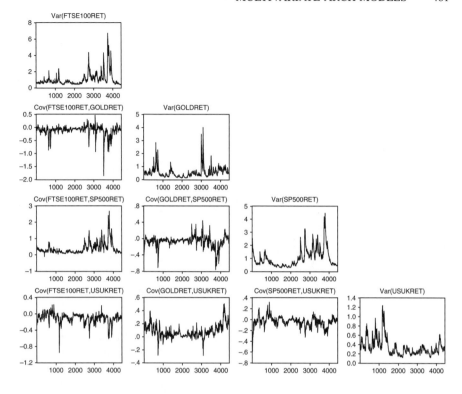

Figure 11.1   Conditional variances and conditional covariances of the Diag-BEKK(1,1) model, in EViews 6.0.

$$
vech(\mathbf{H}_t) = \begin{pmatrix} \sigma_{1,t}^2 \\ \sigma_{1,2,t} \\ \sigma_{1,3,t} \\ \sigma_{1,4,t} \\ \sigma_{2,t}^2 \\ \sigma_{3,2,t} \\ \sigma_{4,2,t} \\ \sigma_{3,t}^2 \\ \sigma_{4,3,t} \\ \sigma_{4,t}^2 \end{pmatrix} = \begin{pmatrix} a_{1,1} + \tilde{a}_{1,1}\varepsilon_{1,t-1}^2 + \gamma_{1,1}\varepsilon_{1,t-1}^2 d_{1,t-1} + \tilde{b}_{1,1}\sigma_{1,t-1}^2 \\ a_{1,2} + \tilde{a}_{2,2}\varepsilon_{1,t-1}\varepsilon_{2,t-1} + \gamma_{2,2}\varepsilon_{1,t-1}d_{1,t-1}\varepsilon_{2,t-1}d_{2,t-1} + \tilde{b}_{2,2}\sigma_{1,2,t-1} \\ a_{1,3} + \tilde{a}_{3,3}\varepsilon_{1,t-1}\varepsilon_{3,t-1} + \gamma_{3,3}\varepsilon_{1,t-1}d_{1,t-1}\varepsilon_{3,t-1}d_{3,t-1} + \tilde{b}_{3,3}\sigma_{1,3,t-1} \\ a_{1,4} + \tilde{a}_{4,4}\varepsilon_{1,t-1}\varepsilon_{4,t-1} + \gamma_{4,4}\varepsilon_{1,t-1}d_{1,t-1}\varepsilon_{4,t-1}d_{4,t-1} + \tilde{b}_{4,4}\sigma_{1,4,t-1} \\ a_{2,2} + \tilde{a}_{5,5}\varepsilon_{2,t-1}^2 + \gamma_{5,5}\varepsilon_{2,t-1}^2 d_{2,t-1} + \tilde{b}_{5,5}\sigma_{2,t-1}^2 \\ a_{2,3} + \tilde{a}_{6,6}\varepsilon_{2,t-1}\varepsilon_{3,t-1} + \gamma_{6,6}\varepsilon_{2,t-1}d_{2,t-1}\varepsilon_{3,t-1}d_{3,t-1} + \tilde{b}_{6,6}\sigma_{3,2,t-1} \\ a_{2,4} + \tilde{a}_{7,7}\varepsilon_{2,t-1}\varepsilon_{4,t-1} + \gamma_{7,7}\varepsilon_{2,t-1}d_{2,t-1}\varepsilon_{4,t-1}d_{4,t-1} + \tilde{b}_{7,7}\sigma_{4,2,t-1} \\ a_{3,3} + \tilde{a}_{8,8}\varepsilon_{3,t-1}^2 + \gamma_{8,8}\varepsilon_{3,t-1}^2 d_{3,t-1} + \tilde{b}_{8,8}\sigma_{3,t-1}^2 \\ a_{3,4} + \tilde{a}_{9,9}\varepsilon_{3,t-1}\varepsilon_{4,t-1} + \gamma_{9,9}\varepsilon_{3,t-1}d_{3,t-1}\varepsilon_{4,t-1}d_{4,t-1} + \tilde{b}_{9,9}\sigma_{4,3,t-1} \\ a_{4,4} + \tilde{a}_{10,10}\varepsilon_{4,t-1}^2 + \gamma_{10,10}\varepsilon_{4,t-1}^2 d_{4,t-1} + \tilde{b}_{10,10}\sigma_{4,t-1}^2 \end{pmatrix}
$$

$$(11.38)$$

Table 11.4 provides the estimated parameters of the Diag-VECH(1,1) model and Figure 11.2 presents plots of the conditional variances and conditional covariances.

Table 11.3 The conditional variance matrix in the Diag-VECH(1,1) model, with the matrix of constants estimated without restrictions and GJR-type asymmetric effects. $d_{i,t-1} = 1$ if $\varepsilon_{i,t-1} < 0$ and $d_{i,t-1} = 0$ otherwise

$$
vech \begin{pmatrix} \sigma_{1,t}^2 & \sigma_{1,2,t} & \sigma_{1,3,t} & \sigma_{1,4,t} \\ \sigma_{1,2,t} & \sigma_{2,t}^2 & \sigma_{3,2,t} & \sigma_{4,2,t} \\ \sigma_{1,3,t} & \sigma_{3,2,t} & \sigma_{3,t}^2 & \sigma_{4,3,t} \\ \sigma_{1,4,t} & \sigma_{4,2,t} & \sigma_{4,3,t} & \sigma_{4,t}^2 \end{pmatrix} = vech \begin{pmatrix} a_{1,1} & a_{1,2} & a_{1,3} & a_{1,4} \\ a_{1,2} & a_{2,2} & a_{2,3} & a_{2,4} \\ a_{1,3} & a_{2,3} & a_{3,3} & a_{3,4} \\ a_{1,4} & a_{2,4} & a_{3,4} & a_{4,4} \end{pmatrix}
$$

$$
+ vech \begin{pmatrix} \tilde{a}_{1,1} & 0 & 0 & 0 \\ 0 & \tilde{a}_{2,2} & 0 & 0 \\ 0 & 0 & \tilde{a}_{3,3} & 0 \\ 0 & 0 & 0 & \tilde{a}_{4,4} \end{pmatrix} \circ vech \begin{pmatrix} \varepsilon_{1,t-1}^2 & \varepsilon_{1,t-1}\varepsilon_{2,t-1} & \varepsilon_{1,t-1}\varepsilon_{3,t-1} & \varepsilon_{1,t-1}\varepsilon_{4,t-1} \\ \varepsilon_{1,t-1}\varepsilon_{2,t-1} & \varepsilon_{2,t-1}^2 & \varepsilon_{2,t-1}\varepsilon_{3,t-1} & \varepsilon_{2,t-1}\varepsilon_{4,t-1} \\ \varepsilon_{1,t-1}\varepsilon_{3,t-1} & \varepsilon_{2,t-1}\varepsilon_{3,t-1} & \varepsilon_{3,t-1}^2 & \varepsilon_{3,t-1}\varepsilon_{4,t-1} \\ \varepsilon_{1,t-1}\varepsilon_{4,t-1} & \varepsilon_{2,t-1}\varepsilon_{4,t-1} & \varepsilon_{3,t-1}\varepsilon_{4,t-1} & \varepsilon_{4,t-1}^2 \end{pmatrix}
$$

$$
+ vech \begin{pmatrix} \gamma_{1,1} & 0 & 0 & 0 \\ 0 & \gamma_{2,2} & 0 & 0 \\ 0 & 0 & \gamma_{3,3} & 0 \\ 0 & 0 & 0 & \gamma_{4,4} \end{pmatrix} \circ vech \begin{pmatrix} \varepsilon_{1,t-1}^2 d_{1,t-1} & \varepsilon_{1,t-1}d_{1,t-1}\varepsilon_{2,t-1}d_{2,t-1} & \varepsilon_{1,t-1}d_{1,t-1}\varepsilon_{3,t-1}d_{3,t-1} & \varepsilon_{1,t-1}d_{1,t-1}\varepsilon_{4,t-1}d_{4,t-1} \\ \varepsilon_{1,t-1}d_{1,t-1}\varepsilon_{2,t-1}d_{2,t-1} & \varepsilon_{2,t-1}^2 d_{2,t-1} & \varepsilon_{2,t-1}d_{2,t-1}\varepsilon_{3,t-1}d_{3,t-1} & \varepsilon_{2,t-1}d_{2,t-1}\varepsilon_{4,t-1}d_{4,t-1} \\ \varepsilon_{1,t-1}d_{1,t-1}\varepsilon_{3,t-1}d_{3,t-1} & \varepsilon_{2,t-1}d_{2,t-1}\varepsilon_{3,t-1}d_{3,t-1} & \varepsilon_{3,t-1}^2 d_{3,t-1} & \varepsilon_{3,t-1}d_{3,t-1}\varepsilon_{4,t-1}d_{4,t-1} \\ \varepsilon_{1,t-1}d_{1,t-1}\varepsilon_{4,t-1}d_{4,t-1} & \varepsilon_{2,t-1}d_{2,t-1}\varepsilon_{4,t-1}d_{4,t-1} & \varepsilon_{3,t-1}d_{3,t-1}\varepsilon_{4,t-1}d_{4,t-1} & \varepsilon_{4,t-1}^2 d_{4,t-1} \end{pmatrix}
$$

$$
+ vech \begin{pmatrix} \tilde{b}_{1,1} & 0 & 0 & 0 \\ 0 & \tilde{b}_{2,2} & 0 & 0 \\ 0 & 0 & \tilde{b}_{3,3} & 0 \\ 0 & 0 & 0 & \tilde{b}_{4,4} \end{pmatrix} \circ vech \begin{pmatrix} \sigma_{1,t-1}^2 & \sigma_{1,2,t-1} & \sigma_{1,3,t-1} & \sigma_{1,4,t-1} \\ \sigma_{1,2,t-1} & \sigma_{2,t-1}^2 & \sigma_{2,3,t-1} & \sigma_{2,4,t-1} \\ \sigma_{1,3,t-1} & \sigma_{2,3,t-1} & \sigma_{3,t-1}^2 & \sigma_{3,4,t-1} \\ \sigma_{1,4,t-1} & \sigma_{2,4,t-1} & \sigma_{3,4,t-1} & \sigma_{4,t-1}^2 \end{pmatrix}
$$

Table 11.4 Estimated parameters of the Diag-VECH(1,1) model, in EViews 6.0. The coefficient to standard error ratios are reported in brackets

$$
\mathbf{y}_t =
\begin{pmatrix}
y_{FTSE100,t} \\
y_{Gold,t} \\
y_{SP500,t} \\
y_{\$/\pounds,t}
\end{pmatrix}
=
\begin{pmatrix}
\beta_{0,1} \\
\beta_{0,2} \\
\beta_{0,3} \\
\beta_{0,4}
\end{pmatrix}
+ \boldsymbol{\varepsilon}_t =
\begin{pmatrix}
0.036[3.07] \\
-0.002[-0.19] \\
0.044[3.95] \\
0.009[1.31]
\end{pmatrix}
+ \boldsymbol{\varepsilon}_t
$$

$$
vech(\mathbf{H}_t) =
\begin{pmatrix}
\sigma^2_{1,t} \\
\sigma_{1,2,t} \\
\sigma_{1,3,t} \\
\sigma_{1,4,t} \\
\sigma^2_{2,t} \\
\sigma_{3,2,t} \\
\sigma_{4,2,t} \\
\sigma^2_{3,t} \\
\sigma_{4,3,t} \\
\sigma^2_{4,t}
\end{pmatrix}
=
\begin{pmatrix}
a_{1,1} + \tilde{a}_{1,1}\varepsilon^2_{1,t-1} + \gamma_{1,1}\varepsilon^2_{1,t-1}d_{1,t-1} + \tilde{b}_{1,1}\sigma^2_{1,t-1} \\
a_{1,2} \\
a_{1,3} \\
a_{1,4} \\
a_{2,2} + \tilde{a}_{2,2}\varepsilon^2_{2,t-1} + \gamma_{2,2}\varepsilon^2_{2,t-1}d_{2,t-1} + \tilde{b}_{2,2}\sigma^2_{2,t-1} \\
a_{2,3} \\
a_{2,4} \\
a_{3,3} + \tilde{a}_{3,3}\varepsilon^2_{3,t-1} + \gamma_{3,3}\varepsilon^2_{3,t-1}d_{3,t-1} + \tilde{b}_{3,3}\sigma^2_{3,t-1} \\
a_{3,4} \\
a_{4,4} + \tilde{a}_{4,4}\varepsilon^2_{4,t-1} + \gamma_{4,4}\varepsilon^2_{4,t-1}d_{4,t-1} + \tilde{b}_{4,4}\sigma^2_{4,t-1}
\end{pmatrix}
$$

$$
=
\begin{pmatrix}
\begin{array}{l}
0.024 + 0.017 \quad \varepsilon^2_{1,t-1} + 0.065 \quad \varepsilon^2_{1,t-1}d_{1,t-1} + 0.923 \quad \sigma^2_{1,t-1} \\
{[5.70]} \quad [2.37] \qquad\quad [5.85] \qquad\qquad\qquad [88.86] \\
\qquad\qquad\qquad\qquad -0.069 \\
\qquad\qquad\qquad\qquad [-7.71] \\
\qquad\qquad\qquad\qquad 0.222 \\
\qquad\qquad\qquad\qquad [17.21] \\
\qquad\qquad\qquad\qquad -0.083 \\
\qquad\qquad\qquad\qquad [-10.68]
\end{array} \\
\begin{array}{l}
0.007 + 0.081 \quad \varepsilon^2_{2,t-1} - 0.051 \quad \varepsilon^2_{2,t-1}d_{2,t-1} + 0.928 \quad \sigma^2_{2,t-1} \\
{[6.74]} \quad [10.93] \qquad\; [-6.71] \qquad\qquad\qquad [158.6] \\
\qquad\qquad\qquad\qquad -0.041 \\
\qquad\qquad\qquad\qquad [-4.75] \\
\qquad\qquad\qquad\qquad 0.060 \\
\qquad\qquad\qquad\qquad [11.18]
\end{array} \\
\begin{array}{l}
0.017 + 0.006 \quad \varepsilon^2_{3,t-1} + 0.082 \quad \varepsilon^2_{3,t-1}d_{3,t-1} + 0.932 \quad \sigma^2_{3,t-1} \\
{[2.37]} \quad [0.91] \qquad\quad [7.56] \qquad\qquad\qquad [117.2] \\
\qquad\qquad\qquad\qquad -0.026 \\
\qquad\qquad\qquad\qquad [-3.56]
\end{array} \\
\begin{array}{l}
0.002 + 0.020 \quad \varepsilon^2_{4,t-1} + 0.013 \quad \varepsilon^2_{4,t-1}d_{4,t-1} + 0.964 \quad \sigma^2_{4,t-1} \\
{[4.46]} \quad [4.42] \qquad\quad [2.26] \qquad\qquad\qquad [225.7]
\end{array}
\end{pmatrix}
$$

Student t distribution degrees of freedom, $v = 6.442[21.9]$.

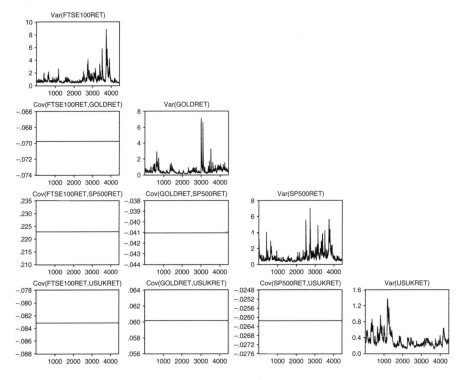

Figure 11.2 Conditional variances and conditional covariances of the Diag-VECH(1,1) model, in EViews 6.0.

Finally, the CCC-ARCH model is estimated as

$$\mathbf{y}_t = \begin{pmatrix} y_{FTSE100,t} \\ y_{Gold,t} \\ y_{SP500,t} \\ y_{\$/\pounds,t} \end{pmatrix} + \boldsymbol{\varepsilon}_t,$$

$$\boldsymbol{\varepsilon}_t | I_{t-1} \sim t[0, \mathbf{H}_t],$$
$$\mathbf{H}_t = \boldsymbol{\Sigma}_t^{1/2} \mathbf{C} \boldsymbol{\Sigma}_t^{1/2},$$
(11.39)

where the diagonal elements of the $\boldsymbol{\Sigma}_t^{1/2}$ matrix are univariate GJR(1,1) specifications, or

$$\sigma_{i,t} = \left(a_{i,0} + \alpha_{i,1} \varepsilon_{i,t-1}^2 + \gamma_{i,1} d_{i,t-1} \varepsilon_{i,t-1}^2 + b_{i,1} \sigma_{i,t-1}^2 \right)^{1/2},$$
(11.40)

and the covariance matrix $\mathbf{H}_t$ is computed as

$$\mathbf{H}_t = \begin{pmatrix} \sigma_{1,t} & 0 & 0 & 0 \\ 0 & \sigma_{2,t} & 0 & 0 \\ 0 & 0 & \sigma_{3,t} & 0 \\ 0 & 0 & 0 & \sigma_{4,t} \end{pmatrix} \begin{pmatrix} 1 & c_{1,2} & c_{1,3} & c_{1,4} \\ c_{1,2} & 1 & c_{2,3} & c_{2,4} \\ c_{1,3} & c_{2,3} & 1 & c_{3,4} \\ c_{1,4} & c_{2,4} & c_{3,4} & 1 \end{pmatrix} \begin{pmatrix} \sigma_{1,t} & 0 & 0 & 0 \\ 0 & \sigma_{2,t} & 0 & 0 \\ 0 & 0 & \sigma_{3,t} & 0 \\ 0 & 0 & 0 & \sigma_{4,t} \end{pmatrix}.$$
(11.41)

Thus, the conditional variance of the log-return $y_{i,t}$ is given by (11.40), whereas the conditional covariance between the log-returns $y_{i,t}$ and $y_{j,t}$ is the (i,j) th element of $\mathbf{H}_t$,

$$\sigma_{i,j,t} = c_{i,j}\sigma_{i,t}\sigma_{j,t}. \qquad (11.42)$$

For the CCC-ARCH specification, where the conditional variances are defined as univariate GJR(1,1) specifications, the estimation of $(n(n-1)/2)+n(1+2q+p)$ parameters is required. So, for $n=4$, $\mathbf{H}_t$ requires the estimation of 22 parameters.

Table 11.5 provides the estimated parameters of the CCC-ARCH model and Figure 11.3 depicts plots of the conditional variances and conditional covariances.

11.4   Estimating multivariate ARCH models using G@RCH 5.0

G@RCH 5.0 incorporates the MGarch class, which allows the estimation of the Scalar-BEKK, the Diag-BEKK, the CCC-ARCH, the DCC-ARCH, the VC-ARCH[3], the O-ARCH and the GO-ARCH model specifications assuming multivariate normal or multivariate Student t distributed innovations. G@RCH 5.0 allows the choice of asymmetric and fractionally integrated volatility specifications rather than just restricting to the symmetric GARCH model. The IGARCH, EGARCH, GJR, APARCH, FIGARCH, FIGARCHC, FIEGARCH, FIAPARCH, FIAPARCHC and HYGARCH specifications of the conditional variance are available for the CCC-ARCH, the DCC-ARCH, the O-ARCH and the GO-ARCH models. In earlier versions of G@RCH, the estimation of multivariate ARCH models is not available. Technical details on the estimation of the models are provided in Laurent's (2007) G@RCH 5.0 manual.

For the same data set, three model specifications are estimated: the Diag-BEKK(1,1) and CCC-ARCH models as in EViews 6.0 and the GO-ARCH model. In OxMetrics 5, the estimation of multivariate ARCH models is available via the menu-driven G@RCH module. However, a program which makes use of the MGarch class code of G@RCH 5.0 can be written. The program named *chapter11.diagBEKK11_ student.ox*, given below, estimates the Diag-BEKK(1,1) model:[4]

```
#include <oxstd.h>
#include <oxdraw.h>
#import <packages/MGarch1/mgarch>
main()
{
```

---

[3] G@RCH 5.0 allows the estimation of Engle's (2002b) DCC-ARCH model as well as of Tse and Tsui's (2002) DCC-ARCH, or VC-ARCH, model.

[4] The programs for the estimation of the CCC-ARCH and GO-ARCH models (*chapter11. CCC_CGR11_student.1step.ox* and *chapter11.GOARCH_MLE.ox*) are given in the Appendix to this chapter.

Table 11.5   Estimated parameters of the CCC-ARCH model, in EViews 6.0. The coefficient to standard error ratios are reported in brackets

$$
\mathbf{y}_t = \begin{pmatrix} y_{FTSE100,t} \\ y_{Gold,t} \\ y_{SP500,t} \\ y_{\$/£,t} \end{pmatrix} = \begin{pmatrix} \beta_{0,1} \\ \beta_{0,2} \\ \beta_{0,3} \\ \beta_{0,4} \end{pmatrix} + \varepsilon_t = \begin{pmatrix} 0.034[2.96] \\ -0.001[-0.08] \\ 0.046[4.20] \\ 0.011[1.48] \end{pmatrix} + \varepsilon_t
$$

$$
vech(\mathbf{H}_t) = \begin{pmatrix} \sigma^2_{1,t} \\ \sigma_{1,2,t} \\ \sigma_{1,3,t} \\ \sigma_{1,4,t} \\ \sigma^2_{2,t} \\ \sigma_{3,2,t} \\ \sigma_{4,2,t} \\ \sigma^2_{3,t} \\ \sigma_{4,3,t} \\ \sigma^2_{4,t} \end{pmatrix} = \begin{pmatrix} a_{1,0} + \alpha_{1,1}\varepsilon^2_{1,t-1} + \gamma_{1,1}d_{1,t-1}\varepsilon^2_{1,t-1} + b_{1,1}\sigma^2_{1,t-1} \\ c_{1,2}\sigma_{1,t}\sigma_{2,t} \\ c_{1,3}\sigma_{1,t}\sigma_{3,t} \\ c_{1,4}\sigma_{1,t}\sigma_{4,t} \\ a_{2,0} + \alpha_{2,1}\varepsilon^2_{i,t-1} + \gamma_{2,1}d_{2,t-1}\varepsilon^2_{2,t-1} + b_{2,1}\sigma^2_{2,t-1} \\ c_{2,3}\sigma_{2,t}\sigma_{3,t} \\ c_{2,4}\sigma_{2,t}\sigma_{4,t} \\ a_{3,0} + \alpha_{3,1}\varepsilon^2_{3,t-1} + \gamma_{3,1}d_{3,t-1}\varepsilon^2_{3,t-1} + b_{3,1}\sigma^2_{3,t-1} \\ c_{3,4}\sigma_{3,t}\sigma_{4,t} \\ a_{4,0} + \alpha_{4,1}\varepsilon^2_{4,t-1} + \gamma_{4,1}d_{4,t-1}\varepsilon^2_{4,t-1} + b_{4,1}\sigma^2_{4,t-1} \end{pmatrix}
$$

$$
= \begin{pmatrix}
0.013 + 0.016\ \varepsilon^2_{1,t-1} + 0.071\ d_{1,t-1}\varepsilon^2_{1,t-1} + 0.935\ \sigma^2_{1,t-1} \\
\text{[5.03]}\quad\text{[2.33]}\qquad\text{[6.50]}\qquad\qquad\text{[117]} \\
-0.137\ \sigma_{1,t}\sigma_{2,t} \\
\text{[-8.26]} \\
0.371\ \sigma_{1,t}\sigma_{3,t} \\
\text{[25.88]} \\
-0.191\ \sigma_{1,t}\sigma_{4,t} \\
\text{[-11.7]} \\
0.004 + 0.085\ \varepsilon^2_{2,t-1} - 0.054\ d_{2,t-1}\varepsilon^2_{2,t-1} + 0.932\ \sigma^2_{2,t-1} \\
\text{[5.62]}\quad\text{[11.5]}\qquad\text{[-7.12]}\qquad\qquad\text{[177]} \\
-0.092\ \sigma_{2,t}\sigma_{3,t} \\
\text{[-5.36]} \\
0.228\ \sigma_{2,t}\sigma_{4,t} \\
\text{[11.25]} \\
0.009 + 0.006\ \varepsilon^2_{3,t-1} + 0.083\ d_{3,t-1}\varepsilon^2_{3,t-1} + 0.943\ \sigma^2_{3,t-1} \\
\text{[5.47]}\quad\text{[0.98]}\qquad\text{[7.92]}\qquad\qquad\text{[153]} \\
-0.068\ \sigma_{3,t}\sigma_{4,t} \\
\text{[-3.98]} \\
0.002 + 0.029\ \varepsilon^2_{4,t-1} + 0.010\ d_{4,t-1}\varepsilon^2_{4,t-1} + 0.960\ \sigma^2_{4,t-1} \\
\text{[3.54]}\quad\text{[5.44]}\qquad\text{[1.51]}\qquad\qquad\text{[211]}
\end{pmatrix}
$$

Student t distribution degrees of freedom, $v = 6.574[22.2]$.

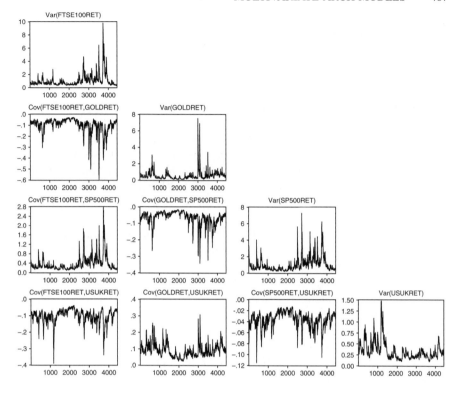

Figure 11.3   Conditional variances and conditional covariances of the CCC-ARCH(1,1) model, in EViews 6.0.

```
decl model = new MGarch();
model.Load("Chapter11Data.xls");
model.Deterministic(-1);
model.Select(Y_VAR, {"FTSE100RET", 0, 0});
model.Select(Y_VAR, {"GOLDRET", 0, 0});
model.Select(Y_VAR, {"SP500RET", 0, 0});
model.Select(Y_VAR, {"USUKRET", 0, 0});
model.CSTS(1,1);
model.DISTRI(STUDENT);
model.ARMA_ORDERS(0,0);
model.GARCH_ORDERS(1,1);
model.MODEL(DIAG_BEKK);
model.MLE(QMLE);
 model.SetSelSampleByDates(dayofcalendar(1988,4,5),
   dayofcalendar(2005,4, 5));
model.Initialization(<>);
model.PrintOutput(1);
```

```
model.DoEstimation();
decl cov_vec,cor_vec,Varf,covf_vec,corf_vec;
delete model;
}
```

There is one more way of estimating either univariate or multivariate ARCH models with OxMetrics based on its batch language.[5]

The Diag-BEKK(1,1) model is estimated in the form of (11.32), where A_0 is a lower triangular matrix. G@RCH 5.0 also provides the variance targeting technique for the computation of the constant matrix, $A_0 A_0'$. Table 11.6 provides the estimated parameters of the Diag-BEKK(1,1) model in G@RCH 5.0.[6]

Both packages provide similar parameter estimates. G@RCH presents the lower triangular matrix A_0, instead of the matrix with the constant coefficients, $A_0 A_0'$. Thus, the constant term in the conditional variance and the conditional covariance equations must be computed. In particular, for the conditional variance of the FTSE100 index, the constant term is equal to $a_{0,1,1}^2$, whereas, for the conditional variance of the \$/£ exchange rate, the constant term is equal to $a_{0,1,4}^2 + a_{0,2,4}^2 + a_{0,3,4}^2 + a_{0,4,4}^2$. Figure 11.4 presents plots of the conditional variances and conditional covariances, which are almost identical to those estimated in EViews, except for the initial values in some covariance estimates (e.g. in the case of the covariance between the S&P500 index returns and the \$/£ exchange rate).

The CCC-ARCH model is estimated in the form of (11.39)–(11.41). G@RCH provides the estimation of the CCC-ARCH model (as well as for the DCC-ARCH model) either with the classical maximum likelihood approach presented in Section 11.2 (called one-step maximum likelihood) or with the two-step maximum likelihood estimation method proposed by Engle and Sheppard (2001).[7] Table 11.7 provides the parameters of the CCC-ARCH model as estimated in G@RCH 5.0 with the classical maximum likelihood.[8] They are almost identical to those estimated in EViews 6.0.

Finally, the GO-ARCH model is estimated in the form of (11.18) to (11.20). G@RCH provides the estimation of the GO-ARCH model with the maximum likelihood approach proposed by Van der Weide (2002), or with the non-linear least

---

[5] Whenever a model is estimated with the menu-driven interface, the batch code is automatically generated in the background, and can be saved using the batch editor. For details, see Laurent (2007, pp. 83 and 205).

[6] The d*iagBEKK11_student.out* file on the accompanying CD-ROM is the G@RCH estimation output (folder *chapter11.Ox.Output*).

[7] When the one-step maximum likelihood method is selected, the G@RCH package gets as initial values the two-step maximum likelihood parameter estimates.

[8] The *ccc_cgr11_student.1step.out* and *ccc_cgr11_student.2step.out* files on the accompanying CD-ROM are the G@RCH outputs with the one-step and two-step maximum likelihood estimation methods, respectively.

Table 11.6 Estimated parameters of the Diag-BEKK(1,1) model, in G@RCH 5.0

$$\mathbf{y}_t = \begin{pmatrix} y_{FTSE100,t} \\ y_{Gold,t} \\ y_{SP500,t} \\ y_{\$/\pounds,t} \end{pmatrix} = \begin{pmatrix} \beta_{0,1} \\ \beta_{0,2} \\ \beta_{0,3} \\ \beta_{0,4} \end{pmatrix} + \boldsymbol{\varepsilon}_t = \begin{pmatrix} 0.048[4.24] \\ -0.009[-1.13] \\ 0.063[5.74] \\ 0.010[1.42] \end{pmatrix} + \boldsymbol{\varepsilon}_t$$

$$vech(\mathbf{H}_t) = \begin{pmatrix} \sigma_{1,t}^2 \\ \sigma_{1,2,t} \\ \sigma_{1,3,t} \\ \sigma_{1,4,t} \\ \sigma_{2,t}^2 \\ \sigma_{3,2,t} \\ \sigma_{4,2,t} \\ \sigma_{3,t}^2 \\ \sigma_{4,3,t} \\ \sigma_{4,t}^2 \end{pmatrix} =$$

$$\begin{pmatrix} a_{0,1,1}^2 + a_{1,1,1}\varepsilon_{1,t-1}^2 a_{1,1,1} + b_{1,1,1}\sigma_{1,t-1}^2 b_{1,1,1} \\ a_{0,1,1}a_{0,1,2} + a_{1,1,1}\varepsilon_{1,t-1}\varepsilon_{2,t-1}a_{1,2,2} + b_{1,1,1}\sigma_{1,2,t-1}b_{1,2,2} \\ a_{0,1,1}a_{0,1,3} + a_{1,1,1}\varepsilon_{1,t-1}\varepsilon_{3,t-1}a_{1,3,3} + b_{1,1,1}\sigma_{1,3,t-1}b_{1,3,3} \\ a_{0,1,1}a_{0,1,4} + a_{1,1,1}\varepsilon_{1,t-1}\varepsilon_{4,t-1}a_{1,4,4} + b_{1,1,1}\sigma_{1,4,t-1}b_{1,4,4} \\ a_{0,1,2}^2 + a_{0,2,2}^2 + a_{1,2,2}\varepsilon_{2,t-1}^2 a_{1,2,2} + b_{1,2,2}\sigma_{2,t-1}^2 b_{1,2,2} \\ a_{0,1,2}a_{0,1,3} + a_{0,2,2}a_{0,2,3} + a_{1,2,2}\varepsilon_{2,t-1}\varepsilon_{3,t-1}a_{1,3,3} + b_{1,2,2}\sigma_{2,3,t-1}b_{1,3,3} \\ a_{0,1,2}a_{0,1,4} + a_{0,2,2}a_{0,2,4} + a_{1,2,2}\varepsilon_{2,t-1}\varepsilon_{4,t-1}a_{1,4,4} + b_{1,2,2}\sigma_{2,4,t-1}b_{1,4,4} \\ a_{0,1,3}^2 + a_{0,2,3}^2 + a_{0,3,3}^2 + a_{1,3,3}\varepsilon_{3,t-1}^2 a_{1,3,3} + b_{1,3,3}\sigma_{3,t-1}^2 b_{1,3,3} \\ a_{0,1,3}a_{0,1,4} + a_{0,2,3}a_{0,2,4} + a_{0,3,3}a_{0,3,4} + a_{1,3,3}\varepsilon_{3,t-1}\varepsilon_{4,t-1}a_{1,4,4} + b_{1,3,3}\sigma_{3,4,t-1}b_{1,4,4} \\ a_{0,1,4}^2 + a_{0,2,4}^2 + a_{0,3,4}^2 + a_{0,4,4}^2 + a_{1,4,4}\varepsilon_{4,t-1}^2 a_{1,4,4} + b_{1,4,4}\sigma_{4,t-1}^2 b_{1,4,4} \end{pmatrix}$$

$$= \begin{pmatrix} 0.011 + 0.194\varepsilon_{1,t-1}^2 0.194 + 0.976\sigma_{1,t-1}^2 0.976 \\ -0.001 + 0.194\varepsilon_{1,t-1}\varepsilon_{2,t-1}0.182 + 0.976\sigma_{1,2,t-1}0.978 \\ 0.002 + 0.194\varepsilon_{1,t-1}\varepsilon_{3,t-1}0.138 + 0.976\sigma_{1,3,t-1}0.989 \\ -0.0008 + 0.194\varepsilon_{1,t-1}\varepsilon_{4,t-1}0.162 + 0.976\sigma_{1,4,t-1}0.984 \\ 0.004 + 0.182\varepsilon_{2,t-1}^2 0.182 + 0.978\sigma_{2,t-1}^2 0.978 \\ -0.0005 + 0.182\varepsilon_{2,t-1}\varepsilon_{3,t-1}0.138 + 0.978\sigma_{2,3,t-1}0.989 \\ 0.0005 + 0.182\varepsilon_{2,t-1}\varepsilon_{4,t-1}0.162 + 0.978\sigma_{2,4,t-1}0.984 \\ 0.003 + 0.138\varepsilon_{3,t-1}^2 0.138 + 0.989\sigma_{3,t-1}^2 0.989 \\ -0.00007 + 0.138\varepsilon_{3,t-1}\varepsilon_{4,t-1}0.162 + 0.989\sigma_{3,4,t-1}0.984 \\ 0.002 + 0.162\varepsilon_{4,t-1}^2 0.162 + 0.984\sigma_{4,t-1}^2 0.984 \end{pmatrix}$$

Student t distribution degrees of freedom, $\nu = 6.185[20.1]$.

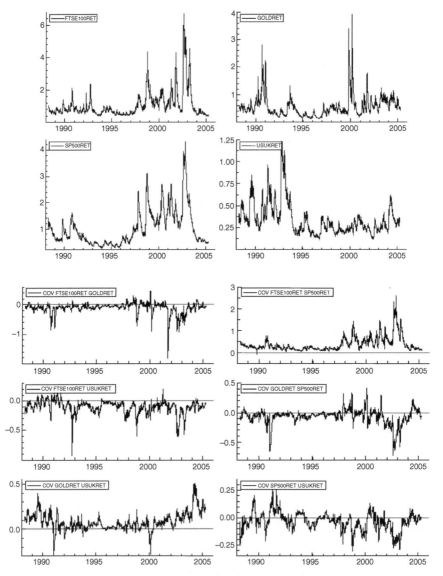

Figure 11.4   Conditional variances and conditional covariances of the Diag-BEKK(1,1) model, in G@RCH 5.0.

squares (NLS) method proposed by Boswijk and van der Weide (2006). The *goarch_mle.out* and *goarch_nls.out* files on the accompanying CD-ROM are the G@RCH outputs with the maximum likelihood estimation method and the non-linear least squares method, respectively. The GO-ARCH model with the maximum

Table 11.7   Estimated parameters of the CCC-ARCH model, in G@RCH 5.0. The coefficient to standard error ratios are reported in brackets

$$
\mathbf{y}_t = \begin{pmatrix} y_{FTSE100,t} \\ y_{Gold,t} \\ y_{SP500,t} \\ y_{\$/\pounds,t} \end{pmatrix} = \begin{pmatrix} \beta_{0,1} \\ \beta_{0,2} \\ \beta_{0,3} \\ \beta_{0,4} \end{pmatrix} + \varepsilon_t = \begin{pmatrix} 0.034[2.91] \\ -0.001[-0.10] \\ 0.047[4.23] \\ 0.011[1.55] \end{pmatrix} + \varepsilon_t
$$

$$
vech(\mathbf{H}_t) = \begin{pmatrix} \sigma_{1,t}^2 \\ \sigma_{1,2,t} \\ \sigma_{1,3,t} \\ \sigma_{1,4,t} \\ \sigma_{2,t}^2 \\ \sigma_{3,2,t} \\ \sigma_{4,2,t} \\ \sigma_{3,t}^2 \\ \sigma_{4,3,t} \\ \sigma_{4,t}^2 \end{pmatrix} = \begin{pmatrix} a_{1,0} + \alpha_{1,1}\varepsilon_{1,t-1}^2 + \gamma_{1,1}d_{1,t-1}\varepsilon_{1,t-1}^2 + b_{1,1}\sigma_{1,t-1}^2 \\ c_{1,2}\sigma_{1,t}\sigma_{2,t} \\ c_{1,3}\sigma_{1,t}\sigma_{3,t} \\ c_{1,4}\sigma_{1,t}\sigma_{4,t} \\ a_{2,0} + \alpha_{2,1}\varepsilon_{i,t-1}^2 + \gamma_{2,1}d_{2,t-1}\varepsilon_{2,t-1}^2 + b_{2,1}\sigma_{2,t-1}^2 \\ c_{2,3}\sigma_{2,t}\sigma_{3,t} \\ c_{2,4}\sigma_{2,t}\sigma_{4,t} \\ a_{3,0} + \alpha_{3,1}\varepsilon_{3,t-1}^2 + \gamma_{3,1}d_{3,t-1}\varepsilon_{3,t-1}^2 + b_{3,1}\sigma_{3,t-1}^2 \\ c_{3,4}\sigma_{3,t}\sigma_{4,t} \\ a_{4,0} + \alpha_{4,1}\varepsilon_{4,t-1}^2 + \gamma_{4,1}d_{4,t-1}\varepsilon_{4,t-1}^2 + b_{4,1}\sigma_{4,t-1}^2 \end{pmatrix}
$$

$$
= \begin{pmatrix} \begin{array}{llll} 0.013 + 0.017 & \varepsilon_{1,t-1}^2 + 0.072 & d_{1,t-1}\varepsilon_{1,t-1}^2 + 0.935 & \sigma_{1,t-1}^2 \\ {[4.63]} \quad [2.57] & [6.46] & [114] \end{array} \\ \begin{array}{ll} -0.137 & \sigma_{1,t}\sigma_{2,t} \\ {[-8.89]} \end{array} \\ \begin{array}{ll} 0.371 & \sigma_{1,t}\sigma_{3,t} \\ {[26.64]} \end{array} \\ \begin{array}{ll} -0.191 & \sigma_{1,t}\sigma_{4,t} \\ {[-12.5]} \end{array} \\ \begin{array}{llll} 0.005 + 0.085 & \varepsilon_{2,t-1}^2 - 0.054 & d_{2,t-1}\varepsilon_{2,t-1}^2 + 0.932 & \sigma_{2,t-1}^2 \\ {[2.87]} \quad [5.49] & [-3.94] & [86.5] \end{array} \\ \begin{array}{ll} -0.092 & \sigma_{2,t}\sigma_{3,t} \\ {[-6.14]} \end{array} \\ \begin{array}{ll} 0.228 & \sigma_{2,t}\sigma_{4,t} \\ {[15.20]} \end{array} \\ \begin{array}{llll} 0.009 + 0.006 & \varepsilon_{3,t-1}^2 + 0.080 & d_{3,t-1}\varepsilon_{3,t-1}^2 + 0.945 & \sigma_{3,t-1}^2 \\ {[3.15]} \quad [1.27] & [4.54] & [95] \end{array} \\ \begin{array}{ll} -0.068 & \sigma_{3,t}\sigma_{4,t} \\ {[-4.41]} \end{array} \\ \begin{array}{llll} 0.002 + 0.029 & \varepsilon_{4,t-1}^2 + 0.010 & d_{4,t-1}\varepsilon_{4,t-1}^2 + 0.960 & \sigma_{4,t-1}^2 \\ {[1.79]} \quad [4.36] & [1.09] & [104] \end{array} \end{pmatrix}
$$

Student t distribution degrees of freedom, $v = 6.552[19.5]$.

likelihood estimation method is

$$
\mathbf{y}_t = \begin{pmatrix} y_{FTSE100,t} \\ y_{Gold,t} \\ y_{SP500,t} \\ y_{\$/£,t} \end{pmatrix} = \begin{pmatrix} \beta_{0,1} \\ \beta_{0,2} \\ \beta_{0,3} \\ \beta_{0,4} \end{pmatrix} + \varepsilon_t = \begin{pmatrix} 0.023 \\ -0.002 \\ 0.034 \\ -0.0001 \end{pmatrix} + \varepsilon_t,
$$
(11.43)

$$
\varepsilon_t | I_{t-1} \sim N[0, \mathbf{H}_t],
$$
$$
\varepsilon_t = \mathbf{V}^{1/2} \mathbf{W} \mathbf{f}_t,
$$
$$
\mathbf{H}_t \equiv V_{t-1}(\varepsilon_t) = \mathbf{V}^{1/2} \mathbf{W} \mathbf{H}_{\mathbf{f},t} \mathbf{W}' \mathbf{V}'^{1/2}.
$$

The random vector process $\mathbf{f}_t = (f_{1,t}, f_{2,t}, f_{3,t}, f_{4,t})'$ has zero conditional mean and GARCH(1,1)-type conditional variance:

$$
\mathbf{H}_{\mathbf{f},t} = V_{t-1}(\mathbf{f}_t) = diag\left(\sigma^2_{f_{1,t}}, \sigma^2_{f_{2,t}}, \sigma^2_{f_{3,t}}, \sigma^2_{f_{4,t}}\right),
$$
$$
\sigma^2_{f_{1,t}} = 0.012 + 0.073 f^2_{1,t-1} + 0.914 \sigma^2_{f_{1,t-1}},
$$
$$
\sigma^2_{f_{2,t}} = 0.005 + 0.042 f^2_{2,t-1} + 0.953 \sigma^2_{f_{2,t-1}},
$$
(11.44)
$$
\sigma^2_{f_{3,t}} = 0.008 + 0.038 f^2_{3,t-1} + 0.954 \sigma^2_{f_{3,t-1}},
$$
$$
\sigma^2_{f_{4,t}} = 0.007 + 0.053 f^2_{4,t-1} + 0.940 \sigma^2_{f_{4,t-1}}.
$$

The matrix $\mathbf{V}^{1/2} = diag(v_1, v_2, v_3, v_4)$ consists of the standard deviations of ε_t. The matrix $\mathbf{W}$ is computed as

$$
\mathbf{W} = \mathbf{P}\mathbf{\Lambda}^{1/2}\mathbf{U} = \begin{bmatrix} 0.993 & 0.025 & 0.017 & 0.113 \\ -0.112 & -0.992 & -0.053 & -0.032 \\ 0.293 & 0.024 & -0.052 & 0.954 \\ -0.149 & -0.129 & -0.978 & -0.074 \end{bmatrix},
$$
(11.45)

where $\mathbf{P}$ contains the four eigenvectors of the unconditional correlation matrix of ε_t,

$$
\mathbf{P} = \begin{bmatrix} 0.621 & -0.287 & 0.082 & -0.724 \\ -0.385 & -0.580 & 0.717 & -0.018 \\ 0.552 & -0.499 & -0.091 & 0.662 \\ -0.401 & -0.576 & -0.686 & -0.194 \end{bmatrix},
$$
(11.46)

$\mathbf{\Lambda}^{1/2}$ is the diagonal matrix with the square roots of the corresponding eigenvalues,

$$
\mathbf{\Lambda}^{1/2} = \begin{bmatrix} 1.247 & 0 & 0 & 0 \\ 0 & 1.027 & 0 & 0 \\ 0 & 0 & 0.897 & 0 \\ 0 & 0 & 0 & 0.765 \end{bmatrix},
$$
(11.47)

and $\mathbf{U}$ is computed as the product of six rotation matrices,

$$\mathbf{U} = \prod_{j=i+1}^{4} \prod_{i=1}^{3} \mathbf{G}_{i,j}(\delta_{i,j}) = \begin{bmatrix} 0.707 & -0.274 & 0.086 & -0.646 \\ 0.371 & 0.614 & -0.695 & 0.053 \\ 0.316 & 0.599 & 0.712 & 0.187 \\ 0.512 & -0.436 & -0.055 & 0.738 \end{bmatrix}. \tag{11.48}$$

The rotation matrices $\mathbf{G}_{i,j}(\delta_{i,j})$ are (4×4) identity matrices, where the (i,i) th and (j,j) th elements are replaced by $\cos(\delta_{i,j})$, whereas the (i,j) th and (j,i) th elements are replaced by $\sin(\delta_{i,j})$ and $-\sin(\delta_{i,j})$, respectively. For example, the matrices $\mathbf{G}_{1,2}(\delta_{1,2})$ and $\mathbf{G}_{3,4}(\delta_{3,4})$ have the form

$$\mathbf{G}_{1,2}(\delta_{1,2}) = \begin{bmatrix} \cos(\delta_{1,2}) & \sin(\delta_{1,2}) & 0 & 0 \\ -\sin(\delta_{1,2}) & \cos(\delta_{1,2}) & 0 & 0 \\ 0 & 0 & 1 & 0 \\ 0 & 0 & 0 & 1 \end{bmatrix} = \begin{bmatrix} 0.933 & 0.361 & 0 & 0 \\ -0.361 & 0.933 & 0 & 0 \\ 0 & 0 & 1 & 0 \\ 0 & 0 & 0 & 1 \end{bmatrix}, \tag{11.49}$$

$$\mathbf{G}_{3,4}(\delta_{3,4}) = \begin{bmatrix} 1 & 0 & 0 & 0 \\ 0 & 1 & 0 & 0 \\ 0 & 0 & \cos(\delta_{3,4}) & \sin(\delta_{3,4}) \\ 0 & 0 & -\sin(\delta_{3,4}) & \cos(\delta_{3,4}) \end{bmatrix} = \begin{bmatrix} 1 & 0 & 0 & 0 \\ 0 & 1 & 0 & 0 \\ 0 & 0 & 0.969 & -0.246 \\ 0 & 0 & 0.246 & 0.969 \end{bmatrix}. \tag{11.50}$$

The $\delta_{i,j}$ stand for the rotation parameters, which are estimated (with coefficient to standard error ratios in brackets) as

$$\begin{aligned} \delta_{1,2} &= 0.369[4.7], \\ \delta_{1,3} &= -0.113[-2.03], \\ \delta_{1,4} &= 0.703[2.88], \\ \delta_{2,3} &= 0.784[20.10], \\ \delta_{2,4} &= -0.070[-1.52], \\ \delta_{3,4} &= -0.249[-3.32]. \end{aligned} \tag{11.51}$$

11.5   Evaluation of multivariate ARCH models

The evaluation criteria presented in the previous chapter for the univariate ARCH models can be extended to take into consideration cross-interactions, such as time-varying conditional correlations. Based on Rosenblatt's (1952) dynamic probability integral transforms, Diebold et al. (1999) proposed a framework for multivariate density forecast evaluation as a generalization of the univariate procedure proposed in

Diebold et al. (1998). Ishida (2005) extended Diebold's et al. (1999) study and proposed new ways to construct probability integral transforms, which provide more powerful tests. Laurent et al. (2008) extended Hansen's (2005) SPA test to a multivariate framework by providing four alternative matrix distance metrics, namely the Frobenius, eigenvalue, Forstner and Moonen and cosinus mass metrics.

Recently, an increasing number of studies have been concerned with evaluating multivariate ARCH models in the context of various financial applications, such as VaR forecasting, option pricing, portfolio optimization and hedging. Morimoto and Kawasaki (2008) made a comparative evaluation of the forecasting performance of univariate and multivariate ARCH models in estimating the intraday VaR. McAleer and da Veiga (2008) compared multivariate ARCH models in a VaR forecasting framework and noted that VaR forecasts appear to be insensitive to the inclusion of spillover effects.[9] Specht and Gohout (2003) and Specht and Winker (2008) evaluated principal component ARCH models, such as the O-ARCH model, in portfolio selection and VaR estimation, respectively.

---

[9] We referred to volatility spillover effects in Section 2.10.

Appendix

EViews 6

- *chapter11.multivariate.prg*

```
load chapter11_data.wf1
smpl @all
'_____Diag-BEKK(1,1) with Student t distributed
  innovations____
system bekk
bekk.append ftse100ret = c(1)
bekk.append goldret = c(2)
bekk.append sp500ret = c(3)
bekk.append usukret = c(4)
bekk.arch(deriv=aa, tdist) @diagbekk c(indef) arch(1,diag)
  garch(1,diag)
'_____Diag-VECH(1,1) with GJR-type asym effects and Student
  t innovations____
system vech
vech.append ftse100ret = c(1)
vech.append goldret = c(2)
vech.append sp500ret = c(3)
vech.append usukret = c(4)
vech.arch(deriv=aa, tdist) @diagvech c(indef) arch(1,diag)
  tarch(1,diag) garch(1,diag)
'_____CCC-ARCH with GJR-type asym effects and Student
  t innovations____
system ccc
ccc.append ftse100ret = c(1)
ccc.append goldret = c(2)
ccc.append sp500ret = c(3)
ccc.append usukret = c(4)
ccc.arch(deriv=aa, tdist) @ccc c arch(1) tarch(1) garch(1)

save chapter11_data.output.wf1
```

G@RCH 5.0

- *chapter11.CCC_CGR11_student.1step.ox*

```
#include <oxstd.h>
#include <oxdraw.h>
#import <packages/MGarch1/mgarch>
main()
```

```
{
  decl model = new MGarch();
  model.Load("Chapter11Data.xls");
  model.Deterministic(-1);
  model.Select(Y_VAR, {"FTSE100RET", 0, 0});
  model.Select(Y_VAR, {"GOLDRET", 0, 0});
  model.Select(Y_VAR, {"SP500RET", 0, 0});
  model.Select(Y_VAR, {"USUKRET", 0, 0});
  model.CSTS(1,1);
  model.DISTRI(STUDENT);
  model.ARMA_ORDERS(0,0);
  model.GARCH_ORDERS(1,1);
  model.MODEL(CCC);
  model.MLE(QMLE);
  model.UGARCH_MODELS(GJR);
  model.UGARCH_TRUNC(1000);
  model.UGARCH_PrintOutput(1);
  model.UGARCH_ARFIMA(0);
  model.ONE_STEP(1);
  model.TSE_LAGS(0);
  model.SetSelSampleByDates(dayofcalendar(1988, 4, 5),
  dayofcalendar(2005, 4, 5));
  model.Initialization(<>);
  model.PrintOutput(1);
  model.DoEstimation();
  decl cov_vec,cor_vec,Varf,covf_vec,corf_vec;
  delete model;
}
```

- *chapter11.GOARCH_MLE.ox*

```
#include <oxstd.h>
#include <oxdraw.h>
#import <packages/MGarch1/mgarch>
main()
{
  decl model = new MGarch();
  model.Load("Chapter11Data.xls");
  model.Deterministic(-1);
  model.Select(Y_VAR, {"FTSE100RET", 0, 0});
  model.Select(Y_VAR, {"GOLDRET", 0, 0});
  model.Select(Y_VAR, {"SP500RET", 0, 0});
  model.Select(Y_VAR, {"USUKRET", 0, 0});
  model.CSTS(1,2);
```

```
model.DISTRI(NORMAL);
model.ARMA_ORDERS(0,0);
model.GARCH_ORDERS(1,1);
model.MODEL(GOGARCH_ML);
model.MLE(QMLE);
model.UGARCH_MODELS(GARCH);
model.UGARCH_TRUNC(1000);
model.UGARCH_PrintOutput(2);
model.UGARCH_ARFIMA(0);
model.OGARCH_M(4);
model.SetSelSampleByDates(dayofcalendar(1988, 4, 5),
  dayofcalendar(2005, 4, 5));
model.Initialization(<>);
model.PrintOutput(1);
model.DoEstimation();
decl cov_vec,cor_vec,Varf,covf_vec,corf_vec;
delete model;
}
```

References

Abhyankar, A.H. (1995). Trading-round-the clock: Return, volatility and volume spillovers in the Eurodollar futures market. *Pacific-Basin Finance Journal*, 3, 75–92.

Abrahamson, I.G. (1967). Exact Bahadur efficiencies for the Kolmogorov–Smirnov and Kuiper one- and two-sample statistics. *Annals of Mathematical Statistics*, 38(5), 1475–1490.

Acerbi, C. (2002). Spectral measures of risk: A coherent representation of subjective risk aversion. *Journal of Banking and Finance*, 26(7), 1505–1518.

Acerbi, C., Nordio, C. and Sirtori, C. (2001). Expected shortfall as a tool for financial risk management. Working Paper, http://www.gloriamundi.org/var/wps.html.

Adrangi, B. and Chatrath, A. (2003). Non-linear dynamics in futures prices: Evidence from the Coffee, Sugar and Cocoa Exchange. *Applied Financial Economics*, 13, 245–256.

Aguilar, O. and West, M. (2000). Bayesian dynamic factor models and portfolio allocation. *Journal of Business and Economic Statistics*, 18, 338–57.

Ahrens, R. and Reitz, S. (2004). Heterogeneous expectations in the foreign exchange market: Evidence from the daily dollar/DM exchange rate. *Journal of Evolutionary Economics*, 15(1), 65–82.

Aiolfi, M. and Favero, C.A. (2005). Model uncertainty, thick modelling and the predictability of stock returns. *Journal of Forecasting*, 24(4), 233–254.

Aït-Sahalia, Y. (2001). Telling from discrete data whether the underlying continuous time model is a diffusion. *NBER Working Paper*, 8504, October 2001.

Aït-Sahalia, Y. (2002). Maximum likelihood estimation of discretely sampled diffusions: A closed-form approximation approach. *Econometrica*, 70(1), 223–262.

Aït-Sahalia, Y., Mykland, P.A. and Zhang, L. (2005). How often to sample a continuous-time process in the presence of market microstructure noise. *Review of Financial Studies*, 18, 351–416.

Akaike, H. (1973). Information theory and an extension of the maximum likelihood principle. in B.N. Petrov and F. Csáki (eds), *Proceedings of the Second International Symposium on Information Theory*, pp. 267–281. Budapest: Akadémiai Kiadó. Reproduced in S. Kotz and

N.L. Johnson (eds), *Breakthroughs in Statistics Vol I Foundations and Basic Theory*, pp. 610–624. New York: Springer-Verlag 1992.

Alexander, C.O. (1998). Volatility and correlation: Methods, models and applications. In C.O. Alexander (ed.), *Risk Management and Analysis: Measuring and Modelling Financial Risk*, pp. 125–172. Chichester: John Wiley & Sons, Ltd.

Alexander, C.O. (2000). A primer on the orthogonal GARCH model. ISMA Centre, University of Reading, Working Paper.

Alexander, C.O. (2001). Orthogonal GARCH. In C.O. Alexander (ed.), *Mastering Risk*, Vol. 2, pp. 21–38. London: Financial Times-Prentice Hall.

Alexander, C.O. (2002). Principal component models for generating large GARCH covariance matrices. *Economic Notes*, 31(2), 337–359.

Alexander, C.O. (2008). *Market Risk Analysis, Volume 1: Quantitative Methods in Finance*. Chichester: John Wiley & Sons Ltd.

Alexander, C.O. and Chibumba, A.M. (1997). Multivariate orthogonal factor GARCH. University of Sussex, Discussion Papers in Mathematics.

Alexander, C.O. and Leigh, C.T. (1997). On the covariance models used in value at risk models. *Journal of Derivatives*, 4, 50–62.

Alizadeh, S., Brandt, M.W. and Diebold, F.X. (2002). Range-based estimation of stochastic volatility models. *Journal of Finance*, 57, 1047–1091.

Allen, L. and Bali, T.G. (2007). Cyclicality in catastrophic and operational risk measurements. *Journal of Banking and Finance*, 31(4), 1191–1235.

Andersen, T. (2000). Some reflections on analysis of high-frequency data. *Journal of Business and Economic Statistics*, 18(2), 146–153.

Andersen, T. and Bollerslev, T. (1997). Intraday periodicity and volatility persistence in financial markets. *Journal of Empirical Finance*, 4, 115–158.

Andersen, T. and Bollerslev, T. (1998a). Answering the skeptics: Yes, standard volatility models do provide accurate forecasts. *International Economic Review*, 39, 885–905.

Andersen, T. and Bollerslev, T. (1998b). DM-dollar volatility: Intraday activity patterns, macroeconomic announcements and longer-run dependencies. *Journal of Finance*, 53, 219–265.

Andersen, T. and Bollerslev, T. (1998c). ARCH and GARCH models. In S. Kotz, C.B. Read and D. L. Banks (eds), *Encyclopedia of Statistical Sciences Vol. II*, New York: John Wiley & Sons, Inc.

Andersen, T., Bollerslev, T. and Lange, S. (1999a). Forecasting financial market volatility: Sample frequency vis-à-vis forecast horizon. *Journal of Empirical Finance*, 6, 457–477.

Andersen, T., Bollerslev, T., Diebold, F.X. and Labys, P. (1999b). Understanding, optimizing, using and forecasting realized volatility and correlation. Northwestern University, Duke University and University of Pennsylvania, Working Paper. Published in revised form as: Great realizations, *Risk*, 2000, 105–108.

Andersen, T., Bollerslev, T. and Cai, J. (2000a). Intraday and interday volatility in the Japanese stock market. *Journal of International Financial Markets, Institutions and Money*, 10, 107–130.

Andersen, T., Bollerslev, T., Diebold, F.X. and Labys, P. (2000b). Market microstructure effects and the estimation of integrated volatility. Northwestern University, Duke University, and the University of Pennsylvania, Working Paper.

Andersen, T., Bollerslev, T., Diebold, F.X. and Labys, P. (2000c). Exchange rate returns standardized by realized volatility are (nearly) Gaussian. *Multinational Finance Journal*, 4, 159–179.

Andersen, T., Bollerslev, T., Diebold, F.X. and Ebens, H. (2001a). The distribution of realized stock return volatility. *Journal of Financial Economics*, 61, 43–76.

Andersen, T., Bollerslev, T., Diebold, F.X. and Labys, P. (2001b). The distribution of realized exchange rate volatility. *Journal of the American Statistical Association*, 96, 42–55.

Andersen, T., Bollerslev, T., Diebold, F.X. and Labys, P. (2003). Modeling and forecasting realized volatility. *Econometrica*, 71, 529–626.

Andersen, T., Bollerslev, T. and Diebold, F.X. (2005a). Parametric and nonparametric volatility measurement. In Y. Aït-Sahalia and L.P. Hansen (eds), *Handbook of Financial Econometrics*, Amsterdam: North Holland.

Andersen, T., Bollerslev, T. and Meddahi, N. (2005b). Correcting the errors: Volatility forecast evaluation using high-frequency data and realized volatilities. *Econometrica*, 73(1), 279–296.

Andersen, T., Bollerslev, T., Christoffersen, P. and Diebold, F.X. (2006). Volatility and correlation forecasting. In G. Elliott, C.W.J. Granger and A. Timmermann (eds), *Handbook of Economic Forecasting*, Amsterdam: North Holland.

Andersen, T., Bollerslev, T. and Diebold, F.X. (2007). Roughing it up: Including jump components in the measurement, modeling and forecasting of return volatility. *Review of Economics and Statistics*, 89(4), 701–720.

Anderson, T.W. and Darling, D.A. (1952). Asymptotic theory of certain goodness of fit criteria based on stochastic processes. *Annals of Mathematical Statistics*, 23, 193–212.

Anderson, T.W. and Darling, D.A. (1954). A test of goodness of fit. *Journal of the American Statistical Association*, 49, 765–769.

Ang, A. and Bekaert, G. (2002). International asset allocation with regime shifts. *Review of Financial Studies*, 15(4), 1137–1187.

Angelidis, T. and Benos, A. (2008). Value-at-risk for Greek stocks. *Multinational Finance Journal*, 12(1/2), 67–105.

Angelidis, T. and Degiannakis, S. (2005). Modeling risk for long and short trading positions. *Journal of Risk Finance*, 6(3), 226–238.

Angelidis, T. and Degiannakis, S. (2007). Backtesting VaR models: A two-stage procedure. *Journal of Risk Model Validation*, 1(2), 1–22.

Angelidis, T. and Degiannakis, S. (2008). Volatility forecasting: Intra-day vs. inter-day models. *Journal of International Financial Markets, Institutions and Money*, 18, 449–465.

Angelidis, T., Benos, A. and Degiannakis, S. (2004). The use of GARCH models in VaR estimation. *Statistical Methodology*, 1(2), 105–128.

Angelidis, T., Benos, A. and Degiannakis, S. (2007). A robust VaR model under different time periods and weighting schemes. *Review of Quantitative Finance and Accounting*, 28(2), 187–201.

Aragó, V. and Fernández, A. (2002). Expiration and maturity effect: Empirical evidence from the Spanish spot and futures stock index. *Applied Economics*, 34, 1617–1626.

Areal, N.M.P.C. and Taylor, S.J. (2002). The realised volatility of FTSE-100 future prices. *Journal of Futures Markets*, 22, 627–648.

Artzner, P., Delbaen, F., Eber, J.-M. and Heath, D. (1997). Thinking coherently. *Risk*, 10, 68–71.

Artzner, P., Delbaen, F., Eber, J.-M. and Heath, D. (1999). Coherent measures of risk. *Mathematical Finance*, 9, 203–228.

Arvanitis, S. and Demos, A. (2004). Time dependence and moments of a family of time-varying parameter GARCH in mean models. *Journal of Time Series Analysis*, 25(1), 1–25.

Assoe, K.G. (1998). Regime-switching in emerging stock market returns. *Multinational Finance Journal*, 2, 101–132.

Asteriou, D. (2006). *Applied Econometrics, A Modern Approach Using Eviews and Microfit*. New York: Palgrave Macmillan.

Atchison, M., Butler, K. and Simonds, R. (1987). Non-synchronous security trading and market index autocorrelation. *Journal of Finance*, 42, 111–118.

Attanasio, O.P. and Wadhwani, S. (1989). Risk and the predictability of stock market returns. Manuscript, Stanford University.

Awartani, B. and Corradi, V. (2005). Predicting the volatility of the S&P-500 stock index via GARCH models: The role of asymmetries. *International Journal of Forecasting*, 21(1), 167–183.

Baba, Y., Engle, R.F., Kraft, D. and Kroner, K.F. (1990). Multivariate simultaneous generalized ARCH. Mimeo, Department of Economics, University of California, San Diego.

Bachellier, L. (1900). Theory of speculation. In P. Cootner (ed.), *The Random Character of Stock Market Prices*, MIT Press, Cambridge, MA (1964).

Backus, D.K., Gregory, A.W. and Zin, S.E. (1989). Risk premiums in the term structure: Evidence from artificial economics. *Journal of Monetary Economics*, 24, 371–399.

Bai, J. and Ng, S. (2001). A consistent test for conditional symmetry in time series models. *Journal of Econometrics*, 103, 225–258.

Baillie, R.T. and Bollerslev, T. (1989). The message in daily exchange rates: A conditional variance tale. *Journal of Business and Economic Statistics*, 7, 297–305.

Baillie, R.T. and Bollerslev, T. (1990). A multivariate generalized ARCH approach to modeling risk-premia in forward foreign exchange rate markets. *Journal of International Money and Finance*, 16, 109–124.

Baillie, R.T. and Morana, C. (2009). Modelling long memory and structural breaks in conditional variances: An adaptive FIGARCH Approach. *Journal of Economic Dynamics and Control*, 33(8), 1577–1592.

Baillie, R.T. and Myers, R.J. (1991). Bivariate GARCH estimation of optimal commodity futures hedge. *Journal of Applied Econometrics*, 16, 109–124.

Baillie, R.T., Bollerslev, T. and Mikkelsen, H.O. (1996a). Fractionally integrated generalized autoregressive conditional heteroskedasticity. *Journal of Econometrics*, 74, 3–30.

Baillie, R.T., Chung, C.F. and Tieslau, M.A. (1996b). Analysing inflation by the fractionally integrated ARFIMA-GARCH model. *Journal of Applied Econometrics*, 11, 23–40.

Bali, T.G. and Lu, Y. (2004). Forecasting stock market volatility: The empirical performance of SGED-GARCH, implied, and realized volatility models. Working Paper.

Bali, T.G. and Theodossiou, P. (2007). A conditional-SGT-VaR approach with alternative GARCH models. *Annals of Operations Research*, 151(1), 241–267.

Bams, D., Lehnert, T. and Wolff, C.C.P. (2005). An evaluation framework for alternative VaR-models. *Journal of International Money and Finance*, 24, 944–958.

Bandi, F.M. and Russell, J.R. (2005). Microstructure noise, realized volatility, and optimal sampling. Technical Report, Graduate School of Business, University of Chicago.

Bandi, F.M. and Russell, J.R. (2006). Separating microstructure noise from volatility. *Journal of Financial Economics*, 79, 655–692.

Bao, Y., Lee, T.-H. and Saltoglu, B. (2006). Evaluating predictive performance of value-at-risk models in emerging markets: A reality check. *Journal of Forecasting*, 25, 101–128.

Barndorff-Nielsen, O.E. and Shephard, N. (2001). Non-Gaussian Ornstein–Uhlenbeck based models and some of their uses in financial economics. *Journal of the Royal Statistical Society, Series B*, 63, 197–241.

Barndorff-Nielsen, O.E. and Shephard, N. (2002a). Econometric analysis of realised volatility and its use in estimating stochastic volatility models. *Journal of the Royal Statistical Society, Series B*, 64, 253–280.

Barndorff-Nielsen, O.E. and Shephard, N. (2002b). Estimating quadratic variation using realized variance. *Journal of Applied Econometrics*, 17, 457–477.

Barndorff-Nielsen, O.E. and Shephard, N. (2003). Realized power variation and stochastic volatility models. *Bernoulli*, 9, 243–265.

Barndorff-Nielsen, O.E. and Shephard, N. (2004a). Econometric analysis of realized covariation: High frequency based covariance, regression, and correlation in financial economics. *Econometrica*, 72, 885–925.

Barndorff-Nielsen, O.E. and Shephard, N. (2004b). Power and bipower variation with stochastic volatility and jumps. *Journal of Financial Econometrics*, 2, 1–37.

Barndorff-Nielsen, O.E. and Shephard, N. (2005). How accurate is the asymptotic approximation to the distribution of realised volatility? In D. Andrews, J. Powell, P. Ruud and J. Stock (eds), *Identification and Inference for Econometric Models*, Cambridge: Cambridge University Press.

Barndorff-Nielsen, O.E. and Shephard, N. (2006). Econometrics of testing for jumps in financial economics using bipower variation. *Journal of Financial Econometrics*, 4(1), 1–30.

Barndorff-Nielsen, O.E., Nicolato, E. and Shephard, N. (2002). Some recent developments in stochastic volatility modelling. *Quantitative Finance*, 2, 11–23.

Barndorff-Nielsen, O.E., Hansen, P.R., Lunde, A. and Shephard, N. (2004). Regular and modified kernel-based estimators of integrated variance: The case of independent noise. Economics Series, 2004-FE-20, Oxford Financial Research Centre.

Barone-Adesi, G. and Giannopoulos, K. (2001). Non-parametric VaR techniques. Myths and realities. Economic Notes by Banca Monte dei Paschi di Siena SpA, 30, 167–181.

Barone-Adesi, G., Giannopoulos, K. and Vosper, L. (1999). VaR without correlations for nonlinear portfolios. *Journal of Futures Markets*, 19, 583–602.

Barone-Adesi, G., Rasmussen, H. and Ravanelli, C. (2004). An option pricing formula for the GARCH diffusion model. *Computational Statistics and Data Analysis*, 49(2), 287–310.

Basak, S. and Shapiro, A. (2001). Value-at-risk based risk management: Optimal policies and asset prices. *Review of Financial Studies*, 14(2), 371–405.

Basle Committee on Banking Supervision. (1995a). An internal model-based approach to market risk capital requirements. Basle Committee on Banking Supervision, Basle, Switzerland.

Basle Committee on Banking Supervision. (1995b). Planned supplement to the capital accord to incorporate market risks. Basle Committee on Banking Supervision, Basle, Switzerland.

Basle Committee on Banking Supervision. (1998). *International Convergence of Capital Measurement and Capital Standards.*

Bates, C. and White, H. (1988). Efficient instrumental variables estimation of systems of implicit heterogeneous nonlinear dynamic equations with non-spherical errors. In W.A. Barnett, E.R. Berndt and H., White (eds), *Dynamic Econometric Modeling*, Cambridge: Cambridge University Press.

Bates, J.M. and Granger, C.W.J. (1969). The combination of forecasts. *Operational Research Quarterly*, 20, 451–468.

Baumol, W.J. (1963). An expected gain confidence limit criterion for portfolio selection. *Management Science*, 10, 174–182.

Bauwens, L. and Giot, P. (2000). The logarithmic ACD model: An application to the bid-ask quote process of three NYSE stocks. *Annales d'Economie et de Statistique*, 60, 117–149.

Bauwens, L. and Laurent, S. (2005). A new class of multivariate skew densities, with application to GARCH models. *Journal of Business and Economic Statistics*, 23(3), 346–354.

Bauwens, L. and Lubrano, M. (1998). Bayesian inference on GARCH models using the Gibbs sampler. *Econometrics Journal*, 1, 23–46.

Bauwens, L. and Lubrano, M. (1999). *Bayesian Dynamic Econometrics*. Oxford: Oxford University Press.

Bauwens, L., Giot, P., Grammig, J. and Veredas, D. (2004). A comparison of financial duration models via density forecasts. *International Journal of Forecasting*, 20(4), 589–609.

Bauwens, L., Laurent, S. and Rombouts, J.V.K. (2006). Multivariate GARCH models: A survey. *Journal of Applied Econometrics*, 21, 79–109.

Becker, R., Clements, A.E. and White, S.I. (2007). Does implied volatility provide any information beyond that captured in model-based volatility forecasts? *Journal of Banking and Finance*, 31, 2535–2549.

Beckers, S. (1981). A note on estimating the parameters in the jump-diffusion model of stock returns. *Journal of Financial and Quantitative Analysis*, 26, 127–140.

Beckers, S. (1983). Variances of security price returns based on high, low and closing prices. *Journal of Business*, 56, 97–109.

Beder, T. (1995). VaR: Seductive but dangerous. *Financial Analysts Journal*, 51, 12–24.

Beine, M., Bénassy-Quéré, A. and Lecourt, C. (2002). Central bank intervention and foreign exchange rates: New evidence from FIGARCH estimations. *Journal of International Money and Finance*, 21, 115–144.

Beine, M., Laurent, S. and Lecourt, C. (2003). Official central bank interventions and exchange rate volatility: Evidence from a regime-switching analysis. *European Economic Review*, 47, 891–911.

Bekaert, G. and Wu, G. (2000). Asymmetric volatility and risk in equity markets. *NBER. Review of Financial Studies*, 13(1), 1–42.

Bera, A.K. and Higgins, M.L. (1993). ARCH models: Properties, estimation and testing. *Journal of Economic Surveys*, 7, 305–366.

Bera, A.K. and Jarque, C.M. (1982). Model specification tests: A simultaneous approach. *Journal of Econometrics*, 20, 59–82.

Bera, A.K. and Kim, S. (1996). Testing constancy of correlation with an application to international equity returns. Mimeo, Center for International Business Education and Research (CIBER), working paper 96–107, University of Illinois, Urbana-Champaign.

Bera, A.K. and Lee, S. (1990). On the formulation of a general structure for conditional heteroskedasticity. Mimeo, University of Illinois.

Bera, A.K., Garcia, P. and Roh, J.S. (1991). Estimation of time varying hedge ratios for agricultural commodities: BGARCH and random coefficient approaches. Mimeo, Department of Economics, University of Illinois at Urbana Champaign.

Bera, A.K., Higgins, M.L. and Sangkyu, L. (1992). Interaction between autocorrelation and conditional heteroscedasticity: A random-coefficient approach. *Journal of Business and Economic Statistics*, 10(2), 133–142.

Beran, J. (1995). Maximum likelihood estimation of the differencing parameter for invertible short- and long-memory ARIMA models. *Journal of the Royal Statistical Society, Series B*, 57(4), 672–695.

Berkes, I. and Horváth, L. (2003). The rate of consistency of the quasi-maximum likelihood estimator. *Statistics and Probability Letters*, 61, 133–143.

Berkes, I., Horváth, L. and Kokoszka, P. (2003). GARCH processes: Structure and estimation. *Bernoulli*, 9, 201–207.

Barkoulas, J.T. and Travlos, N. (1998). Chaos in an emerging capital market? The case of the Athens Stock Exchange. *Applied Financial Economics*, 8, 231–243.

Barkoulas, J.T., Baum, C.F. and Travlos, N. (2000). Long memory in the Greek stock market. *Applied Financial Economics*, 10, 177–184.

Berkowitz, J. (2001). Testing density forecasts, with applications to risk management. *Journal of Business and Economic Statistics*, 19, 465–474.

Berndt, E.R., Hall, B.H., Hall, R.E. and Hausman, J.A. (1974). Estimation and inference in nonlinear structural models. *Annals of Economic and Social Measurement*, 3, 653–665.

Bessembinder, H. and Seguin, P. (1993). Price volatility, trading volume and market depth: Evidence from the futures market. *Journal of Financial and Quantitative Analysis*, 28, 21–39.

Best, P. (1999). *Implementing Value at Risk*. John Wiley & Sons Ltd, Chichester.

Bhar, R. and Hamori, S. (2004). Empirical characteristics of the permanent and transitory components of stock return: Analysis in a Markov switching heteroskedasticity framework. *Economics Letters*, 82, 157–165.

Bhattacharya, P.S. and Thomakos, D.D. (2008). Forecasting industry-level CPI and PPI Inflation: Does exchange rate pass-through matter? *International Journal of Forecasting*, 24(1), 134–150.

Bickel, P.J. and Doksum, K.A. (1981). An Analysis of transformations revisited. *Journal of American Statistical Association*, 76, 296–311.

Billio, M. and Caporin, M. (2006). A generalized dynamic conditional correlation model for portfolio risk evaluation. Economics Research Paper 53/06, University Ca' Foscari of Venice. http://ssrn.com/abstract=948405.

Billio, M. and Pelizzon, L. (2000). Value-at-risk: A multivariate switching regime approach. *Journal of Empirical Finance*, 7, 531–554.

Billio, M., Caporin, M. and Gobbo, M. (2006). Flexible dynamic conditional correlation multivariate GARCH for asset allocation. *Applied Financial Economics Letters*, 2, 123–130.

Black, F. (1975). Fact and fantasy in the use of options. *Financial Analysts Journal*, 31(4), 36–41, 61–72.

Black, F. (1976). Studies of stock market volatility changes. *Proceedings of the American Statistical Association, Business and Economic Statistics Section*, 177–181.

Black, F. and Scholes, M. (1972). The valuation of option contracts and a test of market efficiency. *Journal of Finance*, 27(2), 399–418.

Black, F. and Scholes, M. (1973). The pricing of options and corporate liabilities. *Journal of Political Economy*, 81(3), 637–654.

Blackman, J. (1958). Correction to 'An extension of the Kolmogorov distribution'. *Annals of Mathematical Statistics*, 29, 318–324.

Blair, B.J., Poon, S.H. and Taylor, S.J. (2001). Forecasting S&P100 volatility: The incremental information content of implied volatilities and high frequency returns. *Journal of Econometrics*, 105, 5–26.

Blattberg, R. and Gonedes, N. (1974). A comparison of stable and Student's *t* distributions as statistical models for stock prices. *Journal of Business*, 47, 244–280.

Bod, P., Blitz, D., Franses, P.H. and Kluitman, R. (2002). An unbiased variance estimator for overlapping returns. *Applied Financial Economics*, 12, 155–158.

Bollen, N.P.B., Gray, S.F. and Whaley, R. (2000). Regime switching in foreign exchange rates: Evidence from currency option prices. *Journal of Econometrics*, 94, 239–276.

Bollerslev, T. (1986). Generalized autoregressive conditional heteroskedasticity. *Journal of Econometrics*, 31, 307–327.

Bollerslev, T. (1987). A conditional heteroskedastic time series model for speculative prices and rates of return. *Review of Economics and Statistics*, 69, 542–547.

Bollerslev, T. (1990). Modeling the coherence in short-run nominal exchange rates: A multivariate generalized ARCH approach. *Review of Economics and Statistics*, 72, 498–505.

Bollerslev, T. and Domowitz, I. (1991). Price volatility, spread variability and the role of alternative market mechanisms. *Review of Futures Markets*, 10(1), 78–102.

Bollerslev, T. and Engle, R.F. (1993). Common persistence in conditional variances. *Econometrica*, 61, 167–186.

Bollerslev, T. and Ghysels, E. (1996). Periodic autoregressive conditional heteroskedasticity. *Journal of Business and Economic Statistics*, 14, 139–157.

Bollerslev, T. and Mikkelsen, H.O. (1996). Modeling and pricing long-memory in stock market volatility. *Journal of Econometrics*, 73, 151–184.

Bollerslev, T. and Wooldridge, J.M. (1992). Quasi-maximum likelihood estimation and inference in dynamic models with time-varying covariances. *Econometric Reviews*, 11, 143–172.

Bollerslev, T. and Wright, J.H. (2000). Semiparametric estimation of long-memory volatility dependencies: The role of high-frequency data. *Journal of Econometrics*, 98, 81–106.

Bollerslev, T. and Wright, J.H. (2001). Volatility forecasting, high-frequency data and frequency domain inference. *Review of Economics and Statistics*, 83, 596–602.

Bollerslev, T., Engle, R.F. and Wooldridge, J.M. (1988). A capital asset pricing model with time-varying covariances. *Journal of Political Economy*, 96, 116–131.

Bollerslev, T., Chou, R. and Kroner, K.F. (1992). ARCH modeling in finance: A review of the theory and empirical evidence. *Journal of Econometrics*, 52, 5–59.

Bollerslev, T., Engle, R.F. and Nelson, D. (1994). ARCH models. In R.F. Engle and D. McFadden (eds), *Handbook of Econometrics*, Volume 4, pp. 2959–3038. Amsterdam: Elsevier Science.

Boswijk, H.P. and van der Weide, R. (2006). Wake me up before you GO-GARCH. Tinbergen Institute Discussion Paper, TI2006–079/4.

Boudoukh, J., Richardson, M. and Whitelaw, R. (1997). Investigation of a class of volatility estimators. *Journal of Derivatives*, 4, 63–71.

Boudoukh, J., Richardson, M. and Whitelaw, R. (1998). The best of both worlds. *Risk*, 11, 64–67.

Bougerol, P. and Picard, N. (1992). Stationarity of GARCH processes and of some non-negative time series. *Journal of Econometrics*, 52, 115–128.

Box, G.E.P. and Cox, D.R. (1964). An analysis of transformations. *Journal of the Royal Statistical Society, Series B*, 26, 211–243.

Box, G.E.P. and Pierce, D.A. (1970). Distribution of residual correlations in autoregressive-integrated moving average time series models. *Journal of the American Statistical Association*, 65, 1509–1526.

Box, G.E.P. and Tiao, G.C. (1962). A further look at robustness via Bayes's theorem. *Biometrika*, 49, 419–432.

Box, G.E.P. and Tiao, G.C. (1973). *Bayesian Inference in Statistical Analysis*. Reading, MA: Addison-Wesley.

Brailsford, T.J. (1996). The empirical relationship between trading volume, returns and volatility. *Accounting and Finance*, 35, 89–111.

Brailsford, T.J. and Faff, R.W. (1996). An evaluation of volatility forecasting techniques. *Journal of Banking and Finance*, 20, 419–438.

Brandt, M.W. and Diebold, F.X. (2006). A no-arbitrage approach to range-based estimation of return covariances and correlations. *Journal of Business*, 79(1), 61–73.

Braun, P.A., Nelson, D. and Sunier, A.M. (1995). Good news, bad news, volatility and betas. *Journal of Finance*, 50, 1575–1603.

Brenner, M., Shu, J. and Zhang, J.E. (2007). The market for volatility trading: VIX futures. Working Paper, 07-003, Stern School of Business, New York University.

Breusch, T.S. and Pagan, A.R. (1979). A simple test for heteroskedasticity and random coefficient variation. *Econometrica*, 48, 1287–1294.

Brillinger, D.R. (2007). Volatility for point and marked point processes. Invited paper, *Stochastic Volatility Modeling: Reflections, Recent Developments and the Future, E. Xekalaki (organizer)*, 56th Session of the International Statistical Institute, 22–29 August 2007, Lisbon, Portugal.

Brillinger, D.R. (2008). Extending the volatility concept to point processes. *Journal of Statistical Planning and Inference*, 138, 9(1), 2607–2614.

Brock, W.A. (1986). Distinguishing random and deterministic systems: Abridged version. *Journal of Economic Theory*, 40, 168–195.

Brock, W.A. and Kleidon, A.W. (1990). Exogenous demand stocks and trading volume: A model of intraday bids and asks. Manuscript, University of Wisconsin.

Brock, W.A., Dechert, W.D. and Scheinkman, J.A. (1987). A test for independence based on the correlation dimension. SSRI, Working Paper no. 8702, Department of Economics, University of Wisconsin, Madison.

Brock, W.A., Hsieh, D. and LeBaron, B. (1991). *Nonlinear Dynamics, Chaos and Instability: Statistical Theory and Economic Evidence*. MIT Press, Cambridge, MA.

Brock, W.A., Dechert, W.D., Scheinkman, J.A. and LeBaron, B. (1996). A test for independence based on the correlation dimension. *Econometric Reviews*, 15(3), 197–235.

Brockwell, P.J. and Davis, R.A. (1991). *Time Series: Theory and Methods*. New York: Springer.

Brooks, C. (2002). *Introductory Econometrics for Finance*. Cambridge: Cambridge University Press.

Brooks, C. and Burke, S.P. (2003). Information criteria for GARCH model selection. *European Journal of Finance*, 9, 557–580.

Brooks, C. and Persand, G. (2003a). The effect of asymmetries on stock index return value-at-risk estimates. *Journal of Risk Finance*, 4(2), 29–42.

Brooks, C. and Persand, G. (2003b). Volatility forecasting for risk management. *Journal of Forecasting*, 22, 1–22.

Brooks, C., Faff, R.W., McKenzie, M.D. and Mitchell, H. (2000). A multi-country study of power ARCH models and national stock market returns. *Journal of International Money and Finance*, 19, 377–397.

Brooks, C., Burke, S.P. and Persand, G. (2001). Benchmarks and the accuracy of GARCH model estimation. *International Journal of Forecasting*, 17, 45–56.

Brooks, C., Clare, A.D., Dalle Molle, J.W. and Persand, G. (2005). A comparison of extreme value theory approaches for determining value at risk. *Journal of Empirical Finance*, 12, 339–52.

Brooks, S., Dellaportas, P. and Roberts, G.O. (1997). An approach to diagnosing total variation convergence of MCMC algorithms. *Journal of Computational and Graphical Statistics*, 6, 251–265.

Brorsen, B.W. and Yang, S.R. (1994). Nonlinear dynamics and the distribution of daily stock index returns. *Journal of Financial Research*, 17(2), 187–203.

Brown, B.W. and Ligeralde, A.V. (1990). Conditional heteroskedasticity in overlapping prediction models. Manuscript, Rice University.

Broyden, C.G. (1965). A class of methods for solving nonlinear simultaneous equations. *Mathematics of Computation*, 19, 577–593.

Broyden, C.G. (1967). Quasi-Newton methods and their application to function minimization. *Mathematics of Computation*, 21, 368–381.

Broyden, C.G. (1970). The convergence of a class of double rank minimization algorithms, 2: The new algorithm. *Journal of the Institute for Mathematics and Applications*, 6, 222–231.

Bruneau, C., De Bandt, O., Flageollet, A. and Michaux, E. (2007). Forecasting inflation using economic indicators: The case of France. *Journal of Forecasting*, 26(1), 1–22.

Brunetti, C. and Gilbert, C.L. (1998). A bivariate FIGARCH model of crude oil price volatility. Working Paper, 390, Department of Economics, QMW, London.

Brunetti, C. and Gilbert, C.L. (2000). Bivariate FIGARCH and fractional cointegration. *Journal of Empirical Finance*, 7, 509–530.

Brunetti, C., Mariano, R., Scotti, C. and Tan, A.H.H. (2003). Markov switching GARCH models of currency crises in Southeast Asia. Working Paper, 03-008, Penn Institute for Economic Research.

Busch, T. (2005). A robust LR test for the GARCH model. *Economics Letters*, 88(3), 358–364.

Bühlmann, P. and McNeil, A.J. (2002). An algorithm for non-parametric GARCH modelling. *Computational Statistics and Data Analysis*, 40, 665–683.

Byström, H.N.E. (2004). Managing extreme risks in tranquil and volatile markets using conditional extreme value theory. *International Review of Financial Analysis*, 13(2), 133–152.

Cabedo, D.J. and Moya, I. (2003). Estimating oil price 'value at risk' using the historical simulation approach. *Energy Economics*, 25, 239–253.

Cai, J. (1994). A Markov model of switching-regime ARCH. *Journal of Business and Economic Statistics*, 12, 309–316.

Calzorari, G. and Fiorentini, G. (1998). A Tobit model with GARCH errors. *Econometric Reviews*, 17, 85–104.

Campbell, J., Lo, A. and MacKinlay, A.C. (1997). *The Econometrics of Financial Markets*. Princeton, NJ: Princeton University Press.

Cappiello, L., Engle, R.F. and Sheppard, K. (2006). Asymmetric dynamics in the correlations of global equity and bond returns. *Journal of Financial Econometrics*, 4, 537–572.

Carr, P. and Madan, D. (1998). Towards a theory of volatility trading. In R. Jarrow (ed.), *Volatility Estimation Techniques for Pricing Derivatives*, pp. 417–427. London: Risk Books.

Cecchetti, S.G., Cumby, R.E. and Figlewski, S. (1988). Estimation of the optimal futures hedge. *Review of Economics and Statistics*, 70, 623–630.

Chan, K.C. (1988). On the contrarian investment strategy. *Journal of Business*, 61, 147–163.

Chan, K.F. and Gray, P. (2006). Using extreme value theory to measure value-at-risk for daily electricity spot prices. *International Journal of Forecasting*, 22(2), 283–300.

Chan, N.H. (2002). *Time Series: Applications to Finance*. New York: John Wiley & Sons Inc.

Chaudhuri, K. and Klaassen, F. (2000). Have East Asian stock markets calmed down? Evidence from a regime switching model. Mimeo, University of Amsterdam.

Chen, A.-S. and Leung, M.T. (2005). Modeling time series information into option prices: An empirical evaluation of statistical projection and GARCH option pricing model. *Journal of Banking and Finance*, 29(12), 2947–2969.

Chib, S., Kim, S. and Shephard, N. (1998). Stochastic volatility: Likelihood inference and comparison with ARCH models. *Review of Economic Studies*, 65, 361–393.

Chong, Y.Y. and Hendry, D.F. (1986). Econometric evaluation of linear macroeconomic models. *Review of Economic Studies*, 53, 671–690.

Chopra, N., Lakonishok, J. and Ritter, J.R. (1992). Measuring abnormal performance: Do stocks overreact? *Journal of Financial Economics*, 31, 235–268.

Choudhry, T. (1995). Integrated-GARCH and non-stationary variances: Evidence from European stock markets during the 1920s and 1930s. *Economics Letters*, 48, 55–59.

Christensen, B.J. and Prabhala, N.R. (1998). The relation between implied and realised volatility. *Journal of Financial Economics*, 50, 125–150.

Christie, A.A. (1982). The stochastic behavior of common stock variances: Value, leverage and interest rate effects. *Journal of Financial Economics*, 10, 407–432.

Christodoulakis, G.A. and Satchell, S.E. (2002). Correlated ARCH (CorrARCH): Modelling the time-varying conditional correlation between financial asset returns. *European Journal of Operational Research*, 139, 351–370.

Christoffersen, P. (1998). Evaluating interval forecasts. *International Economic Review*, 39, 841–862.

Christoffersen, P. (2003). *Elements of Financial Risk Management*. New York: Academic Press.

Christoffersen, P. and Jacobs, K. (2004). Which GARCH model for option evaluation? *Management Science*, 50(9), 1204–1221.

Chung, C.F. (1999). Estimating the fractionally integrated GARCH model. National Taiwan University, Working paper.

Clark, P.K. (1973). A subordinated stochastic process model with finite variance for speculative prices. *Econometrica*, 41, 135–156.

Clark, T.E. (1999). Finite-sample properties of tests for equal forecast accuracy. *Journal of Forecasting*, 18, 489–504.

Clark, T.E. and McCracken, M.W. (2001). Tests of equal forecast accuracy and encompassing for nested models. *Journal of Econometrics*, 105, 85–110.

Clemen, R.T. (1989). Combining forecasts: A review and annotated bibliography. *International Journal of Forecasting*, 5, 559–583.

Clements, M.P. (2005). *Evaluating Econometric Forecasts of Economic and Financial Variables*. Basingstoke: Palgrave Macmillan.

Clements, M.P. and Hendry, D.F. (1993). On the limitations of comparing mean square forecast errors. *Journal of Forecasting*, 12, 617–637.

Cohen, K., Maier, S., Schwartz, R. and Whitcomb, D. (1978). The returns generation process, returns variance and the effect of thinness in securities markets. *Journal of Finance*, 33, 149–167.

Cohen, K., Hawawini, G., Maier, S., Schwartz, R. and Whitcomb, D. (1979). On the existence of serial correlation in an efficient securities market. *TIMS Studies in the Management Sciences*, 11, 151–168.

Cohen, K., Hawawini, G., Maier, S., Schwartz, R. and Whitcomb, D. (1983). Friction in the trading process and the estimation of systematic risk. *Journal of Financial Economics*, 12, 263–278.

Conrad, J. and Kaul, G. (1993). Long-term market overreaction or biases in computed returns. *Journal of Finance*, 48, 39–63.

Conrad, J., Gultekin, M.N. and Kaul, G. (1991). Asymmetric predictability of conditional variances. *Review of Financial Studies*, 4, 597–662.

Corrado, C. and Truong, C. (2007). Forecasting stock index volatility: Comparing implied volatility and the intraday high-low price range. *Journal of Financial Research*, XXX(2), 201–215.

Corsi, F. (2004). A simple long memory model of realized volatility. Technical Report, University of Southern Switzerland.

Corsi, F., Zumbach, G., Müller, U.A. and Dacorogna, M.M. (2001). Consistent high-precision volatility from high-frequency data. *Economic Notes*, 30, 183–204.

Corsi, F., Kretschmer, U., Mittnik, S. and Pigorsch, C. (2005). The volatility of realised volatility. Working Paper 33, Center for Financial Studies.

Cowles, A. and Jones, H. (1937). Some a posteriori probabilities in stock market action. *Econometrica*, 5, 280–294.

Cox, G.V. (1930). Evaluation of economic forecasts. *Journal of the American Statistical Association*, 25(169), 31–35.

Cox, J. and Ross, S.A. (1976). The valuation of options for alternative stochastic processes. *Journal of Financial Economics*, 3, 145–166.

Cramér, H. (1928). On the composition of elementary errors. *Skandinavisk Aktuarietidskrift*, 11, 13–74, 141–180.

Cramér, J.S. (1986). *Econometric Applications of Maximum Likelihood Methods*. Cambridge: Cambridge University Press.

Csörgő, S. and Faraway, J. (1996). The exact and asymptotic distributions of Cramér–von Mises statistics. *Journal of the Royal Statistical Society, Series B*, 58(1), 221–234.

Cuoco, D., He, H. and Issaenko, S. (2001). Optimal dynamic trading strategies with risk limits. Manuscript, The Wharton School, University of Pennsylvania.

Cuthbertson, K. and Nitzsche, D. (2004). *Quantitative Financial Economics: Stocks, Bonds and Foreign Exchange*, 2nd edition. Chichester: John Wiley & Sons Ltd.

Dacorogna, M.M., Müller, U.A., Nagler, R.J., Olsen, R.B. and Pictet, O.V. (1993). A geographical model for the daily and weekly seasonal volatility in the foreign exchange market. *Journal of International Money and Finance*, 12(4), 413–438.

Dacorogna, M.M., Müller, U.A., Dav, R., Olsen, R.B. and Pictet, O.V. (1998). Modelling Short-term Volatility with GARCH and HARCH Models. In C. Dunis and B. Zhou (eds), *Nonlinear Modelling of High Frequency Financial Time Series*, pp. 161–176. Chichester: John Wiley & Sons, Ltd.

D'Agostino, R.B. and Stephens, M.A. (1986). *Goodness-of-Fit Techniques*. New York: Markel Dekker.

D'Agostino, R.B., Belanger, A. and D'Agostino, R.B. Jr. (1990). A suggestion for using powerful and informative tests of normality. *American Statistician*, 44(4), 316–321.

Daigler, R.T. (1994). *Advanced Option Trading*. Chicago: Probus.

Dallal, G.E. and Wilkinson, L. (1986). An analytic approximation to the distribution of Lilliefor's test statistic for normality. *American Statistician*, 40(4), 294–296.

Danielsson, J. and Zigrand, J.-P. (2004). On time-scaling of risk and the square-root-of-time rule. EFA 2004 Maastricht Meetings Paper No. 5339. http://ssrn.com/abstract=567123.

Darling, D.A. (1957). The Kolmogorov–Smirnov, Cramér–von Mises tests. *Annals of Mathematical Statistics*. 28(4), 823–838.

Davidon, W.C. (1959). Variable metric method for minimization. Research and Development Report ANL-5990, Argonne National Laboratory.

Davidson, J. (2004). Moment and memory properties of linear conditional heteroscedasticity models, and a new model. *Journal of Business and Economic Statistics*, 22(1), 16–29.

Davis, P.J. (1965). The gamma function and related functions. In M. Abramowitz and N. Stegun (eds), *Handbook of Mathematical Functions*, pp. 253–294. New York: Dover.

Day, T.E. and Lewis, C.M. (1988). The behaviour of the volatility implicit in the prices of stock index options. *Journal of Financial Economics*, 22, 103–122.

Day, T.E. and Lewis, C.M. (1992). Stock market volatility and the information content of stock index options. *Journal of Econometrics*, 52, 267–287.

de Vries, C.G. (1991). On the relation between GARCH and stable processes. *Journal of Econometrics*, 48, 313–324.

DeBondt, W.F.M. and Thaler, R.H. (1985). Does the stock market overreact? *Journal of Finance*, 40, 793–805.

DeBondt, W.F.M. and Thaler, R.H. (1987). Further evidence on investor overreaction and stock market seasonality. *Journal of Finance*, 42, 557–581.

DeBondt, W.F.M. and Thaler, R.H. (1989). A mean-reverting walk down Wall Street. *Journal of Economic Perspectives*, 3, 189–202.

Degiannakis, S. (2004). Volatility forecasting: Evidence from a fractional integrated asymmetric power ARCH skewed-t model. *Applied Financial Economics*, 14, 1333–1342.

Degiannakis, S. (2008a). ARFIMAX and ARFIMAX-TARCH realized volatility modeling. *Journal of Applied Statistics*, 35(10), 1169–1180.

Degiannakis, S. (2008b). Forecasting VIX. *Journal of Money, Investment and Banking*, 4, 5–19.

Degiannakis, S. and Floros, C. (2010). VIX index in interday and intraday volatility models. *Journal of Money, Investment and Banking*, 13, forthcoming.

Degiannakis, S. and Xekalaki, E. (2004). Autoregressive conditional heteroscedasticity models: A review. *Quality Technology and Quantitative Management*, 1(2), 271–324.

Degiannakis, S. and Xekalaki, E. (2005). Predictability and model selection in the context of ARCH models. *Journal of Applied Stochastic Models in Business and Industry*, 21, 55–82.

Degiannakis, S. and Xekalaki, E. (2007a). Assessing the performance of a prediction error criterion model selection algorithm in the context of ARCH models. *Applied Financial Economics*, 17, 149–171.

Degiannakis, S. and Xekalaki, E. (2007b). Simulated evidence on the distribution of the standardized one-step-ahead prediction errors in ARCH processes. *Applied Financial Economics Letters*, 3, 31–37.

Degiannakis, S. and Xekalaki, E. (2008). SPEC model selection algorithm for ARCH models: An options pricing evaluation framework. *Applied Financial Economics Letters*, 4(6), 419–423.

Degiannakis, S., Livada, A. and Panas, E. (2008). Rolling-sampled parameters of ARCH and Levy-stable models. *Journal of Applied Economics*, 40(23), 3051–3067.

Delbaen, F. (2002). Coherent risk measures on general probability spaces. In K. Sandmann and P.J. Schönbucher (eds), *Advances in Finance and Stochastics: Essays in Honour of Dieter Sondermann*, pp. 1–38. Berlin: Springer.

Dellaportas, P. and Roberts, G.O. (2003). An introduction to MCMC. In J. Muller (ed.), *Spatial Statistics and Computational Methods*, pp. 1–42. New York: Springer-Verlag.

Demeterfi, K., Derman, E., Kamal, M. and Zhou, J. (1999). More than you ever wanted to know about volatility swaps. *Goldman Sachs Quantitative Strategies*, Research Notes.

Demos, A. (2002). Moments and dynamic structure of a time-varying-parameter stochastic volatility in mean model. *Econometrics Journal*, 5, 345–357.

Demos, A. and Sentana, E. (1998). Testing for GARCH effects: A one-sided approach. *Journal of Econometrics*, 86(1), 97–127.

Dempster, A.P., Laird, N.M. and Rubin, D.R. (1977). Maximum likelihood from incomplete data via the EM algorithm. *Journal of the Royal Statistical Society, Series B*, 39, 1–22.

Devaney, M. (2001). Time varying risk premia for real estate investment trusts: A GARCH-M model. *Quarterly Review of Economics and Finance*, 41, 335–346.

Dewachter, H. (2001). Can markov switching models replicate chartist profits in the foreign exchange market? *Journal of International Money and Finance*, 20, 25–41.

Diebold, F.X. (1986). Modeling the persistence of conditional variances: A comment. *Econometric Reviews*, 5, 51–56.

Diebold, F.X. and Lopez, J.A. (1995). Modeling volatility dynamics. In K. Hoover (ed.), *Macroeconometrics: Developments, Tensions and Prospects*, pp. 427–472. Boston: Kluwer Academic.

Diebold, F.X. and Lopez, J. (1996). Forecast evaluation and combination. In G.S. Maddala and C.R. Rao (eds), *Handbook of Statistics*, pp. 241–268. Amsterdam: North-Holland.

Diebold, F.X. and Mariano, R. (1995). Comparing predictive accuracy. *Journal of Business and Economic Statistics*, 13(3), 253–263.

Diebold, F.X. and Nerlove, M. (1989). The dynamics of exchange rate volatility: A multivariate latent factor ARCH model. *Journal of Applied Econometrics*, 4, 1–21.

Diebold, F.X. and Pauly, P. (1988). Endogenous risk in a portfolio balance rational expectations model of the Deutschmark-dollar rate. *European Economic Review*, 32, 27–53.

Diebold, F.X., Hickman, A., Inoue, A. and Schuermann, T. (1996). Converting 1-day volatility to h-day volatility: Scaling by $\sqrt{h}$ is worse than you think. Working paper, Department of Economics, University of Pennsylvania.

Diebold, F.X., Gunther, T.A. and Tay, A.S. (1998). Evaluating density forecasts with applications to financial risk management. *International Economic Review*, 39(4), 863–883.

Diebold, F.X., Hahn, J. and Tay, A.S. (1999). Multivariate density forecast evaluation and calibration in financial risk management: High-frequency returns on foreign exchange. *Review of Economics and Statistics*, 81, 661–673.

Dimson, E. (1979). Risk measurement when shares are subject to infrequent trading. *Journal of Financial Economics*, 7, 197–226.

Ding, Z. and Engle, R.F. (2001). Large scale conditional covariance matrix modeling, estimation and testing. Working paper, Department of Finance, Stern School of Business, New York University.

Ding, Z. and Granger, C.W.J. (1996). Modeling volatility persistence of speculative returns: A new approach. *Journal of Econometrics*, 73, 185–215.

Ding, Z., Granger, C.W.J. and Engle, R.F. (1993). A long memory property of stock market returns and a new model. *Journal of Empirical Finance*, 1, 83–106.

Domowitz, I. and Hakkio, C.S. (1985). Conditional variance and the risk premium in the foreign exchange market. *Journal of International Economics*, 19, 47–66.

Doob, J.L. (1949). Heuristic approach to the Kolmogorov–Smirnov theorems. *Annals of Mathematical Statistics*, 20, 393–403.

Doornik, J.A. (2001). *Ox 3.0 – An Object Oriented Matrix Programming Language*. London: Timberlake Consultants.

Doornik, J.A. and Ooms, M. (2001). A package for estimating, forecasting and simulating Arfima models: Arfima Package 1.01 for Ox. Working Paper, Nuffield College, Oxford.

Doornik, J.A. and Ooms, M. (2006). A package for estimating, forecasting and simulating Arfima models: Arfima Package 1.04 for Ox. Working Paper, Nuffield College, Oxford.

Dowd, K. (2005). *Measuring Market Risk*, 2nd edition. Chichester: John Wiley & Sons Ltd.

Dowd, K., Blake, D. and Cairns, A. (2004). Long-term value-at-risk. *Journal of Risk Finance*, 5(2), 52–57.

Drost, F.C. and Werker, B.J.M. (1996). Closing the GARCH gap: Continuous time GARCH modeling. *Journal of Econometrics*, 74, 31–57.

Drost, F.C., Nijman, T.E. and Werker, B.J.M. (1998). Estimation and testing in models containing both jumps and conditional heteroskedasticity. *Journal of Business and Economic Statistics*, 16(2), 237–243.

Duan, J. (1995). The Garch option pricing model. *Mathematical Finance*, 5(1), 31–32.

Duan, J. (1996). A unified theory of option pricing under stochastic volatility – from Garch to diffusion. Working Paper, Hong Kong University of Science and Technology.

Duan, J. (1997). Augmented GARCH(p,q) process and its diffusion limit. *Journal of Econometrics*, 79, 97–127.

Duan, J. and Simonato, J.G. (1998). Empirical martingale simulation for asset prices. *Management Science*, 44, 1218–1233.

Duan, J. and Simonato, J.G. (2001). American option pricing under GARCH by a Markov chain approximation. *Journal of Economic Dynamics and Control*, 25, 1689–1718.

Duan, J. and Wei, J. (1999). Pricing foreign currency and cross-currency options under GARCH. *Journal of Derivatives*, 7, 51–63.

Duan, J., Gauthier, G. and Simonato, J. (1999a). An analytical approximation for the Garch option pricing model. *Journal of Computational Finance*, 2, 75–116.

Duan, J., Gauthier, G. and Simonato, J.G. (1999b). Fast valuation of derivative contracts by simulation. Research report, HEC Montréal.

Duan, J., Gauthier, G., Sasseville, C. and Simonato, J.G. (2003). Approximating American option prices in the GARCH framework. *Journal of Futures Markets*, 23, 915–929.

Duan, J., Gauthier, G., Sasseville, C. and Simonato, J.G. (2006a). Approximating the GJR-GARCH and EGARCH option pricing models analytically. *Journal of Computational Finance*, 9(3).

Duan, J., Ritchken, P. and Sum, Z. (2006b). Approximating GARCH-jump models, jump-diffusion processes, and option pricing. *Mathematical Finance*, 16, 21–52.

Dudley, R.M. (1989). *Real Analysis and Probability*. Pacific Grove, CA: Wadsworth & Brooks/Cole.

Dueker, M.J. (1997). Markov switching in Garch processes and mean-reverting stock market volatility. *Journal of Business and Economic Statistics*, 15, 26–34.

Duffie, D. and Pan, J. (1997). An overview of value at risk. *Journal of Derivatives*, 4, 7–49.

Dumas, B., Fleming, J. and Whaley, R. (1998). Implied volatility functions: Empirical tests. *Journal of Finance*, 53, 2059–2106.

Ebens, H. (1999). Realized stock volatility. Working Paper 420, Department of Economics, Johns Hopkins University.

Einstein, A. (1905). Ueber die von der molecular-kinetischen Theorie der Wärme geforderte Bewegung von in ruhenden Flüssigkeiten suspendierten Teilchen. *Annalen der Physik*, 17, 549–560.

Elyasiani, E. and Mansur, I. (1998). Sensitivity of bank stock returns distribution to changes in the level and volatility of interest rates: A GARCH-M model. *Journal of Banking and Finance*, 22, 535–563.

Embrechts, P. (ed.) (2000). *Extremes and Integrated Risk Management*. London: Risk Books.

Enders, W. (2003). *Applied Econometric Time Series*, 2nd edition. Hoboken, NJ: John Wiley & Sons Inc.

Engel, C. (1994). Can the Markov switching model forecast exchange rates? *Journal of International Economics*, 36, 151–165.

Engel, C. and Hamilton, J.D. (1990). Long swings in the dollar: Are they in the data and do markets know it? *American Economic Review*, 80, 689–713.

Engle, R.F. (1982). Autoregressive conditional heteroskedasticity with estimates of the variance of U.K. inflation. *Econometrica*, 50, 987–1008.

Engle, R.F. (1983). Estimates of the Variance of US inflation based on the ARCH model. *Journal of Money Credit and Banking*, 15, 286–301.

Engle, R.F. (1984). Wald, likelihood ratio and Lagrange multiplier tests in econometrics. In Z. Griliches and M.D. Intriligator (eds), *Handbook of Econometrics*, Volume II. Amsterdam: North Holland.

Engle, R.F. (1987). Multivariate ARCH with factor structures – cointegration in variance. Discussion Paper 87–27, University of California, San Diego.

Engle, R.F. (1990). Discussion: Stock market volatility and the crash of '87. *Review of Financial Studies*, 3, 103–106.

Engle, R.F. (ed.) (1995). *ARCH: Selected Readings*. Oxford: Oxford University Press.

Engle, R.F. (2001a). GARCH 101: The use of ARCH/GARCH models in applied econometrics. *Journal of Economic Perspectives*, 15(4), 157–168.

Engle, R.F. (2001b). Financial econometrics – a new discipline with new methods. *Journal of Econometrics*, 100, 53–56.

Engle, R.F. (2002a). New frontiers for ARCH models. *Journal of Applied Econometrics*, 17, 425–446.

Engle, R.F. (2002b). Dynamic conditional correlation: A simple class of multivariate GARCH models. *Journal of Business and Economic Statistics*, 20, 339–350.

Engle, R.F. (2004). Risk and volatility: Econometric models and financial practice. *American Economic Review*, 94(3), 405–420.

Engle, R.F. and Bollerslev, T. (1986). Modelling the persistence of conditional variances. *Econometric Reviews*, 5(1), 1–50.

Engle, R.F. and González-Rivera, G. (1991). Semiparametric ARCH models. *Journal of Business and Economic Statistics*, 9, 345–359.

Engle, R.F. and Kraft, D. (1983). Multiperiod forecast error variances of inflation estimated from ARCH models. In A. Zellner (ed.), *Applied Time Series Analysis of Economic Data*, pp. 293–302. Washington, DC: US Dept. of Commerce, Bureau of the Census.

Engle, R.F. and Kroner, K.F. (1995). Multivariate simultaneous generalized ARCH. *Econometric Theory*, 11, 122–150.

Engle, R.F. and Lee, G.G.J. (1993). A permanent and transitory component model of stock return volatility. Discussion Paper 9244, Department of Economics, University of California, San Diego.

Engle, R.F. and Manganelli, S. (2004). CAViaR: Conditional autoregressive value at risk by regression quantiles. *Journal of Business and Economic Statistics*, 22(4), 367–381.

Engle, R.F. and Mustafa, C. (1992). Implied ARCH models from options prices. *Journal of Econometrics*, 52, 289–311.

Engle, R.F. and Ng, V.K. (1993). Measuring and testing the impact of news on volatility. *Journal of Finance*, 48, 1749–1778.

Engle, R.F. and Russell, J.R. (1998). Autoregressive conditional duration: A new model for irregularly spaced transaction data. *Econometrica*, 66, 1127–1162.

Engle, R.F. and Sheppard, K. (2001). Theoretical and empirical properties of dynamic conditional correlation multivariate GARCH. Working paper, NYU Stern School of Business and UCSD.

Engle, R.F. and Sun, Z. (2005). Forecasting volatility using tick by tick data. Paper presented to the European Finance Association, 32th Annual Meeting, Moscow.

Engle, R.F. and Susmel, R. (1990). Intraday mean and volatility relations between US and UK stock market returns. Manuscript, University of California, San Diego.

Engle, R.F., Granger, C.W.J. and Kraft, D. (1986). Combining competing forecasts of inflation using a bivariate ARCH model. *Journal of Economic Dynamics and Control*, 8, 151–165.

Engle, R.F., Lilien, D.M. and Robins, R.P. (1987). Estimating time varying risk premia in the term structure: The ARCH-M model. *Econometrica*, 55, 391–407.

Engle, R.F., Ito, T. and Lin, W.L. (1990a). Meteor showers or heat waves? Heteroskedastic intra-daily volatility in the foreign exchange market. *Econometrica*, 58, 525–542.

Engle, R.F., Ng, V.K. and Rothschild, M. (1990b). Asset pricing with a factor ARCH Covariance structure: Empirical estimates for Treasury bills. *Journal of Econometrics*, 45, 213–238.

Engle, R.F., Hong, C.H., Kane, A. and Noh, J. (1993). Arbitrage valuation of variance forecasts with simulated options. *Advances in Futures and Options Research*, 6, 393–415.

Engle, R.F., Kane, A. and Noh, J. (1997). Index-option pricing with stochastic volatility and the value of accurate variance forecasts. *Review of Derivatives Research*, 1, 120–144.

Ericsson, N.R. (1992). Parameter constancy, mean square forecast errors, and measuring forecast performance: An exposition, extensions, and illustration. *Journal of Policy Modelling*, 4, 465–495.

Espasa, A. and Albacete, R. (2007). Econometric modelling for short-term inflation forecasting in the euro area. *Journal of Forecasting*, 26(5), 303–316.

Fama, E.F. (1965). The behaviour of stock market prices. *Journal of Business*, 38, 34–105.

Fama, E.F. (1970). Efficient capital markets: A review of theory and empirical work. *Journal of Finance*, 25, 383–417.

Feller, W. (1948). On the Kolmogorov-Smirnov limit theorems for empirical distributions. *Annals of Mathematical Statistics*, 19, 177–189.

Feller, W. (1951). The asymptotic distribution of the range of sums of random variables. *Annals of Mathematical Statistics*, 22, 427–432.

Feng, Y., Beran, J. and Yu, K. (2007). Modelling financial time series with SEMIFAR–GARCH model. *Journal of Management Mathematics*, 18, 395–412.

Fernandez, C. and Steel, M. (1998). On Bayesian modeling of fat tails and skewness. *Journal of the American Statistical Association*, 93, 359–371.

Ferreira, M.A. and Lopez, J.A. (2005). Evaluating interest rate covariance models within a value-at-risk framework. *Journal of Financial Econometrics*, 3(1), 126–168.

Ferson, W.E. (1989). Changes in expected security returns, risk and the level of interest rates. *Journal of Finance*, 44, 1191–1218.

Fiorentini, G., Sentana, E. and Calzolari, G. (2003). Maximum likelihood estimation and inference in multivariate conditionally heteroskedastic dynamic regression models with Student t innovations. *Journal of Business and Economic Statistics*, 21, 532–546.

Fisher, L. (1966). Some new stock market indices. *Journal of Business*, 39, 191–225.

Fisher, R.A. (1924). The conditions under which X^2 measures the discrepancy between observations and hypothesis. *Journal of the Royal Statistical Society*, 87, 442–450.

Fleming, J. (1998). The quality of market volatility forecast implied by S&P 100 index option prices. *Journal of Empirical Finance*, 5, 317–345.

Fleming, J., Ostdiek, B. and Whaley, R. (1995). Predicting stock market volatility: A new measure. *Journal of Futures Markets*, 15(3), 265–302.

Fletcher, R. (1970). A new approach to variable metric algorithms. *Computer Journal*, 13(3), 317–322.

Fletcher, R. (1987). *Practical Methods of Optimization*, 2nd edition. Chichester: John Wiley & Sons Ltd.

Fletcher, R. and Powell, M.J.D. (1963). A rapidly convergent descent method for minimization. *Computer Journal*, 6, 163–168.

Fofana, N'Z.F. and Brorsen, B.W. (2001). GARCH option pricing with implied volatility. *Applied Economics Letters*, 8, 335–340.

Fong, W.M. (1998). The dynamics of DM/£ exchange rate volatility: A SWARCH analysis. *International Journal of Finance and Economics*, 3, 59–71.

Fornari, F. and Mele, A. (1995). Sign- and volatility-switching ARCH models: Theory and applications to international stock markets. Banca d'Italia Discussion Paper 251.

Fornari, F. and Mele, A. (1996). Modeling the changing asymmetry of conditional variances. *Economics Letters*, 50, 197–203.

Fornari, F., Monticelli, C., Pericoli, M. and Tivegna, M. (2002). The impact of news on the exchange rate of the lira and long-term interest rates. *Economic Modelling*, 19, 611–639.

Francq, C., Roussignol, M. and Zakoïan, J.-M. (2001). Conditional heteroskedasticity driven by hidden Markov chains. *Journal of Time Series Analysis*, 22, 197–220.

Franses, P.H. and Hafner, C.M. (2003). A generalised dynamic conditional correlation model for many asset returns. Econometric Institute Report EI 2003–18, Erasmus University, Rotterdam.

Franses, P.H. and Homelen, P.V. (1998). On forecasting exchange rates using neural networks. *Applied Financial Economics*, 8, 589–596.

Franses, P.H. and van Dijk, D. (2000). *Nonlinear Time Series Models in Empirical Finance*. Cambridge: Cambridge University Press.

French, K.R. and Roll, R. (1986). Stock return variances: The arrival of information and the reaction of traders. *Journal of Financial Economics*, 17, 5–26.

French, K.R., Schwert, G.W. and Stambaugh, R.F. (1987). Expected stock returns and volatility. *Journal of Financial Economics*, 19, 3–29.

Frennberg, P. and Hansson, B. (1996). An evaluation of alternative models for predicting stock volatility. Evidence from a small stock market. *Journal of International Financial Markets Institutions and Money*, 5, 117–134.

Frey, R. and Michaud, P. (1997). The effect of GARCH-type volatilities on prices and payoff-distributions of derivative assets – a simulation study. Working Paper, ETH Zurich.

Friedmann, R. and Sanddorf-Köhle, W.G. (2002). Volatility clustering and non-trading days in Chinese stock markets. *Journal of Economics and Business*, 54, 193–217.

Gallant, A.R. and Tauchen, G. (1989). Semi non-parametric estimation of conditional constrained heterogeneous processes: Asset pricing applications. *Econometrica*, 57, 1091–1120.

Gallant, A.R., Hsieh, D. and Tauchen, G. (1991). On fitting a recalcitrant series: The pound/dollar exchange rate 1974-83. In W.A. Barnett, J. Powell and G. Tauchen (eds), *Nonparametric and Semiparametric Methods in Econometrics and Statistics*, Cambridge: Cambridge University Press.

Gallant, A.R., Rossi, P. and Tauchen, G. (1993). Nonlinear dynamic structures. *Econometrica*, 61, 871–907.

Garman, M.B. and Klass, M.J. (1980). On the estimation of security price volatilities from historical data. *Journal of Business*, 53(1), 67–78.

Gau, Y.-F. and Tang, W.-T. (2004). Forecasting value-at-risk using the markov-switching ARCH model. Econometric Society 2004 Far Eastern Meetings.

Gençay, R. and Selçuk, F. (2004). Extreme value theory and value-at- risk: Relative performance in emerging markets. *International Journal of Forecasting*, 20(2), 287–303.

Geweke, J. (1986). Modeling the persistence of conditional variances: A comment. *Econometric Reviews*, 5, 57–61.

Geweke, J. (1988a). Comments on Poirier: Operational Bayesian methods in econometrics. *Journal of Economic Perspectives*, 2, 159–166.

Geweke, J. (1988b). Exact inference in models with autoregressive conditional heteroskedasticity. In W.A. Barnett, E.R. Berndt and H., White (eds), *Dynamic Econometric Modeling*, Cambridge: Cambridge University Press.

Geweke, J. (1989). Exact predictive densities in linear models with Arch distrubances. *Journal of Econometrics*, 44, 307–325.

Geweke, J. (1996). Monte Carlo simulation and numerical integration. In H. Amman, D. Kendrick and J., Rust (eds), *Handbook of Computational Economics*, pp. 731–800. Amsterdam: North-Holland.

Ghose, D. and Kroner, K.F. (1995). The relationship between GARCH and symmetric stable processes: Finding the source of fat tails in financial data. *Journal of Empirical Finance*, 2, 225–251.

Ghysels, E., Harvey, A.C. and Renault, E. (1996). Stochastic volatility. In G.S. Maddala (ed.), *Handbook of Statistics, Vol. 14: Statistical Methods in Finance*, pp. 119–191. Amsterdam: North Holland.

Gill, P.E., Murray, W. and Wright, M.H. (1981). *Practical Optimization*. New York: Academic Press.

Giot, P. (2005). Market risk models for intraday data. *European Journal of Finance*, 11, 309–324.

Giot, P. and Laurent, S. (2003a). Value-at-risk for long and short trading positions. *Journal of Applied Econometrics*, 18, 641–664.

Giot, P. and Laurent, S. (2003b). Market risk in commodity markets: A VaR approach. *Energy Economics*, 25, 435–457.

Giot, P. and Laurent, S. (2004). Modelling daily value-at-risk using realized volatility and ARCH type models. *Journal of Empirical Finance*, 11, 379–398.

Giovannini, A. and Jorion, P. (1989). The time variation of risk and return in the foreign exchange and stock markets. *Journal of Finance*, 44, 307–325.

Giraitis, L. and Robinson, P.M. (2000). Whittle estimation of ARCH models. *Econometric Theory*, 17, 608–631.

Glejser, H. (1969). A new test for heteroscedasticity. *Journal of the American Statistical Association*, 64, 316–323.

Glosten, L., Jagannathan, R. and Runkle, D. (1993). On the relation between the expected value and the volatility of the nominal excess return on stocks. *Journal of Finance*, 48, 1779–1801.

Gnedenko, B.V. (1952). Some results on the maximum discrepancy between two empirical distributions. in *Selected Translations in Mathematical Statistics and Probability, 1961*, pp. 73–75. Providence, RI: American Mathematical Society.

Gnedenko, B.V. and Rvaceva, E.R. (1952). On a problem of the comparison of two distributions. In *Selected Translations in Mathematical Statistics and Probability, 1961*, pp. 69–72. Providence, RI: American Mathematical Society.

Godfrey, L.G. (1978). Testing for multiplicative heteroscedasticity. *Journal of Econometrics*, 8, 227–236.

Goffe, W.L., Ferrier, G.D. and Rogers, J. (1994). Global optimization of statistical functions with simulated annealing. *Journal of Econometrics*, 60(1/2), 65–99.

Goldfarb, D. (1970). A family of variable metric methods derived by variational means. *Mathematics of Computation*, 24(109), 23–26.

González-Rivera, G. (1996). Smooth transition GARCH models. Working paper, Department of Economics, University of California, Riverside.

González-Rivera, G., Lee, T.-H. and Mishra, S. (2004). Forecasting volatility: A reality check based on option pricing, utility function, value-at-risk and predictive likelihood. *International Journal of Forecasting*, 20, 629–645.

Gouriéroux, C. (1997). *ARCH Models and Financial Applications*. New York: Springer,

Gouriéroux, C. and Jasiak, J. (2001). *Financial Econometrics: Problems, Models, and Methods*. Princeton, NJ: Princeton University Press.

Gouriéroux, C. and Monfort, A. (1992). Qualitative threshold ARCH models. *Journal of Econometrics*, 52, 159–199.

Grammig, J. and Maurer, K.-O. (2000). Non-monotonic hazard functions and the autoregressive conditional duration model. *Econometrics Journal*, 3, 16–38.

Granger, C.W.J. (1980). Long memory relationships and the aggregation of dynamic models. *Journal of Econometrics*, 14, 227–238.

Granger, C.W.J. (1981). Some properties of time series data and their use in econometric model specification. *Journal of Econometrics*, 16, 121–130.

Granger, C.W.J. (1989). Combining forecasts – twenty years later. *Journal of Forecasting*, 8, 167–173.

Granger, C.W.J. and Joyeux, R. (1980). An introduction to long memory time series models and fractional differencing. *Journal of Time Series Analysis*, 1, 15–39.

Granger, C.W.J. and Newbold, P. (1973). Some comments on the evaluation of economic forecasts. *Applied Economics*, 5, 35–47.

Granger, C.W.J. and Newbold, P. (1977). *Forecasting Economic Time Series*. New York: Academic Press,

Granger, C.W.J. and Pesaran, M.H. (2000a). Economic and statistical measures of forecasting accuracy. *Journal of Forecasting*, 19, 537–560.

Granger, C.W.J. and Pesaran, M.H. (2000b). A decision theoretical approach to forecast evaluation. In W.S. Chow, W.K. Li and K., Tong (eds), *Statistics and Finance: An Interface*, pp. 261–278. London: Imperial College Press.

Gray, S.F. (1996). Modeling the conditional distribution of interest rates as a regime-switching process. *Journal of Financial Economics*, 42, 27–62.

Greene, W.H. (1997). *Econometric Analysis*, 3rd edition. Upper Saddle River, NJ: Prentice Hall.

Groenendijk, P.A., Lucas, A. and de Vries, C.G. (1995). A note on the relationship between GARCH and symmetric stable processes. *Journal of Empirical Finance*, 2, 253–264.

Guermat, C. and Harris, R.D.F. (2002). Forecasting value at risk allowing for time variation in the variance and kurtosis of portfolio returns. *International Journal of Forecasting*, 18, 409–419.

Guidolin, M. and Timmermann, A. (2003). Value at risk and expected shortfall under regime switching. Working paper, University of Virginia and University of California at San Diego.

Gultekin, B., Rogalski, R. and Tinic, S. (1982). Option pricing model estimates: Some empirical results. *Financial Management*, 11, 58–69.

Haas, M., Mittnik, S. and Paolella, M.S. (2004). A new approach to Markov switching GARCH models. *Journal of Financial Econometrics*, 2(4), 493–530.

Hacker, S.R. and Abdulnasser, H.-J. (2005). A test for multivariate ARCH effects. *Applied Economics Letters*, 12(7), 411–417.

Hagerman, R. (1978). More evidence on the distribution of security returns. *Journal of Finance*, 33, 1213–1220.

Hagerud, G.E. (1996). A smooth transition Arch model for asset returns. Working Paper Series in Economics and Finance 162, Department of Finance, Stockholm School of Economics.

Hall, P. and Yao, Q. (2003). Inference in ARCH and GARCH models with heavy-tailed errors. *Econometrica*, 71, 285–317.

Hamilton, J.D. (1988). Rational expectations econometric analysis of changes in regime. An investigation of the term structure of interest rates. *Journal of Economic Dynamics and Control*, 12, 385–423.

Hamilton, J.D. (1989). A new approach to the economic analysis of non-stationary time series and the business cycle. *Econometrica*, 57, 357–384.

Hamilton, J.D. (1994). *Time Series Analysis*. Princeton, NJ: Princeton University Press.

Hamilton, J.D. and Susmel, R. (1994). Autoregressive conditional heteroskedasticity and changes in regime. *Journal of Econometrics*, 64, 307–333.

Hamilton, M.A. (1991). Model validation: An annotated bibliography. *Communications in Statistics – Theory and Methods*, 20(7), 2207–2266.

Hannan, E.J. and Quinn, B.G. (1979). The determination of the order of an autoregression. *Journal of the Royal Statistical Society, Series B*, 41, 190–195.

Hansen, B.E. (1994). Autoregressive conditional density estimation. *International Economic Review*, 35, 705–730.

Hansen, C.B., McDonald, J.B. and Theodossiou, P. (2003). Flexible parametric distributions for financial data. Paper presented at the Annual Meetings of the American Statistical Association, San Francisco.

Hansen, C.B., McDonald, J.B. and Newey, W. (2007a). Instrumental variables estimation with flexible distributions. CeMMAP Working Paper CWP21/07, Centre for Microdata Methods and Practice, Institute for Fiscal Studies.

Hansen, C.B., McDonald, J.B. and Theodossiou, P. (2007b). Some flexible parametric models for partially adaptive estimators of econometric models. *Economics: The Open-Access, Open-Assessment E-Journal*, 1, 2007–7 http://www.economics-ejournal.org/economics/journalarticles/2007–7.

Hansen, L. and Hodrick, R.J. (1980). Forward exchange rates as optimal predictors of future spot rates: An econometric analysis. *Journal of Political Economy*, 88, 829–853.

Hansen, P.R. (2005). A test for superior predictive ability. *Journal of Business and Economic Statistics*, 23, 365–380.

Hansen, P.R. and Lunde, A. (2005a). A forecast comparison of volatility models: Does anything beat a GARCH(1,1)? *Journal of Applied Econometrics*, 20(7), 873–889.

Hansen, P.R. and Lunde, A. (2005b). A realized variance for the whole day based on intermittent high-frequency data. *Journal of Financial Econometrics*, 3(4), 525–554.

Hansen, P.R. and Lunde, A. (2006). Consistent ranking of volatility models. *Journal of Econometrics*, 131, 97–121.

Hansen, P.R., Kim, J. and Lunde, A. (2003). Testing for superior predictive ability using Ox. A manual for SPA for Ox. Technical Document. http://www.stanford.edu./~prhansen/software/SPA.html.

Hansen, P.R., Lunde, A. and Nason, J.M. (2005). Model confidence sets for forecasting models. Working Paper 2005–7, Federal Reserve Bank of Atlanta. http://ssrn.com/abstract=522382.

Härdle, W. and Hafner, C.M. (2000). Discrete time option pricing with flexible volatility estimation. *Finance and Stochastics*, 4(2), 189–207.

Harmantzis, C.F., Miao, L. and Chien, Y. (2006). Empirical study of value-at-risk and expected shortfall models with heavy tails. *Journal of Risk Finance*, 7(2), 117–135.

Harris, R.D.F. and Sollis, R. (2003). *Applied Time Series Modelling and Forecasting*. Chichester: John Wiley & Sons Ltd.

Harris, R.D.F. and Küçüközmen, C.C. (2001). The empirical distribution of UK and US stock returns. *Journal of Business Finance and Accounting*, 28(5–6), 715–740.

Harvey, A.C. (1976). Estimating regression models with multiplicative heteroscedasticity. *Econometrica*, 44, 461–465.

Harvey, A.C. (1981). *The Econometric Analysis of Time Series*. Oxford: Philip Allan.

Harvey, A.C. and Shephard, N. (1993). The econometrics of stochastic volatility. Discussion Paper 166, London School of Economics.

Harvey, A.C., Ruiz, E. and Sentana, E. (1992). Unobserved component time series models with ARCH disturbances. *Journal of Econometrics*, 52, 129–157.

Harvey, D.I., Leybourne, S.J. and Newbold, P. (1997). Testing the equality of prediction mean squared errors. *International Journal of Forecasting*, 13, 281–291.

Harvey, D.I., Leybourne, S.J. and Newbold, P. (1998). Tests for forecast encompassing. *Journal of Business and Economic Statistics*, 16(2), 254–259.

Hauksson, H.A. and Rachev, S.T. (2001). The GARCH-stable option pricing model. *Mathematical and Computer Modelling*, 34, 1199–1212.

He, C. and Teräsvirta, T. (1999a). Properties of moments of a family of GARCH processes. *Journal of Econometrics*, 92, 173–192.

He, C. and Teräsvirta, T. (1999b). Fourth moment structure of the GARCH(p,q) model. *Econometric Theory*, 15, 824–846.

He, C., Teräsvirta, T. and Malmsten, H. (2002). Fourth moment structure of a family of first-order exponential GARCH models. *Econometric Theory*, 18, 868–885.

Hecq, A. (1996). IGARCH effect on autoregressive lag length selection and causality tests. *Applied Economics Letters*, 3, 317–323.

Hentschel, L. (1995). All in the family: Nesting symmetric and asymmetric GARCH models. *Journal of Financial Economics*, 39, 71–104.

Hertz, J., Krogh, A. and Palmer, R. (1991). *Introduction to the Theory of Neural Computation*. Reading, MA: Addison-Wesley.

Heston, S.L. and Nandi, S. (2000). A closed-form GARCH option valuation model. *Review of Financial Studies*, 13(3), 585–625.

Heynen, R. and Kat, H.M. (1994). Volatility prediction: A comparison of the stochastic volatility, Garch(1,1) and Egarch(1,1) models. *Journal of Derivatives*, 94, 50–65.

Higgins, M.L. and Bera, A.K. (1992). A class of nonlinear ARCH models. *International Economic Review*, 33, 137–158.

Hill, R.C., Griffiths, W.E. and Judge, G.G. (2000). *Undergraduate Econometrics*, 2nd edition. New York: John Wiley & Sons Inc.

Ho, L.-C., Burridge, P., Cadle, J. and Theobald, M. (2000). Value-at-risk: Applying the extreme value approach to Asian markets in the recent financial turmoil. *Pacific-Basin Finance Journal*, 88, 249–275.

Hodrick, R.J. (1989). Risk, uncertainty and exchange rates. *Journal of Monetary Economics*, 23, 433–459.

Hol, E. and Koopman, S. (2000). Forecasting the variability of stock index returns with stochastic volatility models and implied volatility. Tinbergen Institute Discussion Paper No. 104,4.

Holden, A. (1986). *Chaos*, Princeton, NJ: Princeton University Press.

Holton, G.A. (2003). *Value-at-Risk: Theory and Practice*. San Diego, CA: Academic Press.

Hong, Y. and Lee, J. (2001). One-sided testing for ARCH effects using wavelets. *Econometric Theory*, 17(6), 1051–1081.

Hoppe, R. (1998). VAR and the unreal world. *Risk*, 11, 45–50.

Hoppe, R. (1999). Finance is not physics. *Risk Professional*, 1(7).

Hosking, J.R.M. (1981). Fractional differencing. *Biometrica*, 68, 165–176.

Hsieh, D. (1988). The statistical properties of daily foreign exchange rates: 1974–1983. *Journal of International Economics*, 24, 129–145.

Hsieh, D. (1989). Modeling heteroscedasticity in daily foreign-exchange rates. *Journal of Business and Economic Statistics*, 7(3), 307–317.

Hsieh, D. (1991). Chaos and nonlinear dynamics: application to financial markets. *Journal of Finance*, 46, 1839–1877.

Hsieh, K.C. and Ritchken, P. (2005). An empirical comparison of GARCH option pricing models. *Review of Derivatives Research*, 8(3), 129–150.

Huang, Y.C. and Lin, B.-J. (2004). Value-at-risk analysis for Taiwan stock index futures: Fat tails and conditional asymmetries in return innovations. *Review of Quantitative Finance and Accounting*, 22, 79–95.

Hull, J. and White, A. (1987). The pricing of options on assets with stochastic volatility. *Journal of Finance*, 42, 281–300.

Hull, J. and White, A. (1998). Incorporating volatility updating into the historical simulation method for VaR. *Journal of Risk*, 1, 5–19.

Hurst, H. (1951). The long term storage capacity of reservoirs. *Transactions of the American Society of Civil Engineers*, 116, 770–799.

Hutchinson, J., Lo, A. and Poggio, T. (1994). A nonparametric approach to the pricing and hedging of derivative securities via learning networks. *Journal of Finance*, 49, 851–889.

Hwang, Y. (2001). Asymmetric long memory GARCH in exchange return. *Economics Letters*, 73, 1–5.

Inui, K. and Kijima, M. (2005). On the significance of expected shortfall as a coherent risk measure. *Journal of Banking and Finance*, 29, 853–864.

Ioannides, M., Kat, H.M. and Theodossiou, P. (2002). Fitting distributions to hedge fund index returns. Paper presented to the Tenth Annual Conference of the Multinational Finance Society, Montreal, Canada, MFC104.

Ishida, I. (2005). Scanning multivariate conditional densities with probability integral transforms. CIRJE F-Series 369, Faculty of Economics, University of Tokyo.

Jackson, P., Maude, D.J. and Perraudin, W. (1998). Testing value-at-risk approaches to capital adequacy. *Bank of England Quarterly Bulletin*, 38, 256–266.

Jacquier, E., Polson, N. and Rossi, P. (1994). Bayesian analysis of stochastic volatility models. *Journal of Business and Economic Statistics*, 12(4), 371–417.

Jacquier, E., Polson, N. and Rossi, P. (1999). Stochastic volatility: Univariate and multivariate extensions. *CIRANO, Scientific Series*, 99s-26.

Jacquier, E., Polson, N. and Rossi, P. (2004). Bayesian analysis of stochastic volatility models with fat tails and correlated errors. *Journal of Econometrics*, 122, 185–212.

Jarque, C.M. and Bera, A.K. (1980). Efficient tests for normality, heteroscedasticity and serial independence of regression residuals. *Economic Letters*, 6, 255–259.

Jarque, C.M. and Bera, A.K. (1987). A test for normality of observations and regression residuals. *International Statistical Review*, 55(2), 163–172.

Jasic, T. and Wood, D. (2004). The profitablity of daily stock market indices trades based on neural network predictions: Case study for the S&P500, the DAX, the TOPIX and the FTSE in the period 1965–1999. *Applied Financial Economics*, 14, 285–297.

Jeantheau, T. (1998). Strong consistency of estimators for multivariate ARCH models. *Econometric Theory*, 14, 70–86.

Johnson, N.L., Kotz, S. and Balakrishnan, N. (1995). *Continuous Univariate Distributions*, Vol. 2, 2nd edition. New York: John Wiley & Sons Inc.

Jondeau, E. and Rockinger, M. (2001). Gram–Charlier densities. *Journal of Economic Dynamics & Control*, 25, 1457–1483.

Jondeau, E. and Rockinger, M. (2003). Testing for differences in the tails of stock-market returns. *Journal of Empirical Finance*, 10, 559–581.

Jorion, P. (1988). On jump processes in the foreign exchange and stock markets. *Review of Financial Studies*, 1, 427–445.

Jorion, P. (1995). Predicting volatility in the foreign exchange market. *Journal of Finance*, 50, 507–582.

Jorion, P. (2006). *Value at Risk. The New Benchmark for Managing Financial Risk*. New York: McGraw-Hill.

Judge, G.G., Griffiths, W.E., Hill, R.C., Lütkepohl, H. and Lee, T.-C. (1985). *The Theory and Practice of Econometrics*. New York: John Wiley & Sons Inc.

Kallsen, J. and Taqqu, M.S. (1998). Option pricing in Arch-type models. *Mathematical Finance*, 8(1), 13–26.

Kapetanios, G., Labhard, V. and Price, S. (2006). Forecasting using predictive likelihood model averaging. *Economics Letters*, 91, 373–379.

Karanasos, M. (1996). A theoretical and empirical comparison of the GARCH(1,1), the two-state Markov switching ARCH models. Mimeo, Department of Economics, Birkbeck College, London.

Karanasos, M. (1999). The second moment and the autoregressive function of the squared errors of the GARCH model. *Journal of Econometrics*, 90, 63–76.

Karanasos, M. and Kim, J. (2003). Moments of the ARMA–EGARCH model. *Econometrics Journal*, 6, 146–166.

Karanasos, M. and Kim, J. (2006). A re-examination of the asymmetric power ARCH model. *Journal of Empirical Finance*, 13, 113–128.

Kaufmann, S. and Frühwirth-Schnatter, S. (2002). Bayesian analysis of switching ARCH models. *Journal of Time Series Analysis*, 23(4), 425–458.

Kawakatsu, H. (2006). Matrix exponential GARCH. *Journal of Econometrics*, 134, 95–128.

Kayahan, B., Saltoglu, B. and Stengos, T. (2002). Intra-day features of realized volatility: Evidence from an emerging market. *International Journal of Business and Economics*, 1(1), 17–24.

Kendal, M.G. and Stuart, A. (1969). *The Advanced Theory of Statistics*, Vol. 1, 3rd edition. London: Charles Griffin.

Kibble, W.F. (1941). A two variate gamma type distribution. *Sankhyā*, 5, 137–150.

Kim, C.M. (1989). Volatility effect on time series behaviour of exchange rate changes. Working paper, Korea Institute for International Economic Policy.

Kim, D. and Kon, S.J. (1999). Structural change and time dependence of stock returns. *Journal of Empirical Finance*, 6, 283–308.

Kim, J. (2000). The relationship between the monetary regime and output volatility: A multivariate GARCH-M model of the Japanese experience, 1919–1996. *Japan and the World Economy*, 12, 49–69.

Kim, Y.H., Rachev, S.T. and Chung, D.M. (2007). The modified tempered stable distribution, GARCH models and option pricing. Technical report, Chair of Econometrics, Statistics and Mathematical Finance School of Economics and Business Engineering, University of Karlsruhe.

King, M., Sentana, E. and Wadhwani, S. (1994). Volatility and links between national stock markets. *Econometrica*, 62(4), 901–933.

Klaassen, F. (2002). Improving GARCH volatility forecasts with regime-switching GARCH. *Empirical Economics*, 27(2), 363–394.

Knight, J. and Satchell, S.E. (2002a). GARCH processes – some exact results, some difficulties and a suggested remedy. In J. Knight and S.E. Satchell (eds), *Forecasting Volatility in the Financial Markets*, Oxford: Butterworth-Heinemann.

Knight, J. and Satchell, S.E. (2002b). GARCH predictions and the predictions of option prices. In J. Knight and S.E. Satchell (eds), *Forecasting Volatility in the Financial Markets*, Oxford: Butterworth-Heinemann.

Kodres, L.E. (1988). Tests of unbiasedness in foreign exchange futures markets: The effects of price limits. *Review of Futures Markets*, 7, 138–166.

Kodres, L.E. (1993). Tests of unbiasedness in the foreign exchange futures markets: An examination of price limits and conditional heteroskedasticity. *Journal of Business*, 66(3), 463–490.

Koenker, R. and Bassett, G. (1978). Regression quantiles. *Econometrica*, 46(1), 33–50.

Kolmogorov, A.N. (1933). Sulla deternimazione empirica di una legge di distribuzione. *Giornale dell'Istituto Italiano degli Attuari*, 4, 83–91.

König, H. and Gaab, W. (1982). *The Advanced Theory of Statistics, Vol. 2 of Inference and Relationships*. New York: Haffner.

Koopman, S., Jungbacker, B. and Hol, E. (2005). Forecasting daily variability of the S&P100 stock index using historical, realised and implied volatility measurements. *Journal of Empirical Finance*, 12, 445–475.

Kroner, K.F. and Claessens, S. (1991). Optimal currency composition of external debt: Applications to Indonesia and Turkey. *Journal of International Money and Finance*, 10, 131–148.

Kroner, K.F. and Lastrapes, W.D. (1991). The impact of exchange rate volatility on international trade: Estimates using the GARCH-M Model. Manuscript, University of Arizona.

Kroner, K.F. and Ng, V.K. (1998). Modeling asymmetric comovements of asset returns. *Review of Financial Studies*, 11(4), 817–844.

Kroner, K.F. and Sultan, J. (1991). Exchange rate volatility and time varying hedge ratios. In S.G. Rhee and R.P. Change (eds), *Pacific-Basin Capital Markets Research*, Vol. II. Amsterdam: North Holland.

Kroner, K.F. and Sultan, J. (1993). Time varying distribution and dynamic hedging with foreign currency futures. *Journal of Financial and Quantitative Analysis*, 28, 535–551.

Kuester, K., Mittnik, S. and Paolella, M.S. (2006). Value-at-risk prediction: A comparison of alternative strategies. *Journal of Financial Econometrics*, 4(1), 53–89.

Kuiper, N.H. (1960). Tests concerning random points on a circle. *Koninklijke Nederlandse Akademie van Wetenschappen, Proceedings Series A*, 63, 38–47.

Kullback, S. and Leibler, R.A. (1951). On information and sufficiency. *Annals of Mathematical Statistics*, 22, 79–86.

Kunitomo, N. (1992). Improving the Parkinson method of estimating security price volatilities. *Journal of Business*, 65, 295–302.

Kupiec, P.H. (1995). Techniques for verifying the accuracy of risk measurement models. *Journal of Derivatives*, 3, 73–84.

Kwan, C.K., Li, W.K. and Ng, K. (2010). A multivariate threshold GARCH model with time-varying correlations. *Econometric Reviews*, 29(1), 20–38.

Lai, K.S. and Pauly, P. (1988). Time series properties of foreign exchange rates re-examined. Manuscript, University of Pennsylvania.

Lambadiaris, G., Papadopoulou, L., Skiadopoulos, G. and Zoulis, Y. (2003). VAR: History or simulation? *Risk*, 16(9), 122–127.

Lambert, P. and Laurent, S. (2000). Modeling skewness dynamics in series of financial data. Discussion Paper, Institut de Statistique, Louvain-la-Neuve.

Lambert, P. and Laurent, S. (2001). Modeling financial time series using GARCH-type models and a skewed Student density. Mimeo, Université de Liège.

Lamoureux, G.C. and Lastrapes, W.D. (1990a). Heteroskedasticity in stock return data: Volume versus GARCH effects. *Journal of Finance*, 45, 221–229.

Lamoureux, G.C. and Lastrapes, W.D. (1990b). Persistence in variance, structural change, and the GARCH model. *Journal of Business and Economic Statistics*, 8, 225–234.

Lanne, M. (2006). A mixture multiplicative error model for realized volatility. *Journal of Financial Econometrics*, 4(4), 594–616.

Lanne, M. and Saikkonen, P. (2007). A multivariate generalized orthogonal factor GARCH model. *Journal of Business and Economic Statistics*, 25, 61–75.

Latane, H.A. and Rendleman, R.J. (1976). Standard deviations of stock price ratios implied in option prices. *Journal of Finance*, 31, 369–381.

Laurent, S. (2007). *Estimating and Forecasting ARCH Models Using G@RCH 5*. London: Timberlake Consultants Press.

Laurent, S. and Peters, J.-P. (2002a). G@RCH 2.2: An Ox package for estimating and forecasting various ARCH models. *Journal of Economic Surveys*, 16, 447–485.

Laurent, S. and Peters, J.-P. (2002b). A tutorial for G@RCH 2.3, a complete Ox package for estimating and forecasting ARCH models. Mimeo, Department of Economics, University of Liège, Belgium, and Department of Quantitative Economics, University of Amsterdam.

Laurent, S. and Peters, J.-P. (2004). *G@RCH 4.0, Estimating and Forecasting ARCH Models*. London: Timberlake Consultants Press.

Laurent, S. and Peters, J.-P. (2006). *G@RCH 4.2, Estimating and Forecasting ARCH Models*. London: Timberlake Consultants Press.

Laurent, S., Rombouts, J.V.K., Silvennoinen, A. and Violante, F. (2008). Comparing and ranking covariance stractures of M-GARCH volatility models. Working paper, HEC Montreal.

Laux, P.A. and Ng, L.K. (1993). The sources of GARCH: Empirical evidence from an intraday returns model incorporating systematic and unique risks. *Journal of International Money and Finance*, 12, 543–560.

Lawrence, C. and Tits, A. (2001). A computationally efficient feasible sequential quadratic programming algorithm. *SIAM Journal on Optimisation*, 11(4), 1092–1118.

LeBaron, B. (1992). Some relations between volatility and serial correlations in stock market returns. *Journal of Business*, 65(2), 199–219.

Lebedev, N.N. (1972). *Special Functions and their Applications*. New York: Dover.

Lee, J. (1999). The inflation and output variability trade-off: evidence from a GARCH model. *Economics Letters*, 62, 63–67.

Lee, L.F. (1999). Estimation of dynamic and ARCH Tobit models. *Journal of Econometrics*, 92, 355–390.

Lee, S.W. (1991). Asymptotic properties of the maximum likelihood estimator of the GARCH-M and I-GARCH-M models. Mimeo, Department of Economics, University of Rochester.

Lee, S.W. and Hansen, B.E. (1991). Asymptotic properties of the maximum likelihood estimator and test of the stability of parameters of the GARCH and I-GARCH models. Mimeo, Department of Economics, University of Rochester.

Lee, S.W. and Hansen, B.E. (1994). Asymptotic theory for the GARCH(1,1) quasi-maximum likelihood estimator. *Econometric Theory*, 10, 29–52.

Lee, T.K.Y. and Tse, Y.K. (1991). Term structure of interest rates in the Singapore Asian dollar market. *Journal of Applied Econometrics*, 6, 143–152.

Lewis, P.A.W. (1961). Distribution of the Anderson-Darling statistic. *Annals of Mathematical Statistics*, 32(4), 1118–1124.

Li, J. (2007). Stock return and higher conditional moments: Evidence from 30 Dow-Jones Industrial stocks. Working paper, Department of Finance, Drexel University.

Li, M.-Y.L. and Lin, H.-W.W. (2004). Estimating value-at-risk via Markov switching ARCH models. An empirical study on stock index returns. *Applied Economics Letters*, 11, 679–691.

Li, W.K. and Mak, T.K. (1994). On the squared residual autocorrelations in non-linear time series with conditional heteroscedasticity. *Journal of Time Series Analysis*, 15, 627–636.

Li, W.K., Ling, S. and McAleer, M. (2001). A survey of recent theoretical results for time series models with GARCH errors. Discussion Paper 545, Institute of Social and Economic Research, Osaka University, Japan.

Lien, D. and Tse, Y.K. (1998). Hedging time varying downside risk. *Journal of Futures Markets*, 18, 705–722.

Lilliefors, H.W. (1967). On the Kolmogorov-Smirnov test for normality with mean and variance unknown. *Journal of the American Statistical Association*, 62, 399–402.

Lilliefors, H.W. (1969). On the Kolmogorov-Smirnov test for the exponential distribution with mean unknown. *Journal of the American Statistical Association*, 64, 387–389.

Lin, B.H. and Yeh, S.-K. (2000). On the distribution and conditional heteroscedasticity in Taiwan stock prices. *Journal of Multinomial Financial Management*, 10, 367–395.

Lin, W.L. (1992). Alternative estimators for factor GARCH Models – a Monte Carlo comparison. *Journal of Applied Econometrics*, 7, 259–279.

Ling, S. and Li, W.K. (1997). Diagnostic checking of nonlinear multivariate time series with multivariate ARCH errors. *Journal of Time Series Analysis*, 18, 447–464.

Ling, S. and McAleer, M. (2001). Necessary and sufficient moment conditions for the GARCH (r,s) and asymmetric power GARCH(r,s) models. *Econometric Theory*, 18, 722–729.

Liu, S. and Brorsen, B.W. (1995). Maximum likelihood estimation of a GARCH-stable model. *Journal of Applied Econometrics*, 10, 273–285.

Liu, Q. and Morimune, K. (2005). A modified GARCH model with spells of shocks. *Asia-Pacific Financial Markets*, 12(1), 29–44.

Ljung, G.M. and Box, G.E.P. (1978). On a measure of lack of fit in time series models. *Biometrica*, 65, 297–303.

Lo, A. and MacKinlay, A.C. (1988). Stock market prices do not follow random walks: Evidence from a simple specification test. *Review of Financial Studies*, 1, 41–66.

Lo, A. and MacKinlay, A.C. (1990a). An econometric analysis of non-synchronous trading. *Journal of Econometrics*, 45, 181–212.

Lo, A. and MacKinlay, A.C. (1990b). When are contrarian profits due to stock market overreaction? *Review of Financial Studies*, 3, 175–205.

Locke, P.R. and Sayers, C.L. (1993). Intra-day futures prices volatility: Information effects and variance persistence. *Journal of Applied Econometrics*, 8, 15–30.

Longin, F. (2000). From value-at-risk to stress testing: The extreme value approach. *Journal of Banking and Finance*, 24, 1097–1130.

Longin, F. and Solnik, B. (1995). Is the correlation in international equity returns constant: 1960–1990? *Journal of International Money and Finance*, 14(1), 3–26.

Lopez, J.A. (1999). Methods for evaluating value-at-risk estimates. *Economic Policy Review*, Federal Reserve Bank of New York, 2, 3–17.

Lubrano, M. (1998). Smooth transition GARCH models: A Bayesian perspective. CORE Discussion Paper 9866.

Lumsdaine, R.L. (1991). Asymptotic properties of the quasi-maximum likelihood estimator in GARCH(1,1) and IGARCH(1,1) models, Mimeo, Department of Economics, Princeton University.

Lumsdaine, R.L. (1996). Consistency and asymptotic normality of the quasi-maximum likelihood estimator in IGARCH(1,1) and covariance stationary GARCH(1,1) models. *Econometrica*, 64(3), 575–596.

Lumsdaine, R.L. and Ng, S. (1999). Testing for ARCH in the presence of a possibly misspecified conditional mean. *Journal of Econometrics*, 93, 257–279.

Lunde, A. (1999). A generalized gamma autoregressive conditional duration model. Working paper, Department of Economics, Politics and Public Administration, Aalborg University, Denmark.

Lütkepohl, H. (2005). *New Introduction to Multiple Time Series Analysis*. New York: Springer.

Lütkepohl, H. and Krätzig, M. (2004). *Applied Time Series Econometrics*. Cambridge: Cambridge University Press.

Luu, J.C. and Martens, M. (2003). Testing the mixture of distributions hypothesis using 'realized' volatility. *Journal of Futures Markets*, 23(7), 661–679.

Luukkonen, R., Saikkonen, P. and Teräsvirta, T. (1988). Testing linearity against smooth transition autoregressive models. *Biometrika*, 75, 491–499.

Maag, U.R. (1973). The asymptotic expansion of the distribution of the goodness of fit statistic V_{NN}. *Annals of Statistics*, 1(6), 1185–1188.

MacBeth, J.D. and Merville, L.J. (1979). An empirical examination of the Black–Scholes call option pricing model. *Journal of Finance*, 34(5), 1173–1186.

MacBeth, J.D. and Merville, L.J. (1980). Test of the Black–Scholes and Cox call option valuation models. *Journal of Finance*, 35(2), 285–300.

Madhavan, A. (2000). Market microstructure: A survey. *Journal of Financial Markets*, 3, 205–258.

Maheu, J.M. and McCurdy, T.H. (2000). Identifying bull and bear markets in stock returns. *Journal of Business and Economic Statistics*, 18, 100–112.

Mandelbrot, B. (1963). The variation of certain speculative prices. *Journal of Business*, 36, 394–419.

Mandelbrot, B. (1964). New methods in statistical economics. *Journal of Political Economy*, LXXI, 421–440.

Mandelbrot, B. (1982). *The Fractal Geometry of Nature*. W. H. Freeman, San Francisco.

Mandelbrot, B. and Taylor, H. (1967). On the distribution of stock price differences. *Operational Research*, 15, 1057–1062.

Mandelbrot, B. and Van Ness, J. (1968). Fractional Brownian motion, fractional noises and applications. *SIAM Review*, 10, 422–437.

Mann, H.B. and Wald, A. (1942). On the choice of the number of class intervals in the application of the chi-square test. *Annals of Mathematical Statistics*, 13(3), 306–317.

Mark, N. (1988). Time varying betas and risk premia in the pricing of forward foreign exchange contracts. *Journal of Financial Economics*, 22, 335–354.

Marquardt, D.W. (1963). An algorithm for least squares estimation of nonlinear parameters. *Journal of the Society for Industrial and Applied Mathematics*, 11, 431–441.

Marshall, C. and Siegel, M. (1997). Value at risk: Implementing a risk measurement standard. *Journal of Derivatives*, 4(3), 91–110.

Martens, M. (2002). Measuring and forecasting S&P 500 index-futures volatility using high-frequency data. *Journal of Futures Markets*, 22, 497–518.

Martens, M. and van Dijk, D. (2007). measuring volatility with the realized range. *Journal of Econometrics*, 138, 181–207.

McAleer, M. (2005). Automated inference and learning in modeling financial volatility. *Econometric Theory*, 21, 232–261.

McAleer, M., Chan, F., Hoti, S. and Lieberman, O. (2005). Generalised autoregressive conditional correlation. Working paper, School of Economics and Commerce, University of Western Australia.

McAleer, M. and da Veiga, B. (2008). Forecasting value-at-risk with a parsimonious portfolio spillover GARCH (PS-GARCH) model. *Journal of Forecasting*, 27, 1–19.

McCracken, M.W. (2000). Robust out-of-sample inference. *Journal of Econometrics*, 99, 195–223.

McDonald, J.B. and Newey, W. (1988). Partially adaptive estimation of regression models via the generalized *t* distribution. *Econometric Theory*, 4, 428–457.

McDonald, J.B. and Xu, Y.J. (1995). A generalization of the beta distribution with applications. *Journal of Econometrics*, 66, 133–152.

McNeil, A.J. (1997). Estimating the tails of loss severity distributions using extreme value theory. *ASTIN Bulletin*, 27(1), 117–137.

McNeil, A.J. (1998). Calculating quantile risk measures for financial return series using extreme value theory. ETH E-Collection, Department of Mathematics, Swiss Federal Technical University, Zurich.

McNeil, A.J. (1999). Extreme value theory for risk managers. In *Internal Modelling and CAD II*, pp. 93–113. London: Risk Books.

McNeil, A.J. and Frey, R. (2000). Estimation of tail-related risk measures for heteroskedasticity financial time series: An extreme value approach. *Journal of Empirical Finance*, 7, 271–300.

Meitz, M. and Teräsvirta, T. (2006). Evaluating models of autoregressive conditional duration. *Journal of Business and Economic Statistics*, 24(1), 104–124.

Mencia, F.J. and Sentana, E. (2005). Estimation and testing of dynamic models with generalised hyperbolic innovations. CEPR Discussion Paper 5177, Centre for Economic Policy Research, London.

Menn, C. and Rachev, S.T. (2005). A GARCH option pricing model with α-stable innovations. *European Journal of Operational Research*, 163, 201–209.

Merton, R.C. (1973). Rational theory of option pricing. *Bell Journal of Economics and Management Science*, 4(1), 141–183.

Merton, R.C. (1976). Option pricing when underlying stock returns are discontinuous. *Journal of Financial Economics*, 3, 125–144.

Merton, R.C. (1980). On estimating the expected return on the market: An explanatory investigation. *Journal of Financial Economics*, 8, 323–361.

Michelfelder, R.A. (2005). Volatility of stock returns: Emerging and mature markets. *Managerial Finance*, 31(2), 66–86.

Mikosch, T. and Stărică, C. (2004). Nonstationarities in financial time series, the long-range dependence, and the IGARCH effects. *Review of Economics and Statistics*, 86, 378–390.

Mikosch, T. and Straumann, D. (2002). Whittle estimation in a heavy-tailed GARCH(1,1) model. *Stochastic Processes and their Applications*, 100, 187–222.

Milhøj, A. (1987). A multiplicative parameterization of ARCH models. Mimeo, Department of Statistics, University of Copenhagen.

Mills, T.C. (1996). Non-linear forecasting of financial time series: An overview and some new models. *Journal of Forecasting*, 15, 127–135.

Mills, T.C. (1999). *The Econometric Modelling of Financial Time Series*, 2nd edition. Cambridge: Cambridge University Press.

Mincer, J. and Zarnowitz, V. (1969). The Evaluation of Economic Forecasts. In J. Mincer (ed.), *Economic Forecasts and Expectations*, pp. 3–46. New York: National Bureau of Economic Research.

Mittnik, S. and Paolella, M.S. (2000). Conditional density and value-at-risk prediction of Asian currency exchange rates. *Journal of Forecasting*, 19, 313–333.

Mittnik, S. and Paolella, M.S. (2003). Prediction of financial downside-risk with heavy tailed conditional distributions. In S.T. Tachev (ed.), *Handbook of Heavy Tailed Distributions in Finance*, Amsterdam: North-Holland.

Mittnik, S., Paolella, M.S. and Rachev, S.T. (2000). Diagnosing and treating the fat tails in financial return data. *Journal of Empirical Finance*, 7, 389–416.

Mittnik, S., Paolella, M.S. and Rachev, S.T. (2002). Stationarity of stable power-GARCH processes. *Journal of Econometrics*, 106, 97–107.

Mittnik, S., Paollela, M.S. and Rachev, S.T. (1998a). Unconditional and conditional distributional models for the Nikkei index. *Asia-Pacific Financial Markets*, 5, 99–128.

Mittnik, S., Rachev, S.T. and Paolella, M.S. (1998b). Stable Paretian modeling in finance: Some empirical and theoretical aspects. In R. Adler, R. Feldman and M.S. Taqqu (eds), *A Practical Guide to Heavy Tails*, pp. 79–110. Boston: Birkhäuser.

Mittnik, S., Rachev, S.T., Doganoglu, T. and Chenyao, D. (1999). Maximum likelihood estimation of stable Paretian models. *Mathematical and Computer Modelling*, 29, 275–293.

Miyakoshi, T. (2002). ARCH versus information based variances: Evidence from the Tokyo stock market. *Japan and the World Economy*, 14, 215–231.

Mizrach, B. (1990). Learning and conditional heteroskedasticity in asset returns. Mimeo, Department of Finance, The Wharton School, University of Pennsylvania.

Mood, A. (1940). The distribution theory of runs. *Annals of Mathematical Statistics*, 11, 367–392.

Moore, D.S. (1986). Tests of the chi-squared type. In R.B. D'Agostino and M.A. Stephens (eds), *Goodness-of-Fit Techniques*, pp. 63–95. New York: Marcel Dekker.

Morana, C. (2007). Multivariate modelling of long memory processes with common components. *Computational Statistics and Data Analysis*, 52, 919–934.

Morgan, J.P. (1996). *Riskmetrics™*, 4th edition. Technical Document, New York.

Morgan, W.A. (1940). A test for significance of the difference between the two variances in a sample form a normal bivariate population. *Biometrica*, 31, 13–19.

Morgan, I. and Trevor, R. (1999). Limit moves as censored observations of equilibrium futures price in GARCH processes. *Journal of Business and Economic Statistics*, 17(4), 397–408.

Morimoto, T. and Kawasaki, Y. (2008). Empirical comparison of multivariate GARCH models for estimation of intraday value at risk. Working paper, Hitotsubashi University. http://ssrn.com/abstract=1090807.

Moschini, G. and Myers, R.J. (2002). Testing for constant hedge ratios in commodity markets: A multivariate GARCH approach. *Journal of Empirical Finance*, 9, 589–603.

Moser, G., Rumler, F. and Scharler, J. (2007). Forecasting Austrian inflation. *Economic Modelling*, 24, 470–480.

Müller, U.A., Dacorogna, M.M., Davé, R.D., Olsen, R.B., Pictet, O.V. and VonWeizsäcker, J.E. (1997). Volatilities of different time resolutions – analyzing the dynamics of market components. *Journal of Empirical Finance*, 4, 213–239.

Müller, U.A., Dacorogna, M.M., Davé, R.D., Pictet, O.V., Olsen, R.B. and Ward, J.R. (1993). Fractals and intrinsic time – a challenge to econometricians. Invited presentation at the 39th International AEA Conference on Real Time Econometrics, 14–15 October 1993 in Luxembourg and the 4th International PASE Workshop, 22–26 November 1993 in Ascona (Switzerland). Also in B. Lüthje (ed.), *Erfolgreiche Zinsprognose*. Bonn: Verband öffentlicher Banken (1994).

Muñoz, M.P., Marquez, M.D. and Acosta, L.M. (2007). Forecasting volatility by means of threshold models. *Journal of Forecasting*, 26(5), 343–363.

Myers, R.J. (1991). Estimating time-varying optimal hedge ratio on futures markets. *Journal of Futures Markets*, 11, 39–53.

Najand, M. and Yung, K. (1991). A GARCH examination of relationship between volume and price variability in futures markets. *Journal of Futures Markets*, 11, 613–621.

Nakatsuma, T. (2000). Bayesian analysis of ARMA-GARCH models: A Markov chain sampling approach. *Journal of Econometrics*, 95, 57–69.

Nam, K., Pyun, C.S. and Arize, A.C. (2002). Asymmetric mean reversion and contrarian profits: ANST-GARCH approach. *Journal of Empirical Finance*, 9, 563–588.

Natenberg, S. (1994). *Option Volatility and Pricing. Advanced Trading Strategies and Techniques*. New York: McGraw Hill.

Nelson, D. (1990a). Stationarity and persistence in the GARCH(1,1) model. *Econometric Theory*, 6, 318–334.

Nelson, D. (1990b). ARCH models as diffusion approximations. *Journal of Econometrics*, 45, 7–38.

Nelson, D. (1991). Conditional heteroskedasticity in asset returns: A new approach. *Econometrica*, 59, 347–370.

Nelson, D. and Cao, C.Q. (1992). Inequality constraints in the univariate Garch model. *Journal of Business and Economic Statistics*, 10, 229–235.

Nelson, D. and Foster, D. (1994). Asymptotic filtering theory for univariate ARCH models. *Econometrica*, 62, 1–41.

Newbold, P. and Granger, C.W.J. (1974). Experience with forecasting univariate time series and the combination of forecasts. *Journal of the Royal Statistical Society, Series A*, 137, 131–165.

Newey, W. and West, K.D. (1987). A simple positive semi-definite, heteroskedasticity and autocorrelation consistent covariance matrix. *Econometrica*, 55, 703–708.

Ng, V.K. (1988). Equilibrium stock return distributions in a simple log-linear factor economy. Manuscript, University of California, San Diego.

Ng, V.K., Engle, R.F. and Rothschild, M. (1992). A multi-dynamic factor model for stock returns. *Journal of Econometrics*, 52, 245–265.

Níguez, T.-M. and Rubia, A. (2006). Forecasting the conditional covariance matrix of a portfolio under long-run temporal dependence. *Journal of Forecasting*, 25(6), 439–458.

Nocedal, J. and Wright, S.J. (1999). *Numerical Optimization*. New York: Springer.

Noh, J., Engle, R.F. and Kane, A. (1994). Forecasting volatility and option prices of the S&P500 index. *Journal of Derivatives*, 2, 17–30.

Nowicka-Zagrajek, J. and Weron, A. (2001). Dependence structure of stable R-GARCH processes. *Probability and Mathematical Statistics*, 21, 371–380.

Oedegaard, B.A. (1991). Empirical tests of changes in autocorrelation of stock index returns. Mimeo, Graduate School of Industrial Administration, Carnegie Mellon University.

Officer, R. (1972). The distribution of stock returns. *Journal of the American Statistical Association*, 67, 807–812.

Oomen, R. (2001). Using high frequency stock market index data to calculate, model and forecast realized volatility. Manuscript, Department of Economics, European University Institute.

Ozaki, T. (1980). Non-linear time series models for non-linear random vibrations. *Journal of Applied Probability*, 17, 84–93.

Pagan, A.R. and Hong, Y. (1991). Non-parametric estimation and the risk premium. In W.A. Barnett, J. Powell and G. Tauchen (eds), *Nonparametric and Semiparametric Methods in Econometrics and Statistics*, Cambridge: Cambridge University Press.

Pagan, A.R. and Sabau, H. (1987). On the inconsistency of the MLE in certain heteroskedastic regression models. Mimeo, University of Rochester, Department of Economics

Pagan, A.R. and Schwert, G.W. (1990). Alternative models for conditional stock volatility. *Journal of Econometrics*, 45, 267–290.

Palm, F.C. (1996). GARCH models of volatility. In G. Maddala and C. Rao (eds), *Handbook of Statistics*, pp. 209–240. Amsterdam: Elsevier.

Palm, F.C. and Vlaar, P. (1997). Simple diagnostics procedures for modelling financial time series. *Allgemeines Statistisches Archiv*, 81, 85–101.

Panorska, A., Mittnik, S. and Rachev, S.T. (1995). Stable GARCH models for financial time series. *Applied Mathematics Letters*, 8, 33–37.

Pantula, S.G. (1986). Modeling the persistence of conditional variances: A comment. *Econometric Reviews*, 5, 71–73.

Park, T.H. and Switzer, L.N. (1995). Bivariate GARCH estimation of the optimal hedge ratios for stock index futures: A note. *Journal of Futures Markets*, 15, 61–67.

Parke, W.R. (1999). What is a fractional unit root? *Review of Economics and Statistics*, 81(4), 632–638.

Parke, W.R. and Waters, G.A. (2007). An evolutionary game theory explanation of ARCH effects. *Journal of Economic Dynamic and Control*, 31(7), 2234–2262.

Parkinson, M. (1980). The extreme value method for estimating the variance of the rate of return. *Journal of Business*, 53(1), 61–65.

Patton, A.J. (2005). Volatility forecast evaluation and comparison using imperfect volatility proxies. Working paper, London School of Economics.

Patton, A.J. (2006). Volatility forecast comparison using imperfect volatility proxies. Working paper, London School of Economics.

Pearson, K. (1900). On the criterion that a given system of deviations from the probable in the case of a correlated system of variables is such that it can be reasonably supposed to have arisen from random sampling. *Philosophical Magazine*, (5), 50, 157–175. Reprinted in K. Pearson (1956) *Karl Pearson's Early Statistical Papers*, 339–357. Cambridge: Cambridge University Press.

Peel, D.A. and Speight, A.E.H. (1996). Is the US business cycle asymmetric? Some further evidence. *Applied Economics*, 28, 405–415.

Pelletier, D. (2006). Regime switching for dynamic correlations. *Journal of Econometrics*, 131, 445–473.

Perez-Quiros, G. and Timmermann, A. (2001). Business cycle asymmetries in stock returns: Evidence from higher order moments and conditional densities. *Journal of Econometrics*, 103, 259–306.

Perez-Rodriguez, J.V., Torra, S. and Andrada-Felix, J. (2005). Are Spanish Ibex35 Stock Future Index Returns Forecasted with Non-Linear Models? *Applied Financial Economics*, 15, 963–975.

Pesaran, M.H. and Timmermann, A. (1992). A simple nonparametric test of predictive performance. *Journal of Business and Economic Statistics*, 10, 461–465.

Plasmans, J., Verkooijen, W. and Daniels, H. (1998). Estimating structural exchange rate models by artificial neural networks. *Applied Financial Economics*, 8, 541–551.

Poggio, T. and Girosi, F. (1990). Networks for approximation and learning. *Proceedings of the IEEE, special issue: Neural Networks I: Theory and Modeling*, 78, 1481–1497.

Politis, D.N. (2003). Model-free volatility prediction. Paper 2003-16, Department of Economics, University of California, San Diego.

Politis, D.N. (2004). A heavy-tailed distribution for ARCH residuals with application to volatility predictions. *Annals of Economics and Finance*, 5, 283–298.

Politis, D.N. (2006). A multivariate heavy-tailed distribution for ARCH/GARCH residuals. *Advances in Econometrics*, 20, 105–124.

Politis, D.N. (2007a). Model free versus model based volatility prediction. *Journal of Financial Econometrics*, 5(3), 358–389.

Politis, D.N. (2007b). Model free prediction. Invited paper, *Stochastic Volatility Modeling: Reflections, Recent Developments and the Future, E. Xekalaki (organizer)*, 56th Session of the International Statistical Institute, 22–29 August 2007, Lisbon, Portugal.

Politis, D.N. and Romano, J.P. (1994). The stationary bootstrap. *Journal of the American Statistical Association*, 89, 1303–1313.

Poon, S.H., Rockinger, M. and Tawn, J. (2001). New extreme-value dependence measures and finance applications. Manuscript, HEC Paris.

Poon, S.H. and Granger, C.W.J. (2003). Forecasting volatility in financial markets: A review. *Journal of Economic Literature*, XLI, 478–539.

Praetz, P. (1972). The distribution of share price changes. *Journal of Business*, 45, 49–55.

Press, W.H., Teukolsy, S.A., Vetterling, W.T. and Flannery, B.P. (1992a). *Numerical Recipes in Fortran*. Cambridge: Cambridge University Press.

Press, W.H., Teukolsy, S.A., Vetterling, W.T. and Flannery, B.P. (1992b). *Numerical Recipes in C: The Art of Scientific Computing*. Cambridge: Cambridge University Press.

Priestley, M. (1988). *Non-linear and Non-stationary Time Series Analysis*. London: Academic Press.

Pyun, C.S., Lee, S.Y. and Nam, K. (2000). Volatility and information flows in emerging equity market: A case of the Korean Stock Exchange. *International Review of Financial Analysis*, 9, 405–420.

Quandt, R.E. (1983). Computational problems and methods. In Z. Griliches and M.D. Intriligator (eds), *Handbook of Econometrics*, Volume I. Amsterdam: North Holland.

Rabemananjara, R. and Zakoian, J.M. (1993). Threshold ARCH models and asymmetries in volatility. *Journal of Applied Econometrics*, 8, 31–49.

Rachev, S.T. and Mittnik, S. (2000). *Stable Paretian Models in Finance*. Chichester: John Wiley & Sons Ltd.

Rich, R.W., Raymond, J. and Butler, J.S. (1991). Generalized instrumental variables estimation of autoregressive conditional heteroskedastic models. *Economics Letters*, 35, 179–185.

Richardson, M. and Stock, J.H. (1989). Drawing inference from statistics based on multiyear asset returns. *Journal of Financial Economics*, 25, 323–348.

Ritchken, P. and Trevor, R. (1999). Pricing options under generalized GARCH and stochastic volatility processes. *Journal of Finance*, 54, 377–402.

Rogers, L.C.G. and Satchell, S.E. (1991). Estimating variance from high, low and closing prices. *Annals of Applied Probability*, 1, 504–512.

Rosenblatt, M. (1952). Remarks on a multivariate transformation. *Annals of Mathematical Statistics*, 23, 470–472.

Ross, S.A. (1989). Information and volatility: The no-arbitrage martingale approach to timing and resolution irrelevancy. *Journal of Finance*, 44, 1–17.

Rothman, E.D. and Woodroofe, M. (1972), A Cramér–von Mises type statistic for testing symmetry. *Annals of Mathematical Statistics*, 43(6), 2035–2038.

Ruiz, E. and Pérez, A. (2003). Asymmetric long memory GARCH: A reply to Hwang's model. *Economics Letters*, 78(3), 415–422.

Rydén, T., Teräsvirta, T. and Asbrink, S. (1998). Stylized facts of daily return series and the hidden Markov model. *Journal of Applied Econometrics*, 13, 217–244.

Sabbatini, M. and Linton, O. (1998). A GARCH model of the implied volatility of the Swiss market index from option prices. *International Journal of Forecasting*, 14, 199–213.

Sadorsky, P. (2005). Stochastic volatility forecasting and risk management. *Applied Financial Economics*, 15, 121–135.

Sadorsky, P. (2006). Modeling and forecasting petroleum futures volatility. *Energy Economics*, 28(4), 467–488.

Saez, M. (1997). Option pricing under stochastic volatility and stochastic interest rate in the Spanish case. *Applied Financial Economics*, 7, 379–394.

Saltoglu, B. (2003). Comparing forecasting ability of parametric and non-parametric methods: An application with Canadian monthly interest rates. *Applied Financial Economics*, 13, 169–179.

Sarkar, N. (2000). Arch model with Box-Cox transformed dependent variable. *Statistics and Probability Letters*, 50, 365–374.

Sarma, M., Thomas, S. and Shah, A. (2003). Selection of VaR models. *Journal of Forecasting*, 22(4), 337–358.

Schmalensee, R. and Trippi, R.R. (1978). Common stock volatility expectations implied by option premia. *Journal of Finance*, 33(1), 129–147.

Scholes, M. and Williams, J. (1977). Estimating betas from non-synchronous data. *Journal of Financial Economics*, 5, 309–328.

Schorr, B. (1974). On the choice of the class intervals in the application of the chi-square test. *Optimization*, 5, 357–377.

Schwarz, G. (1978). Estimating the dimension of a model. *Annals of Statistics*, 6, 461–464.

Schwert, G.W. (1989a). Why does stock market volatility changes over time. *Journal of Finance*, 44, 1115–1153.

Schwert, G.W. (1989b). Business cycles, financial crisis and stock volatility. *Carnegie Rochester Conference Series on Public Policy*, 39, 83–126.

Schwert, G.W. (1990). Stock volatility and the crash of '87. *Review of Financial Studies*, 3, 77–102.

Schwert, G.W. and Seguin, P. (1990). Heteroskedasticity in stock returns. *Journal of Finance*, 45, 1129–1155.

Sentana, E. (1995). Quadratic ARCH models. *Review of Economic Studies*, 62, 639–661.

Sentana, E. and Wadhwani, S. (1991). Feedback traders and stock returns autocorrelations: Evidence from a century of daily data. *Economic Journal*, 58, 547–563.

Seymour, A.J. and Polakow, D.A. (2003). A coupling of extreme-value theory and volatility updating with value-at-risk estimation in emerging markets: A South African test. *Multinational Finance Journal*, 7, 3–23.

Shanno, D.F. (1970). Conditioning of quasi-Newton methods for function minimization. *Mathematics of Computation*, 24(111), 647–656.

Shanno, D.F. (1985). An example of numerical nonconvergence of a variable-metric method. *Journal of Optimization Theory and Applications*, 46(1), 87–94.

Shapiro, S.S. and Francia, R.S. (1972). Approximate analysis of variance test for normality. *Journal of the American Statistical Association*, 67, 215–225.

Shapiro, S.S. and Wilk, M.B. (1965). An analysis of variance test for normality. *Biometrika*, 42, 591–611.

Shapiro, S.S., Wilk, M.B. and Chen, N.J. (1968). A comparative study of various tests for normality. *Journal of the American Statistical Association*, 63, 1343–1372.

Sharma, J.L., Mougoue, M. and Kamath, R.R. (1996). Heteroskedasticity in stock market indicator return data: Volume versus GARCH effects. *Applied Financial Economics*, 6, 337–342.

Shephard, N. (1996). Statistical aspects of ARCH and stochastic volatility models. In D.R. Cox, D.V. Hinkley and O.E. Barndorff-Nielsen (eds) *Time Series Models in Econometrics, Finance and Other Fields*, pp. 1–67. London: Chapman & Hall.

Shibata, R. (1980). Asymptotically efficient selection of the order of the model for estimating parameters of a linear process. *Annals of Statistics*, 8, 147–164.

Silvennoinen, A. and Teräsvirta, T. (2005). Multivariate autoregressive conditional hetero-skedasticity with smooth transitions in conditional correlations. Working Paper 577, Series in Economics and Finance, SSE/EFI.

Silvennoinen, A. and Teräsvirta, T. (2007). Modelling multivariate autoregressive conditional heteroskedasticity with the double smooth transition conditional correlation GARCH model. Working Paper 652, Series in Economics and Finance, SSE/EFI.

Silvennoinen, A. and Teräsvirta, T. (2008). Mulivariate GARCH models. In T.G. Andersen, R. A. Davis, J-.P. Kreiss and T. Mikosch (eds), *Handbook of Financial Time Series*, New York: Springer.

Silverman, B.W. (1986). *Density Estimation for Statistics and Data Analysis*. London: Chapman & Hall.

Simon, D.P. (1989). Expectations and risk in the Treasury bill market: An instrumental variables approach. *Journal of Financial and Quantitative Analysis*, 24, 357–366.

Smirnov, N.V. (1937). On the distribution of the von Mises' ω^2-criterion (in Russian). *Matematicheskii Sbornik*, 5, 973–993.

Smirnov, N.V. (1939). On the estimation of the discrepancy between empirical curves of distribution of two independent samples. *Bulletin Mathématique de l'Université de Moscou*, 2.

Smith, G.W. (1987). Endogenous conditional heteroskedasticity and tests of menu-cost pricing theory. Manuscript, Queen's University.

So, M.K.P. and Yu, P.L.H. (2006). Empirical analysis of GARCH models in value at risk estimation. *Journal of International Markets, Institutions and Money*, 16(2), 180–197.

Sollis, R. (2005). Predicting returns and volatility with macroeconomic variables: Evidence from tests of encompassing. *Journal of Forecasting*, 24, 221–231.

Specht, K. and Gohout, W. (2003). Portfolio selection using the principal components Garch model. *Financial Markets and Portfolio Management*, 17(4), 450–458.

Specht, K. and Winker, P. (2008). Portfolio optimization under VaR constraints based on dynamic estimates of the variance-covariance matrix. In E.J. Kontoghiorghes, B. Rustem and P. Winker (eds), *Computational Methods in Financial Engineering: Essays in Honour of Manfred Gilli*, pp. 73–94. Berlin: Springer.

Stephens, M.A. (1965). The goodness-of-fit statistic V_N: Distribution and significance points. *Biometrika*, 52, 309–321.

Stephens, M.A. (1969). Results from the relation between two statistics of the Kolmogorov–Smirnov type. *Annals of Mathematical Statistics*, 40(5), 1833–1837.

Stephens, M.A. (1970). Use of Kolmogorov–Smirnov, Cramér–von Mises and related statistics without extensive tables. *Journal of the Royal Statistical Society, Series B*, 32, 115–122.

Stephens, M.A. (1986). Tests based on EDF statistics. In R.B. D'Agostino and M.A. Stephens (eds), *Goodness-of-Fit Techniques*, New York: Marcel Dekker.

Stock, J.H. (1987). Measuring business cycle time. *Journal of Political Economy*, 95, 1240–1261.

Stock, J.H. (1988). Estimating continuous time process subject to time deformation. *Journal of the Statistical Association*, 83, 77–85.

Stuart, A. and Ord, K. (1994). *Kendall's Advanced Theory of Statistics: Distribution Theory*, Vol. 1, 6th edition. London: Edward Arnold.

Subbotin, M.T.H. (1923). On the law of frequency of error. *Mathematicheskii Sbornik*, 31, 296–301.

Sullivan, R., Timmermann, A. and White, H. (1999). Data-snooping, technical trading rule performance, and the bootstrap. *Journal of Finance*, 54(5), 1647–1691.

Susmel, R. (2000). Switching volatility in international equity markets. *International Journal of Finance and Economics*, 5, 265–283.

Tai, C.-S. (2001). A multivariate GARCH in mean approach to testing uncovered interest parity: Evidence from Asia-Pacific foreign exchange markets. *Quartely Review of Economics and Finance*, 41, 441–460.

Taleb, N. (1997a). The world according to Nassim Taleb. *Derivatives Strategy*, December/January.

Taleb, N. (1997b). Against VaR. *Derivatives Strategy*, April.

Tauchen, G. and Pitts, M. (1983). The price variability-volume relationship on speculative markets. *Econometrica*, 51, 485–505.

Tavares, A.B., Curto, J.D. and Tavares, G.N. (2008). Modelling heavy tails and asymmetry using ARCH-type models with stable Paretian distributions. *Nonlinear Dynamics*, 51, 231–243.

Taylor, N. (2004). Modeling discontinuous periodic conditional volatility: Evidence from the Commodity Futures Market. *Journal of Futures Markets*, 24(9), 805–834.

Taylor, S.J. (1986). *Modelling Financial Time Series*. Chichester: John Wiley & Sons Ltd.

Taylor, S.J. (1994). Modelling stochastic volatility: A review and comparative study. *Mathematical Finance*, 4, 183–204.

Taylor, S.J. (2005). *Asset Price Dynamics, Volatility, and Prediction*. Princeton, NJ: Princeton University Press.

Taylor, S.J. and Xu, X. (1997). The incremental volatility information in one million foreign exchange quotations. *Journal of Empirical Finance*, 4, 317–340.

Teräsvirta, T. (1994). Specification, estimation and evaluation of smooth transition auto-regressive models. *Journal of American Statistical Association*, 89, 208–218.

Teräsvirta, T., Tjøstheim, D. and Granger, C.W.J. (1994). Aspects of modeling nonlinear time series. In R.F. Engle and D. McFadden (eds), *Handbook of Econometrics*, Volume 4. Amsterdam: Elsevier Science.

Teyssiere, G. (1997). Modelling exchange rates volatility with multivariate long-memory ARCH processes. Working Paper 97B03, DT-GREQAM, Marseille.

Thadewald, T. and Büning, H. (2007). Jarque–Bera Test and its competitors for testing normality – a power comparison. *Journal of Applied Statistics*, 34(1), 87–105.

Theodossiou, P. (1998). Financial data and the skewed generalized *t* distribution. *Management Science*, 44, 1650–1661.

Theodossiou, P. (2002). Skewness and kurtosis in financial data and the pricing of options. Working paper, Rutgers University.

Theodossiou, P. and Trigeorgis, L. (2003). Option pricing when log-returns are skewed and leptokurtic. Paper presented to the *Eleventh Annual Conference of the Multinational Finance Society, Istanbul*, MFC223.

Thisted, R.A. (1988). *Elements of Statistical Computing. Numerical Computation*. New York: Chapman & Hall.

Thomaidis, N.S. and Dounias, G. (2008). A general class of neural network-GARCH models for financial time series analysis. Working paper, School of Business Studies, University of the Aegean.

Thomakos, D.D. and Wang, T. (2003). Realized volatility in the futures markets. *Journal of Empirical Finance*, 10, 321–353.

Thompson, J. and Stewart, H. (1986). *Nonlinear Dynamics and Chaos*. Chichester: John Wiley & Sons Inc.

Thum, F. (1988). Economic foundations of ARCH exchange rate process. Manuscript, University of Texas.

Tong, H. (1990). *Nonlinear Time Series: A Dynamic System Approach*. Oxford: Oxford University Press.

Tong, H. (2007). Exploring volatility from a dynamical system perspective. Invited paper, *Stochastic Volatility Modeling: Reflections, Recent Developments and the Future*, E. Xekalaki (organizer), 56th Session of the International Statistical Institute, 22–29 August 2007, Lisbon, Portugal.

Tsay, R. (1987). Conditional heteroscedastic time series models. *Journal of the American Statistical Association*, 82, 590–604.

Tsay, R. (2002). *Analysis of Financial Time Series*. New York: John Wiley & Sons Inc.

Tse, Y.K. (1998). The conditional heteroskedasticity of the yen-dollar exchange rate. *Journal of Applied Econometrics*, 193, 49–55.

518   REFERENCES

Tse, Y.K. (2000). A test for constant correlations in a multivariate GARCH model. *Journal of Econometrics*, 98, 107–127.

Tse, Y.K. (2002). Residual-based diagnostics for conditional heteroscedasticity models. *Econometrics Journal*, 5, 358–373.

Tse, Y.K. and Tsui, K.C. (1999). A note on diagnosing multivariate conditional heteroscedasticity models. *Journal of Time Series Analysis*, 20(6), 679–691.

Tse, Y.K. and Tsui, K.C. (2002). A multivariate generalized autoregressive conditional heteroscedasticity model with time-varying correlations. *Journal of Business and Economic Statistics*, 20, 351–362.

Tse, Y.K. and Zuo, X.L. (1997). Testing for conditional heteroscedasticity: Some Monte Carlo results. *Journal of Statistical Computation and Simulation*, 58, 237–253.

Tsionas, E.G. (1999). Monte Carlo inference in econometric models with symmetric stable disturbances. *Journal of Econometrics*, 88, 2, 365–401.

Tsionas, E.G. (2002). Likelihood-based comparison of stable Paretian and competing models: Evidence from daily exchange rates. *Journal of Statistical Computation and Simulation*, 72(4), 341–353.

Tsui, A.K. (2004). Diagnostics for conditional heteroscedasticity models: Some simulation results. *Mathematics and Computers in Simulation*, 64, 113–119.

Tsui, A.K. and Yu, Q. (1999). Constant conditional correlation in a bivariate GARCH model: Evidence from the stock markets of China. *Mathematics and Computers in Simulation*, 48, 503–509.

Turner, C.M., Startz, R. and Nelson, C.R. (1989). A Markov model of heteroskedasticity, risk, and learning in the stock market. *Journal of Financial Economics*, 25, 3–22.

Urzúa, C. (1996). On the correct use of omnibus tests for normality. *Economics Letters*, 53, 247–251.

Van Den Goorbergh, R.W.J. and Vlaar, P. (1999). Value-at-risk analysis of stock returns. Historical simulation, variance techniques or tail index estimation? DNB Staff Reports 40, Netherlands Central Bank.

van der Weide, R. (2002). GO-GARCH: A multivariate generalized orthogonal GARCH model. *Journal of Applied Econometrics*, 17, 549–564.

van Dijk, D. and Franses, P.H. (2003). Selecting a nonlinear time series model using weighted tests of equal forecast accuracy. *Oxford Bulletin of Economics and Statistics*, 65(1), 727–744.

Veronesi, P. (1999). Stock market overreaction to bad news in good times: A rational expectations equilibrium model. *Review of Financial Studies*, 12, 975–1007.

Vigfusson, R. (1997). Switching between chartists and fundamentalists: A Markov regime-switching approach. *International Journal of Finance and Economics*, 2, 291–305.

Vilasuso, J. (2002). Forecasting exchange rate volatility. *Economics Letters*, 76, 59–64.

Vlaar, P. and Palm, F.C. (1993). The message in weekly exchange rates in the European monetary system: Mean reversion, conditional heteroscedasticity and jumps. *Journal of Business and Economic Statistics*, 11(3), 351–360.

Von Mises, R. (1931). *Wahrscheinlichkeitsrechnung und Ihre Anwendung in der Statistik und Theoretischen Psysik*. Leipzig and Vienna: Deiticke.

Vrontos, I.D., Dellaportas, P. and Politis, D.N. (2003). A full-factor multivariate GARCH model. *Econometrics Journal*, 6, 311–333.

Walsh, D.M. and Tsou, G.Y.-G. (1998). Forecasting index volatility: Sampling interval and non-trading effects. *Applied Financial Economics*, 8, 477–485.

Wang, K.-L., Fawson, C., Barrett, C.B. and McDonald, J.B. (2001). A flexible parametric GARCH model with an application to exchange rates. *Journal of Applied Econometrics*, 16, 521–536.

Wei, K.C.J. and Chiang, R. (1997). GMM and MLE approaches for estimation of volatility and regression models when daily prices are subjected to price limits. Department of Finance School of Business and Management, Hong Kong University of Science and Technology.

Wei, S.X. (1999). A Bayesian approach to dynamic Tobit models. *Econometric Reviews*, 18, 417–439.

Wei, S.X. (2002). A censored GARCH model of asset returns with price limits. *Journal of Empirical Finance*, 9, 197–223.

Weiss, A. (1986). Asymptotic theory for ARCH models: Estimation and testing. *Econometric Theory*, 2, 107–131.

West, K.D. (2006). Forecast evaluation. In G. Elliott, C.W.J. Granger and A., Timmermann (eds), *Handbook of Economic Forecasting*, Amsterdam: North Holland.

West, K.D. and Cho, D. (1995). The predictive ability of several models of exchange rate volatility. *Journal of Econometrics*, 69, 367–391.

West, K.D., Edison, H.J. and Cho, D. (1993). A utility based comparison of some models for exchange rate volatility. *Journal of International Economics*, 35, 23–45.

Westerfield, R. (1977). The distribution of common stock price changes: An application of transactions time and subordinated stochastic models. *Journal of Financial and Quantitative Analysis*, 12, 743–765.

Whaley, R. (1982). Valuation of American call options on dividend paying stocks. *Journal of Financial Economics*, 10, 29–58.

Whaley, R. (1993). Derivatives on market volatility: Hedging tools long overdue. *Journal of Derivatives*, 1, 71–84.

White, H. (1980). A heteroskedastic-consistent covariance matrix estimator and a direct test for heteroskedasticity. *Econometrica*, 48, 817–838.

White, H. (1992). *Artificial Neural Networks: Approximation and Learning Theory*. Oxford: Blackwell.

White, H. (2000). A reality check for data snooping. *Econometrica*, 68, 1097–1126.

Whittle, P. (1953). Estimation and information in stationary time series. *Arkiv för Matematik*, 2, 423–434.

Wilhelmsson, A. (2006). Garch forecasting performance under different distribution assumptions. *Journal of Forecasting*, 25(8), 561–578.

Xekalaki, E. and Degiannakis, S. (2005). Evaluating volatility forecasts in option pricing in the context of a simulated options market. *Computational Statistics and Data Analysis*, 49(2), 611–629.

Xekalaki, E., Panaretos, J. and Psarakis, S. (2003). A predictive model evaluation and selection approach – the correlated gamma ratio distribution. In J. Panaretos (ed.), *Stochastic Musings: Perspectives from the Pioneers of the Late 20th Century*, pp. 188–202. Mahwah, NJ: Lawrence Erlbaum Associates.

Yamai, Y. and Yoshiba, T. (2005). Value-at-risk versus expected shortfall: A practical perspective. *Journal of Banking and Finance*, 29(4), 997–1015.

Yang, L. (2006). A semiparametric GARCH model for foreign exchange volatility. *Journal of Econometrics*, 130, 365–384.

Yu, J. (2002). Forecasting volatility in the New Zealand stock market. *Applied Financial Economics*, 12, 193–202.

Zakoian, J.M. (1990). Threshold heteroskedastic models. Manuscript, CREST, INSEE, Paris.

Zarowin, P. (1990). Size, seasonality and stock market over-reaction. *Journal of Financial and Quantitative Analysis*, 25, 113–125.

Zhang, G.P. and Berardi, V.L. (2001). Time series forecasting with neural network ensembles: An application for exchange rate prediction. *Journal of the Operational Research Society*, 52 (6), 652–664.

Zhang, J.E. and Zhu, Y. (2006). VIX futures. *Journal of Futures Markets*, 26(6), 521–531.

Zhang, M.Y., Russell, J.R. and Tsay, R. (2001). A nonlinear autoregressive conditional duration model with applications to financial transaction data. *Journal of Econometrics*, 104, 179–207.

Zhang, L., Mykland, P.A. and Ait-Sahalia, Y. (2005). A tale of two time scales: Determining integrated volatility with noisy high-frequency data. *Journal of the American Statistical Association*, 100, 1394–1411.

Zivot, E. and Jiahui, W. (2006). *Modeling Financial Time Series with S-PLUS*, 2nd edition. New York: Springer.

Author Index

Subject Index